ELASTIC-PLASTIC FRACTURE MECHANICS

ON MATERIALS, ENGINEERING AND MECHANICAL SCIENCE

A series devoted to the publication of courses and educational seminars given at the Joint Research Centre, Ispra Establishment, as part of its education and training program. Published for the Commission of the European Communities, Directorate-General Information Market and Innovation.

The publisher will accept continuation orders for this series which may be cancelled at any time and which provide for automatic billing and shipping of each title in the series upon publication. Please write for details.

ELASTIC-PLASTIC FRACTURE MECHANICS

Proceedings of the 4th Advanced Seminar on Fracture Mechanics,
Joint Research Centre, Ispra, Italy,
24–28 October 1983
in collaboration with the
EUROPEAN GROUP ON FRACTURE

Edited by

LARS HANNES LARSSON

Commission of the European Communities,
Joint Research Centre, Ispra Establishment,
Ispra, Italy

D. REIDEL PUBLISHING COMPANY

A MEMBER OF THE KLUWER ACADEMIC PUBLISHERS GROUP

DORDRECHT / BOSTON / LANCASTER

Library of Congress Cataloging in Publication Data

Advanced Seminar on Fracture Mechanics (4th: 1983: Joint Research Centre,
   Ispra, Italy)
Elastic-plastic fracture mechanics.

   (Ispra courses on materials, engineering, and mechanical science)
   Including index.
   1.   Fracture mechanics–Congresses.   2.   Elasticity–Congresses.
3.   Plasticity–Congresses.   I.   Larsson, Lars Hannes.   II.   Title.   III.   Series.
TA409.A38   1983   620.1'126   85–2399

ISBN-13: 978-94-010-8874-9   e-ISBN-13: 978-94-009-5380-2
DOI: 10.1007/978-94-009-5380-2

Commission of the European Communities Joint Research Centre Ispra (Varese), Italy

Publication arrangements by
Commission of the European Communities
Directorate-General Information Market and Innovation, Luxembourg

EUR 9830
© 1985, ECSC, EEC, EAEC, Brussels and Luxembourg

Softcover reprint of the hardcover 1st edition 1985

LEGAL NOTICE
Neither the Commission of the European Communities nor any person acting on behalf of the
Commission is responsible for the use which might be made of the following information.

Published by D. Reidel Publishing Company
P.O. Box 17, 3300 AA Dordrecht, Holland

Sold and distributed in the U.S.A. and Canada
by Kluwer Academic Publishers,
190 Old Derby Street, Hingham, MA 02043, U.S.A.

In all other countries, sold and distributed
by Kluwer Academic Publishers Group,
P.O. Box 322, 3300 AH Dordrecht, Holland

CONTENTS

FOREWORD

This book contains the texts of the lectures and workshops of the 4th
Advanced Seminar on Fracture Mechanics (ASFM 4) held at the Joint Research Centre, Ispra, on 24-28 October 1983. The birth of this series
of advanced seminars is contemporary and closely connected with the
birth of the European Group on Fracture (EGF): both events can be
traced back to the mid-seventies when a group of European experts met
and decided to start cooperation in the field of fracture mechanics.
The organisation of seminars was suggested as one of the ways of en-
hancing European collaboration. The JRC Ispra offered to host such
seminars because the European character of this establishment and the
existing facilities of the Ispra Courses offered an excellent frame-
work. Indeed, one of the aims of the Ispra Courses is to contribute
to exchanges and ties between European scientific workers in subjects
related to EEC research programmes.

The first ASFM took place in 1975. ASFM 2 was organized in 1979 and
was devoted to elastic-plastic fracture mechanics. Thereafter the
ASFMs have been biennial, alternating with the other regularly sche-
duled event of the EGF, namely the European Conference on Fracture
which is organized at different places on even years. ASFM 3 in 1981
had as its theme subcritical crack growth due to fatigue, stress corrosion and creep.

As before, the EGF nominated for ASFM 4 an Advisory Board with the
following members:
  Prof. J. Carlsson, The Royal Institute of Technology, Sweden
  Prof. D. François, Ecole Centrale des Arts et Manufactures, France
  Dr. F. Sommer, Fraunhofer Institut für Werkstoffmechanik, Germany
  Prof. C.E. Turner, Imperial College, U.K.
  Dr. H. van Elst, Metaalinstituut TNO, The Netherlands.

This group played an essential role in drafting the programme of
ASFM 4 and in taking the first contacts with the proposed lecturers.
Right from the early discussions of the Advisory Board it became
evident that Elastic-Plastic Fracture Mechanics was again the favoured
theme. In order to allow an in-depth treatment, it was decided to con-
centrate on EPFM under monotonic loading and time-independent material
properties, excluding cyclic loading and creep. However, dynamic
loading was considered to be an important matter and two lectures were
inserted on this topic. One of them discussed the problems encountered
when modelling rate-dependent material behaviour.

The lectures and workshops presented here have been revised and up-
dated to make them better suited to book form. In order to remind
readers of the origin of the different contributions making up the

book, the words lecture and workshop are used where appropriate. The four workshops illustrate the four design methods discussed at ASFM 4: the R6 method of CEGB, the EPRI method, the EnJ method and the COD design curve method. Each workshop consists of a general presentation of the method followed by example problems (which the reader might try to solve himself as an exercise), and a discussion of the solutions.

Fifteen lectures are presented. They cover experimental, theoretical and numerical aspects, micromechanisms and continuum mechanics and also the concept of damage and its use in local fracture criteria. My introductory lecture gives a general outline of the problem area that is tackled in the following lectures and how they are related. It finishes with a list of questions to which an answer is expected from the various contributions. At ASFM 4 Prof. Carlsson had the difficult task of drawing the conclusions and expanding on some questions not fully covered. His written contribution concludes this book.

In order to shorten the publication time camera-ready texts delivered by the authors were used. Though as much as humanly possible was done to have a uniform presentation, some minor variations in typing standards and details were unavoidable, I hope the reader finds them acceptable.

As the course-coordinator of ASFM 4 and as the Editor of these proceedings it is my pleasure to express my thanks to:
   Mr. B. Henry, Manager of the Ispra Training and Education Programme
   Mr. Porrini and the secretariat of Ispra Courses who took care of
       the practical organisation
   The members of the EGF Advisory Board for their help
   The lecturers for their contributions and their patience during
       the editing process.

                                        L.H. Larsson

INTRODUCTION TO THE ADVANCED SEMINAR ASFM4 ON ELASTIC-PLASTIC FRACTURE
MECHANICS

L.H. Larsson
Commission of the European Communities
Joint Research Centre - Ispra Establishment
21020 Ispra (Va) - Italy

ABSTRACT. Apart from the introduction in which a short discussion of
the boundary area of non-destructive examination is given, this lecture
is an overview of the problem areas within EPFM to be treated in the
lectures and workshops of the seminar. It starts with a presentation of
elastic-plastic crack behaviour: different regimes depending on the
spread of plasticity, stable crack growth, failure modes and failure
paths, constraints and the transferability of specimen results to structural response, local phenomena at the crack tip. Crack characterizing
parameters (J or COD) or, alternatively, the local criteria approach,
are then presented. The need for and the criteria to be satisfied by
design methods are discussed. Follows a short presentation of the problems to be studied in the field of dynamic effects. Finally, a list
of questions is given to which answers are expected from the lectures
and discussions during the seminar.

1. INTRODUCTION

It is hardly thinkable today to envisage the performance of any significant assessments of structural integrity without having recourse to
fracture mechanics. Whenever the safety of structures is an important
issue, fracture mechanics concepts find an application right from the
stage of material selection and design. Although the discipline of fracture mechanics is traditionally considered to be an amalgam of continuum mechanics and materials science, the application of fracture mechanics technology encompasses a much wider range of activities such as
- failure analysis and case histories;
- non-destructive testing and inspection;
- codes and standards;
- legal aspects (product liability, insurance).
   Of particular interest in connection with the subject of this seminar is the boundary area of non-destructive examination (NDE) techniques needed for pre-service and in-service inspection. Sophistications

1

L. H. Larsson (ed.), Elastic-Plastic Fracture Mechanics, 1–11.
© *1985 ECSC, EEC, EAEC, Brussels and Luxembourg.*

in fracture mechanics would be worthless without a parallel development of NDE methodologies leading to a reliable detection and characterization of real defects in structures. A few words on this area seem therefore relevant, taking as an example nuclear pressure vessels.

In order to evaluate the effectiveness of NDE, international co-operative efforts have been organized in three successive steps:

1) Verification of the effectiveness of existing procedures. This was the object of PISC I (1974-1979) in which 34 organizations from ten European countries applied the ASME XI procedure to three samples containing defects. Fig.1 shows the obtained /1/ defect detection probability (DDP) as a function of defect size 2a. Clearly, the performances are not acceptable for reactor reliability standards: to have DDP > 0.95 one must have 2a > 50 mm, i.e. 20 - 25% of the plate thickness (the PISC I samples were 200 - 250 mm thick).

2) Identification of performant procedures or techniques. PISC I contained limited information on European procedures showing as a trend that for DDP > 0.95 the lower limit of defect size could be reduced to 10% of the plate thickness. A mini-PISC (PISC = Programme for Inspection of Steel Components) called DDT, organized in the U.K./2/ gave very positive results showing that NDE can be efficient. A large programme PISC II /3/ started by OECD/CSNI and CEC in 1982 and continuing until the end of 1984, involves the participation of 55 teams from Europe (12 countries), the USA, Canada and Japan. The JRC Ispra Establishment is acting as the operating agent and referee laboratory to manage the programme, make the final destructive testing of four samples containing artificial defects in the welds (one is shown in Fig.2) and evaluate the results. Improved or advanced procedures (X-rays, ultrasonic techniques and eddy currents) will be applied.

3) Validation of these procedures on real structures and real defects. Work on real structures is performed in national laboratories in Europe, USA and Japan. Measurements on real defects (such as those existing in retired pressure vessels) are the object of the proposed PISC III programme.

Various parametric studies are performed in support of the PISC programme. One of the problems under investigation is the improvement of ultrasonic techniques in austenitic stainless steels in which these techniques have a poor performance due to the microstructural features of the material.

This introductory lecture to ASFM4 is intended as an overview of the problem areas connected with EPFM and which will be treated in the various lectures and workshops. At the end a list of questions is presented to which an answer is expected to be obtained during the seminar. It will be the task of Prof. Carlsson to discuss in the conclusion of the seminar, whether the answers are complete or whether they reflect the fact that much further research is needed in this advanced area of fracture mechanics.

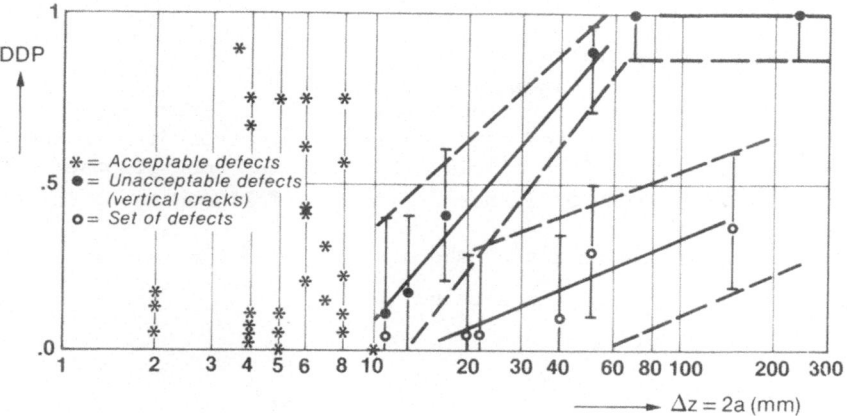

Figure 1. Defect detection probability as a function of defect size (PISC I Programme /1/).

Figure 2. PISC II: one of the samples.

2. ELASTIC-PLASTIC CRACK BEHAVIOUR

2.1 Regimes of a Cracked Body

Fig.3 is a schematic representation of the different regimes that a (two-dimensional) cracked body can experience depending on the spread of plasticity. This kind of classification has been proposed by Soete /4/; the definitions of Turner /5/ will be followed here. The notations used in Fig.3 are:

σ_1 : local stress at a very short distance from the crack tip;

σ_n : average net section stress;

σ_y : uniaxial yield stress;

σ : uniform applied stress.

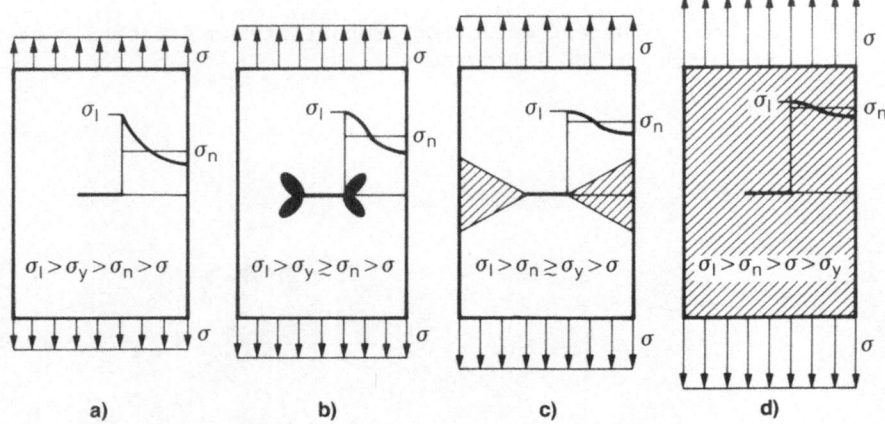

Figure 3. Different regimes of a cracked elastic-plastic body.

A schematic stress distribution in the net section is also shown in Fig.3, as well as the extent of plastic zones (black or shadowed areas). The four different regimes are the following:
a) LEFM regime: yielding limited to a small zone in the immediate vicinity of the crack tip;
b) elastic-plastic regime: a yield zone develops ahead of the crack tip but it does not reach the lateral boundary of the structure (contained yielding);
c) gross yield (or full ligament yielding): yielding spreads to the lateral boundaries and is thus uncontained;
d) general yield: the applied stress is larger than the yield stress; extensive plasticity develops in the whole structure.

Regimes b), c) and d) are those that require a treatment by EPFM concepts, although in regime b) acceptable answers may be obtained by
LEFM using a crack length corrected for the extent of the plastic zone.
Regime a) is supposed to be well-known throughout this seminar but it
will be evoked now and then in order to show the deviation of elastic-
plastic behaviour from LEFM behaviour.

The boundaries between the four regimes are more or less arbitrary.
Under a monotonically increasing load these regimes will merge gradually one into the other. However, the structure will reach general yield
only if the material is sufficiently ductile and crack growth resistent
to avoid failure within one of the prior regimes with less extended
plasticity. Failure in the LEFM regime is by unstable rapid crack propagation. The same failure mode usually occurs in the elastic-plastic
regime. In the gross yield regime the structure may fail by plastic
collapse of the net section but in less tough materials unstable crack
growth may be the governing process. Finally, in the general yield regime, plastic collapse tends to be the dominant failure mode.

2.2 Stable Crack Growth and Tearing Instability

Fig.3 gives only a simplified picture of stress redistribution with
the spread of plasticity, supposing a stationary crack. The real situation is more complex due to slow stable crack growth which may appear
in regimes b), c) and d) and which depends on the material crack growth
resistance properties. Thus, methods are needed for the prediction of
the extent of stable crack growth and of the conditions under which
this process may become unstable (tearing instability).

Crack growth resistance properties of a material are generally represented by crack growth resistance curves which describe the variation
of a crack characterizing parameter with crack extension Δa, Fig.4.
Usually the crack characterizing parameter is chosen to be J (J-resistance curve), sometimes the crack opening displacement (COD-resistance
curve). The theoretical basis for J-R (resistance) curves and their
use for predicting stable crack extension and tearing instability are
discussed in the lecture by Bakker. The problems of measurement of
J-R curves are treated in the lecture by Blauel, whereas Lamain will
discuss the possibilities of numerical analysis.

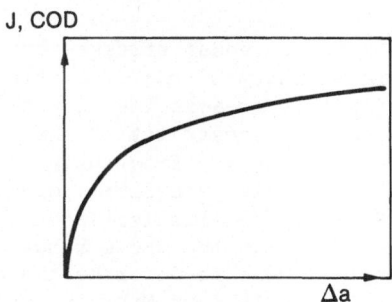

Figure 4. Crack growth resistance curves.

2.3 Constraint and Transferability

The two-dimensional structure shown in Fig.3 will behave differently if it is thin or thick. In thin plates the plane stress situation is approached where the out-of-plane deformation is free and the corresponding stress zero. In the interior of very thick plates the out-of-plane deformation is prevented by the constraint produced by the surrounding material and a triaxial stress state is created. The amount of constraint influences greatly the process of stable crack growth and the failure mode of the component.

Fig.5 illustrates the influence of constraint on global response curves. It shows typical load-displacement curves (of e.g. a three-point bend specimen) obtained by finite element analysis assuming plane strain or plane stress conditions. A more realistic analysis requires three-dimensional modelling with several finite element layers across the thickness. Indeed, experimental results tend to follow the calculated 3-D curve. Thus the real behaviour of the specimen is close to the plane strain situation for small loads but approaches more and more the plane stress situation with increasing load. The degree of constraint at a crack tip is thus essentially dependent on the spread of plasticity.

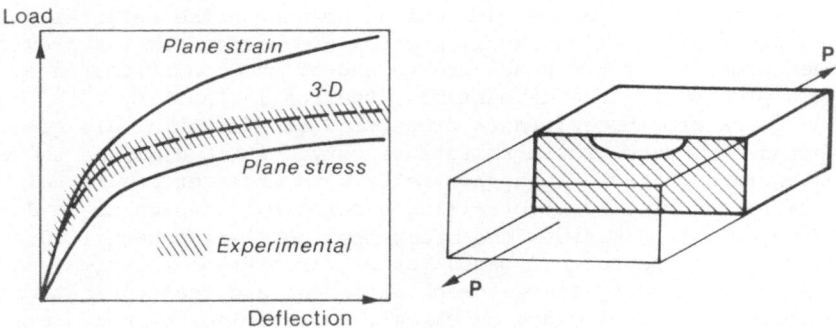

Figure 5.   Global response curves.

Figure 6.   Surface crack.

Referring to Fig.3, when plasticity extends, stresses and strains are less localized in the immediate vicinity of the crack tip and therefore the out-of-plane deformation in this area is less constrained. It is not so easy to describe the effects of constraint in the case of a surface crack (Fig.6). However, such cracks are frequently encountered in real structures, e.g. in pressure vessels, and their correct treatment is of paramount importance. Recently, Pellissier-Tanon /6/ has discussed the behaviour of surface cracks in the whole range from LEFM to general yield and presented simple models to describe the elastic-plastic response. In this seminar, the lecture by Munz is devoted to surface cracks.

One of the difficult problems in EPFM is that of the transferability of experimental results from specimens to real structures. For obvious reasons of economy, laboratory specimens are often of small size and offer less constraint for crack tip deformation than real structures do. Therefore the structural component may fail in the elastic-plastic regime b) of Fig.3, whereas a small specimen may easily be pushed in the gross or general yield regime where it will fail by a different mechanism. Hints to our present understanding of how to treat the problem of transferability will be found in various lectures of ASFM4.

2.4 Local Phenomena

Looking more closely at what happens in the crack tip region in ductile materials, one can observe the phenomena depicted in Fig.7. The letters a) to d) are not in relation to the four regimes described in Fig.3. Under increasing loading, the crack tip starts blunting to accommodate the high strains arising in its vicinity. When the local strain reaches a certain level, microcracks or voids are formed. With increasing strain these voids grow until they reach a critical size. At this point the crack tip advances through coalescence of the voids. New voids are formed ahead of the new crack tip and after a further increase of load these coalesce, etc., producing the stable crack growth phenomenon.

It is evident that microstructural features play a key role in this process as they do in fast fracture. Micromechanisms are discussed in the lecture by Dahl and Dormagen. They also describe another type of ductile fracture called shear decohesion, occurring e.g. in quenched and tempered steels.

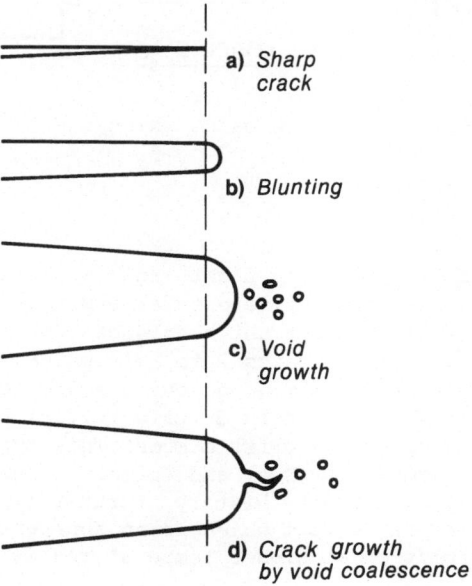

a) *Sharp crack*

b) *Blunting*

c) *Void growth*

d) *Crack growth by void coalescence*

Figure 7. Crack tip behaviour.

A particular environmental effect on microstructure is that produced by neutron irradiation in nuclear reactors. Some basic explanations on irradiation damage are given in the lecture by Bernard and Verzeletti who discuss some special precautions to be taken and methods used when dealing with J-R curve measurements on irradiated materials. The reader not interested in irradiated materials might, however, find the described key-curve method attractive whenever he needs to obtain as much information as possible from very few and small specimens.

2.5 Failure Paths Followed by a Cracked Body

Adding to the failure modes summarized in 2.1, the effects of stable crack growth presented in 2.2, all possible failure paths that a cracked body can follow can be schematically represented as in Fig.8.

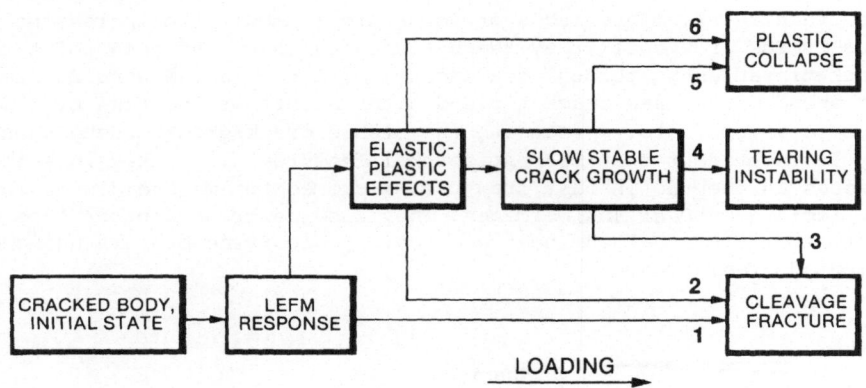

Figure 8. The different failure paths which a cracked body can follow.

Loading increases from the left to the right, and is scaled in such a way that the end points in the three different failure modes are placed on the same vertical line at the right hand side.

ASFM4 is designed to treat failure paths 2 to 6 in which elastic-plastic effects are present. Particularly dangerous are the descending lines 2 and 3. These lead, after a seemingly ductile initial behaviour, to a catastrophic final fracture in the brittle mode. This type of problems are treated in the lecture by Dahl and Dormagen. When the temperature is varying during the loading history, further complications are introduced. The warm prestress effect which is an apparent enhancement of toughness after deformation at higher temperatures is the object of the second lecture by Mudry.

3. CHARACTERIZING PARAMETERS

In order to be accepted for use in quantitative fracture control, any variable proposed as a parameter characterizing crack tip deformation fields must satisfy the following criteria /5/:
- it must be based on a sound physical model;
- it must be accessible both experimentally and numerically by well defined testing methods and computer analysis;
- the range of applicability to practical problems should be as wide as possible.

Undoubtedly the J-integral has reached today a much more widespread use than the COD. Correspondingly, a considerable space has been reserved to J in this seminar. The COD-concept is, however, applied in the COD design method which has reached the status of British Standard. This method will be treated in the workshop by Turner.

The theoretical basis of the path integral J, its extension to 3-D situations and the limitations of this concept are dealt with by Bakker. Experimentally J is evaluated from the energy equivalent of J (the area under the load-load point displacement curve of the specimen). In fracture assessments critical values of J and COD are needed; their definition and measurement is the object of the lecture by Garwood. Lamain will discuss the numerical aspects of J- and COD evaluation.

Recently, a quite different approach to ductile fracture problems, based on local criteria has been developed. In this method no characterizing parameter is needed. Continuum mechanics are applied to calculate the elastic-plastic deformation and stress fields. Ductile fracture damage is evaluated by simulating the growth of cavities by a model which accounts for the triaxial stress state. The damage can be introduced in the constitutive relations /7/ to modify the behaviour of the material or, alternatively, the cavity growth prediction can serve directly in a fracture criterion. Thus, this method allows predicting crack growth initiation and slow stable crack growth in specimens and structures, including 3-D situations like surface cracks. The local criteria are discussed in the first lecture by Mudry. The concepts of fracture and damage are further elaborated in the lecture by François.

4. DESIGN METHODS

Engineers need practical rules for applying fracture mechanics technology to design problems and for assessing the safety of existing structures. Such rules should be assembled in a code format in which each step is documented. The results coming out from advanced research are either implemented in the design methods only after careful testing and when consensus has been reached, or even ignored because considered to be too complex. In the latter case one has of course to demonstrate that the simplified approaches adopted in the design methods are conservative.

Another requirement for a design method is that it should be complete:

- tell how a defect or a group of defects is to be characterized by a crack;
- indicate how crack growth due to creep, fatigue and environment should be incorporated in the analysis;
- give protection against all possible failure paths (Fig. 8).

Four design methods are treated in ASFM4 from the point of view of how EPFM is handled by them (two of them are limited to EPFM and do not have the pretension of codes):
- the R6 (CEGB) procedure by Dowling et al.;
- the EPRI method by Rousselier;
- the J-based procedure (En J) by Turner;
- the COD method by Turner.

5. DYNAMIC EFFECTS

The interest in dynamic fracture mechanics stems from two considerations:
a) dynamic loadings which create high deformation rates and wave propagation effects;
b) the crack arrest problem.

For safety reasons it is important to know under which conditions a crack which for some reason has started to grow rapidly (accidental loads, local embrittlements in the material) can be arrested. To cope with these problems one needs experimental and analytical methods which differ from those used in static conditions. For instance, crack arrest specimens should be designed in such a way that the measurements are not perturbed by wave propagation effects (otherwise complex numerical analysis is mandatory for the evaluation of such tests). Numerical analysis is much more involved than in static problems not only due to inertia loads and wave propagation but also due to the strain rate effect on material properties. These problems are discussed in two lectures of ASFM4, the first by Brickstad and Nilsson and the second by Kalthoff.

6. QUESTIONS TO BE DISCUSSED DURING ASFM4

1) Do we fully understand the behaviour of elastic-plastic cracked bodies in all regimes (increasing degree of yielding)?
   - stable crack growth and tearing instability
   - plastic collapse
   - failure paths
   - correlation of macroscopic behaviour with microstructure
   - description and prediction of the effect of constraint
     . 3-dimensional problems
     . transferability of specimen results to structures.
2) Are there acceptable characterizing parameters?
   - possibilities and limitations of J and COD
   - problems in the experimental determination
     . R-curves
     . critical values
   - problems in the numerical analysis

- use of local criteria (no J- or COD-like parameters needed)
- damage.
3) What is the state-of-the-art of design methods?
   - conservatism
   - protection against all types of failure (completeness)
   - ease of use.
4) How much is known about dynamic effects?
   - crack arrest methodology
   - dynamic tearing
   - material strain rate dependence
   - problems of measurement and analysis.

REFERENCES

1. Crutzen, S.J. and Parry, G.W., 'Method of evaluation and the re-
    sults of the PISC trials', Post Conference Seminar on NDE in Rela-
    tion to Structural Integrity, SMiRT-5, Berlin, 22nd August 1979.
2. International Symposium on the UK Defect Detection Trial, Risley,
    7-8 October 1982.
3. Crutzen, S.J., 'PISC I and PISC II. Working for effective and re-
    liable inspection procedures', IAEA Int. Symp. on Reliability of
    Reactor Pressure Components, Stuttgart, 21-25 March 1983.
4. Soete, W., 'An experimental approach to fracture initiation in
    structural steel', *Advances in Research on the Strength and Frac-
    ture of Materials*, D.M.R. Taplin, Ed., Vol.I, pp.775-804, Pergamon
    Press, 1978.
5. Turner, C.E., 'Methods for post-yield fracture safety assessment',
    Post-yield Fracture Mechanics, D.G.H. Latzko, Ed., pp.23-210,
    Applied Science Publishers, 1979.
6. Pellissier-Tanon, A., 'Fracture mechanics in structure design',
    Application of Fracture Mechanics to Materials and Structures,
    G.C. Sih, E. Sommer and W. Dahl, Eds., pp.213-234, Martinus Nij-
    hoff Publishers, 1984.
7. Rousselier, G., 'Finite deformation constitutive relations inclu-
    ding ductile fracture damage', *Three-dimensional Constitutive Re-
    lations and Ductile Fracture*, S. Nemat-Nasser, Ed., North-Holland
    Publishing Co., 1981, pp.331-355.

THE J-CONCEPT: THEORETICAL BASIS AND ITS USE IN EPFM

Ad Bakker
Delft University of Technology
Laboratory for Thermal Power Engineering
P.O. box 5055, 2600 GA Delft
The Netherlands

ABSTRACT. The first part of this paper summarizes the original (two-dimensional) definition of the J-integral as introduced by Rice and the physical interpretation of J in terms of potential energy release and crack tip stress/strain field characterization. The derivation of a conservation law is discussed, from which the J-theory is generalized to plane defects in three-dimensional bodies and loadings not covered by the original concept, such as thermally induced initial strains, body forces and crack surface pressure.
The second part deals with the application of the J-concept to elastic-plastic materials. It is argued that, as the theoretical concept is only valid for linear or non-linear elastic materials, or materials obeying a deformation theory of plasticity, practical applications are limited to the prediction of crack growth initiation and (to within certain limits) stable crack growth and final instability in structures with monotonically increasing proportional loads. Some important backgrounds of these specific applications of the J-concept are discussed briefly.

1. INTRODUCTION

The impact of the presence of a crack on the load bearing capacity of a structure is an important issue of safety assessments with regard to both design and fitness for purpose validations of operating structures. For structures behaving pre-dominantly in a linear elastic fashion the Irwin stress intensity factor concept [1] has become an important tool for this purpose, as reflected by the existence of an ASTM [2] and BSI [3] standard for the determination of critical material stress intensity factors (K_{Ic}) and its use by the ASME [4] pressure vessel codes for the design (section III) and fitness for purpose validation (section XI) of pressurized components in nuclear power stations. This concept, however, loses its sense when fracture is preceded by a significant amount of inelastic (plastic) deformation, as generally occurs in structures made of low- and intermediate strength materials

13

L. H. Larsson (ed.), Elastic-Plastic Fracture Mechanics, 13–53.
© *1985 ECSC, EEC, EAEC, Brussels and Luxembourg.*

of high toughness and operating above the transition temperature.

The first proposals for the application of the Rice [5] J-integral as a fracture criterion in elastic-plastic materials were made by Broberg [6] and Begley & Landes [7]. Numerous subsequent studies have revealed J as a suitable parameter to predict the onset of crack growth in elastic-plastic materials and even (though in a fairly restricted sense) to describe the process of stable crack growth and final tearing instability in such materials, as first proposed by Paris et al. [8]. The confidence in the J-concept of at least an important part of the engineering community is reflected by the introduction of national standards in the USA (ASTM [9]) and Japan (JSME [10]) for the experimental determination of the critical J-value at the onset of crack growth, J_{Ic}.

The first part (section 2) of this lecture gives a brief survey of the original definition of J, i.e. for cracks in two-dimensional bodies without initial strains, body forces and crack face tractions, and its physical interpretation in *linear or non-linear elastic* materials in terms of potential energy release and crack tip stress/strain field characterization. As a general acceptance of the concept also requires it to be applicable to general (three-dimensional) bodies with loading conditions mentioned above, the extension of the theoretical concept to such situations, based on an extended version of a conservation law derived by Knowles & Sternberg [11], is described in section 3. Finally, some aspects of the application of the J-concept in the field of *Elastic-Plastic* Fracture Mechanics (EPFM) are discussed in section 4. By the scope of this lecture, i.e. to give some theoretical backgrounds of the J-concept which must lead to a better understanding of subsequent lectures at this seminar dealing with specific aspects of the concept, this discussion is almost completely restricted to the limitations of the use of J in EPFM imposed by the fact that the theoretical basis as discussed in sections 2 and 3 is essentially valid for linear and non-linear elastic material behaviour only.

Reference to original work is made as much as possible, without, however, having the intention of being complete, as this would render this paper useless as a seminar text. A more comprehensive treatise and literature survey of the subject is given by Turner [12].

2. ORIGINAL CONCEPT

2.1. Definitions

The original definition of the J-integral by Rice [5] refers to a homogeneous body of a linear or non-linear elastic material, free of body forces and initial (stress free) strains, subjected to a two-dimensional stress- and deformation field in a Cartesian $\underline{x} = \{x_i\}$, (i = 1,2) space.

When:

$$\underline{u} = \{u_i\} = \text{Cartesian displacement vector field,}$$

$$\underline{\varepsilon} = \{\varepsilon_{ij}\} = \text{infinitesimal strain tensor field:}$$

$$\varepsilon_{ij} = \frac{1}{2} \left(\frac{\partial u_i}{\partial x_j} + \frac{\partial u_j}{\partial x_i} \right)$$

$$\underline{\sigma} = \{\sigma_{ij}\} = \text{stress tensor field}$$

are defined to be functions of the Cartesian coordinates $\underline{x}$ only, the strain energy density W is defined as ★):

$$W = W(\underline{\varepsilon}) = \int_0^{\varepsilon} \sigma_{ij} \, d\varepsilon_{ij} \tag{1}$$

Rice showed that for any *closed* curve $\Gamma^{\bigstar}$ (cf. figure 1):

$$\oint_{\Gamma^{\bigstar}} \left(W \, dx_2 - T_i \frac{\partial u_i}{\partial x_1} \, ds \right) = 0 \tag{2}$$

where $\underline{T}$ is the traction vector defined according to the outward normal $\underline{n}$ along $\Gamma^{\bigstar}$, $T_i = \sigma_{ij} n_j$, and ds is an element of arc length along $\Gamma^{\bigstar}$.

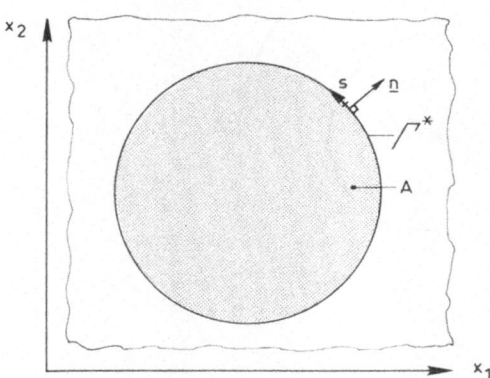

Figure 1. Closed contour in a two-dimensional body. Enclosed area A is free of singularities and in equilibrium.

★) *Note that the summation convention of repeated indices of vector and tensor components is used throughout this paper.*

The proof of equation (2) is subject to the condition that the area A enclosed by Γ^{*} is free of singularities and in equilibrium, i.e. in the absence of body forces:

$$\frac{\partial \sigma_{ij}}{\partial x_j} = 0 \tag{3}$$

For a traction free notch having flat surfaces parallel to the x_1-axis and a rounded tip, denoted by the arc Γ_t in figure 2, the J-integral is defined as the integral of equation (2) for an *open* path Γ surrounding the notch tip, starting and ending at the lower and upper flat notch surfaces, respectively:

$$J = \int_{\Gamma} \left(W \, dx_2 - T_i \frac{\partial u_i}{\partial x_1} \, ds \right) \tag{4}$$

By application of equation (2) to a closed contour surrounding the shaded area A in figure 2, i.e. formed by Γ, any other arbitrary contour inside Γ (denoted as Γ' in figure 2) and the parts of the bottom and top notch surfaces connecting Γ and Γ', it can easily be verified that, since the integrand of equation (2) vanishes at the flat notch surfaces ($dx_2 = 0$ and $\underline{T} = 0$), the J-integral according to

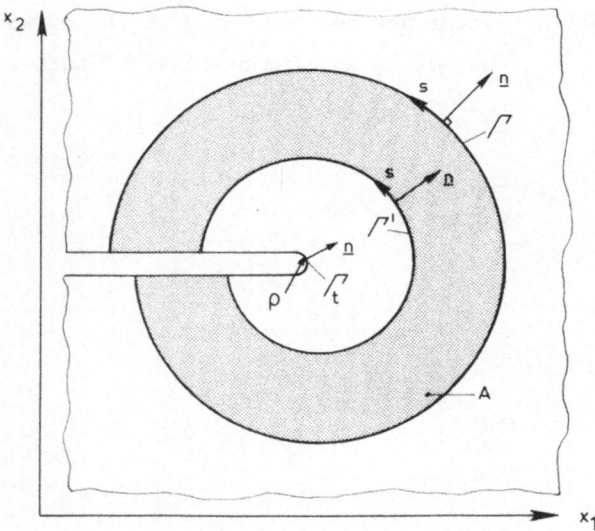

Figure 2. Contours surrounding the notch tip of a notch in a two-dimensional body. The flat notch surfaces are parallel to the x_1-axis. The curved notch tip contour is denoted as Γ_t.

equation (4) is identical for the paths Γ and Γ' provided that identical positive normal directions are used for both paths. Hence, J according to equation (4) is a *path independent* contour integral. By shrinking the path to the notch tip, $\Gamma \rightarrow \Gamma_t$ and since $\underline{T} = 0$ for the notch surfaces:

$$J = \int_{\Gamma_t} W \, dx_2 \tag{5}$$

which implies that, since $W = W(\underline{\varepsilon})$, J is an averaged measure of the strain on the notch tip.

This reflects the basic utility of the method, i.e. that the state at a notch tip according to equation (5) can be determined by evaluating the integral of equation (4) for an arbitrary contour Γ away from the notch tip. In certain cases, of which some are discussed by Rice [5], this permits a direct evaluation of J by a suitable choice of Γ.

Equation (5) is not valid for the limiting case of a straight *crack*, for which the notch tip radius $\rho \rightarrow 0$. However, the proof of path independence as outlined before still applies. Hence, the J-value evaluated for an arbitrary contour Γ is still equal to that for a small circular contour of radius r, as indicated in figure 3:

$$J = \int_{\Gamma} \left(W \, dx_2 - T_i \frac{\partial u_i}{\partial x_1} \, ds \right) = r \int_{-\pi}^{\pi} \left(W \cos \theta - T_i \frac{\partial u_i}{\partial x_1} \right) d\theta \tag{6}$$

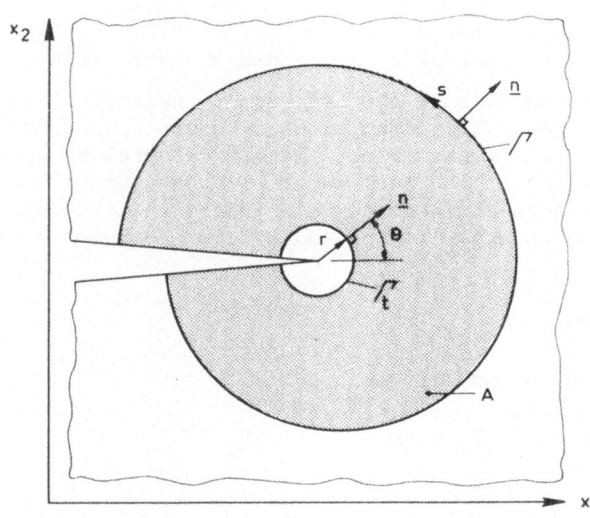

Figure 3. Contours surrounding the crack tip of a crack in a two-dimensional body. The flat crack surfaces are parallel to the x_1-axis. Γ_t is a circular contour with radius r ($ds = rd\theta$, $dx_2 = n_1 \, ds = r\cos\theta d\theta$).

This holds for any arbitrarily small value of r, which implies that the limit r → 0 exists. The J-value for a crack is then defined as:

$$J = \lim_{r \to 0} \left[r \int_{-\pi}^{\pi} \left(W \cos \theta - T_i \frac{\partial u_i}{\partial x_1} \right) d\theta \right]$$ (7)

By this, J depends on the magnitude of near crack tip 1/r singular terms in the strain energy density W and products of stresses and strains only. The equality of J determined from equation (7) to that for a general, far field contour Γ according to equation (6) is valid for the conditions for which the proof of path independence applies only, i.e. cracks in two-dimensional bodies in the absence of:

- body forces
- initial strains
- surface tractions on crack faces.

Equation (7) is adopted as the *general definition* of J, as it depends on the singularity of stresses and strains only, also in the presence of the above effects. The impact of these effects on the *far field* definition of J is discussed in section 3, together with the extension of the method to plane cracks in three-dimensional bodies.

2.2. Physical interpretation

Two important physical interpretations can be given to the J-integral as defined in the previous subsection, viz.:

- in terms of potential energy release
- in terms of crack tip stress/strain field characterization.

The short survey given below only applies from a theoretical point of view, i.e. from which the basic assumptions of *small (infinitesimal) deformations* and *elastic* (linear or non-linear) material behaviour are correct. The consequences of the real material behaviour in the neighbourhood of cracks in engineering materials (finite deformations at the blunting tip and plasticity obeying a flow theory) are discussed in section 4.

Potential energy release

The potential energy of a body is given by:

$$P = \int_V W \, dV - \int_{A_T} T_i u_i \, dA - \int_V f_i u_i \, dV$$ (8)

where V is the volume of the body, A_T is that portion of the body surface with prescribed surface tractions $\underline{T}$, and $\underline{f}$ are body forces per unit volume. The first volume integral gives the absorbed energy and the surface and second volume integral give the work done by external tractions and body forces, respectively.

Rice [13] showed that in the absence of body forces ($\underline{f} = 0$) the J-integral is identical to the energy release rate, i.e. the rate of potential energy release with respect to crack length a:

$$J = - \frac{\partial P}{\partial a} \tag{9}$$

where, for a two-dimensional body, P is given per unit thickness. Hence, for *linear*-elastic material behaviour, J is equal to the LEFM energy release rate G and related to the stress intensity factors K, both introduced by Irwin [1]. For a state of plane strain:

$$J = G = \frac{(1-\nu^2)}{E} \left(K_I^2 + K_{II}^2 \right) + \frac{(1+\nu)}{E} K_{III}^2 \tag{10}$$

where the indices I, II and III to K denote the opening, in-plane shear and anti-plane shear modes, respectively, E is Young's modulus and ν is Poisson's ratio. The factor $(1-\nu^2)$ should be removed for a plane stress situation. By the equality of equation (10), the definition of J as a far field contour integral according to equation (6) supplies an attractive way to determine stress intensity factors in pure mode situations.

The proof of equation (9) as given by Rice [13] is limited to two-dimensional bodies free of initial strains, body forces and crack surface tractions, and refers primarily to the far field definition of J. However, it can be deduced from the work of various authors, aimed at the extension of the J-theory, that equation (9) is valid in general, provided the near tip field definition of J according to equation (7) is used (e.g. Sakata et al. [14]: body forces, Ainsworth et al. [15]: initial strains, de Lorenzi [16]: three-dimensional bodies).

Crack tip stress/strain field characterization

In subsection 2.1 it was already argued that only the 1/r singular terms in the near tip distribution of the strain energy W and products of stress- and strain components contribute to the non-zero value of J according to equation (7). Hence, J can be regarded to characterize the singularity in the near tip stress/strain field.

For *linear*-elastic materials, the singular terms of the Westergaard [17] solution for stresses and strains can be expressed in terms of the stress intensity factors. E.g. for the stresses in mode I loading:

$$\sigma_{ij} = \frac{K_I}{\sqrt{2\pi r}} f_{ij} (\theta) \tag{11}$$

where r, θ are polar coordinates with origin at the crack tip (cf.

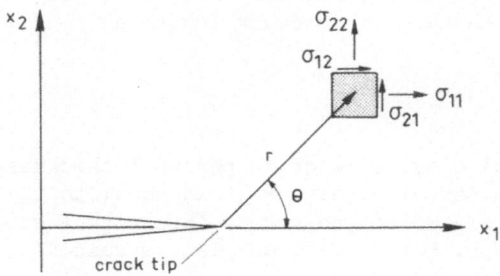

Figure 4. Polar coordinate system with origin at the crack tip.

figure 4) and the set of functions $f_{ij}(\theta)$ are the same for all cracks loaded in mode I. The solutions for modes II and III differ by these functions only. Hence, by the equality of equation (10), the linear-elastic stresses (and strains) for cracks loaded in a pure mode can also be written in terms of J. Rice [5] argued that this also holds for small-scale yielding near the crack tip (plastic zone size small compared to crack length and other relevant dimensions), for which it is anticipated that the elastic singularity still governs the stresses and strains at distances from the crack tip that are large compared to the yield zone but still small compared to crack size and other relevant dimensions. Based on finite element results, Larsson & Carlsson [18] found a geometry dependence of the small scale yielding plastic zone size, which is in contradiction to this. They, however, suggested that this was caused by differences in the non-singular terms in the normal stress parallel to the crack, and Rice [19], who arrived at the same conclusion, argued that these terms have no effect on the near tip J-value which only depends on the singular terms in the stress distribution.

   The singular behaviour of stresses and strains in the vicinity of a crack tip in *inelastic* materials was investigated by Hutchinson [20], [21] and Rice & Rosengren [22]. They employed a Ramberg-Osgood power relation for the material stress-strain curve according to:

$$\frac{\varepsilon}{\varepsilon_y} = \frac{\sigma}{\sigma_y} + \alpha \left(\frac{\sigma}{\sigma_y}\right)^n \tag{12}$$

where σ_y is the yield stress, ε_y the corresponding yield strain ($\varepsilon_y = \sigma_y/E$) and α and n are material constants. Some examples of stress-strain curves according to equation (12) are shown in figure 5.

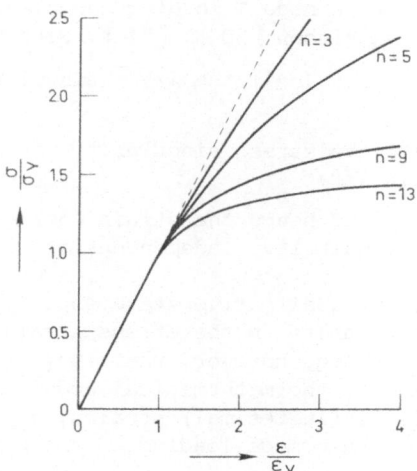

Figure 5. Ramberg-Osgood stress-strain relations for $\alpha = 0.02$ and different values of n.

The analyses of Hutchinson, Rice & Rosengren (often referred to jointly as the HRR solutions) employ a deformation theory of plasticity, which is, however, a special form of non-linear elasticity with imposed incompressibility of the non-linear (plastic) deformations. Hutchinson expressed the singular stress- and strain distribution in terms of a "singularity amplitude" denoted as K. It was pointed out by McClintock [23] that they can also be expressed in terms of J, i.e. in the notation adopted here:

$$\frac{\sigma_{ij}}{\sigma_y} = \left(\frac{1}{\alpha \, \sigma_y \, \varepsilon_y \, I_n} \cdot \frac{J}{r} \right)^{\frac{1}{n+1}} \cdot \tilde{\sigma}_{ij}(\theta)$$

$$\frac{\varepsilon_{ij}}{\varepsilon_y} = \alpha \left(\frac{1}{\alpha \, \sigma_y \, \varepsilon_y \, I_n} \cdot \frac{J}{r} \right)^{\frac{n}{n+1}} \cdot \tilde{\varepsilon}_{ij}(\theta) \tag{13}$$

where:

I_n = constant dependent on n in equation (12), loading mode and stress state (plane stress/plane strain)

$\tilde{\sigma}_{ij}(\theta)$, $\tilde{\varepsilon}_{ij}(\theta)$ = non-dimensional functions of θ, dependent on n, loading mode and stress state.

Numerical values of I_n and graphs of $\tilde{\sigma}_{ij}(\theta)$ and $\tilde{\varepsilon}_{ij}(\theta)$ for several values of n, plane stress and plane strain mode I loading and plane strain mode II loading are given by Hutchinson [20], [21]. Note that:

- for linear elastic material behaviour (n=1) the $1/\sqrt{r}$ singularity of equation (11) is restored;

- for ideal plastic behaviour (n→∞) the stress singularity vanishes and the strains are singular with $1/r$;

- products of stresses and strains (and hence the strain energy W of equation (1)) exhibit a $1/r$ singularity, independent on n.

Equations (13) confirm the previous qualitative statement, i.e. that J characterizes the dominant singularity in the stress/strain field at a crack tip. It should be recalled, however, that they are only correct within the bounds imposed by the mathematical model used to generate the solutions, i.e. small (infinitesimal) strains, a deformation theory of plasticity and a pure mode loading of the crack.

3. EXTENSIONS OF THE THEORETICAL CONCEPT

3.1. Conservation laws

For the extension of the J-integral method to general (three-dimensional) bodies subjected to loadings which induce initial strains (e.g. non-uniform temperature distributions), body forces and/or surface tractions on the crack surfaces, we consider $\underline{u}(\underline{x})$, $\underline{\varepsilon}(\underline{x})$ and $\underline{\sigma}(\underline{x})$ to represent, respectively, the Cartesian displacement vector field, its associated infinitesimal strain tensor field and the equilibrium stress tensor field within a body. Adoption of infinitesimal strains implies ★):

$$\varepsilon_{ij} = \frac{1}{2}\left(\frac{\partial u_i}{\partial x_j} + \frac{\partial u_j}{\partial x_i}\right) \tag{14}$$

$$\sigma_{ij} = \sigma_{ji}$$

and the equilibrium state of the stress field:

$$\frac{\partial \sigma_{ij}}{\partial x_j} - f_i = 0 \tag{15}$$

★) *Note that the dependence of variables on the spatial position $\underline{x}$ within the body is not indicated further, i.e. $\varepsilon_{ij} \equiv \varepsilon_{ij}(\underline{x})$, etc.*

where f_i stands for the components of the body force vector field $\underline{f}$ (per unit volume).

The *total* strains of equation (14) are split into a *mechanical* (stress induced) part $\underline{\varepsilon}^m$ and an *initial* (stress free) part $\underline{\varepsilon}^o$:

$$\varepsilon_{ij} = \varepsilon_{ij}^m + \varepsilon_{ij}^o \tag{16}$$

In case of thermally induced initial strains in isotropic materials:

$$\varepsilon_{ij}^o = \alpha \, \delta_{ij} \, \theta \tag{17}$$

where α is the coefficient of thermal expansion, δ_{ij} is Kronecker's delta and θ is the material temperature relative to a reference temperature at which the body is free of thermally induced stresses. The mechanical strains and stresses within an elastic body may be related by a constitutive relation:

$$\varepsilon_{ij}^m = C_{ijkl} \, \sigma_{kl} \tag{18}$$

For *linear* elastic materials, equation (18) represents Hooke's law and $\underline{C}$ is a tensor of material constants (depending on Young's modulus E and Poisson's ratio ν only if the material is also isotropic). For *non-linear* elastic materials, $\underline{C} \equiv \underline{C}(\underline{\varepsilon})$.

The strain energy density at a point, given by equation (1) in the absence of initial strains, is re-defined according to:

$$W = \int_0^{\underline{\varepsilon}^m} \sigma_{ij} \, d \, \varepsilon_{ij}^m \tag{19}$$

By equation (18) $W \equiv W(\underline{\varepsilon}^m)$, and hence stresses can be derived from W according to:

$$\sigma_{ij} = \frac{\partial W}{\partial \varepsilon_{ij}^m} \tag{20}$$

We now consider Ω to be a regular subregion of the body with closed, piecewise differentiable surface $\Sigma^\star$. Ω' is an image of Ω according to the mapping:

$$x_i' = x_i + \delta x_i \tag{21}$$

where $\delta \underline{x} \equiv \delta \underline{x}(\underline{x})$ is a smooth virtual vector field of infinitesimal translations.

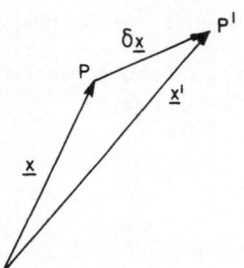

Figure 6. Mapping of a point P on Ω to its image P' on Ω'.

The difference between the strain energy at points P' and P, where P' is the image of P (cf. figure 6) follows from:

$$\delta W = W(\underline{x}') - W(\underline{x}) = \delta x_k \frac{\partial W}{\partial x_k} \tag{22}$$

In Appendix 1 it is shown that from equation (22), using equations (14) through (20) and the divergence theorem of Gauss, the following condition can be derived:

$$\int_{\Sigma^\star} \delta x_k \left(W n_k - T_i \frac{\partial u_i}{\partial x_k} \right) dS -$$

$$- \int_\Omega \left[\frac{\partial \delta x_k}{\partial x_j} \left(W \delta_{jk} - \sigma_{ij} \frac{\partial u_i}{\partial x_k} \right) - \delta x_k \left(\sigma_{ij} \frac{\partial \varepsilon_{ij}^o}{\partial x_k} + f_i \frac{\partial u_i}{\partial x_k} \right) \right] dV = 0 \tag{23}$$

where $\underline{T}$ is the traction vector defined according to the outward normal $\underline{n}$ on $\Sigma^\star$ ($T_i = \sigma_{ij} n_j$) and dV is a volume element of Ω.

For a constant $\delta \underline{x}$ field, i.e. independent of the spatial position $\underline{x}$, the mapping is the result of a rigid translation of Ω to Ω' (cf. figure 7).

Hence, $\partial \delta x_k / \partial x_j = 0$ and the first term in the volume integral of equation (23) vanishes, and the remainder of this equation becomes linear in δx_k. By applying unit translations in the three coordinate directions, the following three independent conditions can then be derived (k = 1, 2 or 3):

$$\int_{\Sigma^\star} \left(W n_k - T_i \frac{\partial u_i}{\partial x_k} \right) dS + \int_\Omega \left(\sigma_{ij} \frac{\partial \varepsilon_{ij}^o}{\partial x_k} + f_i \frac{\partial u_i}{\partial x_k} \right) dV = 0 \tag{24}$$

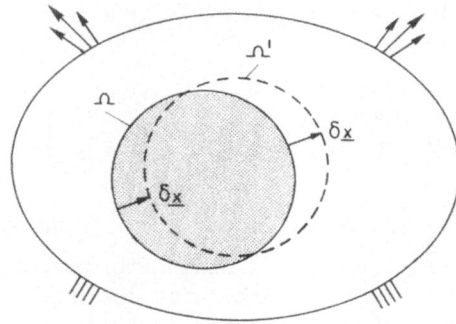

Figure 7. Mapping of subregion Ω to its image Ω' resulting from a rigid translation of Ω.

In the absence of both initial strains and body forces the volume integral vanishes, and equation (24) reduces to a condition derived by Eshelby [24] and stated later by Knowles & Sternberg [11] as the first of a series of three conservation laws in static elastic media.

The derivation of the extended expression for J requires the application of equation (24) to a special subregion formed by a slice of infinitesimal thickness. Appendix 2 gives the derivation according to which equation (24) applied to a slice in the x_1-x_2 plane (cf. figure 8) transforms to (k = 1, 2 or 3):

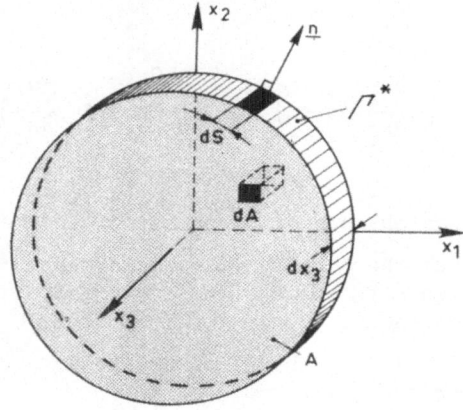

Figure 8. Subregion consisting of a slice of infinitesimal thickness dx_3. Γ^* is a closed contour in the x_1-x_2 plane, A is the enclosed area.

$$\Gamma^\star \int \left(W n_k - T_i \frac{\partial u_i}{\partial x_k} \right) ds +$$

$$+ \int_A \left[\frac{\partial}{\partial x_3} \left(W \delta_{k3} - \sigma_{i3} \frac{\partial u_i}{\partial x_k} \right) + \sigma_{ij} \frac{\partial \varepsilon_{ij}^o}{\partial x_k} + f_i \frac{\partial u_i}{\partial x_k} \right] dA = 0 \quad (25)$$

where $\Gamma^\star$ is the contour of the subregion in the x_1-x_2 plane and A is the enclosed area. Note that for two-dimensional deformation fields and zero initial strains and zero body forces the area integral vanishes, and that the remaining contour integral for $k = 1$ corresponds with Rice's equation (2), as $dx_2 = n_1$ ds.

3.2. Two-dimensional bodies

We consider the two-dimensional crack configuration of figure 3, re-drawn in figure 9. The crack surfaces are now assumed to be loaded by a surface pressure p. The area A enclosed by a far field contour Γ and a near tip circular contour Γ_t with radius r is a subregion to which equation (25) can be applied. The closed contour $\Gamma^\star$ of this sub-region is formed by Γ, Γ_t and the top and bottom parts of the crack surfaces connecting Γ and Γ_t, denoted as Γ_s^+ and Γ_s^-, respectively. By $n_1 = 0$ and $T_1 = 0$ on Γ_s^+ and Γ_s^-, $T_2 = p$ on Γ_s^+ and $T_2 = -p$ on Γ_s^-, application of equation (25) in the x_1 direction ($k = 1$) yields:

$$\int_\Gamma \left(W n_1 - T_i \frac{\partial u_i}{\partial x_1} \right) ds - \int_{\Gamma_t} \left(W n_1 - T_i \frac{\partial u_i}{\partial x_1} \right) ds -$$

$$- \int_{\Gamma_s^+} p \frac{\partial u_2}{\partial x_1} ds + \int_{\Gamma_s^-} p \frac{\partial u_2}{\partial x_1} ds + \int_A \left(\sigma_{ij} \frac{\partial \varepsilon_{ij}^o}{\partial x_1} + f_i \frac{\partial u_i}{\partial x_1} \right) dA = 0 \quad (26)$$

where the integral over Γ_t has changed sign because of the reversed positive normal direction, and the terms involving derivatives to x_3 in the area integral vanish in a two-dimensional deformation field. For the circular contour with radius r, the integral along Γ_t is equal to the second part of equation (6) which is, according to equation (7), defined as the J-integral for the limit $r \rightarrow 0$.

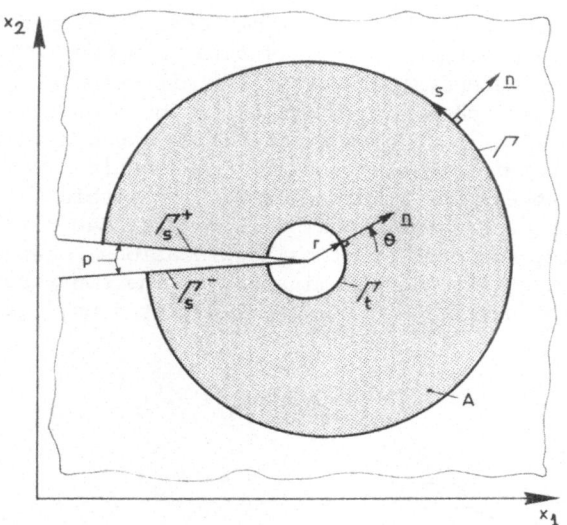

Figure 9. Contours surrounding the crack tip of a crack in a two-dimensional body. Γ_s denotes the parts of the flat crack surfaces parallel to the x_1-axis between the far field contour Γ and a small circular contour Γ_t with radius r.

Hence, we find from equation (26):

$$
J = \lim_{r \to 0} r \int_{-\pi}^{\pi} \left(W \cos \theta - T_i \frac{\partial u_i}{\partial x_1} \right) d\theta
$$

$$
= \int_{\Gamma} \left(W n_1 - T_i \frac{\partial u_i}{\partial x_1} \right) ds - \int_{\Gamma_s^+} p \frac{\partial u_2}{\partial x_1} ds + \int_{\Gamma_s^-} p \frac{\partial u_2}{\partial x_1} ds +
$$

$$
+ \int_{A} \left(\sigma_{ij} \frac{\partial \varepsilon_{ij}^o}{\partial x_1} + f_i \frac{\partial u_i}{\partial x_1} \right) dA \tag{27}
$$

The integral along Γ corresponds with the original definition of J according to equation (6) as $dx_2 = n_1 ds$. The term to include the initial strain effects was given by Ainsworth et al. [15] *), that for

---

*) *For thermally induced initial strains, the initial strain term in the area integral transforms by equation (17) to:*

$$
\sigma_{ij} \frac{\partial \varepsilon_{ij}^o}{\partial x_1} = \sigma_{ii} \, \alpha \, \frac{\partial \theta}{\partial x_1}
$$

the body forces by Sakata et al. [14]. By the terms involving the effects of crack surface pressure, initial strains and body forces, one important utility of the method, i.e. that the limit state at the crack tip in terms of the stress and strain singularity could be determined by evaluating an integral along a contour away from the crack tip, is lost. Evaluation of the complete equation (27) now requires the knowledge of the complete stress and strain distribution within the area A enclosed by the contour Γ. Generally this will require the results of a finite element analysis. For most practical applications, however, the contribution of the integral along Γ dominates, as the contribution of the other integrals vanishes for $\Gamma \to \Gamma_t$, as illustrated by figure 10.

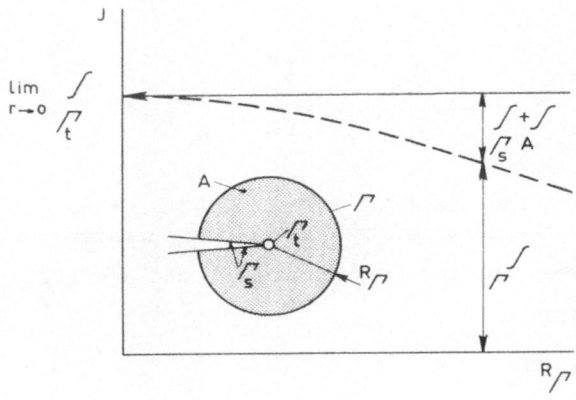

Figure 10. Diminishing effect of area and crack surface integrals on J.

3.3. Three-dimensional bodies

Figure 11 shows the cross section over the crack plane of a three-dimensional body with a plane crack. The global coordinate system is chosen such that the crack lies in the x_1-x_3 plane, with the x_1-axis normal to the crack front at $x_3 = 0$. Figure 9 can now be considered as the cross section through this body at $x_3 = 0$. The *local* J-integral along the crack front at $x_3 = 0$ can now be derived in exactly the same way as for the two-dimensional body in subsection 3.2. Application of equation (25) in the x_1-direction (k = 1) yields the same result as equation (27), except that the terms involving derivatives to x_3 in the area integral do not vanish in the general, three-dimensional situation.

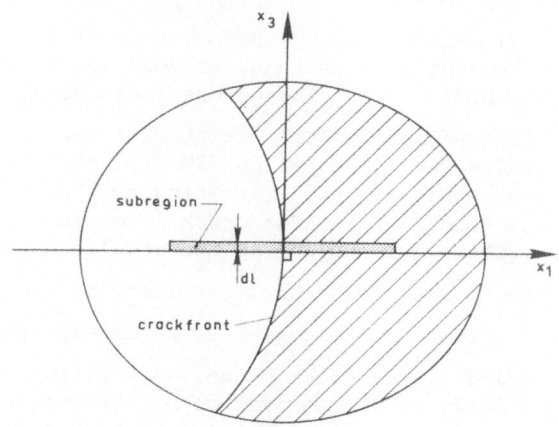

Figure 11. Cross section over crack plane // x_1-x_3 plane of a three-dimensional body with plane crack.

Hence:

$$J = \lim_{r \to 0} r \int_{-\pi}^{\pi} \left(W \cos \theta - T_i \frac{\partial u_i}{\partial x_1} \right) d\theta$$

$$= \int_{\Gamma} \left(W n_1 - T_i \frac{\partial u_i}{\partial x_1} \right) ds - \int_{\Gamma_s^+} p \frac{\partial u_2}{\partial x_1} ds + \int_{\Gamma_s^-} p \frac{\partial u_2}{\partial x_1} ds +$$

$$+ \int_{A} \left[- \frac{\partial}{\partial x_3} \left(\sigma_{i3} \frac{\partial u_i}{\partial x_1} \right) + \sigma_{ij} \frac{\partial \varepsilon_{ij}^o}{\partial x_1} + f_i \frac{\partial u_i}{\partial x_1} \right] dA \qquad (28)$$

The J-value according to equation (28) is in essence the x_1-component of a vector, with components given by:

$$J_k = \lim_{r \to 0} r \int_{-\pi}^{\pi} \left(W n_k - T_i \frac{\partial u_i}{\partial x_k} \right) ds \qquad (29)$$

Expressions similar to equation (28) for $J \equiv J_1$ can also be derived for the J_2 and J_3 components by application of equation (25) for $k = 2$ and 3, respectively. Kikuchi et al. [25] derived these integrals from Eshelby's [24] energy momentum tensor equations, without the terms involving initial strains and body forces. These were included by Kishimoto et al. [26] in their definition of a $\underline{J}$-integral vector which is identical to the J-integral vector of equation (29) for non-linear elastic material behaviour as considered here. Treating the components of the vector as local energy release rates for crack extensions in the three coordinate directions, these authors define the resultant vector, i.e. $\sqrt{\left(J_1^2 + J_2^2 + J_3^2 \right)}$, as the "three-dimensional J-value". In addition to objections against the use of the J_2-component as an energy release rate for crack extensions normal to the crack plane, as e.g. put forward by Labbens [27] on the basis of the derivation of the energetic meaning of the J-vector by Budiansky & Rice [28], such a definition is not in conformity with the two-dimensional definition of J. This is only true for the J_1-component according to equation (28), which is hence denoted as the three-dimensional J-value.

For our further observations it is interesting to note that, in the coordinate system considered here J_3 is equal to zero.

This follows directly from equation (29) for $k = 3$. The term involving W vanishes because $n_3 = 0$ on Γ_t. The strain components $\partial u_i / \partial x_3$ in the second term are non-singular and hence these terms have the same singularity as the stress components in the traction vector. These are singular with r^{-s}, where $s = 1/2$ (linear elastic materials) or less. By that, also the second term vanishes for the limit $r \to 0$, and hence $J_3 = 0$.

By equation (28) we have a general, three-dimensional equation for the local J-distribution along a crack front. However, serious problems arise when this integral has to be evaluated. In general such evaluations will be based on the results of a finite element analysis. The problems are not caused by the special coordinate system adopted for equation (28), as coordinate transformations are easy to perform. They are primarily caused by the derivatives to x_3 of stresses and strains in the area integral. Stresses and strains form part of the normal output of a finite element analyses. Spatial derivations of these quantities at local points cannot, however, be determined accurately. This can most easily be demonstrated by observing the equilibrium condition of equation (15). The finite element method only assures equilibrium of complete elements and not at local points within elements. Hence, as the equilibrium condition involves derivatives of stresses, such terms cannot be calculated accurately. A more practical objection against the use of equation (28) is that it is difficult to define a plane section normal to local points along a crack front within a general three-dimensional finite element mesh.

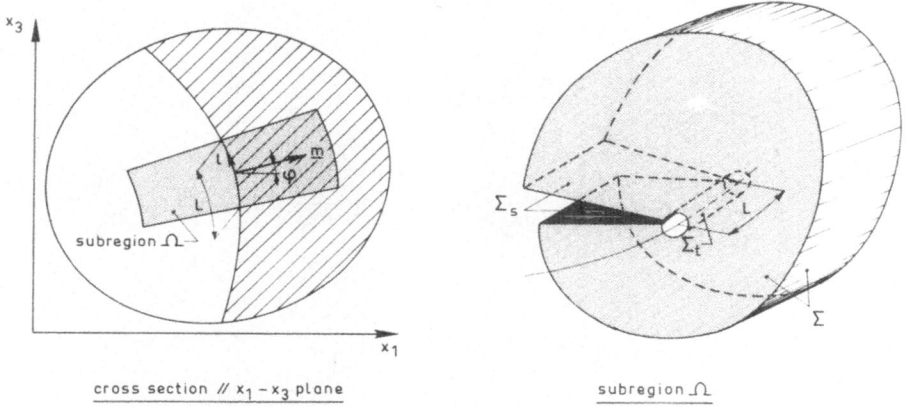

cross section // x_1–x_3 plane   subregion Ω

Figure 12. Subregion cutting a finite arc L from a crack front.
Flat crack surfaces Σ_s are parallel to the x_1-x_3 plane.

      The above arguments formed the motive for an alternative
formulation proposed by Bakker [29]. This involves the application of
the conservation law of equation (24) to a subregion which cuts a finite
arc of the crack front, as shown schematically in figure 12. The total
surface $\Sigma^{\star}$ is divided into a part Σ_t, formed by the surface of a small
circular volume with radius r and the crack front as center line, a
part Σ_s, coinciding with the flat crack surfaces, and the remaining
surface Σ, completely composed of interior points of the body. The
normal direction is defined positive when pointing outward the subregion
on Σ and Σ_s, and when pointing inward on Σ_t. With the x_2-axis normal to
the crack plane, $n_1 = n_3$ and $T_1 = T_3 = 0$ on Σ_s, and $T_2 = p$ on the top
part of $\Sigma_s(\Sigma_s^{+})$ and $T_2 = -p$ on the bottom part Σ_s^{-}. Application of
equation (24) in the x_1 and x_3 directions (k = 1 and 3, note that k = 2
is omitted) yields:

$$\int_{\Sigma} \left(W\, n_k - T_i\, \frac{\partial u_i}{\partial x_k} \right) dS - \int_{\Sigma_t} \left(W\, n_k - T_i\, \frac{\partial u_i}{\partial x_k} \right) dS -$$

$$- \int_{\Sigma_s^{+}} p\, \frac{\partial u_2}{\partial x_k}\, dS + \int_{\Sigma_s^{-}} p\, \frac{\partial u_2}{\partial x_k}\, dS + \int_{\Omega} \left(\sigma_{ij}\, \frac{\partial \varepsilon_{ij}^{o}}{\partial x_k} + f_i\, \frac{\partial u_i}{\partial x_k} \right) dV = 0$$

$$\tag{30}$$

For the limit $r \rightarrow 0$, the integrals over Σ_t are denoted as the x_1 $(k = 1)$ and x_3 $(k = 3)$ components of the vector $\underline{F}$. Hence:

$$F_k = \lim_{r \to 0} \int_{\Sigma_t} \left(W\, n_k - T_i \frac{\partial u_i}{\partial x_k} \right) dS$$

$$= \int_{\Sigma} \left(W\, n_k - T_i \frac{\partial u_i}{\partial x_k} \right) dS - \int_{\Sigma_s^+} p \frac{\partial u_2}{\partial x_k} dS + \int_{\Sigma_s^-} p \frac{\partial u_2}{\partial x_k} dS +$$

$$+ \int_{\Omega} \left(\sigma_{ij} \frac{\partial \epsilon_{ij}^o}{\partial x_k} + f_i \frac{\partial u_i}{\partial x_k} \right) dV = 0 \qquad (k = 1 \text{ or } 3) \qquad (31)$$

The integral over Σ_t can be split into an integral along the arc L of the crack front and one along a local crack tip contour in a plane normal to the local crack tip:

$$F_k = \int_L \left[\lim_{r \to 0} \int_{\Gamma_t} \left(W\, n_k - T_i \frac{\partial u_i}{\partial x_k} \right) ds \right] dl \qquad (k = 1 \text{ or } 3) \qquad (32)$$

The J-vector according to equation (29) refers to the local plane normal to the crack front. Hence, for the limit $r \rightarrow 0$ the integral along Γ_t is equal to $(J_1 \cos \varphi - J_3 \sin \varphi)$ for F_1 $(k = 1)$ and to $(J_1 \sin \varphi + J_3 \cos \varphi)$ for F_3 $(k = 3)$, where φ is the angle between the local and global x_1-coordinate along the crack front (cf. figure 12). According to the arguments given earlier, $J_3 = 0$. Hence, if $m_1 = \cos \varphi$ and $m_3 = \sin \varphi$ are denoted as the x_1 and x_3 components of the local normal to the crack front in the crack plane, equation (32) reduces to:

$$F_k = \int_L J\, m_k\, dl \qquad (k = 1 \text{ or } 3) \qquad (33)$$

where $J \equiv J_1$ is the local J-value along the crack front.

Equation (33) states thet the F_1 and F_3 integrals determined over the surfaces Σ and Σ_s and the volume of the subregion according to equation (31) are a measure for the average J-value along the arc L. For practical applications it is convenient to determine the x_1'-component of the $\underline{F}$-vector in a local x_1'-x_3' system as indicated in figure 13. The mean J-value along the arc L, $\bar{J}$, then follows from:

$$\bar{J} \approx \frac{F_1'}{B} = \frac{\left(F_1 \cos \alpha + F_3 \sin \alpha\right)}{B} \tag{34}$$

where B is the projection of the arc L on the local x_3'-axis in figure 13.

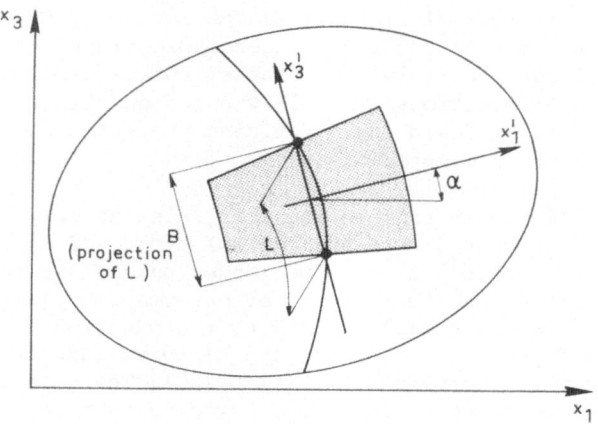

Figure 13. Local coordinates x_1'-x_3' used to calculate mean value of J over arc L.

Equation (34) is exact for an arc length L formed by the edge of an isoparametric finite element of any order with evenly distributed side nodes.

Bakker [29] also described another method to calculate local J-distributions along crack fronts, derived from the condition stated by equation (23). This method is identical to the so-called "virtual crack extension technique" proposed by De Lorenzi [16], which was shown (Bakker [30]) to be identical to similar methods derived by Hellen [31] and Parks [32]. The results of both methods were reported to be identical, although the virtual crack extension method appeared easier to apply. However, it was chosen to discuss the method based on the conservation laws of equation (24) here, as this is in better conformity with the original definition of J in two-dimensional bodies. The virtual crack extension technique is more seen as a numerical tool, which will be discussed by Lamain in his lecture on numerical methods.

4. APPLICATIONS IN EPFM

4.1. General aspects

The physical interpretations of J in terms of potential energy release and in terms of crack tip stress/strain field characterization as discussed in subsection 2.2 were based on the assumption of (linear or non-linear) *elastic* material behaviour and small (infinitesimal) deformations. For elastic-plastic materials, the interpretation in terms of potential energy release is no longer appropriate, as the strain energy is no longer entirely recoverable, but is partly dissipated by plastic deformations. This leaves the question whether the crack tip stress/strain field characterizing property can be maintained in *elastic-plastic* materials.

In general, plasticity in engineering materials can only be described realistically by an incremental (flow) theory of plasticity. Momentary stresses and strains are then no longer uniquely related as in equation (18) for non-linear elasticity or deformation plasticity. Hence, W according to equation (19) is no longer a function of momentary strains, but becomes dependent on the loading history, and stresses cannot be derived from W according to equation (20). The direct consequence is that the conservation laws derived in subsection 3.1 do not exist, and hence that path independence of J cannot be proven. It is anticipated, however, that these properties can be restored in cases where the results of incremental plasticity solutions are almost identical to the results obtained from deformation theory solutions, such that equation (19) is valid in a practical, engineering sense. Although this cannot be assured on beforehand, the conditions for this are best for monotonically *growing* loads (i.e. no unloading) which are in a fixed proportion to a single loading parameter. The above line of thought can be reversed, i.e. if the J-integral determined from an incremental plastic solution appears to be (nearly) path independent, the solution approaches that for deformation theory, and analytical observations based on deformation theory of plasticity or non-linear elasticity also apply to incremental plasticity in a practical sense. It is indeed for this porpose that many investigators studied the path dependent behaviour of J determined from incremental plastic solutions. These are too numerous to list, a review is given by Turner [12] . There seems to be a general consensus that J can be path independent and hence still characterizes the near tip stress/strain field in elastic-plastic materials of bodies loaded proportionally to a monotonically increasing parameter.

The assumption of a sharp crack with zero root radius at the tip cause the stresses and strains to be singular at the crack tip as expressed by the HRR-solution of equation (13). However, these solutions are based on infinitesimal strain theory and hence lose their validity near the crack tip where strains are extremely high. McMeeking [33] studied a more realistic situation, modeling the crack as a notch with an extremely small root radius and accounting for finite deformations.

The results indicated that finite deformation effects are confined to a small region in the direct vicinity of the blunting notch tip and that beyond this region stresses and strains are well approximated by small strain analyses of the sharp crack configuration, as presented by the HRR-solution. The calculated J-integrals were found to be path independent for contours with radius larger than ± 5 times the momentary crack tip opening displacement. This seems to warrant the assumption that, although the definition of J according to equation (7) loses its sense when finite deformations are accounted for, J-values computed from small strain analyses still characterize the intensity of the stress/strain field in the intensely deformed region at the blunting crack tip. Also the slip line solution at the blunting crack tip presented by Rice [5] suggests that this region of large deformations (D in figure 14) is confined to the direct vicinity of the crack tip.

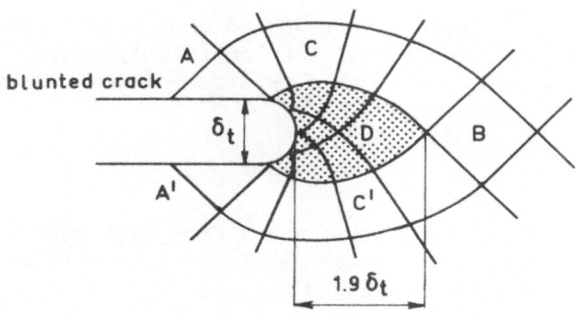

Figure 14. Region of intense deformation (D) created by crack tip blunting illustrated by a perfectly plastic slip line field (Rice [5]).

4.2. Crack growth initiation

The use of J as a single parameter to predict the onset of crack growth initiation in elastic-plastic cracked bodies was first proposed by Broberg [6] and Begley & Landes [7]. A first requisite for such an application is the existence of a geometry independent critical material value of J at which crack growth initiates. The original work of Landes & Begley [34] was aimed at the demonstration of this, and later (Begley & Landes [35]) at the determination of the critical value denoted as J_{Ic}, as all studies were performed for pure mode I loading. Further details are given by Garwood in his lecture on the determination of initiation properties of materials.

A first question arising when J is proposed as a fracture parameter is how it relates with the crack tip opening displacement, δ_t, being another parameter widely used for the characterization of the elastic-plastic fracture toughness of materials, BSI [36] , and the assessment of the integrity of cracked structures, as will be discussed by Turner in the workshop devoted to the COD method.

For perfectly plastic materials and plane stress conditions, Rice [5] finds from a Dugdale approach:

$$J = \sigma_y \delta_t \qquad (35)$$

while an approximate slip line solution of a plane strain deformation state for a Von Mises material leads to [13] :

$$J \approx 1.5 \ \sigma_y \ \delta_t \qquad (36)$$

Many analytical and numerical studies have revealed expressions similar to equation (36), differing by the value of the constant 1.5 only. In general there seems to be a consensus about a relation between J and δ_t of the form:

$$J = m \ \bar{\sigma} \ \delta_t \qquad (37)$$

where m is denoted as the constraint factor (m = 1 for plane stress) and $\bar{\sigma}$ is a material flow stress, often taken as the average of the yield stress σ_y and ultimate tensile stress σ_u, $\bar{\sigma} = \frac{1}{2} \ (\sigma_y + \sigma_u)$.

Accepting equation (37) at least in a qualitative fashion, i.e. that the applied J value for a given δ_t increases with the amount of out-of-plane constraint, or conversely with the relative level of the triaxial state of stress, it reflects the basic difference between the J and δ_t concepts. The maximum strain a material can withstand in a certain direction and hence the critical δ_t of a material, decreases with the out-of-plane constraint. This bounds critical material δ_t values to the material thickness applied in a structure, as prescribed by BSI [36] . The dependence on the out-of-plane constraint makes J a parameter which at least takes the triaxial state of stress at the crack tip into account. Complete transferability of critical J-values to other constraint levels requires J_{Ic} to be independent of thickness. The minimum thickness must of course be large enough to ensure that crack growth initiation takes place in a tearing mode and not by a slant type of fracture. The effect of thickness and other size effects on J_{Ic} are discussed by Garwood at this seminar.

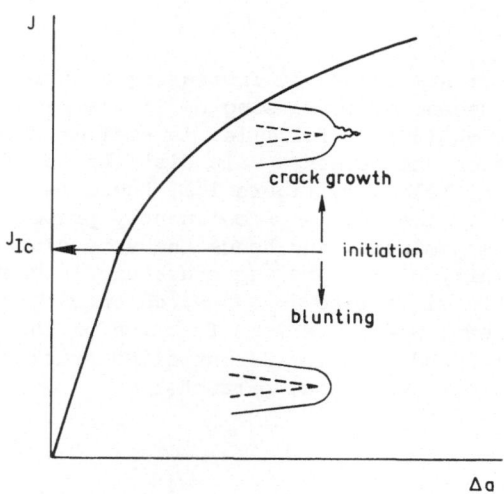

Figure 15. Typical J-R(esistance) curve.

4.3. Stable crack growth and instability

The onset of crack growth in ductile materials is generally followed by some amount of stable crack growth before final instability occurs. Leaving the issue of the validity of J beyond initiation aside for the moment, the progression of crack growth requires a substantial increase of J, as schematically sketched in the J versus crack growth (Δa) graph, denoted as a J-R curve in figure 15. For many materials J has to be increased to several times its initiation value, J_{Ic}, to produce an increase of crack length of only several millimeters, and before final instability occurs. Hence, there is a substantial margin of safety between initiation and instability, and the possibility to quantify this is of great importance for the safety assessment of cracked structures.

Stable Crack Growth

During the process of stable crack growth, a wake of unloaded material is formed behind the crack tip, together with a region surrounding the crack tip where stress- and strain increments will be highly non-proportional. This violates the conditions for deformation theory of plasticity and hence, by the arguments given in subsection 4.1, the conditions for the applicability of the J-concept. However, it was pointed out by Hutchinson and Paris [37], by considering plastic strain increments described by the HRR-model of equation (13) due to increments in J and crack length a, that at distances r from the crack tip large enough to fulfil the condition:

$$\frac{da}{r} \ll \frac{dJ}{J} \qquad (38)$$

the proportional strain increments caused by increasing J dominate the non-proportional increments caused by increasing a. If the radius r for which the condition of equation (38) is fulfilled is smaller than the radius R of the region in which the stress/strain distribution is dominated by the singular HRR field (cf. figure 16), there is an annular region in which plastic loading is predominantly proportional *and* the singular HRR-field is dominant, and hence the stress/strain field beyond the direct vicinity of the *growing* crack tip is completely *controlled* by J. This situation is termed *J-controlled crack growth*. Within fully yielding specimens, R will be some fraction of the uncracked ligament b, and the condition for J-controlled crack growth can be stated in terms of a non-dimensional parameter ω:

$$\omega = \frac{b}{J} \frac{dJ}{da} \gg 1 \qquad (39)$$

as can easily be deduced from equation (38).

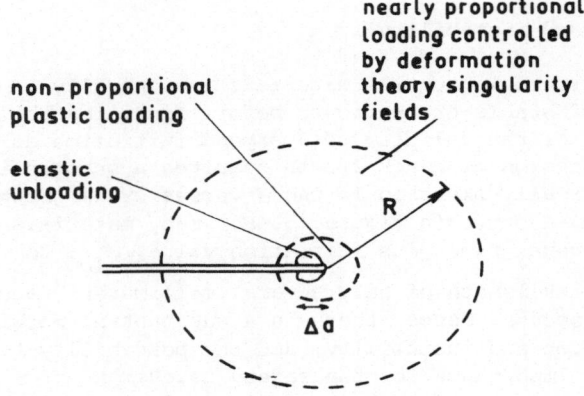

Figure 16. Schematic of crack tip conditions for J-controlled crack growth (Hutchinson & Paris [37]).

Another condition given by Hutchinson and Paris [37] follows from the requirement that the wake of elastic unloading and the region of distinctly non-proportional plastic loading, which will be of the order Δa, must be small compared with the radius R of the singular HRR field, which is some fraction of the uncracked ligament b. Hence:

$$\frac{\Delta a}{b} \ll 1 \qquad (40)$$

In addition it seems obvious that the same limits as used for the determination of J_{Ic} apply, i.e. that the size of the process zone, which is proportional with the crack tip opening displacement and hence with $J/\bar{\sigma}$, is small compared with the remaining ligament b:

$$\alpha = \frac{b \, \bar{\sigma}}{J} \gg 1 \tag{41}$$

and that the thickness B is large enough to result in a plane strain situation.
The above limits for J-controlled crack growth are only qualitative. Various analyses have shown that the actual limits depend on the type of structure, in particular of the loading condition. Based on numerical analyses Shih et al. [38] , [39] suggest $\omega > 10$ and $\alpha > 25$ for bending loads, and $\omega > 80$ and $\alpha > 200$ for tension loads. The limit for maximum crack growth was proposed as $\Delta a/b < 0.06$ and that for plane strain condition as $B > b$, independent of the type of loading.

If the limits for J-controlled crack growth are satisfied *and* the J-R curve (cf. figure 15) may be regarded as a material response, i.e. independent of the type and size of the structure, the crack growth as a function of loading in a structure may be predicted on the basis of the calculation of applied J-values and the material response determined on standard specimens. This approach will be discussed by Rousselier in the workshop devoted to the EPRI estimation scheme [39] , [40]. Analytical studies of the actual process of crack growth e.g. by Rice & Sorensen [41] and Rice, Drugan and Sham [42] show, however, that J-R curves will be dependent on size and/or type of specimen, at least after some amount of crack growth. This is supported by various experimental results, which show e.g. a substantial difference between the slope of J-R curves for tension and bending type of specimens as schematically shown in figure 17.

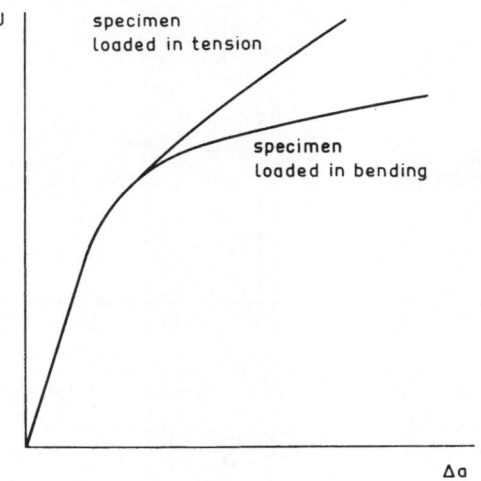

Figure 17. Typical difference between J-R curves of specimens loaded in bending and tension occurring after a limited amount of crack growth.

The analysis of Rice, Drugan and Sham [42] reveals, amongst other
things, the condition that a J-like parameter (i.e. a parameter equal
to J before initiation) has to satisfy in order to be a suitable,
geometry indepent parameter in the presence of crack growth. A parameter
satisfying this condition, i.e. that the rate of change of this parameter must be independent of the rate of increase of crack length, was
recently proposed by Ernst [43] denoting it as the "modified"
J-integral, J_M, according to:

$$J_M = J_D - \int_{a_o}^{a} \frac{\partial J_{pl}}{\partial a} \bigg|_{\Delta_{pl}} da \qquad (42)$$

where:

J_D = original definition of J (based on Deformation theory of
plasticity)

a_o, a = initial and current crack length

J_{pl} = plastic part of J_D ($J_D = J_{el} + J_{pl}$)

Δ_{pl} = plastic part of displacement ($\Delta = \Delta_{el} + \Delta_{pl}$).

As described by Ernst, the determination of applied J_M-values does not
require more information than that required for the original, J_D-values.
Figure 18 shows how the geometry dependence of J_D versus Δa data is
significantly improved when J_M is used instead for correlation.
The comparitive ease of determination makes J_M an attractive parameter
for crack growth studies, but further investigations are needed to
establish its viability for fracture safety assessment purposes.

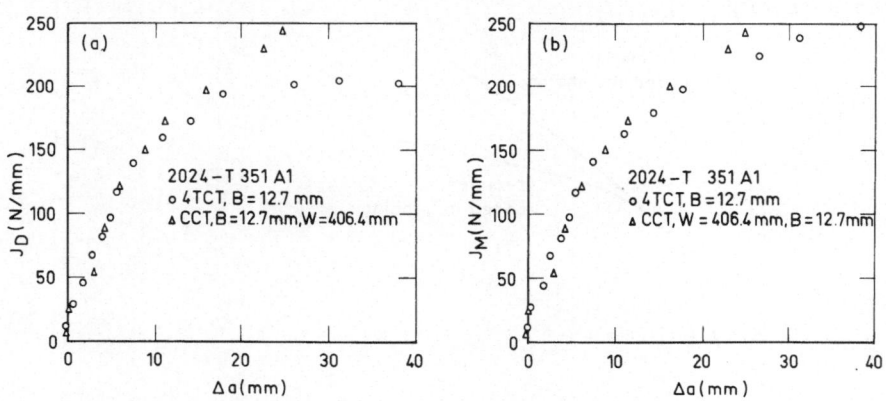

Figure 18. Compact specimen and center cracked panel J-R curves
(Ernst [43]).
(a) using the original (deformation theory) J_D,
(b) using the modified J_M.

Instability

A method for treating the tearing instability in the J-controlled crack growth regime was introduced by Paris et al. [8] and was further elaborated by Hutchinson & Paris [37]. Figure 19 shows a two-dimensional body, representing either a test specimen or a structural member (for simplicity further referred to as the specimen). F and Δ are the load and load point displacement measured at the specimen, respectively. The specimen is loaded in series with a linear spring with compliance C_M (for test specimens this can be the testing machine compliance, for structural members it represents the compliance of the remainder of the structure). The total load point displacement is:

$$\Delta_T = \Delta + C_M F \tag{43}$$

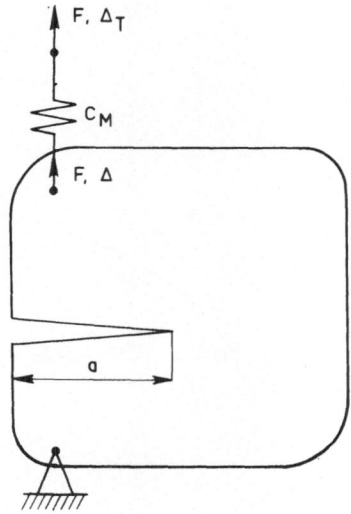

Figure 19. Typical specimen geometry or structural member (Hutchinson & Paris [37]).

It is assumed that crack growth Δa of the specimen remains within the limits for J-controlled crack growth as discussed in the previous sub-section, and that there is a unique material J-R curve. We now distinguish the applied J-value as the J computed within the specimen at a given load F and crack length $a = a_o + \Delta a$, $J_{app} \equiv J (F, a)$, and the material J-value as J at the material resistance curve for a given Δa, $J_{mat} \equiv J(\Delta a)$.

At any F and current crack length a, equilibrium requires:

$$J_{app} = J_{mat}$$

For a fixed total load point displacement Δ_T we may state the stability condition:

$$\left(\frac{\partial J_{app}}{\partial a}\right)_{\Delta_T} < \frac{dJ_{mat}}{da} \qquad (44a)$$

and instability is assumed if:

$$\left(\frac{\partial J_{app}}{\partial a}\right)_{\Delta_T} > \frac{dJ_{mat}}{da} \qquad (44b)$$

as is illustrated by figure 20.

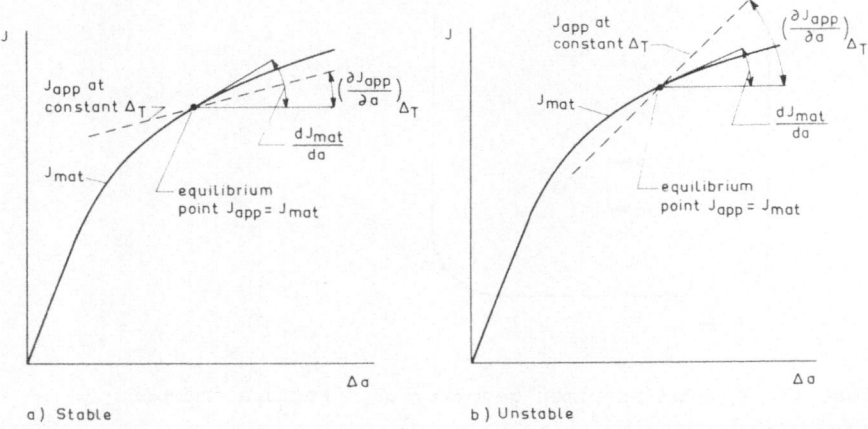

Figure 20. Illustration of the stability condition according to equation (44).

Paris et al. [8] stated the stability condition in terms of the non-dimensional tearing modulus T, defined as:

$$T_{app} = \frac{E}{\bar{\sigma}^2}\left(\frac{\partial J_{app}}{\partial a}\right)_{\Delta_T}$$

$$T_{mat} = \frac{E}{\bar{\sigma}^2}\left(\frac{dJ_{mat}}{da}\right) \tag{45}$$

where E is Young's modulus and $\bar{\sigma}$ is an appropriate material flow stress.

In terms of the tearing modulus, the stability conditions become:

$$T_{app} < T_{mat} \rightarrow stable$$

$$T_{app} > T_{mat} \rightarrow unstable \tag{46}$$

The derivation of the quantity $(\partial J_{app}/\partial a)_{\Delta_T}$ in terms of quantities referring to the specimen and C_M only is given by Hutchinson & Paris [37]:

$$\left(\frac{\partial J_{app}}{\partial a}\right)_{\Delta_T} = \left(\frac{\partial J_{app}}{\partial a}\right)_F - \left(\frac{\partial J_{app}}{\partial F}\right)_a \left(\frac{\partial \Delta}{\partial a}\right)_F \left[C_M + \left(\frac{\partial \Delta}{\partial F}\right)_a\right]^{-1} \tag{47}$$

Note that for $C_M \rightarrow \infty$, the condition applies to a load controlled (dead load) situation, and equation (47) becomes:

$$\left(\frac{\partial J_{app}}{\partial a}\right)_{\Delta_T} = \left(\frac{\partial J_{app}}{\partial a}\right)_F \tag{48}$$

For $C_M = 0$, we obtain a purely displacement controlled situation.

Figure 21 illustrates the large influence of compliance on the applied T values, and hence on the instability of a specimen or structural member.

The determination of applied T values for various deeply cracked configurations is discussed by Hutchinson & Paris [37], together with some suggestions for the experimental determination of T_{mat}, which were further elaborated by Ernst et al. [44], [45]. The tearing modulus approach is also adopted in the EPRI method [39], [40] discussed by Rousselier in this seminar.

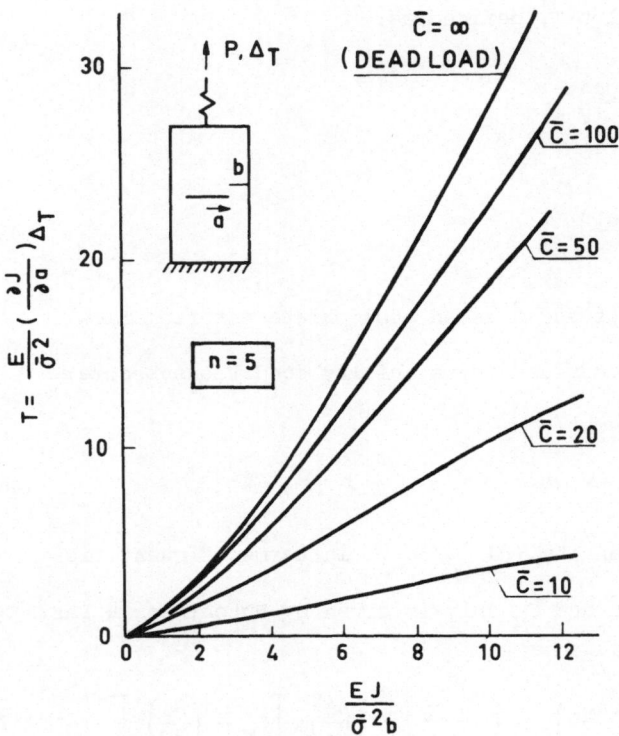

Figure 21. Influence of compliance on T for a deeply cracked plane stress tension specimen. $\bar{C}$ is the combined compliance of the spring and the uncracked specimen, n is the material strain hardening exponent in equation (12) (Hutchinson & Paris [37]).

As T_{mat} is derived from the J-R curve, it will only be a size and geometry independent material property if this also holds for the J-R curve. As discussed previously, this can only be true for limited amounts of crack growth, after which a distinct geometry dependence is observed. Ernst [43] redefined the tearing modulus of equation (45) in terms of his modified J according to equation (42), which yields:

$$T_M = \frac{E}{\bar{\sigma}^2} \frac{\partial J_M}{\partial a} = T - \frac{E}{\bar{\sigma}^2} \frac{\partial J_{pl}}{\partial a}\bigg|_{\Delta_{pl}} \tag{49}$$

such that the stability conditions of equation (46) become:

$$T_{M_{app}} < T_{M_{mat}} \rightarrow stable$$

$$T_{M_{app}} > T_{M_{mat}} \rightarrow unstable \tag{50}$$

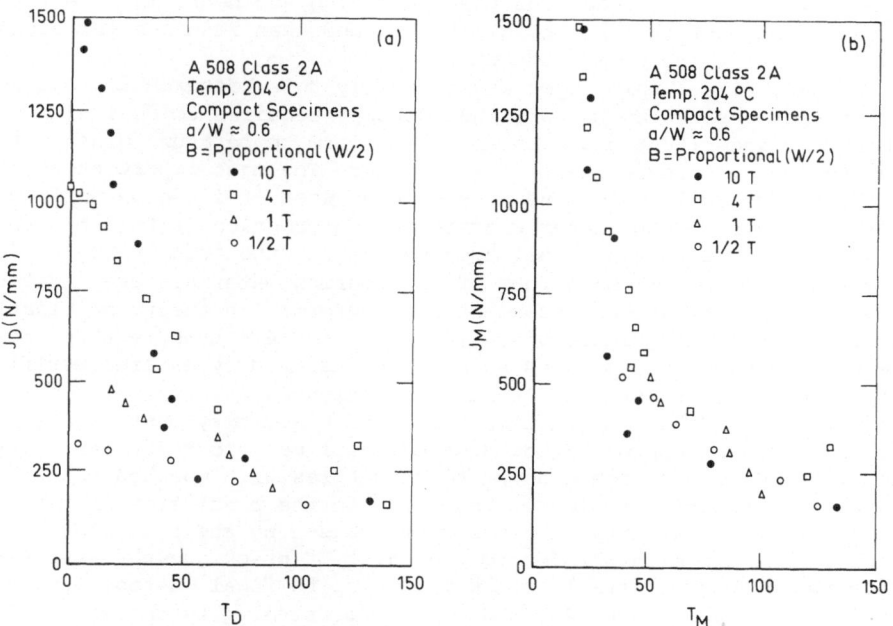

Figure 22. J versus T data for CT specimens of different sizes
Ernst [43]).
(a) using the original (deformation theory),
(b) using the modified J_M and T_M.

Figure 22 illustrates how J versus T data for different sized specimen
show a distinct size dependence in terms of the original J-integral,
but almost collapse into one curve when plotted in terms of J_M and the
associated tearing modulus, T_M.

As remarked earlier, the modified J-integral, J_M, and derived tearing

modulus T_M seem to eliminate at least part of the geometry and size

dependent effects of stable crack growth and instability as observed
when the original definition of J is used. Further explorations of the
method, taking into account a great variety of geometries, sizes and
materials are still required, however.

5. CONCLUDING REMARKS

During the last decade the J-integral has become an important parameter for the safety assessment of cracked structures. However, the method has not yet been formalized in design and fitness for purpose codes; the EPRI EPFM Handbook [40] and the J-design curve [46] can only be considered as first attempts into that direction. The lack of formal rules lays the responsibility for the correct use of the concept completely on the shoulders of the practizing engineer. This requires more knowledge of the theoretical background than required for codified and hence well documented concepts.

The primary aim of this paper was to supply this background, together with the extensions to the original concept required to deal with common loading situations such as non-uniform temperature distributions, body forces and pressurized crack surfaces. The most important aspect of the theoretical concept to keep in mind when it is to be applied, is that it refers to non-linear elastic and deformation (total) plastic material behaviour. Hence, all applications to the actual flow (incremental) plastic behaviour of engineering materials are limited to situations which can be approximated by deformation theory of plasticity. According to the arguments given in subsection 4.1 this requires that the path independence of J is verified or assured by earlier verifications of similar structure/load combinations.

The requirement that deformation theory of plasticity must be able to give a realistic approximation of the actual behaviour further restricts the applicability of the concept to structures with monotonically increasing proportional loads, and hence to the prediction of the loss of load bearing capacity of a structure caused by the presence of cracks. This can either refer to the onset of crack growth initiation or to the process of stable crack growth up to final instability, as discussed in subsections 4.2 and 4.3, respectively. By the scope of this paper, the treatise of these application aspects was limited. More details of important aspects of the application of the J-concept are given in various other lectures at this seminar, viz.:

- Determination of applied J-values (Bakker, Lamain)
- Experimental measurement of J_{Ic} (Garwood) and J-R curves (Blauel)
- Design and fitness for purpose applications (Turner, Rousselier).

ACKNOWLEDGEMENTS

The author's activities in EPFM in the period 1979-1983 were partly sponsored by the Netherlands Ministry of Economic Affairs as part of the co-operative research program BROS, which is greatfully acknowledged. The author is indebted to Professor D.G.H. Latzko for his support and stimulating discussions and to P.A.J.M. Steenkamp for discussions and suggestions after reading the manuscript.

REFERENCES

1. Irwin, G.R.
   Journal of Applied Mechanics, $\underline{\underline{24}}$ (1957), pp. 361/364.

2. Standard Test Method for Plane Strain Fracture Toughness of
   Metallic Materials, *Annual Book of ASTM Standards*,
   ASTM E399-78, American Society for Testing of Materials, USA, 1979.

3. *Methods of Test for Plane Strain Fracture Toughness (K_{Ic}) of
   Metallic Materials*,
   BS 5447, British Standards Institution, UK, 1978.

4. *Boiler and Pressure Vessel Code, Section III, division 1 and
   Section XI, division 1*,
   American Society of Mechanical Engineers, USA, 1983.

5. Rice, J.R.
   Journal of Applied Mechanics, $\underline{\underline{35}}$ (1968), pp. 379/386.

6. Broberg, B.
   Journal of the Mechanics and Physics of Solids, $\underline{\underline{19}}$ (1971),
   pp. 407/418.

7. Begley, J.A. and Landes, J.D.
   In: *Fracture Toughness part II*, ASTM STP 514, pp.1/23,
   American Society for Testing and Materials, USA, 1972.

8. Paris, P., Tada, H., Zahoor, A. and Ernst, H.A.
   A Treatment of the Subject of Tearing Instability,
   U.S. Nuclear Regulatory Commission Report NUREG-0311, USA, 1977.

9. Standard Test for J_{Ic}, a Measure of Fracture Toughness, *Annual
   Book of ASTM Standards*,
   ASTM E813-81, American Society for Testing and Materials,
   USA, 1981.

10. *Standard Method of Test for Elastic-Plastic Fracture Toughness J_{Ic}*,
    Japan Society of Mechanical Engineers, JSME S001-1981, Japan, 1981.

11. Knowles, J.K. and Sternberg, E.
    Archive for Rational Mechanics and Analysis, $\underline{\underline{44}}$ (1972),
    pp. 187/211.

12. Turner, C.E.
    In: *Post-Yield Fracture Mechanics*, D.G.H. Latzko, Chapter 2,
    Applied Science, Publ., London, UK, 1979.

13. Rice, J.R.
    In: *Fracture: an advanced treatise*, 2,
    H. Liebowitz, ed., pp. 191/311, Academic Press, New York,
    USA, 1968.

14. Sakata, M., Aoki, S. and Ishii, K.
    Proc. International Conference on Fracture Mechanics Techn., 1.
    G.C. Sih and C.L. Chow, editors, pp. 515/523, Sijthoff & Noordhoff,
    Leyden, The Netherlands, 1977.

15. Ainsworth, R.A., Neale, B.K. and Price, R.H.
    Proc. Int. Conf. on Tolerance of Flaws in Pressurized Components,
    pp. 171/178,
    Institution of Mechanical Engineers, London, UK, 1978.

16. de Lorenzi, H.G.
    International Journal of Fracture, 19 (1982), pp. 183/193.

17. Westergaard
    Transactions of the ASME, 61 (1939), pp. A49/A53.

18. Larsson, S.G. and Carlsson, A.J.
    Journal of the Mechanics and Physics of Solids, 21 (1973),
    pp. 263/277.

19. Rice, J.R.
    Journal of the Mechanics and Physics of Solids, 22 (1974),
    pp. 17/26.

20. Hutchinson, J.W.
    Journal of the Mechanics and Physics of Solids, 16 (1968),
    pp. 13/31.

21. Hutchinson, J.W.
    Journal of the Mechanics and Physics of Solids, 16 (1968),
    pp. 337/347.

22. Rice, J.R. and Rosengren, G.F.
    Journal of the Mechanics and Physics of Solids, 16 (1968),
    pp. 1/12.

23. McClintock, F.A.
    In: *Fracture: An Advanced Treatise*, 3,
    H. Liebowitz, editor, pp. 48/227, Academic Press, New York,
    USA, 1971.

24. Eshelby, J.D.
    In: *Solid State Physics: Advances in Research and Applications*, 3,
    F. Seitz and D. Turnbull, editors, pp. 79/144, Academic Press,
    New York, USA, 1956.

25. Kikuchi, M., Miyamoto, H. and Sakagichi, Y.
    Proc. 5-th SMiRT Conference, paper G 7/2, Berlin, FRG, 1979.

26. Kishimoto, K., Aoki, S. and Sakata, M.
    Engineering Fracture Mechanics, $\underline{13}$ (1980), pp. 841/850.

27. Labbens, R.
    Introduction à la Mécanique de la Rupture.
    Edition Pluralis, Paris, France, 1980.

28. Budiansky, B. and Rice, J.R.
    Journal of Applied Mechanics, $\underline{40}$ (1973), pp. 201/203.

29. Bakker, A.
    In: *Application of Fracture Mechanics to Materials and Structures*,
    G.C. Sih, E. Sommer and W. Dahl, editors, pp. 657/671, M. Nijhoff
    Publ., The Hague, The Netherlands.

30. Bakker, A.
    International Journal of Pressure Vessels and Piping, $\underline{14}$ (1983),
    pp. 153/179.

31. Hellen, T.K.
    Int. Journal for Numerical Methods in Engineering, $\underline{9}$ (1975),
    pp. 187/207.

32. Parks, D.M.
    International Journal of Fracture, $\underline{10}$ (1974), pp. 487/502.

33. McMeeking, R.M.
    Journal of the Mechanics and Physics of Solids, $\underline{25}$ (1977),
    pp. 357/381.

34. Landes, J.D. and Begley, J.A.
    In: *Fracture Toughness part II*, ASTM STP 514, pp. 24/39,
    American Society for Testing and Materials, USA, 1972.

35. Begley, J.A. and Landes, J.D.
    In: *Fracture Analysis*, ASTM STP 560, pp. 170/186,
    American Society for Testing and Materials, USA, 1974.

36. *Methods for Crack Opening Displacement (COD) Testing*,
    BS 5762, British Standards Institution, London, UK, 1979.

37. Hutchinson, J.W. and Paris, P.C.
    In: *Elastic-Plastic Fracture*, ASTM STP 668, pp. 37/64,
    American Society for Testing and Materials, USA, 1979.

38. Shih, C.F., de Lorenzi, H.G. and Andrews, W.R.
    In: *Elastic-Plastic Fracture*, ASTM STP 668, pp. 65/110,
    American Society for Testing and Materials, USA, 1979.

39. Shih, C.F., German, M.D. and Kumar, V.
    International Journal of Pressure Vessels and Piping, $\underline{9}$ (1981),
    pp. 159/196.

40. Kumar, V., German, M.D. and Shih, C.F.
    An Engineering Approach for Elastic-Plastic Fracture Analysis,
    EPRI NP-1931,
    Electric Power Research Institute, USA, 1981.

41. Rice, J.R. and Sorensen, E.P.
    Journal of the Mechanics and Physics of Solids, $\underline{\underline{26}}$ (1978),
    pp. 163/186.

42. Rice, J.R., Drugan, W.J. and Sham, T.L.
    In: *Fracture Mechanics: Twelfth Conference*,
    ASTM STP 700, pp. 189/200.
    American Society for Testing and Materials, USA, 1980.

43. Ernst, H.A.
    In: *Elastic-Plastic Fracture: Second Symposium*,
    ASTM STP 803, pp. I-191/213,
    American Society for Testing and Materials, USA, 1983.

44. Ernst, H.A., Paris, P.C., Rossow, M. and Hutchinson, J.W.
    In: *Fracture Mechanics*, ASTM STP 677, pp. 581/599,
    American Society for Testing and Materials, USA, 1979.

45. Ernst, H.A., Paris, P.C. and Landes, J.D.
    In: *Fracture Mechanics*, ASTM STP 743, pp. 476/502,
    American Society for Testing and Materials, USA, 1981.

46. Turner, C.E.
    In: *Advances in Elasto-Plastic Fracture Mechanics*,
    L.H. Larsson, editor, pp. 301/318, Applied Science Publ.,
    London, UK, 1979.

APPENDIX 1

Using equation (20), equation (22) can also be written as:

$$\delta w = \delta x_k \frac{\delta W}{\delta x_k} = \delta x_k \frac{\partial W}{\partial \varepsilon_{ij}^m} \frac{\partial \varepsilon_{ij}^m}{\partial x_k} = \delta x_k \, \sigma_{ij} \frac{\partial \varepsilon_{ij}^m}{\partial x_k} \tag{A1-1}$$

By equation (A1-1), the following condition holds for any point P on Ω:

$$\delta x_k \frac{\partial W}{\partial x_k} - \delta x_k \, \sigma_{ij} \frac{\partial \varepsilon_{ij}^m}{\partial x_k} = 0 \tag{A1-2}$$

or, by the definition of $\underline{\varepsilon}^m$ according to equation (16):

$$\delta x_k \frac{\partial W}{\partial x_k} - \delta x_k \, \sigma_{ij} \frac{\partial \varepsilon_{ij}}{\partial x_k} + \delta x_k \, \sigma_{ij} \frac{\partial \varepsilon_{ij}^o}{\partial x_k} = 0 \tag{A1-3}$$

From the definition of $\underline{\varepsilon}$ and the symmetry of $\underline{\sigma}$ by equation (14) it can be shown that:

$$\sigma_{ij} \frac{\partial \varepsilon_{ij}}{\partial x_k} = \sigma_{ij} \frac{\partial^2 u_i}{\partial x_j \, \partial x_k}$$

and by that equation (A1-3) transforms to:

$$\partial x_k \frac{\partial W}{\partial x_k} - \delta x_k \, \sigma_{ij} \frac{\partial^2 u_i}{\partial x_j \, \partial x_k} + \delta x_k \, \sigma_{ij} \frac{\partial \varepsilon_{ij}^o}{\partial x_k} = 0 \tag{A1-4}$$

From the equilibrium condition, equation (15) it follows that:

$$\sigma_{ij} \frac{\partial^2 u_i}{\partial x_j \, \partial x_k} = \frac{\partial}{\partial x_j} \left(\sigma_{ij} \frac{\partial u_i}{\partial x_k} \right) - f_i \frac{\partial u_i}{\partial x_k}$$

and equation (A1-4) may hence be written as:

$$\delta x_k \frac{\partial W}{\partial x_k} - \delta x_k \frac{\partial}{\partial x_j} \left(\sigma_{ij} \frac{\partial u_i}{\partial x_k} \right) + \delta x_k \, f_i \frac{\partial u_i}{\partial x_k} + \delta x_k \, \sigma_{ij} \frac{\partial \varepsilon_{ij}^o}{\partial x_k} = 0$$

$$\tag{A1-5}$$

or also as:

$$\delta x_k \frac{\partial}{\partial x_j} \left(W \, \delta_{jk} - \sigma_{ij} \frac{\partial u_i}{\partial x_k} \right) + \delta x_k \left(f_i \frac{\partial u_i}{\partial x_k} + \sigma_{ij} \frac{\partial \varepsilon_{ij}^o}{\partial x_k} \right) = 0$$

(A1-6)

By partial differentiation the first term can be transformed according to:

$$\frac{\partial}{\partial x_j} \left[\delta x_k \left(W \, \delta_{jk} - \sigma_{ij} \frac{\partial u_i}{\partial x_k} \right) \right] - \frac{\partial \delta x_k}{\partial x_j} \left(W \, \delta_{jk} - \sigma_{ij} \frac{\partial u_i}{\partial x_k} \right)$$

$$+ \delta x_k \left(f_i \frac{\partial u_i}{\partial x_k} + \sigma_{ij} \frac{\partial \varepsilon_{ij}^o}{\partial x_k} \right) = 0$$

(A1-7)

As the condition of equation (A1-7) is valid for any arbitrary point on Ω, it also holds for the integral over the volume of Ω. The divergence theorem of Gauss states:

$$\int_\Omega \frac{\partial}{\partial x_j} \left(g_j(\underline{x}) \right) dV = \int_{\Sigma^\star} g_j(\underline{x}) \, n_j \, dS$$

where g is some function defined on Ω, $\Sigma^\star$ is the (closed) surface of Ω and n is the unit outward normal on $\Sigma^\star$. Hence the divergence theorem can be applied to the first term of equation (A1-7) when this is integrated over the volume of Ω, resulting into equation (23), i.e.:

$$\int_{\Sigma^\star} \delta x_k \left(W \, \delta_{jk} - \sigma_{ij} \frac{\partial u_i}{\partial x_k} \right) n_j \, dS$$

$$- \int_\Omega \left[\frac{\partial \delta x_k}{\partial x_j} \left(W \, \delta_{jk} - \sigma_{ij} \frac{\partial u_i}{\partial x_k} \right) - \delta x_k \left(f_i \frac{\partial u_i}{\partial x_k} + \sigma_{ij} \frac{\partial \varepsilon_{ij}^o}{\partial x_k} \right) \right] dV = 0$$

APPENDIX 2

For the subregion of infinitesimal thickness dx_3 lying in the x_1-x_2 plane as shown in figure 8, the surface Σ^* consists of the contour Γ^* in the x_1-x_2 plane with width dx_3 and the two surfaces A and A' at $x_3 = 0$ and $x_3 = dx_3$, respectively. The surface integral of equation (24) then transforms to:

$$dx_3 \int_{\Gamma^*} \left(W\, n_k - T_i\, \frac{\partial u_i}{\partial x_k} \right) ds + \int_{A+A'} \left(W\, n_k - T_i\, \frac{\partial u_i}{\partial x_k} \right) dA \quad (A2-1)$$

and the volume integral to:

$$dx_3 \int_A \left(\sigma_{ij}\, \frac{\partial \varepsilon_{ij}^o}{\partial x_k} + f_i\, \frac{\partial u_i}{\partial x_k} \right) dA \quad (A2-2)$$

The integrals over A and A' in equation (A2-1) are related according to:

$$\int_{A'} (\ldots) \, dA = - \int_A (\ldots) \, dA - dx_3 \int_A \frac{\partial}{\partial x_3} (\ldots) \, dA \quad (A2-3)$$

where the negative signs are due to the opposite positive normal directions on A and A'.
By introducing (A2-3) into (A2-1), equation (24) becomes:

$$dx_3 \int_{\Gamma^*} \left(W\, n_k - T_i\, \frac{\partial u_i}{\partial x_k} \right) ds - dx_3 \int_A \frac{\partial}{\partial x_3} \left(W\, n_k - T_i\, \frac{\partial u_i}{\partial x_k} \right) dA +$$

$$+ dx_3 \int_A \left(\sigma_{ij}\, \frac{\partial \varepsilon_{ij}^o}{\partial x_k} + f_i\, \frac{\partial u_i}{\partial x_k} \right) dA = 0 \quad (A2-4)$$

This result is linear in dx_3, which can hence be omitted. Furthermore it can be introduced that $n_i = n_2 = 0$ and $n_3 = -1$ on A, by which equation (25) is obtained from equation (A2-4), i.e.:

$$\int_{\Gamma^*} \left(W\, n_k - T_i\, \frac{\partial u_i}{\partial x_k} \right) ds +$$

$$+ \int_A \left[\frac{\partial}{\partial x_3} \left(W\, \delta_{k3} - \sigma_{i3}\, \frac{\partial u_i}{\partial x_k} \right) + \sigma_{ij}\, \frac{\partial \varepsilon_{ij}^o}{\partial x_k} + f_i\, \frac{\partial u_i}{\partial x_k} \right] dx_3 = 0$$

CASE STUDIES ON THE DETERMINATION OF APPLIED J-VALUES

Ad Bakker
Delft University of Technology
Laboratory for Thermal Power Engineering
P.O. box 5055, 2600 GA   Delft
The Netherlands

ABSTRACT. The determination of applied J-values in small test specimens
and structures forms the basis of the experimental determination of
critical material J_{Ic}-values and J-based safety assessments of cracked
structures, respectively. First, the results of approximate J-solutions
for small test specimens (SENB, CT, CNT) are reviewed and compared with
solutions obtained from the formal definition of J and experimental
results. The approximate solutions, which require data obtained from
the experimental records of small specimen tests only and which form
the basis of the J_{Ic} determination of materials, appear to have an
excellent accuracy.

Finally, the results of a J-based initiation analysis of a three-
dimensional (model) structure are compared with experimental results.
Although practical limitations of the experiment prevent a direct
comparison of the predicted and experimental initiation load, it appears
that the (small amount of) crack growth at the end of the test agrees
to within experimental scatter with that predicted from J-computations
and data obtained from small specimen tests.

1.   INTRODUCTION

The determination of applied J-values is an essential issue at two
phases of the application of the J-concept, viz.:

• During the determination of J_{Ic} and J-R curves.

The material value of J_{Ic} and the J-R curve are usually determined
on (preferably small) standard test specimens. Details of these
prodedures [1], [2] are discussed in the lectures by Garwood and
Blauel, respectively. It is essential from an engineering point of
view that the determination of applied J-values during such tests
is as simple as possible, with a minimum number of required
specimens and without needing complicated test arrangements.

55

L. H. Larsson (ed.), Elastic-Plastic Fracture Mechanics, 55–83.
© 1985 ECSC, EEC, EAEC, Brussels and Luxembourg.

• During the analysis of a cracked structure.
The determination of the initiation and/or instability load of a
cracked structure requires the knowledge of the applied Jvalue as
a function of the applied load. As the number of known analytical
solutions is limited, this often requires the execution of finite
element analyses in combination with procedures to calculate J,
which will be discussed later by Lamain. The complexity and high
costs of such analyses cause a strong need for more simple
engineering solutions, like those presented in the EPRI Handbook
[3] and to be discussed in the workshop by Rousselier.

This paper first discusses analytical procedures for the determination
of applied J-values for some common test specimens. Subsequently the
results of these procedures are compared with fully numerical solutions
for Single Edge Notched Bend (SENB), Compact Tension (CT) and Central
Notched Tension (CNT) specimens. Finally, the results of an experimental
and numerical study of the J-based crack growth initiation behaviour of
a model of a structural component are discussed.

Because of the limited scope of this paper, only J-solutions for
stationary (non-growing) cracks are presented. The analytical solutions
for specimen geometries can be extended to growing cracks [4], but
these will not be discussed here.

2. ANALYTICAL SOLUTIONS

2.1. General aspects

As discussed in my first lecture, the J-value in an *elastic* body
can, besides from the usual contour integral, be derived from the rate
of decrease of the potential energy P with cracked area, i.e. for a
two-dimensional body of thickness B and crack length a:

$$J = - \frac{1}{B} \frac{\partial P}{\partial a} \tag{1}$$

It was also stated that the J-theory can only be applied to *inelastic*
(elastic-plastic) bodies when the inelastic behaviour can realistically
be described by a non-linear theory of elasticity (i.e. deformation
theory of plasticity). In that case we can still use equation (1) for
determining J by treating the total absorbed energy as recoverable,
although in reality the plastic energy is dissipated. It is obvious
that this limits the applicability of solutions based on equation (1)
to bodies with monotonically increasing loads, i.e. without unloading.

For bodies loaded in displacement control:

$$P = U = \int_{o}^{\Delta'} F \, d\Delta \tag{2}$$

i.e. the total work done by the external load F with displacement Δ (cf. figure 1).

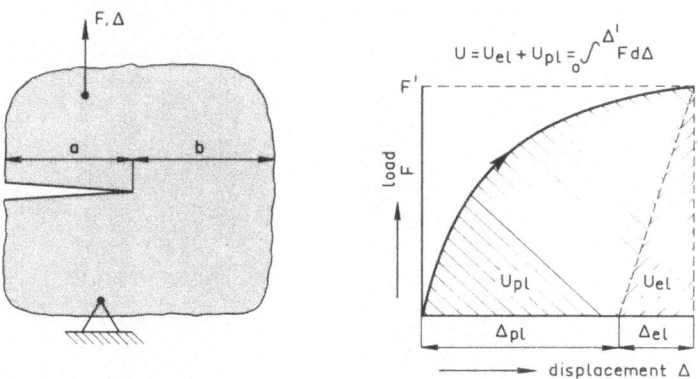

Figure 1. Typical body loaded by a single load.

For bodies loaded in load control:

$$P = U - F'\,\Delta' = \int_0^{\Delta'} F\,d\Delta - F'\,\Delta' = - \int_0^{F'} \Delta\,dF \tag{3}$$

i.e. minus the complementary energy of the work done by the external load. Substitution of equations (2) and (3) into equation (1) yields for J:

$$J = -\frac{1}{B} \int_0^{\Delta'} \left.\frac{\partial F}{\partial a}\right|_\Delta d\Delta = \frac{1}{B} \int_0^{F'} \left.\frac{\partial \Delta}{\partial a}\right|_F dF \tag{4}$$

The total load point displacement, Δ, may be regarded as the sum of the displacement with no crack present, Δ_{nc}, plus the displacement due to the presence of the crack, Δ_c, or:

$$\Delta = \Delta_{nc} + \Delta_c \tag{5}$$

Alternately it may be regarded as the sum of the elastic displacement, Δ_{el}, and that due to plasticity, Δ_{pl}, or:

$$\Delta = \Delta_{el} + \Delta_{pl} \tag{6}$$

By equation (5), the second form of equation (4) may be written as:

$$J = \frac{1}{B} \int_0^{F'} \left.\frac{\partial \Delta_c}{\partial a}\right|_F dF \tag{7}$$

as $\partial \Delta_{nc}/\partial a = 0$.

Using equation (6), J according to equation (1) can be split into an elastic and a plastic component, i.e.:

$$J = J_{el} + J_{pl} \tag{8}$$

with:

$$J_{el} = \frac{1}{B} \int_{o}^{F'} \left. \frac{\partial \Delta_{el}}{\partial a} \right|_{F} dF$$

$$\tag{9}$$

$$J_{pl} = \frac{1}{B} \int_{o}^{F'} \left. \frac{\partial \Delta_{pl}}{\partial a} \right|_{F} dF$$

The elastic displacement is linearly related with the load by the elastic compliance, C, of the body, i.e. $\Delta_{el} = C\,F$. It is then easily verified that the elastic component of J is equal to the well known compliance form of the Griffith energy release rate, G, and by that to the linear elastic stress intensity factor, K, i.e. for mode I loading in plane strain:

$$J_{el} = \frac{1}{2B} \frac{\partial C}{\partial a} F'^2 = G = \frac{(1-\nu^2)}{E} K_I^2 \tag{10}$$

where ν is Poisson's ratio and E is Young's modulus.
Hence, solutions for J_{el} can be obtained from the large variety of elastic K-solutions known from literature, e.g. handbooks like that of Tada et al. [5].

Turner [6] proposed a relation between J and the work done per unit uncracked ligament area according to:

$$J = \eta_{el} \frac{U_{el}}{B\,b} + \eta_{pl} \frac{U_{pl}}{B\,b} \tag{11}$$

where b is the uncracked ligament length and U_{el} and U_{pl} are the elastic and plastic parts of the work done by the external load (cf. figure 1).
With $U_{el} = \frac{1}{2} F'\, \Delta'_{el} = \frac{1}{2} C\, F'^2$ it follows from equation (10) that:

$$\eta_{el} = \frac{b}{C} \frac{\partial C}{\partial a} \tag{12}$$

and hence η_{el} is a dimensionless parameter depending on geometry type and (relative) dimensions only, i.e. independent of material properties and the extent of loading. The general condition for η_{pl} to exist as a factor which is independent of the extent of loading is given by Ernst et al. [4], [7]. For power law hardening materials this condition is met for any two-dimensional configuration.

2.2. Specimens with remaining uncracked ligament in bending

For the derivation of J in bend type specimens it is convenient to replace the load and load point displacement by the moment M and rotation θ, respectively, as shown in figure 2. The derivation of J as given by equation (7) then leads to:

$$J = \frac{1}{B} \int_0^{M'} \left. \frac{\partial \theta_c}{\partial a} \right|_M dM \tag{13}$$

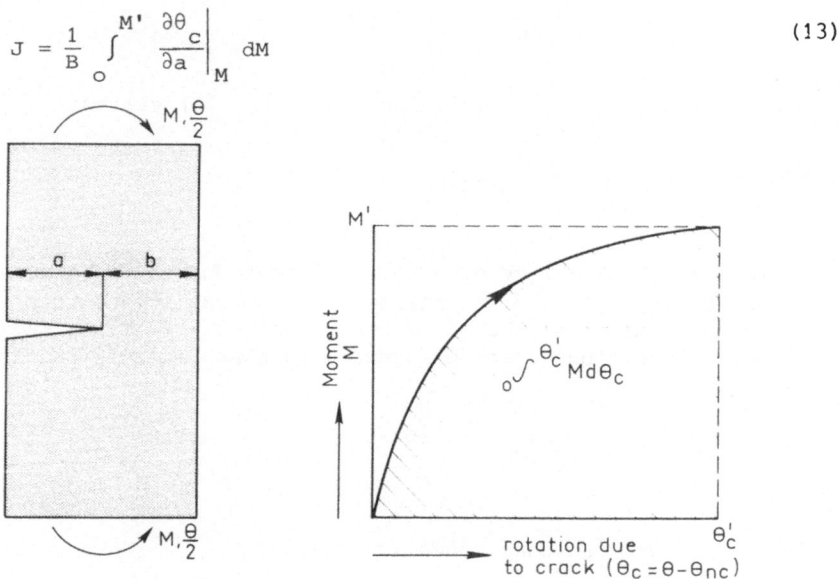

Figure 2. Ligament loaded in pure bending.

Rice et al. [8] state that it follows from dimensional analysis that for *deeply* cracked plates in bending, with plasticity confined to the uncracked ligament, θ_c must have the form:

$$\theta_c = f\left(\frac{M}{B\,b^2}\right) \tag{14}$$

with material constants being the only other parameters involved in the function f.

By equation (14) (with f' denoting the derivative of function f):

$$\left.\frac{\partial\theta_c}{\partial M}\right|_b = \frac{1}{B\,b^2}\,f'\left(\frac{M}{B\,b^2}\right)$$

$$\left.\frac{\partial\theta_c}{\partial a}\right|_M = -\left.\frac{\partial\theta_c}{\partial b}\right|_M = \frac{2M}{B\,b^3}\,f'\left(\frac{M}{B\,b^2}\right)$$

and hence:

$$\left.\frac{\partial\theta_c}{\partial a}\right|_M = \frac{2M}{b}\left.\frac{\partial\theta_c}{\partial M}\right|_b \tag{15}$$

Substitution of equation (15) into equation (13) yields:

$$J = \frac{2}{B\,b}\int_o^{\theta'_c} M\,d\theta_c \tag{16}$$

The integral is the work done by the moment due to the presence of the crack. It can easily be determined from the raw moment-total rotation record by subtracting the displacement without crack. As the deformations without crack are generally elastic, this transforms equation (16) to:

$$J = \frac{2}{B\,b}\int_o^{\theta'} M\,d\theta - \frac{M'\,\theta'_{nc}}{B\,b} \tag{17}$$

Single Edge Notched Bend (SENB) specimen

For an SENB specimen in three point bending equation (17) can easily be reverted to the applied load F and displacement Δ (cf. figure 3):

$$J = \frac{2}{B\,b}\int_o^{\Delta'} F\,d\Delta - \frac{F'\,\Delta'_{nc}}{B\,b} \tag{18}$$

If the plastic displacement becomes very large compared to the elastic contribution, the second term becomes negligible and J is simply related to the total area under the load-displacement record, i.e.:

$$J = \frac{2}{B\,b}\int_o^{\Delta'} F\,d\Delta \tag{19}$$

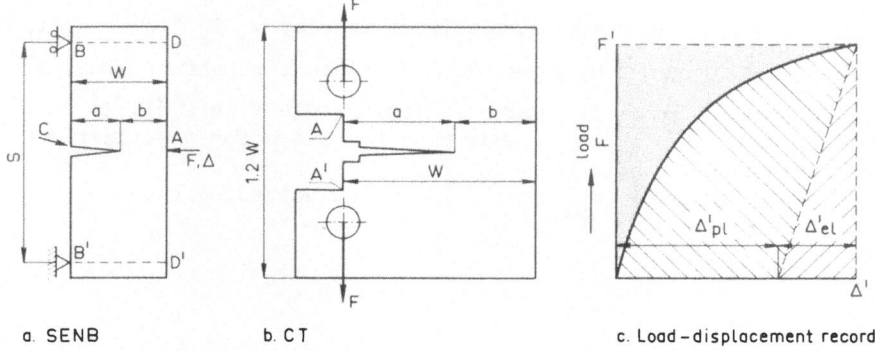

a. SENB   b. CT   c. Load-displacement record

Figure 3. Bend type specimens.

Equations (18) and (19) enable the determination of the applied J-value from the single load-displacement record of a specimen, which is a great advantage with regard to the original procedure used by Begley & Landes [9], which is based on a direct application of equation (1) and requires at least two specimens with slightly different crack length.

As an alternative, the applied J-value of an SENB specimen can be derived from equation (8) and (9). The derivation of J_{pl} follows exactly the same line as that of equation (16), resulting into (again reverting to the use of F and Δ):

$$J_{pl} = \frac{2}{B\,b} \int_{0}^{\Delta'_{pl}} F \; d\Delta_{pl} \qquad (20)$$

The integral is now equal to the plastic part of the work done by the load F (cf. figure 1).
Equation (20) implies that in equation (11) $\eta_{pl} = 2$, which is hence material independent. However, this is only valid for deeply notched specimens (a/W > ≈ 0.5), for which equation (14) for θ_{pl} is valid. For smaller relative crack depths η_{pl} is no longer constant and becomes material dependent [10]. For specimens with a/W > ≈ 0.5, also $\eta_{el} \approx$ 2 for SENB specimens [11], and hence equation (11) with $U = U_{el} + U_{pl}$ and $\eta_{el} = \eta_{pl} = 2$ becomes equal to equation (19), which was adopted by the ASTM standard for J_{Ic} testing [1].

The SENB specimen is also used for the measurement of the critical crack tip opening displacement, δ_t, of a material [12]. However, during these experiments not the load point displacement is measured, but the crack mouth opening displacement (indicated by v_g in figure 4) from which δ_t is estimated by assuming a fictitious rotation point, located at a distance r from the crack tip (point C in figure 4). In the BSI COD standard [12] it is assumed that r = 0.4 b for the plastic component of the displacements.

Using the same model, θ_{pl} can be estimated according to:

$$\theta_{pl} = \frac{v_{g,pl}}{a + z + r} \tag{21}$$

and together with the applied moment M = F S/4, J_{pl} can be obtained from equation (16) with θ_c replaced by θ_{pl}:

$$J_{pl} = \frac{S}{2B \, b(a+z+r)} \int_0^{v_{g,pl}} F \, d \, v_{g,pl} \tag{22}$$

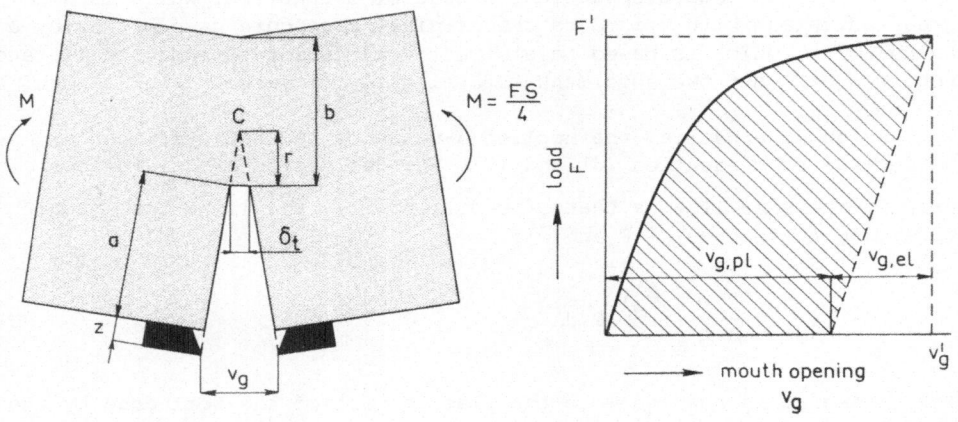

Figure 4. COD measurement arrangement.

Hence, J_{pl} can be determined from the hatched area in the F versus v_g plot in figure 4 (which is usually recorded during a COD test), provided that a good approximation for r can be made. The value of 0.4 b adopted by BSI is in good agreement with finite element analyses [11].

For ideal plastic materials, the integral in equation (22) becomes equal to $F_{lim} \, v'_{g,pl}$, where F_{lim} is the plastic limit load of the specimen.

Equation (22) then becomes equal to the J-solution proposed by Sumpter & Turner [11]. For work hardening materials this would underestimate J, in which case the integral may better be estimated by $\frac{1}{2} (F_{lim} + F') \, v'_{g,pl}$.

Compact Tension (CT) specimen

For the deeply cracked CT specimen (cf. figure 3b), the displacement without crack, Δ_{nc} in equation (18), is always negligibly small compared to the displacement due to the crack, and hence equation (19) would be applicable without any restrictions. However, the uncracked ligament of a CT specimen is loaded in a combination of bending and tension, and not in pure bending as required for the derivation of equation (19). A derivation of J_{pl} which also takes the tension component of the ligament loading into account was made by Merkle & Corten [13], resulting in:

$$J_{pl} = \frac{2(1+\alpha)}{B\,b(1+\alpha^2)} \int_o^{\Delta'_{pl}} F\,d\,\Delta_{pl} + \frac{2\alpha(1-2\alpha-\alpha^2)}{B\,b(1+\alpha^2)^2} \int_o^{F'} \Delta_{pl}\,dF \qquad (23)$$

with:

$$\alpha = \left[\left(\frac{2a}{b}\right)^2 + 2\left(\frac{2a}{b}\right) + 2\right]^{\frac{1}{2}} - \left(\frac{2a}{b} + 1\right)$$

For $a/W \geqslant 0.5$ equation (23) can also be applied for the total J-value, i.e. J_{pl} and Δ_{pl} replaced by J and Δ, respectively. The first integral is then the total work done by the load F (totally hatched area in figure 3c), and the second integral the complementary energy (shaded area in figure 3c). If the plastic deformations are large compared to the elastic deformations, the second integral is small compared to the first one. In addition its multiplication factor is considerably lower (1 : 11 for $a/W = 0.5$, 1 : 13 for $a/W = 0.6$) and hence the second term can usually be neglected, resulting in the approximation:

$$J = \frac{2(1+\alpha)}{B\,b(1+\alpha^2)} \int_o^{\Delta'} F\,d\Delta \qquad (24)$$

which was also adopted by the ASTM standard for J_{Ic} testing [1] (with the factor $2(1+\alpha)/(1+\alpha^2)$ given in tabular form as a function of a/W).

2.3. Specimens with remaining uncracked ligament in tension

Originally the derivation of J for a Double Edge Notched Tension (DENT) specimen was given by Rice et al. [8]. By choosing the correct dimensional and load parameters it also applies for a Centre Notched Tension (CNT) specimen (cf. figure 5). For these configurations the basic form of equations (8) and (9) is used. Again from dimensional analysis it was found [8] that the plastic component of the displacement must have the form:

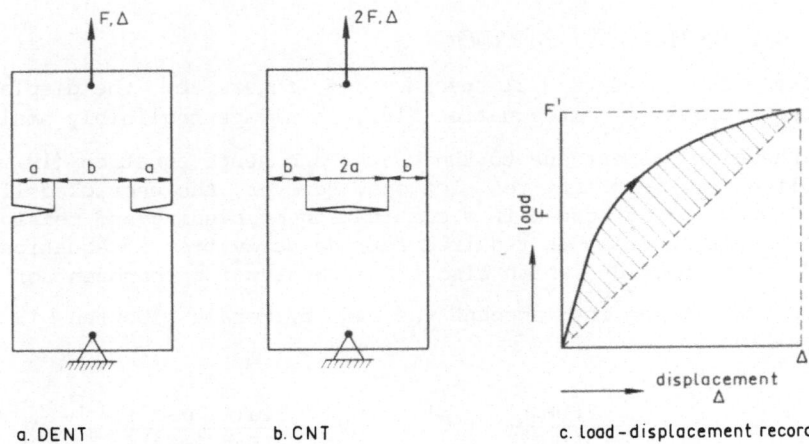

a. DENT   b. CNT   c. load-displacement record

Figure 5. Tension type specimens.

$$\Delta_{pl} = b \ f \left(\frac{F}{B \ b}\right) \tag{25}$$

with material constants being the only other parameters involved in the function f. By equation (25):

$$\left.\frac{\partial \Delta_{pl}}{\partial F}\right|_b = \frac{1}{B} \ f' \left(\frac{F}{B \ b}\right)$$

$$\left.\frac{\partial \Delta_{pl}}{\partial a}\right|_F = \left.\frac{\partial \Delta_{pl}}{\partial b}\right|_F = \frac{F}{b \ B} \ f' \left(\frac{F}{B \ b}\right) - f \left(\frac{F}{B \ b}\right)$$

and hence:

$$\left.\frac{\partial \Delta_{pl}}{\partial a}\right|_F = \frac{1}{b} \left[F \ \left.\frac{\partial \Delta_{pl}}{\partial F}\right|_b - \Delta_{pl} \right] \tag{26}$$

Substitution of equation (26) into equation (9) yields for J_{pl}:

$$J_{pl} = \frac{1}{B \ b} \left(\int_o^{\Delta'_{pl}} F \ d\Delta_{pl} - \int_o^{F'} \Delta_{pl} \ dF \right) \tag{27}$$

or by integrating the second integral by parts:

$$J_{pl} = \frac{2}{B\,b} \left(\int_{o}^{\Delta'_{pl}} F \, d\Delta_{pl} - \frac{1}{2} F' \, \Delta'_{pl} \right) \tag{28}$$

It can easily be verified that by equation (6) this result is equal to:

$$J_{pl} = \frac{2}{B\,b} \left(\int_{o}^{\Delta'} F \, d\Delta - \frac{1}{2} F' \, \Delta' \right) \tag{29}$$

The term in brackets corresponds with the hatched area of the raw load versus (total) displacement record of figure 5c. Hence the applied J-value can be obtained from the load-displacement record of a single specimen by equation (8) with J_{pl} according to equation (29) and the analytical solution for J_{el} [5].

From equations (11) and (28) it follows that:

$$\eta_{pl} = 2 - \frac{F' \, \Delta'_{pl}}{\int_{o}^{\Delta'_{pl}} F \, d\Delta_{pl}} \tag{30}$$

For power law hardening materials, the second term appears to be independent of the extent of the loading [7], but dependent of the hardening exponent. For ideal plastic materials it is easily seen that the second term becomes equal to 1, and hence $\eta_{pl} = 1$.

3. SPECIMEN STUDIES

3.1. SENB Specimen

Figure 6 shows the material stress-strain curve of an SENB specimen with dimensions (cf. figure 3a):

W = 25.4 mm, a/W = 0.5, B/W = 1, S/W = 4

This specimen was used for both an ASTM [14] and (later) a European [15] computational Round Robin exercise, i.e. a comparison between the solutions of a well defined problem solved independently by several analysts. Both programs showed a considerable scatter between the various solutions. It is believed, that this is for an important part caused by the choice of the specimen, as numerical solutions for an SENB specimen are sensitive to the finite element modeling in the vicinity of the load application points (A, B and B' in figure 3a). This can be avoided [16] by considering the relative displacements of points C and D as the load point displacement, instead of that of point A. Only the author's results are discussed here, for a detailed discussion of the Round Robin results one is referred to the respective

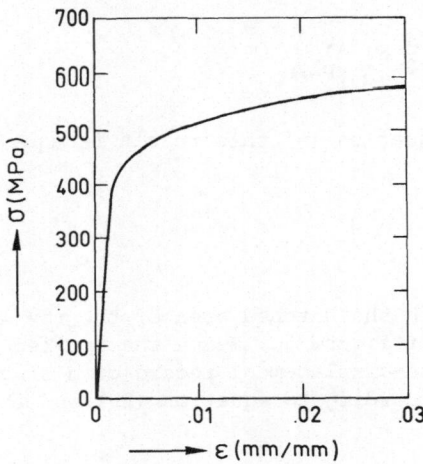

Figure 6. Material stress strain curve (SENB specimen).

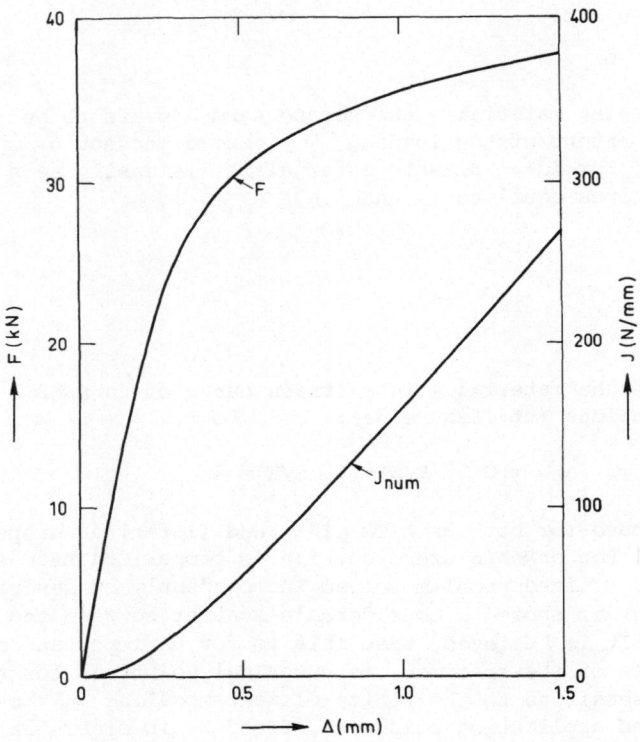

Figure 7. Computational results (SENB specimen).

reports, in particular to that by Larsson [15] , on the European program.
The calculated load F and J-integral as a function of the applied displacement Δ are given in figure 7. The J-value represents the mean value of the contour integral along five different paths [17] , with a maximum path dependence of ± 2.5%. The ratio between the J-values derived from the load-displacement curve of figure 7 according to the procedures discussed in subsection 2.2 and the numerical J-integral (J_{num}) is given in figure 8.

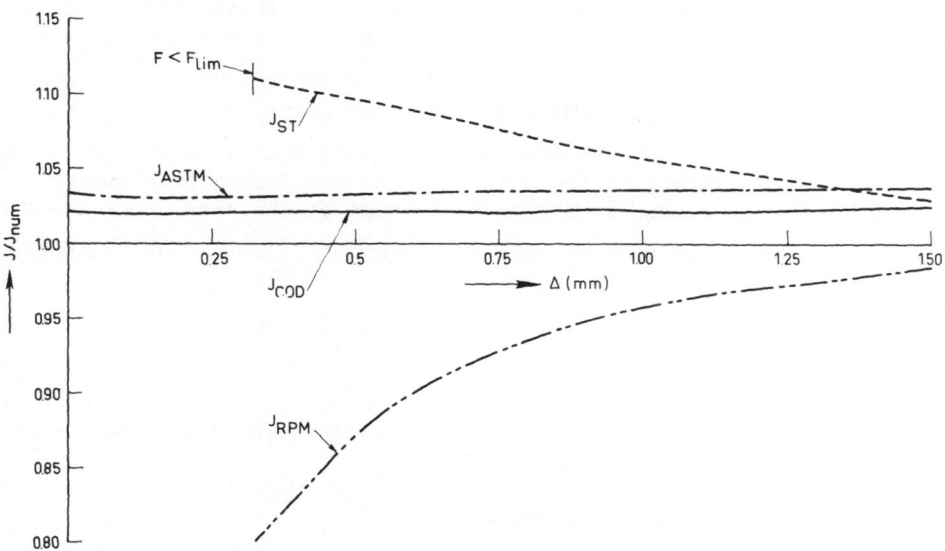

Figure 8. Ratio between estimated and numerical J-values (SENB specimen).

The analytical J-values shown in this figure are:

J_{RPM} : the Rice, Paris & Merkle [8] value with correction for the uncracked deformation according to equation (18);

J_{ASTM}: ASTM [1] value according to equation (19), which corresponds with Turner's [6] procedure according to equation (11) with $\eta_{el} = \eta_{pl} = 2$;

J_{COD} : from the crack mouth opening displacement with J_{pl} according to equation (22) with r = 0.4 b and J_{el} from reference 5;

J_{ST} : Sumpter & Turner's procedure [11] with J_{el} from reference 5 and the integral in equation (22) for J_{pl} estimated by $\frac{1}{2} (F_{lim} + F')$ $v'_{g,pl}$ where the limit load F_{lim} was determined according to reference 3.

Both the J_{ASTM}- and J_{COD}-values show an excellent agreement with the numerical values over the entire range of the displacement. The J_{RPM}-values are lower than J_{ASTM} by the subtraction of the (elastic) uncracked energy. In particular at small load line displacements this results in too low estimates. This confirms observations of Srawley [18], i.e. that the correction for the uncracked energy is valid for very deep cracks (a/W > ≈ 0.8) only. For less deep cracks (0.5 < a/W < 0.8) this correction leads to an underestimate for the elastic component of J. Similar observations are reported by Landes et al. [19] and Sumpter & Turner [11], who argue that for the calculation of J_{el}, the total energy must be used, without correction for the uncracked energy. The good results for J_{COD} indicate that J can accurately be estimated from the crack mouth opening displacement using the BSI rotation factor r = 0.4 b. The accuracy of the simple procedure of Sumpter & Turner, J_{ST}, which does not require the determination of the area under the F versus $v_{g,pl}$ plot, is good at high displacements, i.e. after pronounced plastic deformation.

3.2. CT specimen

Figure 9 shows the material stress-strain curve of a CT specimen [1] with dimensions (cf. figure 3b):

W = 50 mm, B = 25.2 mm, a/W = 0.596

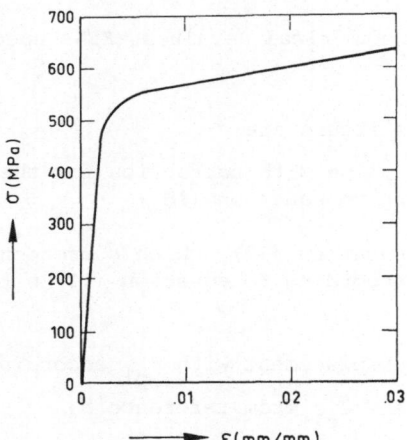

Figure 9. Material stress strain curve (CT specimen).

This specimen was used for both an experimental and computational Round Robin program executed under the auspices of the European Group on Fracture (EGF). The reports on the numerical program have been published [23] , [24]. Only the author's results are presented here. Three finite element analyses were performed [20], viz.:

- two-dimensional, plane stress;

- two-dimensional, plane strain;

- three-dimensional, with one (quadratic) element across half the thickness (with symmetry boundary conditions at the mid plane).

The calculated and experimental load F and J-integral as a function of the applied displacement Δ (taken as the relative displacement between points A and A' in figure 3b) are given by figure 10. The calculated J-integral represents the mean value for seven different paths along which the contour integral was calculated, with a maximum path dependence of ± 2%. The experimental J-value was derived from the experimental load-displacement record by the ASTM equation (24). The agreement between the three-dimensional results and the experiments is surprisingly good, considering the coarse element division in thickness direction. This implies that the pronounced three-dimensional effects as found from three-dimensional analyses with several elements across the thickness [16] are limited to the direct vicinity of the crack tip, whereas the global behaviour is two-dimensional with a lateral constraint somewhere between the limiting cases of plane strain and plane stress, which can be described well by a single layer of the three-dimensional elements.

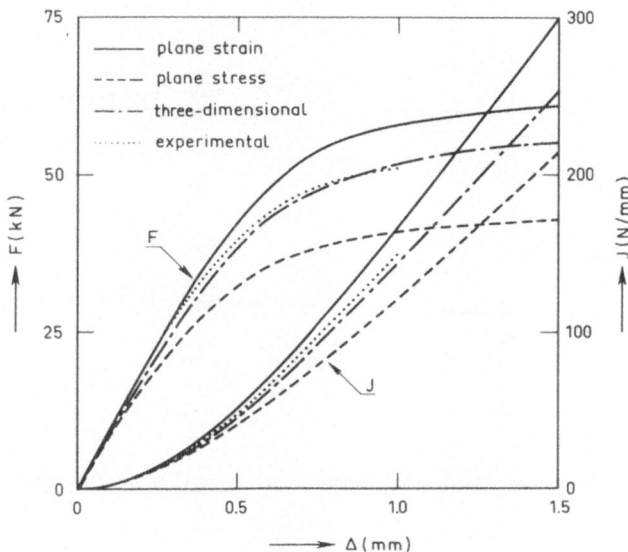

Figure 10. Computational and experimental results (CT specimen).

The ratio between the J-values derived from the plane strain load-displacement curve according to the analytical procedures discussed in subsection 2.3 and J_{num} is given by figure 11. The presented analytical values are:

J_{RPM} : Rice, Paris & Merkle [8] value according to equation (19);

$J_{MC,pl}$: Merkle & Corten value [13] with J_{pl} according to equation (23) and J_{el} taken from reference 5;

$J_{MC,t}$: Merkle & Corten value [13] using the total displacement in equation (23) to calculate the total J-value;

J_{ASTM} : ASTM procedure [1] according to equation (24), i.e. equation (23) using the total displacement and neglecting the complementary energy term.

The Rice, Paris & Merkle procedure, which does not account for the tension component of the ligament loading appears to underestimate J by ≈ 13%, whereas both Merkle & Corten procedures, which account for this effect, give excellent results. The difference between the Merkle & Corten procedures for plastic and total displacement appears to be very small. The neglect of the complementary energy term, resulting in J_{ASTM}, results in a decrease of ≈ 7% at low Δ, and ≈ 3% at high Δ. Beyond the onset of global plasticity ($\Delta > ≈ 0.5$ mm), the error of J_{ASTM} with respect to J_{num} is smaller than 4%, which makes this estimate acceptable for practical use.

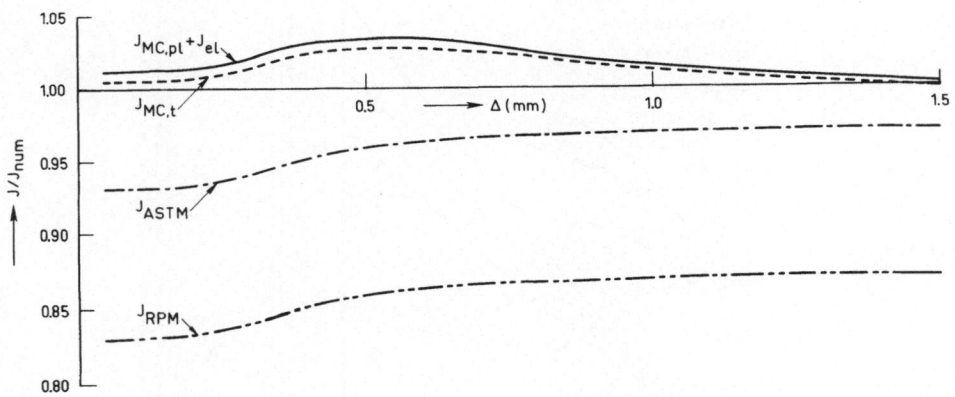

Figure 11. Ratio between estimated and numerical J-values (CT specimen).

3.3. CNT specimen

Figure 12 shows the material stress-strain curve of a CNT specimen with dimensions (cf. figure 5):

a = b = 50 mm, B = 25 mm

This specimen is subject of a combined experimental and numerical study at the author's laboratory, but only the numerical results of a plane strain analysis are discussed here.

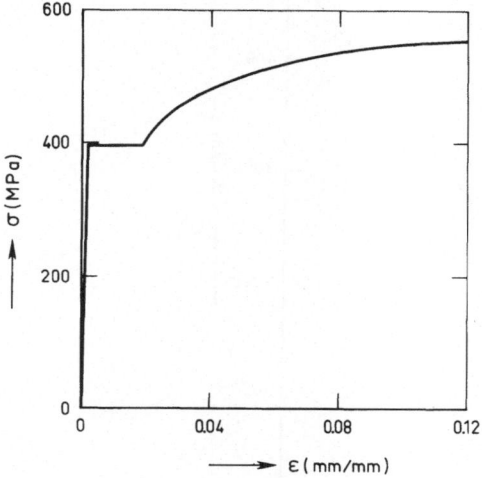

Figure 12. Material stress strain curve (CNT specimen).

The calculated load 2F and J-integral as a function of the applied displacement Δ are shown in figure 13. The J-value represents the mean value of 6 paths along which the contour integral was derived, with a maximum path dependence of $\pm$ 3%. Figure 14 gives the ratio between the J-value derived from the load-displacement record according to the Rice, Paris & Merkle procedure discussed in section 2.3, equation (29) and the numerical J-value (J_{num}) of figure 13. At low Δ, J_{RPM} underestimates J up to a maximum of $\approx$ 4%. Beyond general yield it increases, resulting in a maximum overestimate of 7%. Although this result is reasonable for practical purposes, it is felt that the results will be better for larger relative crack depths, and that a = b is the lower limit for which the basic assumption leading to equation (29), i.e. equation (25), is valid. Analyses with smaller and larger relative crack depths will have to confirm this. The same yields for the statement whether η_{pl} exists for a CNT specimen of a material whose stress strain curve cannot be described accurately by a power law fit (cf. figure 12). The calculated η_{pl} is shown in figure 15 as a function of the applied displacement Δ. Just after the onset of global plasticity

($\Delta \approx 0.5$ mm), η_{pl} is close to 1, i.e. the value for ideal plastic material behaviour as discussed in section 2.3. With increasing plasticity η_{pl} drops to ≈ 0.84. However, on the basis of this single result it cannot be concluded whether η_{pl} really exists, as this variation may have the same origin as that of J_{RPM} in figure 14.

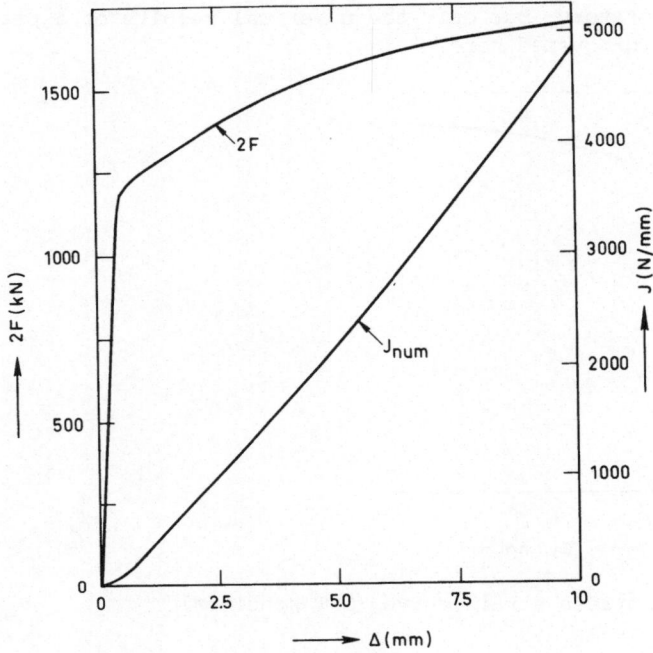

Figure 13. Computational results (CNT specimen).

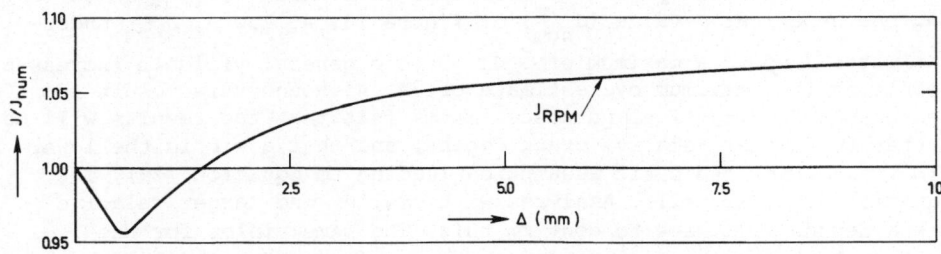

Figure 14. Ratio between estimated and numerical J-values (CNT specimen).

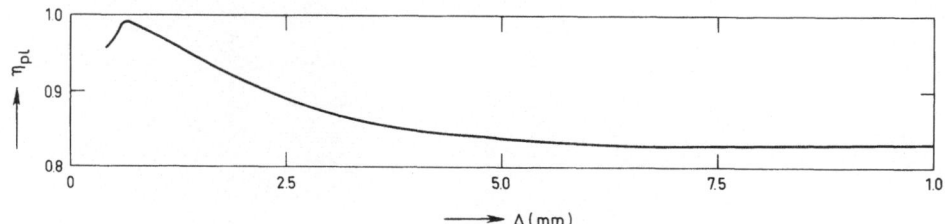

Figure 15. Plastic work factor η_{pl} (CNT specimen).

4. MODEL STRUCTURE STUDY

4.1. Experiments

Figure 16 shows the dimensions of a uni-axially loaded test plate with a quarter elliptical crack emanating from the edge of a central hole. This geometry was chosen because it models a common situation in structures, i.e. a fatigue crack at a location with a high stress concentration.

The crack was generated from a machined notch by fatigue loading. Subsequently the model was loaded once - by displacement control - up to a load for which crack growth could be expected. After unloading the cracked region was cut from the model and put into an oven for some time to colour the ductile crack growth by heat tinting. Finally, it was cooled in liquid nitrogen and broken to enable observations of the crack surfaces. Stable crack growth was found to have occurred along almost the entire crack front. The actual crack growth measurements are shown in figure 17, where the data points represent the mean value of three closely spaced measuring points.

4.2. Computations

A complete three-dimensional elastic-plastic finite element analysis of the model test described in the previous section was performed with the MARC system. For symmetry reasons only one half of the test plate was analyzed as shown in figure 18. Details of the analysis are identical to those of a similar analysis of an SENB specimen reported earlier [16]. The load was applied by prescribing a uniform end displacement equal to the mean value of the displacement transducer readings during the test (cf. figure 16). The entire analysis up to the final experimental end displacement required 14 inelastic (tangent stiffness) displacement increments.

The theoretical basis for J along the curved front of a plane crack in a three-dimensional body was given in my first lecture. An earlier study showed that the most efficient way to calculate this

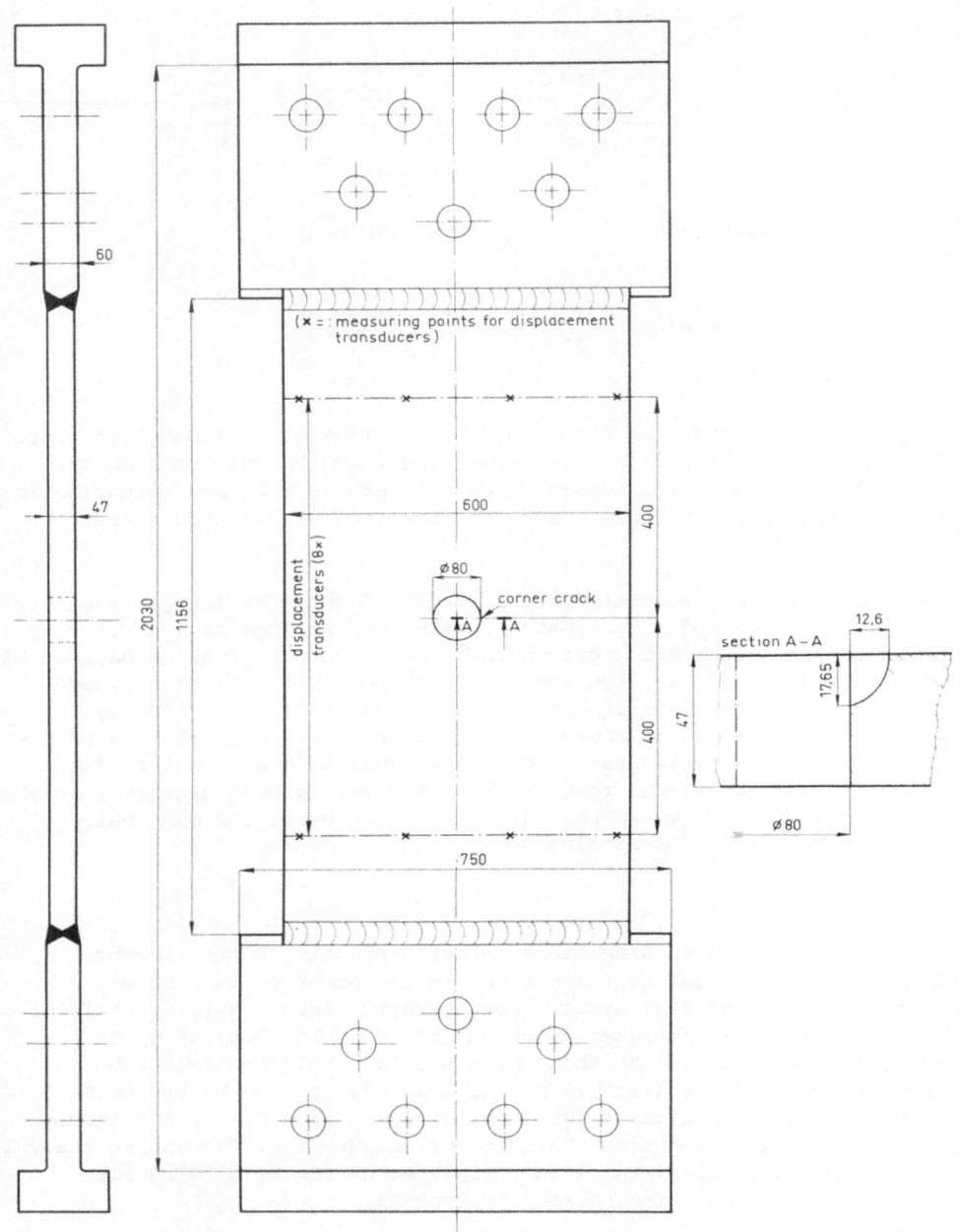

Figure 16. Model plate geometry.

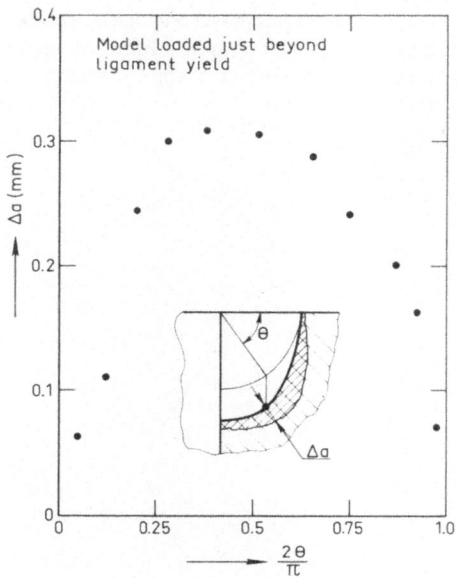

Figure 17. Final crack growth distribution of the model experiment.

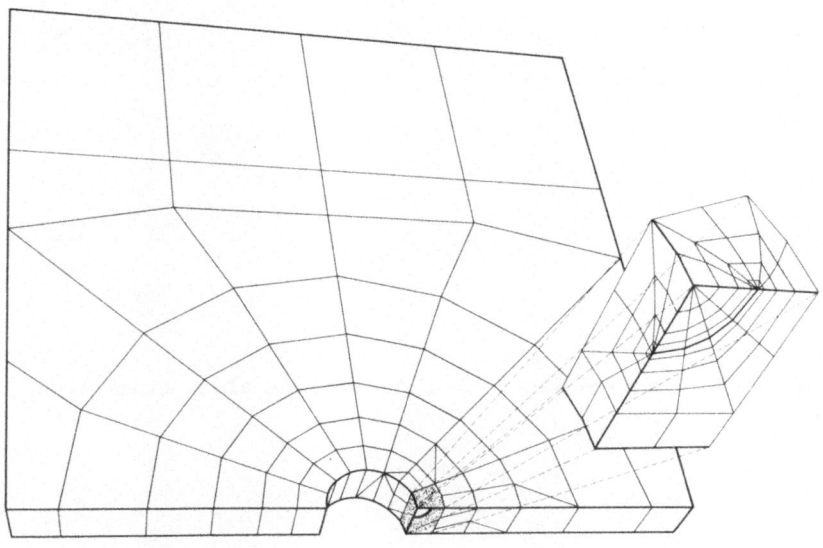

Figure 18. Three-dimensional finite element model.

integral is the virtual crack extension method, in combination with an assumed linear J-distribution per element edge along the crack front [21]. The J-distributions calculated by that procedure at various end displacements, including the final experimental displacement, are shown in figure 19.

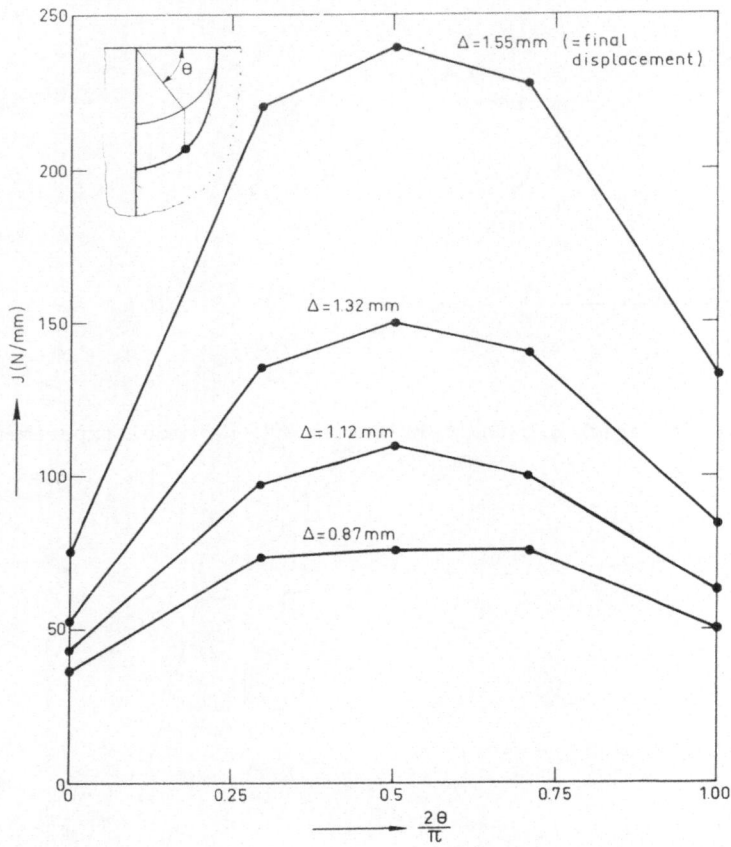

Figure 19. Elastic-plastic J-distributions along crack front at various deformation levels.

4.3. Discussion

Direct comparison between the initiation behaviour of test specimens and the model structure is impossible, because the moment of crack growth initiation in the model could not be detected by the potential drop method. However, the combination of measured crack growth and computed J-distribution, i.e. the measuring points in figure 17 with the computed J-value at the same location along the crack front in figure 19, results in data points in a J-Δa (J-Resistance curve) diagram which can be compared with the J-Δa data obtained from small specimens. This comparison is shown in figure 20. The specimen data in this figure are represented by the 90% confidence limits of a power equation regression ($J = c \, \Delta a^n$, with regression constants c and n) through a series of both side grooved and plane sided (non side grooved) single specimen data points.

Details about the specimen results and a justification for the comparison of mean specimen data with local data along the crack front of the model are given by the author in reference [22].

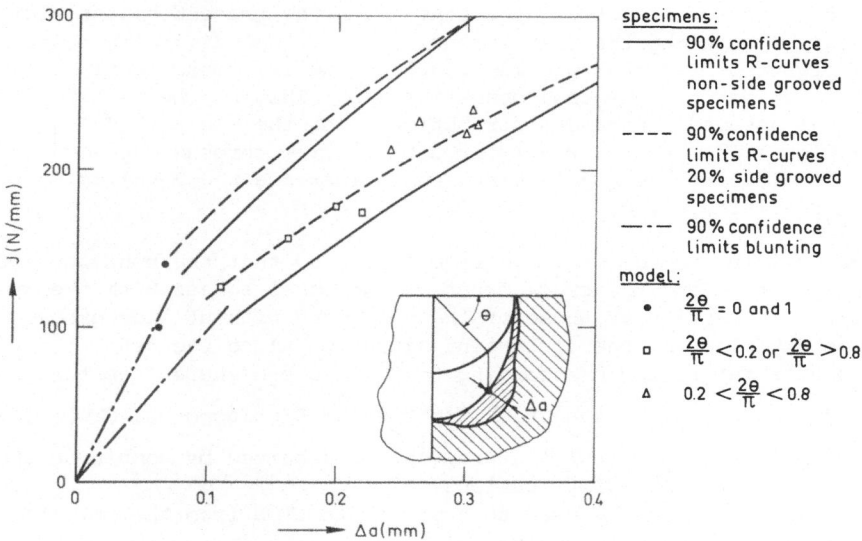

Figure 20. Comparison of specimen and model plate Resistance curves.

The model data points in figure 20 can be divided into three groups:

a. Free surface points $(2\theta/\pi = 0$ and $1)$.
   Although both points fall within the scatter of the specimen
   results, no real significance should be attributed to these points
   because of the coarse element division along the crack front [22].
   As the maximum J-values occur at the interior part of the crack
   front (cf. figure 19), and the surface points did not show real
   (fibrous) crack growth, i.e. blunting only, the surface points
   are of minor interest.

b. Interior points $(\approx 0.2 < 2\theta/\pi < \approx 0.8)$.
   These are the five points on the right in figure 20. The
   J-distribution for this interior part of the crack front is rather
   flat and is hence expected to be accurately described with the
   present element division. The points agree fairly well with the
   specimen results, although they are all in the lower part of the
   confidence interval.

c. Transition points $(2\theta/\pi < \approx 0.2$ and $2\theta/\pi > \approx 0.8)$.
   These four points are located in the transition region between the
   free surfaces and the interior part of the crack front, in which a
   considerable change in constraint takes place. The exact width of
   this region cannot be determined with the present coarse element
   division along the crack front. The corresponding data points in
   figure 20 fit the specimen results, but lie consistently in the
   lower part of the confidence interval. This may be partly due to
   the assumed linear J-distribution along the element edges (cf.
   figure 19). A more realistic smooth curve through the computed
   J-values at the nodal points would increase the J-values in this
   transition region.

In general it can be concluded from figure 20 that the behaviour of
local points along the crack front of the model agrees with the specimen
results, in particular those of the specimens without side grooves. This
may be explained by the constraint situation along the crack front.
The constraint factor m according to $J = m \, \bar{\sigma} \, \delta_t$ (where $\bar{\sigma}$ is the material

flow stress and δ_t is the Crack Tip Opening Displacement, CTOD, cf.

equation (37) in my first lecture) can be obtained by combining the
computed J- and CTOD-distributions.
Two alternatives may be used to approximate CTOD from the results of a
finite element computation, viz.:

   - by extrapolating the deformed far field crack front to the crack
     tip;

   - by taking the opening displacement of the collapsed nodes at the
     crack tip [22].

The first procedure is used for the experimental determination of CTOD
during specimen tests. However, it can only be applied when the
deformed far field crack front is (approximately) straight.

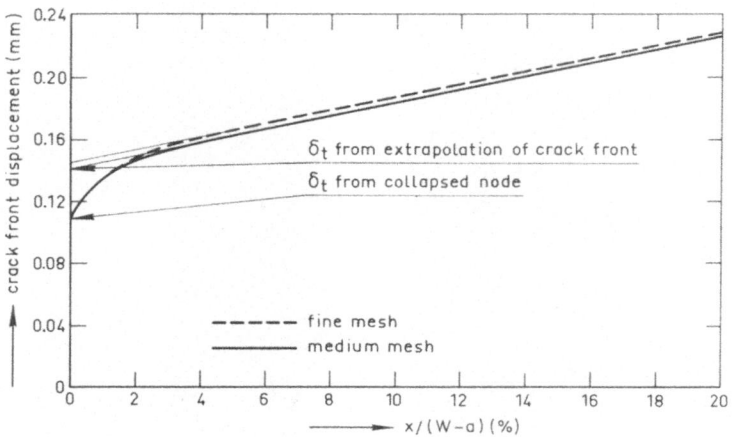

Figure 21. CTOD approximations for two-dimensional SENB analysis at the final displacement $\Delta = 1.5$ mm (cf. subsection 3.1, meshes refer to different mesh configurations used for a mesh convergence study).

Figure 21 compares these approximations for the two-dimensional analyses of an SENB specimen discussed in subsection 3.1. The crack tip elements applied for the present analysis are comparable in size to those of the medium two-dimensional mesh. The results of figure 21 show an adequate convergence of the medium mesh. They also illustrate that the second method (i.e. using the opening displacements of the collapsed crack tip nodes) results in some lower bound estimate of CTOD. This method had to be applied for the present three-dimensional analysis, as it appeared that the deformed crack plane was entirely curved and hence no extrapolations from the deformed far field crack shape could be performed. The use of the collapsed node displacement for CTOD then evidently leads to an upper bound approximation of the constraint factor m. This result is shown in figure 22, together with the constraint factors found for the specimens [22]. As the specimen m factors are obtained from CTOD approximations by application of the extrapolation of the crack front, these may be regarded as lower bound estimates. The maximum constraint for the central part of the crack front approaches, but still is lower than that of the side grooved specimens. For the transition region the constraint is closer to that of the non-side grooved specimens, hence in that region the model behaviour may be expected to show better agreement with the non-side grooved specimen results.
The apparent agreement between model and specimen behaviour is of course also enhanced by the relatively large confidence intervals of the specimen R-curves. There is, however, no way to reduce these confidence intervals, which were found [22] to be caused to a large extent by scatter of the material response. Seen in the light of this

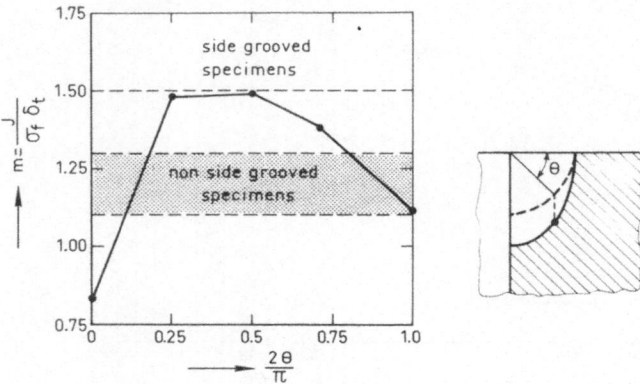

Figure 22. Constraint distribution along crack front.

scatter, the present results indicate that there is a consistent
initiation behaviour in terms of J, with only marginal effect of the
level of constraint. This seems to pave the way for engineering
applications of the J-concept. However, with this conclusion based on
the single result reported here, more experiments are needed to make
it more than tentative.
It should be emphasized that the above conclusion refers to small
amounts of crack growth, i.e. close to initiation, only. For larger
amounts of crack growth there appears to be a structural dependence of
J-controlled crack growth, in particular between structures loaded in
bending and tension, as already discussed in my first lecture.

CONCLUSIONS

• Methods which estimate the applied J-values of SENB, CT and CNT
  specimens from the experimental load-deflection curve have a very
  good accuracy.

  - For SENB specimens, the total energy must be used, without
    correction for the displacement of the body with no crack, as
    originally proposed by Rice, Paris & Merkle [8].

  - For CT specimens, the Rice, Paris & Merkle [8] solution under-
    estimates the applied J as it neglects the axial load component on
    the ligament. The Merkle & Corten [13] procedure, which accounts
    for this load, gives excellent results, as does the ASTM [1]
    approximation.

- The somewhat higher deviations found for the CNT specimen, may be caused by the fact that the relative crack depth a/b = 1 used here, is the lower limit to which the Rice, Paris & Merkle [8] estimate can be applied.
  More results for other relative crack depths are required for a more definite statement. The same yields with regard to the answer to the question whether the plastic work factor η_{pl} exists for a CNT specimen.

• The initiation and crack growth behaviour of a plane curved crack in a three-dimensional body in terms of local J-values is in good agreement with that of SENB test specimens.

ACKNOWLEDGEMENTS

The author's activities in EPFM in the period 1979 - 1983 were partly sponsored by the Netherlands Ministry of Economic Affairs as part of the co-operative research program BROS, which is greatfully acknowledged. The author is indebted to Professor D.G.H. Latzko for his support and stimulating discussions and to P.A.J.M. Steenkamp for discussions and supplying the CNT data used in this paper.

REFERENCES

1.   Standard Test for J_{Ic}, a Measure of Fracture Toughness, *Annual Book of ASTM Standards*, ASTM E813-81, American Society for Testing and Materials, USA, 1981.

2.   Albrecht, P. et.al.
     Journal of Testing and Evaluation, 10 (1982), pp. 245/251.

3.   Kumar, V., German, M.D. and Shih, C.F.,
     An Engineering Approach for Elastic-Plastic Fracture Analysis,
     EPRI NP-1931, Electric Power Research Institute, USA, 1981.

4.   Ernst, H.A., Paris, P.C. and Landes, J.D.,
     In: *Fracture Mechanics, Thirteenth Symposium*,
     ASTM, STP 743, pp. 476/502, American Society for Testing and Materials, USA, 1981.

5.   Tada, H., Paris, P.C. and Irwin, G.,
     The Stress Analysis of Cracks Handbook,
     Del Research Corporation, USA, 1973.

6.   Turner, C.E.,
     In: *Advances in Elastic-Plastic Fracture Mechanics*,
     L.H. Larsson, editor, pp. 301/318, Applied Science Publ., UK, 1979.

7. Paris, P.C., Ernst, H.A. and Turner, C.E.
    In: *Fracture Mechanics, Twelfth Symposium*,
    ASTM STP 700, pp. 338/351, American Society for Testing and
    Materials, USA, 1980.

8. Rice, J.R., Paris, P.C. and Merkle, J.G.,
    In: *Progress in Flaw Growth and Fracture Toughness Testing*,
    ASTM STP 536, pp. 231/245, American Society for Testing and
    Materials, USA, 1973.

9. Begley, J.A. and Landes, J.D.,
    In: *Fracture Toughness*, ASTM STP 514, pp. 1/23 and pp. 24/39,
    American Society for Testing and Materials, USA, 1972.

10. Smith, E.
    Journal of Engineering Materials and Technology, $\underline{104}$ (1982),
    pp. 215/219.

11. Sumpter, J.D.G., and Turner, C.E.,
    In: *Cracks and Fracture*, ASTM STP 601, pp. 3/18,
    American Society for Testing and Materials, USA, 1976.

12. *Methods for Crack Opening Displacement (COD) Testing*, BS 5762,
    British Standards Institution, UK, 1979.

13. Merkle, J.G. and Corten, H.T.,
    Journal of Pressure Vessel Technology, $\underline{96}$ (1974), pp. 286/292.

14. Wilson, W.K. and Osias, J.R.,
    International Journal of Fracture, $\underline{14}$ (1978), pp. R95/R108.

15. Larsson, L.H.,
    International Journal of Pressure Vessels and Piping, $\underline{11}$ (1983),
    pp. 207/228.

16. Bakker, A.,
    International Journal of Pressure Vessels and Piping, $\underline{10}$ (1982),
    pp. 431/449.

17. Bakker, A.,
    International Journal of Pressure Vessels and Piping, $\underline{14}$ (1983),
    pp. 153/179.

18. Srawley, J.E.,
    International Journal of Fracture, $\underline{12}$ (1976), pp. 470/474.

19. Landes, J.D., Walker, H. and Clarke, G.A.,
    In: *Elastic-Plastic Fracture*, ASTM STP 668, pp. 266/287,
    American Society for Testing and Materials, USA, 1979.

20. Bleeker, B.J.,
    Elastic-plastic analysis of a CT specimen,
    EV-1286, Laboratory for Thermal Power Engineering,
    Delft University of Technology, The Netherlands, 1983 (in Dutch).

21. Bakker, A.,
    On the Numerical Evaluation of J in Three Dimensions,
    Proceedings of the International Conference on Application of
    Fracture Mechanics to Materials and Structures,
    Freiburg, FRG, June 20-24, 1983, pp. 657/671.

22. Bakker, A.,
    *The Three-Dimensional J-Integral. An Investigation into Its Use
    for Post-Yield Fracture Safety Assessment*,
    Ph.D. Thesis, Delft University of Technology,
    Delft, The Netherlands, 1984.

23. Larsson, L.H.
    The EGF round robin on numerical EPFM,
    Proceedings of the 5th European Conference on Fracture (EGF-5),
    Lisbon, Portugal, September, 17-21, 1984, pp. 425/446.

24. Larsson, L.H.
    EGF numerical round robins on EPFM,
    Proceedings of the Third International Conference on
    Numerical Methods in Fracture Mechanics,
    University College, Swansea, March 26-30, 1984, pp. 775/814.

THE MEASUREMENT OF CRITICAL VALUES OF CRACK TIP OPENING DISPLACEMENT (CTOD) AND J ON PARENT STEELS AND WELDMENTS FOR USE IN FRACTURE ASSESSMENTS

S. J. Garwood,
The Welding Institute
Abington, Cambridge, CB1 6AL,
United Kingdom

ABSTRACT Following a brief description of the role of elastic-plastic fracture mechanics for fitness for purpose assessments of ferritic materials, current fracture toughness testing standards are reviewed.

Consideration is then given to the measurement of CTOD values for subsequent use in fracture assessments. Particular aspects such as specimen geometry, notch positioning and special test methods for weldments are highlighted.

Estimation procedures for J and CTOD at the initiation of ductile tearing following ASTM and British Standard methods are presented.

Finally the relevance of initiation, maximum load toughness and crack growth resistance curves for the assessment of materials exhibiting fully ductile behaviour is discussed.

1. INTRODUCTION

Elastic-plastic fracture mechanics (EPFM) is gaining increasing acceptance as a tool to aid the designer, fabricator, customer and inspectorate authorities with the assurance of structural integrity. Use of EPFM facilitates rationalisation of material selection, heat treatment, quality control, and non-destructive examination requirements at the design stage and repair considerations during fabrication and service.

In the United Kingdom, the most common approach outside the nuclear industry,is to apply the guidelines outlined in PD 6493 [1] which for the majority of practical cases, recommends adoption of a procedure based on the crack tip opening displacement (CTOD) design curve [2].

L. H. Larsson (ed.), Elastic-Plastic Fracture Mechanics, 85–115.
© *1985 ECSC, EEC, EAEC, Brussels and Luxembourg.*

The CTOD method and its practical application is well documented; the 'state of the art' in 1980 was outlined in reference 3 and more recently, practical problems associated with the use of CTOD procedures have been discussed in reference 4.

The acceptance of the value of the fitness for purpose approach is now being recognised by inclusion in standards such as BS 5500 [5] and BS 4515 [6].

For nuclear and pressure vessel applications use of the CEGB's R6 method [7] and the similar concepts adopted by EPRI [8] are widespread. Toughness estimates for these procedures are generally based on K_{Ic} estimates using J integral procedures.

A new contender in the elastic-plastic fracture analysis field is the EnJ method [9] which is a development of J design curve procedures [10].

All these analysis procedures are dealt with in detail in other lectures.

Within the fracture mechanics fraternity there is considerable current interest in the use of elastic-plastic methods for assessments of fully ductile materials. For the majority of engineering structures which are fabricated from ferritic steels by welding, however, it is the avoidance of brittle fracture which is of vital importance (see Figs. 1, 2 and 3).

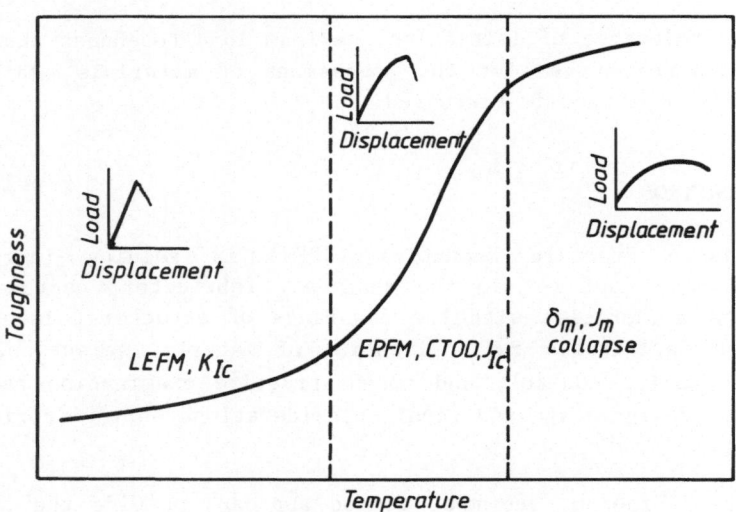

Fig.1. Schematic transition curve showing regions for application of LEFM, EPFM and collapse.

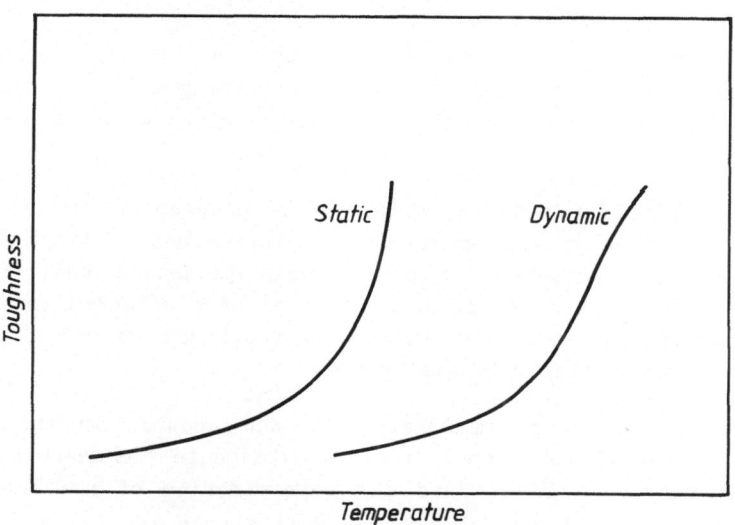

Fig.2. Strain rate effects in structural steels.

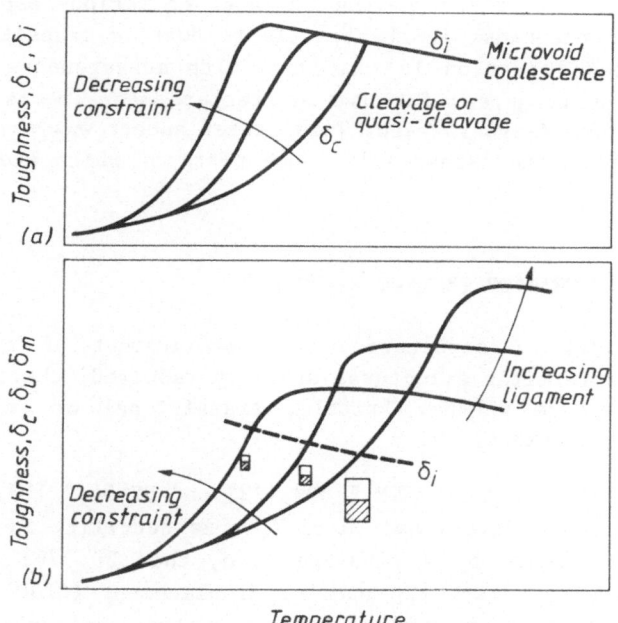

Fig.3. Effects of testpiece geometry on CTOD measured in the transition regime.

Crack arrest capability can only be demonstrated in a very small percentage of structures, therefore avoidance of catastrophic fracture must be assured by preventing initiation of an unstable crack from a fabrication defect.

As by far the majority of fabrication defects are associated with welds it is the understanding of factors affecting weldment toughness and the measurement of appropriate toughness values which is of prime importance.

It is commonplace for material selection procedures to ensure 'upper shelf' behaviour of parent materials at minimum design temperatures, whereas toughness measurements during weld procedure qualification indicate that certain areas of weld metal or heat affected zone (HAZ) exhibit transitional behaviour with CTOD toughness an order of magnitude lower than parent material levels.

In practice the fracture mechanics approach adopted in the safety assessment is usually of second order importance to the measurement of appropriate toughness levels and the determination of the inputs to the assessment such as applied and residual stress levels, and defect dimensions.

The first part of this lecture concentrates on various aspects of the measurement of toughness in the brittle to ductile transition region, whilst the second section is concerned with measurements of appropriate values of toughness for use in fracture assessments when fully ductile behaviour is guaranteed. This latter aspect was the subject of a recent paper by the author [11], some parts of which are reproduced here.

2. FRACTURE MECHANICS TESTING STANDARDS

For the majority of structures, when measurement of the fracture toughness of a ferritic structural steel is required, there is usually no prior knowledge whether brittle, transitional or fully ductile behaviour will be exhibited.

For very brittle behaviour ASTM E 399 [12] and BS 5447 [13] define the test method for the determination of K_{Ic}. As ductility increases, but final fracture occurs by a cleavage mode, then BS 5762 [14] can be employed to measure crack tip opening displacement (CTOD) and is the only formal fracture mechanics test standard applicable for this region. (The COD standard has also been adopted in China as GB 2358-80.) The British Standard was derived from a draft for development (DD19) [15] which employed different CTOD estimation formula. In general the change of formulae to the current procedure outlined below caused a reduction in CTOD of approximately 10% [16, 17]. The background to the development of the existing formulae is detailed in reference 16. ASTM is currently reviewing the British Standard procedure and a draft procedure is being considered by the ASTM E24.08 04 task group.

The current draft suggests that the test procedure is likely to be similar to BS 5762 but will include the compact tension (CT) geometry in addition to the single edge notched three point bend specimens adopted by the British Standard.

The definitions of CTOD (δ) for describing brittle to ductile behaviour follow those of BS5762 i.e:-

δ_c — CTOD at the onset of brittle crack extension or pop-in when the event is not preceded by slow stable crack growth.

δ_u — CTOD at the onset of brittle crack extension or pop-in when the event is preceded by slow stable crack growth.

δ_i — CTOD at which slow stable crack growth commences.

δ_m — CTOD at first attainment of maximum force plateau.

No testing standard for the measurement of J in the transition region exists although there is no reason why J definitions analogous to those for CTOD cannot be adopted using the formulae quoted in the ASTM Standard used for fully ductile behaviour [18]. In addition, consideration is being given by a task group of ASTM E24 to extend the definition of this latter standard to transitional behaviour. All the above fracture testing procedures are currently under review by a British Standard committee with the eventual aim of combining the methods into one definitive standard.

For fully ductile behaviour, BS 5762 describes the measurement of δ_m and also defines a procedure for δ_i in its Appendix A.

When fully ductile behaviour is guaranteed the fracture parameter J_{Ic} can be determined using the procedures described in ASTM E 813 [18] or JSME S001-1981 [19] and various techniques can be employed to measure crack growth (R) curves in terms of CTOD and J as discussed in the lecture by Blauel.

3. TOUGHNESS MEASUREMENTS ON MATERIALS EXHIBITING BRITTLE-DUCTILE BEHAVIOUR

3.1. Crack tip opening displacement (CTOD) Procedures

As noted above the only standard specifically related to the measurements of toughness in the transition region is BS 5762.

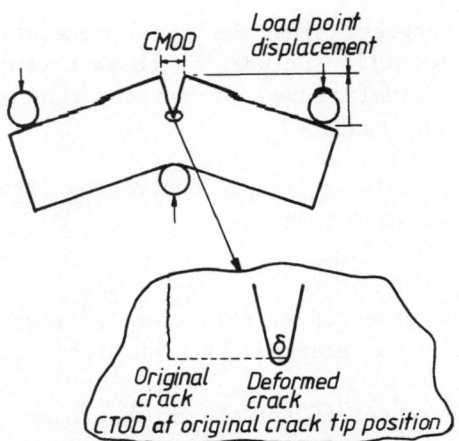

Fig.4. Definition of crack mouth and crack tip opening displacement.

The major aim of the procedures outlined in BS 5762 is to obtain an unambiguous estimate of the displacement at the tip of the fatigue crack in a three point bend test specimen at the onset of unstable fracture (Fig. 4).

To make the test practical CTOD (δ) is estimated using a formula employing the plastic displacement (V_p) measured at or above the crack mouth and an assumed plastic rotational factor ($r_p = 0.4$). The elastic contribution to the CTOD (δ_e) is estimated using the equivalent K value at the failure load and assuming plane strain conditions, i.e.

$$K^2/E' = 2\sigma_Y \delta_e$$

where $\qquad E' = E/(1 - v^2)$ ...(1)
Thus

$$\delta = \frac{K^2(1 - v^2)}{2\sigma_Y E} + \frac{0.4(W - a_o)V_p}{(0.4\ W + 0.6a_o + z)} \qquad ...(2)$$

where $\quad$ W $\quad$ – is the specimen width
$\qquad \sigma_Y$ – is the material yield strength
$\qquad$ z $\quad$ – is the knife edge thickness
$\qquad a_o$ – is the initial crack length
$\qquad$ E $\quad$ – is the Modulus of elasticity
and $\qquad v$ $\quad$ – is Poisson's ratio.

Hence it can be seen that the value of CTOD obtained does not refer necessarily to the physical CTOD at a particular point along the crack front, since the V_p and K values in the formula are influenced by the entire specimen behaviour and the true plastic rotational factor (which can in fact vary between 0.3 and 0.5). Thus comparisons between silicone rubber infiltration methods [20] and the BS 5762 definition of CTOD can show up inaccuracies. These discrepancies are usually small however [21], and the value of having an accepted experimental procedure for CTOD defined in terms of a relatively simple formula is paramount. The idea of defining fracture toughness in terms of the physical opening at the crack tip makes CTOD by far the conceptually simplest elastic-plastic fracture parameter.

For the more ductile ferritic steels, initiation of stable ductile tearing may occur prior to unstable fracture by cleavage. When 'upper shelf' toughness levels are reached tearing will continue until a maximum load is achieved. (see Fig. 5). During the tearing process the crack propagates with an approximately constant crack opening angle and with a relatively sharp tip (see Fig. 6). Measurements of the profile and the displacement at the tip of the growing crack by silicone rubber infiltration methods [22] reveal that, for structural steels, the displacement at the tip of the growing crack (δ_t) is constant at a fraction of δ_i (see Fig. 7).

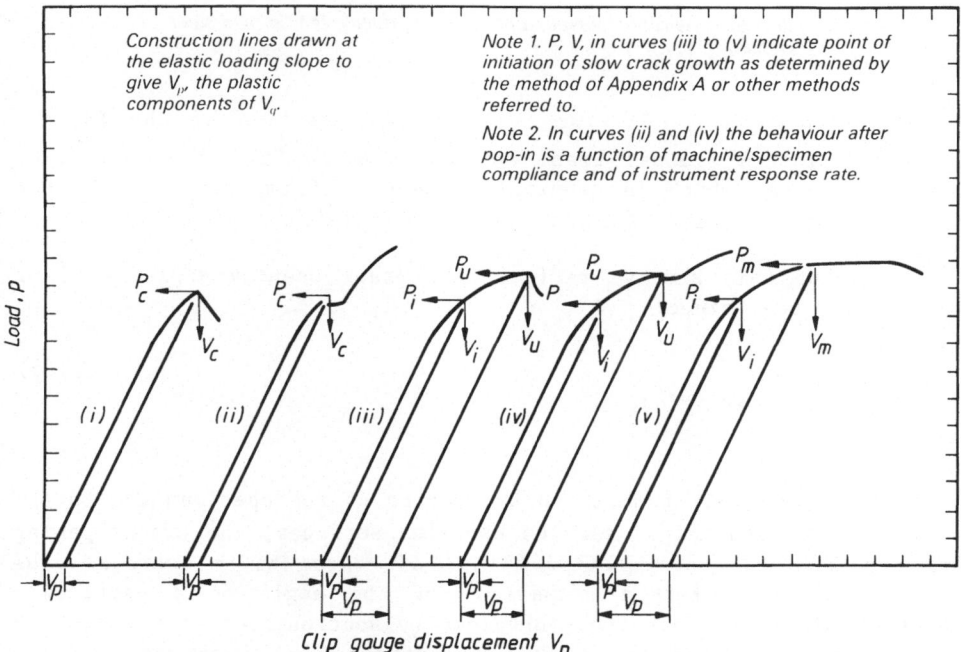

Fig.5. Types of load clip gauge displacement graphs for calculation of CTOD. From BS 5762 amd. 4131.

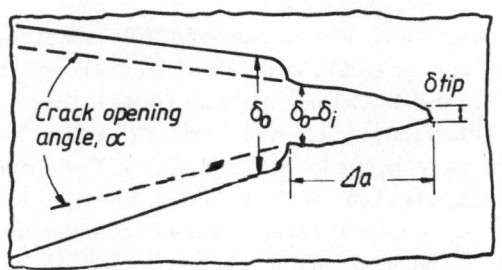

Fig.6. Geometry of a growing ductile tear.

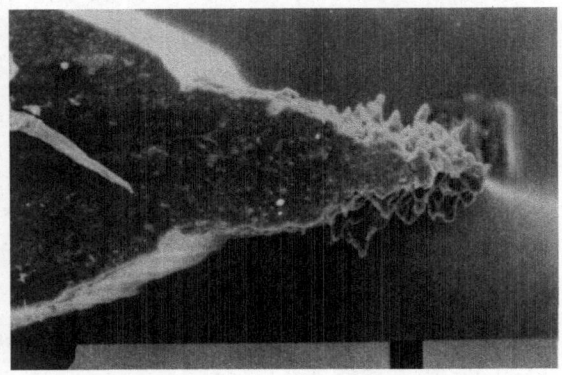

Fig.7. Silicone rubber cast of a propagating ductile tear in mild steel (En32B).

If the displacement (δ_o) at the original position of the fatigue crack is plotted against crack extension as in Fig. 8 a resistance curve is formed which is generally linear in form, particularly for small amounts of crack extension (Δa) [23 - 26].

The slope of this R-curve reflects the crack opening angle (α) of the ductile tear since:-

$$\frac{d\delta_o}{da} \simeq 2 \tan (\alpha/2) \qquad \ldots(3)$$

If the ductile crack tunnels in the centre of the specimen due to the formation of shear lips near the specimen surfaces, the crack opening angle determined from equation 3 will of course be an average value and to obtain a better comparison of the angle at a particular position within the specimen (obtained by sectioning rubber casts for example) with derived values from the physical measurements, the use of the slope of a plot of δ_o against the crack extension at the point under consideration is required.

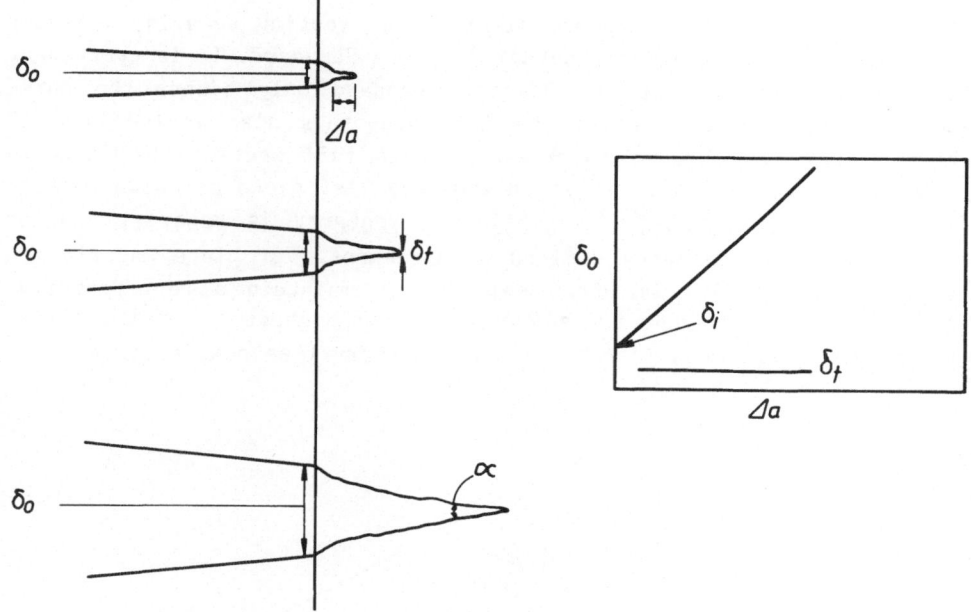

Fig.8. Relationship between CTOD referred to the original crack tip δ_o and the growing crack tip δ_t.

BS 5762 attempts to allow for ductile crack extension by adopting the final crack length measurement (a_f) in place of the initial crack length term (a_o) in equation (2). The result of this is to determine a value of displacement at the growing crack position (δ_r). Unfortunately this displacement no longer has a physical definition since the actual CTOD of the growing crack is very small. The value derived can be likened to the displacement at the tip of a crack, not extended by tearing, in a specimen with an initial crack length of a_f displaced to an equivalent mouth opening displacement (V_g).

However, a far more acceptable definition of CTOD for a growing crack is to use the formula of equation (2) with the initial crack length a_o. Thus this refers the CTOD to the tip of the initial fatigue crack (i.e. δ_o) and therefore has a physical meaning.

This definition is employed for the measurement of δ_m and is adopted in the latest revision of the standard for CTOD resistance curve measurements used for the estimate of δ_i . This procedure is outlined in Appendix A of BS 5762 which was previously ambiguous with regard to the choice of a_o or a_f for the CTOD determination.

At present BS 5762 adopts the final crack length measurement for the determination of δ_u since this gives a lower value than δ_o . However, it is hoped that future revisions to the standard will adopt the δ_o definition for all CTOD measurements.

Specimen dimensions, notching, precracking, testing details, measurement procedures etc are all detailed in the Standard. It is pertinent to note, however, that the British Standard only adopts the three point bend geometry whereas the ASTM proposals also include the CT test piece. For through thickness notches full section thickness is always adopted with the preferred geometry test piece as shown in Fig. 9. For surface notches the subsidiary geometry is generally chosen with the a/W ratio being decided by agreement. For plate materials a value of a/W = 0.3 is often selected to maintain a representative ligament size although a/W = 0.5 gives the highest constraint [16]. For welds special considerations are necessary (see next section).

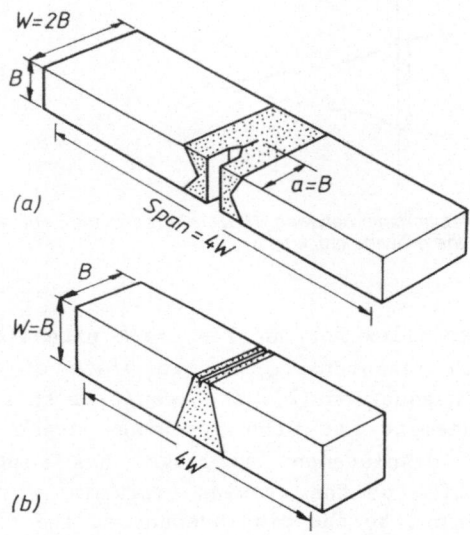

Fig.9. Sketches showing the preferred and subsidiary CTOD testpiece geometries.

3.2.   The Estimation of J Values

ASTM E 813-81 [18] describes the evaluation of J approximating to the initiation of stable tearing, J_{Ic}. As yet no standard employing J equivalent to BS 5762 exists. However, there is no reason why J cannot be measured when carrying out CTOD tests as was originally envisaged by Sumpter and Turner [27].

For the three point bend geometry [28, 29]:-

$$J \simeq \frac{2U_T}{B(W - a_o)} \qquad \qquad ...(4)$$

where U_T is determined by measuring the total area under the load-load point displacement record and B is the specimen thickness. In terms of the area under the load-plastic clip gauge displacement record U_v (see Fig. 10) and assuming a plastic rotational factor of 0.4, a J equivalent to the CTOD formula of equation (2) is given by:-

$$J \simeq \frac{K^2(1 - \nu^2)}{E} + \frac{2U_v S}{4(0.4\ W + 0.6a_o + z)\ B\ (W - a_o)} \qquad \ldots(5)$$

where S is the testpiece span.

In this way J_c, J_u and J_m can be determined for use in analyses where J is more appropriate.

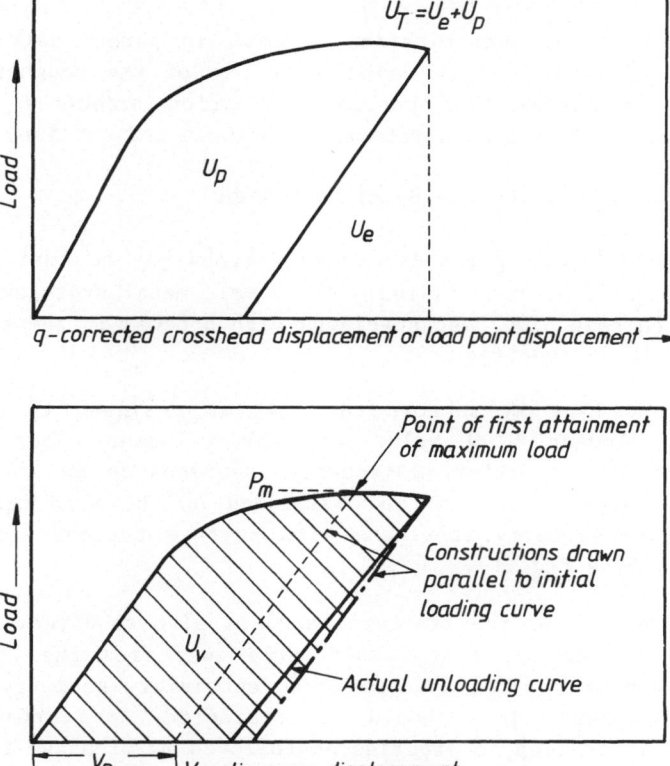

Fig.10. Definitions of values derived from load-displacement records.

4. TESTING WELDMENTS USING CTOD PROCEDURES

4.1. Weldment Toughness

The prediction of the factors giving rise to good weldment toughness is complex, but a number of papers give guidance [30 - 33]. Factors affecting CTOD toughness include: sampling position, parent material, consumable, weld procedure, heat input, number of runs, preheat, restraint and the welder.

It is now common practice to incorporate CTOD testing at the weld procedure qualification stage. The data provided may be used in several different ways, e.g.:

i)   To ensure that the structure has adequate defect tolerance.
ii)   To provide data for subsequent defect assessments.
iii)   For finalising the choice of weld procedure.
iv)   To provide a relevant comparison between CTOD and Charpy (which is the usual quality assurance test during production).

BS 5762 defines the determination of CTOD in parent materials. The testing of weldments and the subsequent use of the toughness values determined for fitness for purpose evaluations requires additional consideration as detailed in references 34 - 36 and outlined below.

4.2. Notch Position and Specimen Design

In some applications, particular defect types may be envisaged (e.g. lack of side wall or root fusion, HAZ or weld metal hydrogen cracking, etc) in which case CTOD test specimens with notches in the appropriate position would be specified.

If the defect type is unknown the test is designed to attempt to measure the lowest CTOD value for the weldment. This generally involves the use of preferred geometry specimens to BS 5762 having a through thickness crack, (often positioned on the weld centre line) and subsidiary geometry specimens with surface notches sampling HAZ regions [36] (see Fig 11).

Careful inspection of the fracture faces is also required. If brittle initiation is seen to occur consistently away from the centre of a through thickness cracked specimen then subsidiary geometry specimens notched into that region should be considered. For surface notched geometries, if tearing is experienced followed by cleavage this may be due to microstructural variations and a deeper notch could result in a lower CTOD value.

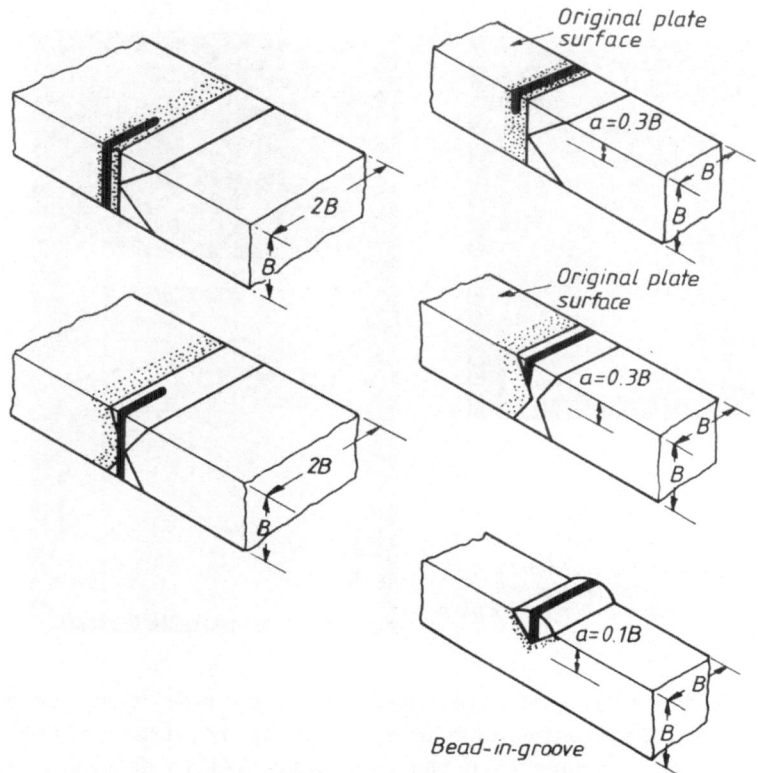

Fig.11. Sketch showing testpiece geometries
and notch orientations used to investigate HAZ toughness.

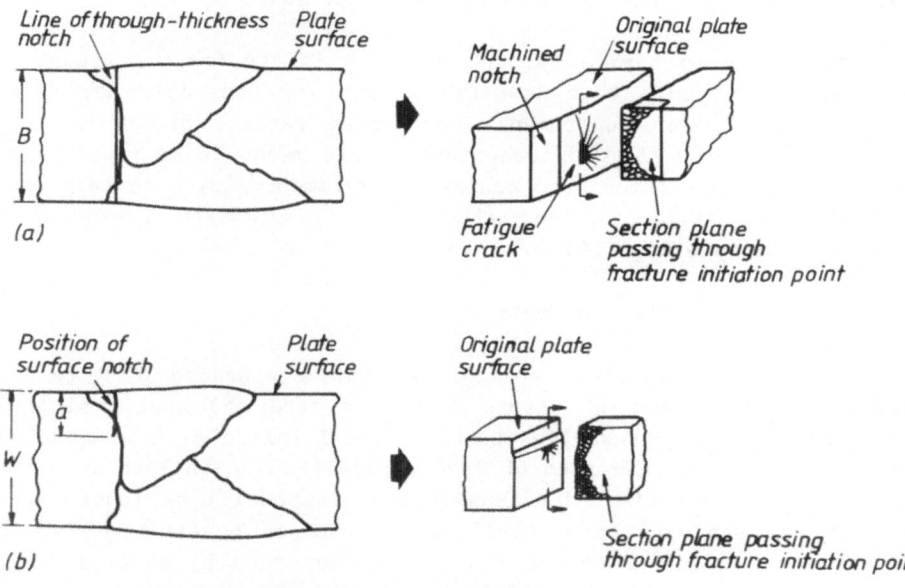

Fig.12. Examples of sectioning techniques for:
a) through-thickness notched, and b) surface notched specimens.

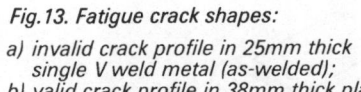

Fig.13. Fatigue crack shapes:

a) invalid crack profile in 25mm thick single V weld metal (as-welded);
b) valid crack profile in 38mm thick plain material.

Heat affected zones require even greater care than weld metal tests [35] and sectioning of the specimen is required at the brittle initiation site subsequent to the test (Fig. 12) to determine in which region of the weld/HAZ initiation occurred.

4.3. Fatigue Precracking and Local Compression

It is necessary to have a fatigue precrack in the fracture toughness test piece to ensure that lowest estimates of toughness are determined. BS 5762 has requirements concerning fatigue crack shape and length to ensure valid CTOD measurements are made. These requirements are particularly difficult to achieve on weldments which contain residual stresses (Fig. 13) and thus local compression treatment is required [36] (see Fig. 14) to ensure valid crack shape.

4.4. Testing Rate and Control

The rate of load application employed for a standard CTOD test is governed by the range of stress intensity rate ($\dot{K}$) specified in BS 5762. This requirement applies to the initial (elastic) loading of the specimen. Therefore the rate of load application at failure in a CTOD test which has experienced extensive plasticity will be lower if the initial $\dot{K}$ is achieved by controlling the rate of movement of the machine crosshead (displacement control) rather than by setting a rate of load application (load control).

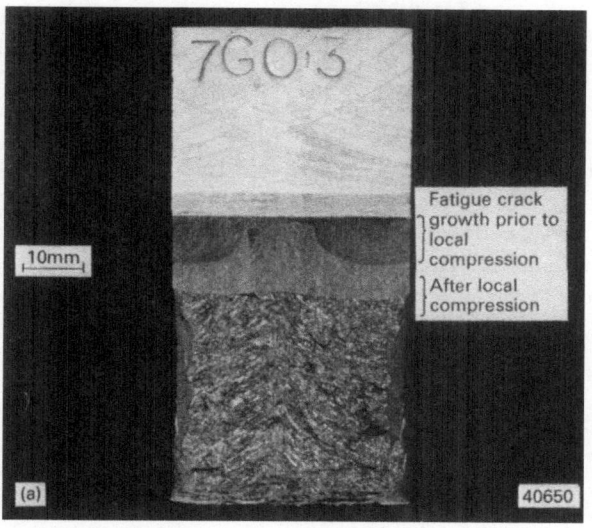

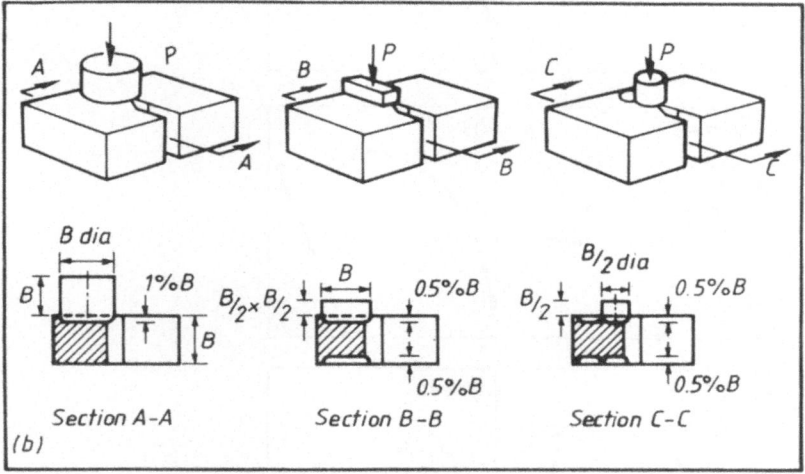

Fig.14. *Local compression treatment is frequently a necessary preliminary to CTOD testing of weldments:*

a) example of fatigue crack front bowing before and after local compression in a 38mm thick MIG weld (as-welded condition).
b) alternative methods of local compression using single and double platens.

Although the $\overset{\bullet}{K}$ range specified generally relates to relatively slow machine cross head displacements, the design of the CTOD specimen ensures that the strain rate achieved at the crack tip is higher than would be relevant to the majority of structural applications. If, however, the structure is expected to experience higher rates of load application, it is strongly advised that tests are carried out at the appropriate strain rate. In certain specialised applications extremely high K̇ rates can be encountered, and guidelines for CTOD testing to model these circumstances have been established [37].

The significance of 'pop-in', a sudden load drop (see Fig. 15) and/or click during the CTOD test, has been discussed by Kamath and Gittos [38]. Some companies are now specifying that CTOD tests should be carried out under 'load control' conditions to assess the significance of short brittle cracks. However, current practice is to treat the 'pop-in' as a critical event (even when followed by a subsequent increase in load as in Type 2b of Fig. 15).

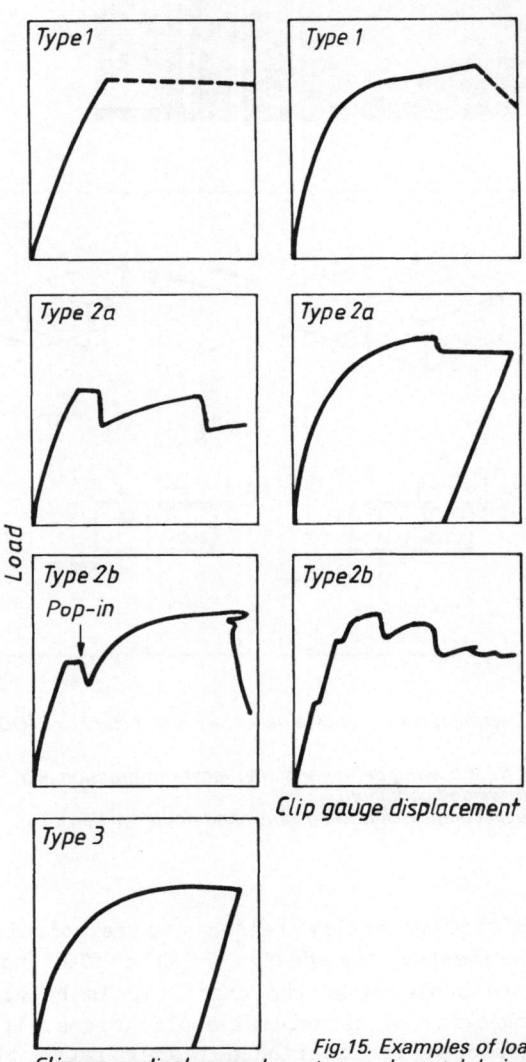

Fig.15. Examples of load v. clip gauge displacement traces observed during CTOD testing.

Load control, as provided by a servo-hydraulic test machine, does not simulate the true elastic 'feedback' of a real structure. Therefore it does not give an accurate guide to the significance of a 'pop-in' in a structure. Also, in many test set ups with conventional instrumentation, the selection of load control can mask a 'pop-in' on the load-displacement diagram, whereas in displacement control the events are more distinctive.

In general, the specification of load control may result in higher strain rates than is necessary for conservatism. This may induce cleavage failures at lower values of CTOD for a given temperature than would be obtained under displacement control. In contrast, for 'upper shelf' behaviour higher CTOD values at first attainment of maximum load may result from load controlled tests [39].

5. INITIATION OF TEARING AND MAXIMUM LOAD

5.1. Detection of Initiation

The initiation point of a stable ductile tear is not usually a critical value in a fracture mechanics test (see Fig. 5). The position of the onset of stable crack growth cannot usually be identified on the load-displacement test record due to the progressive nature of void growth and coalescence and the fact that this is usually accompanied by the development of plasticity. The relationship between initiation, onset of net section yield and maximum load is a function of test geometry and specimen dimensions.

One of the main attractions of measuring the initiation point is that it is a geometry insensitive parameter. Some workers believe that it is a material property, others that it is geometry dependent [40, 41], although this dependence is usually second order to material scatter. It is difficult to resolve this issue due to uncertainties concerning initiation detection techniques and the variation in initiation predictions based on single specimen methods. Techniques have been adopted by various workers (too numerous to reference individually) to detect onset of stable crack growth including; alternating or direct current potential drop (AC or DCPD), ultrasonics, unloading compliance, acoustic emission and stretch zone measurement of the specimen subsequent to the test.

All these techniques have been demonstrated to have been successful in certain applications but as yet none has received universal acceptance. This is mainly due to the need to demonstrate reliability of the particular method on the material and test geometry to be investigated. Thus for the majority of practical situations a multiple specimen procedure (see Fig. 16) is adopted, [7, 11, 14, 18] for the

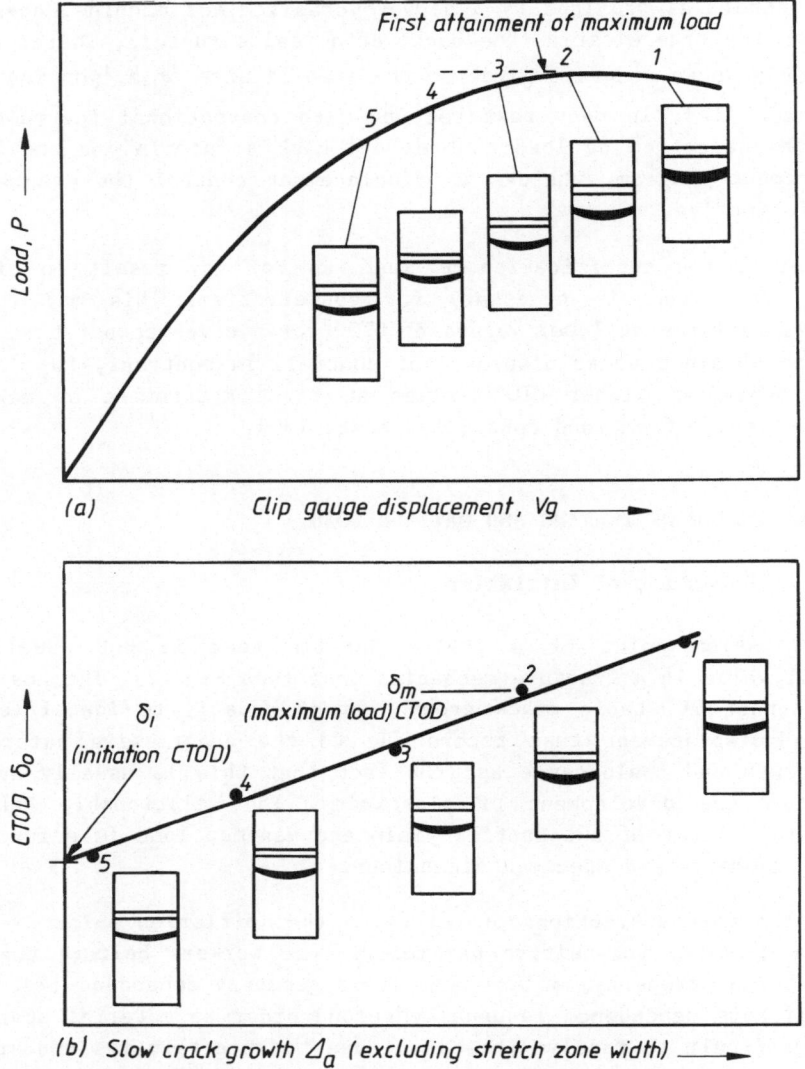

Fig.16. Determination of the initiation of tearing value of CTOD following the procedure of Appendix A, BS 5762

a) load clip gauge displacement record   b) CTOD v. slow crack growth (R-curve).

determination of initiation toughness (although initiation detection techniques are generally used at the same time).

5.2. Measurement of J_{Ic}

The original complicated laboratory procedure for the measurement of J proposed by Begley and Landes [42] based on the $J = dU_T/Bda$ definition has generally been superceded by the Rice, Paris Merkle [28] J estimation formulae for standard toughness evaluation procedures, i.e:-

$$J = \frac{\eta \, U_T}{B(W - a_o)} \qquad \qquad \ldots(6)$$

where η is a geometric factor, which is 2 for bend tests (see equation 4) and a function of a/W for compact tension based on the Merkle-Corten [43] procedures. For the CT specimen this is evaluated using a clip gauge mounted on the load line; for bend specimens other methods such as comparison bar [16] or double clip gauge techniques [11] must be adopted to determine load line displacements.

The multiple specimen method is the approach adopted by ASTM E813, JSME 001-1981 and is permitted by a CEGB draft procedure [44].

Although the original intention of the ASTM E24.08 sub committee, who drafted E813, was undoubtedly to measure the onset of initiation of tearing, the need for a simple standard procedure was also vital. This led to the adoption of a blunting line procedure to simulate stretch zone development (since this causes the initial crack extension, see Fig. 17). This procedure was developed during 'round robin' programmes on selected steels.

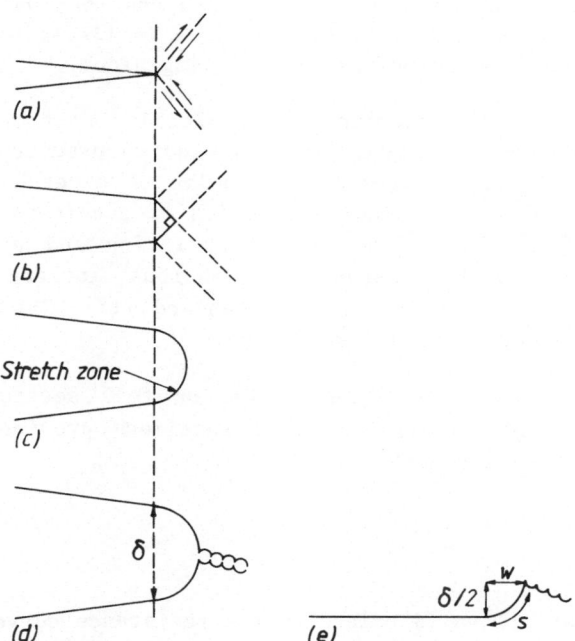

Fig.17. Stretch zone formation and definitions:
a-d) stretch zone formation with increasing load;
e) definitions of stretch zone measurements.

Since the width of the stretch zone is approximately equal to $\delta/2$ and for plane stress,

$$J = \bar{\sigma} \; \delta \qquad\qquad ...(7)$$

where $\bar{\sigma}$ is the material flow stress, then the blunting line was defined as :-

$$J = 2\bar{\sigma} \; \Delta a \qquad\qquad ...(8)$$

It was recognised that this line would not necessarily be universal but it was felt that the value of a consistent definition was paramount. Thus J_{Ic} does not necessarily define the initiation of tearing, in the same way that K_{Ic} does not in ASTM E399 (a maximum of 2% growth being allowable in this latter case).

In fact references 44 and 45 have proposed the adoption of J at 0.2mm of crack extension as a reliable parameter. This parameter $(J_{0.2})$ is likely to be adopted for material characterisation in pressurised water reactor applications in the U.K.

The blunting line procedure has caused much debate at conferences and in the literature, and now that the standard has been used on various materials and in many different countries, it is likely that the whole procedure will be reviewed by the ASTM working group.

The current standard recommends a multiple specimen procedure (although alternative techniques are allowed) to produce a resistance curve of J versus crack extension (inclusive of stretch zone width). This procedure consists of loading a series of specimens to different displacements (Fig. 16a) unloading, heat tinting and breaking open. Details concerning notch and specimen dimensions, fatigue precracking, testing rates etc are covered in the standard [18]. The various procedures are discussed in Blauel's lecture.

Unlike the CTOD procedures which insist on full section thickness, ASTM E813 allows sub-section thickness specimens provided the thickness and ligament requirement

$$B, \; W - a \; > \; 25 \; J_{Ic}/\bar{\sigma} \qquad\qquad ...(9)$$

applies, and that all valid points on the resistance curve comply with

$$B, \; W - a \; > \; 15 \; J/\bar{\sigma} \qquad\qquad ...(10)$$

To date these requirements have been found to be adequate to simulate full constraint on the upper shelf [45, 46]. ASTM E813 also allows side grooving of up to 25% of the thickness in which case the net section thickness is used in equation (6). A minimum of 4 points within the 0.15mm and 1.5mm offset lines parallel to the blunting line, and various requirements on the grouping of data must be adhered to before the J_Q value (which is the intersection of the linear regression of valid data points with the blunting line) can be defined as J_{Ic} (see Fig. 18). Some workers have criticised the linear regression procedure in favour of a power law fit [41, 45].

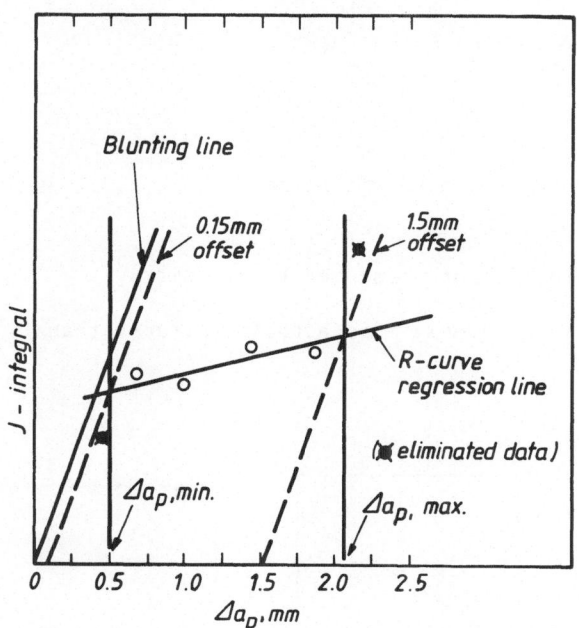

Fig.18. Data elimination after linear regression fit to R-curve data.

The ASTM E813 standard is written with thick pressure vessel applications specifically in mind, and the determination of K_{Ic} from J_{Ic}. This means that the thickness requirements outlined above, and the side grooving concept, are aimed at producing a plane strain resistance curve.

This may mean that it is impossible to measure a 'valid' J_{Ic} on full section thickness specimens for many applications. Also it could be that constraint considerations for upper shelf are not sufficient to ensure complete confidence that cleavage will not occur on thicker section thicknesses if sub section thickness specimens are employed (irrespective of material variability considerations).

5.3. Measurements of δ_i

To determine δ_i, values of δ_o are plotted against ductile crack extension as in Fig. 16 and a linear regression analysis carried out to determine the initiation value at $\Delta a = 0$ as indicated in Appendix A of BS 5762.

The blunting line construction to allow for the development of the stretch zone as outlined in ASTM E813 usually predicts a higher J_{Ic} value than the equivalent J_i following BS 5762 recommendations (see Fig. 19), since the blunting line correction is based on the formula

$$J_i = M \bar{\sigma} \delta_i \qquad \qquad \ldots (11)$$

where

$$\delta_i \simeq 2 \times \text{stretch zone width, and}$$

M is a constraint factor, (assumed to be unity).

For most structural steels M is actually greater than 1.5.

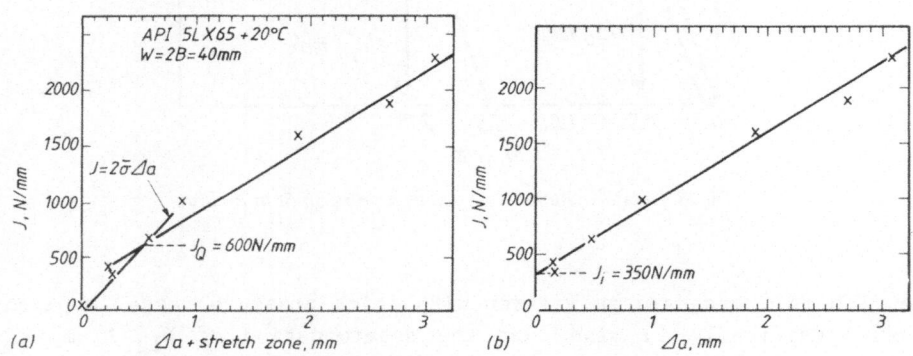

Fig.19. Determination of J initiation;
a) following ASTM 813.81 procedures b) following a BS 5762 type approach.

An interesting experimental observation on plane sided specimens, where crack growth tends to form in a thumb nail shape, is that due to the different procedure for measuring ductile crack extension, the BS 5762 procedure, which excludes stretch zone width, can exceed the ASTM E813 method which includes it.

5.4.   Maximum Load Toughness

In general when ferritic steel plate, forgings, castings and weldments are received for fracture tests it is not known whether they will exhibit brittle, transitional or fully ductile behaviour at the temperature at which the fracture toughness is required. This has necessitated research into the development of test procedures which will enable a unified approach to be adopted over the entire range of material behaviour.

For fully ductile behaviour, the laboratory determination of maximum load toughness using the recommended testpieces of BS 5762 has been discussed in some detail [47 - 49]. The maximum load point in a test specimen loaded under displacement control conditions is coincident with the instability point in a load control test (unless the material is strain rate sensitive) and this point is defined by the tangency condition of the material resistance curve and the maximum load, load controlled, driving force curve (Fig. 20). If the test conditions reproduce those existing in the structure, the maximum load toughness would thus predict the structural instability point. However as bend tests underestimate resistance curves for tension geometries (Fig. 21)

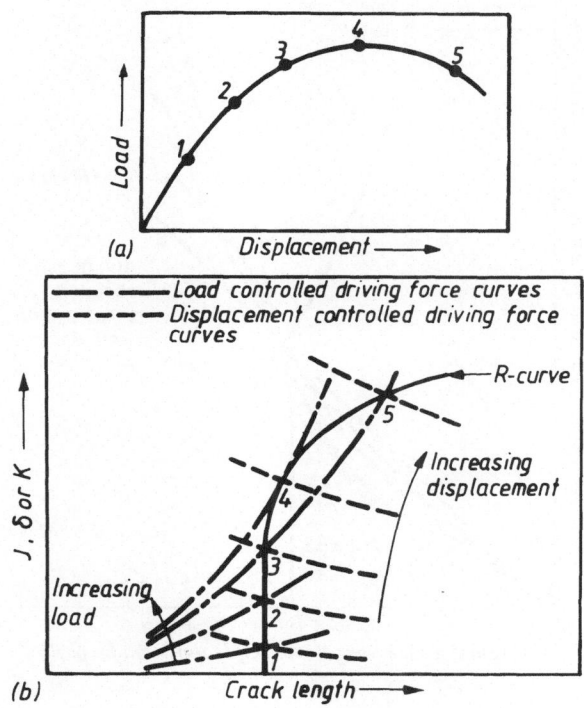

Fig.20. Instability and maximum load behaviour:
a) load/displacement   b) R-curve.

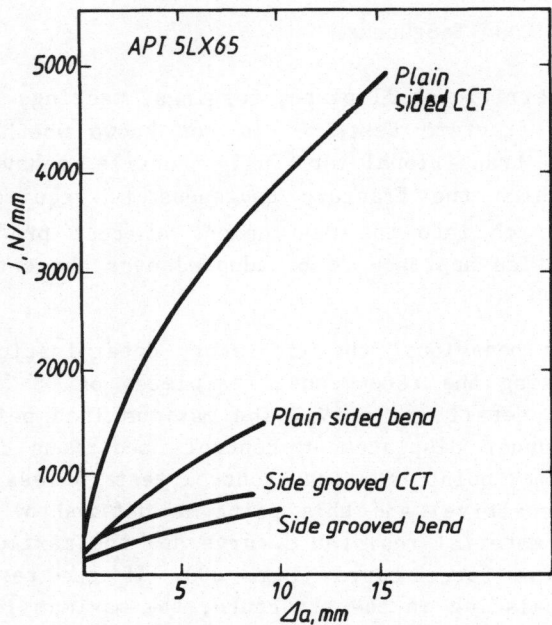

Fig.21. Comparison of bend and CCT J resistance curves.

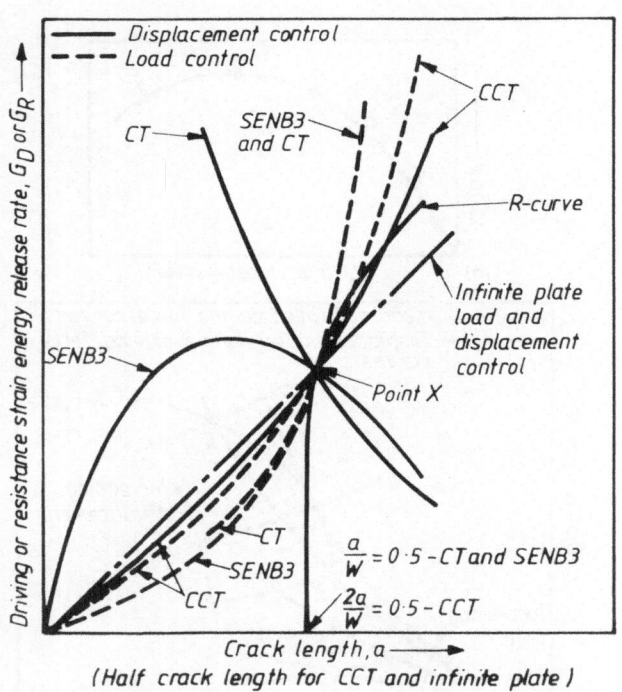

Fig.22. Typical elastic driving force curves, G_D v. a, for four geometries.

and driving force curves are steeper in bend configurations provided certain requirements are met, (Fig. 22), then although the maximum load toughness parameters (δ_m or J_m) are testpiece geometry dependent, conservative estimates of the toughness at which structural ductile instability occurs will be predicted [49].

It is Welding Institute policy that specimen preparation to BS 5762 or BS 5447 should follow identical procedures whether fully ductile or brittle behaviour is anticipated.

For the study of through thickness cracks, full section thickness geometry bend specimens of the preferred geometry of BS 5762, i.e. with W = 2B are recommended. In fact for fully ductile behaviour subsidiary geometry (W = B) specimens may give lower δ_m values due to a reduced ligament size [47]. However, the preferred geometry specimens will still predict toughness values which will be conservative relative to a through crack in a structure [49].

A surface flaw in a structure may effectively be positioned in a section of infinite thickness. Thus to ensure adequate constraint a thickness criterion must be employed to model this type of defect. When the subsidiary geometry does not give adequate thickness constraint, oversquare or side grooved specimens must be employed.

6. MEASUREMENT AND USE OF CRACK GROWTH RESISTANCE CURVES

Resistance curve determinations should only be embarked upon if fully ductile material behaviour can be guaranteed. For most structural steels and weldments the procedures outlined in BS 5762 can be used to derive appropriate values of toughness for the material (whether they be δ_c, δ_u, or δ_m).

In certain specialised circumstances the value of toughness at the initiation of stable tearing and/or the tearing resistance with crack extension, i.e. the slope of the resistance curve, is required. In these circumstances specialised procedures must be adopted to generate the relevant material resistance curve.

For relatively thin sheet, parent material resistance curves can be derived using procedures outlined in references 50 and 51 which employ an equivalant K approach. However, for the section thicknesses employed in general structural steel fabrications, these methods are not applicable and CTOD and J techniques must be employed. These are covered to a limited extent by BS 5762 and ASTM E813 for CTOD and J respectively. However, the R-curve methods described are developed in order to estimate initiation of tearing toughness values. The ASTM E24.08.03 Task Force has carried out a round robin and produced a draft procedure for single specimen R-curve determinations [52], but as yet no detailed R curve procedure has been evolved for 'multiple specimen' testing.

As a general rule, resistance curves are generated for one of four reasons:

a. to establish the initiation of tearing toughness to use as a value of material toughness in a single parameter assessment route such as PD 6493 or R6 where, for various reasons, the possibility of any ductile crack extension cannot be entertained whether stable or not,

b. to establish the 'plane strain' resistance curve of a material to use in a complex structural assessment of a very thick section structure (such as a PWR pressure vessel),

c. to establish structurally representative resistance curves for a ductile instability asessment,

d. to derive resistance curves using a given test geometry to assess the relative resistance to tearing of various materials, with growth in various orientations.

Research [39] into geometry and orientation effects on the crack growth resistance of structural steels and weldments has led to the recommendation that fracture assessments in ductile regimes should be based on the use of maximum load toughness values determined from BS 5762 with design curve type procedures (whether J or CTOD) as in PD 6493. This is also equivalent [53] to determining a K_c value based on J_m for use in the CEGB R6 procedure. Besides providing a relatively simple assessment route, since a single value of fracture toughness is employed, the concept has the great advantage that the test and analysis procedure adopted are essentially identical whether the material behaves in a brittle or a ductile manner.

A valid criticism of the use of a maximum load toughness, (or any value on the resistance curve) is its geometry dependence. However, provided that the recommended test procedures are adopted, the δ_m or J_m values obtained will be conservative estimates of the ductile instability toughness.

This being the case it is difficult to understand why resistance curve methods should be adopted for general toughness testing. Although initiation toughness provides a convenient, geometry insensitive parameter, its relevance for structural assessments must be in doubt and its use generally results in unnecessary over-conservatism. Fundamentally, the determination of relevant resistance curves provides the means for an accurate estimate of ductile instability, but the complexities involved in the prediction of accurate and appropriate driving force curves are not cost effective for the majority of engineering applications.

Where this expense is warranted, as in the Nuclear Power generation industry, the use of R-curve technology is perfectly justifiably well developed. However, the intense research devoted to this field should not dominate the test techniques and standards developed for general engineering where transitional behaviour is often more important.

It would seem much more appropriate to use the knowledge gained by the study of R-curves to develop the existing test procedure and analysis methods in an economical way. This conclusion is reflected by the specification of a J level at 0.2mm of crack growth for material characterisation in the Nuclear Industry and the use of δ_m in North Sea Platform and Piping applications.

REFERENCES

[1] PD 6493:1980. 'Guidance on some methods for the derivation of acceptance levels for defects in fusion welded joints'. British Standards Institution.

[2] HARRISON, J. D. et al. 'The COD approach and its application to welded structures'. ASTM STP 668, 1979, 606-631.

[3] HARRISON, J. D. 'The state-of-the-art in crack tip opening displacement (CTOD) testing and analysis'. Metal Construction, 1980, 12 (9) 413-422 (10) 524-529, (11) 600-605.

[4] GARWOOD, S. J. et. al. 'The use (and abuse) of CTOD'.Metal Construction, 14, No. 5, 1982.

[5] BS 5500:1982. 'Unfired fusion welded pressure vessels'. British Standards Institution.

[6] BS 4515:1969. 'Specifications for field welding of carbon steel pipelines'. British Standards Institution.

[7] HARRISON, R. P., LOOSEMORE, K., MILNE, K., and DOWLING, A. R. CEGB report R/H/R6-Rev2, 1980.

[8] 'An engineering approach for elastic plastic fracture analyses' and 'Procedure for the assessment of the integrity of nuclear pressure vessels and piping containing defects'. EPRI reports NP1931 and NP2431 Projects 1237-1 and 2.

[9]   TURNER, C. E. 'A J based engineering usage of fracture mechanics'. ICF5, 3, Cannes, 1981, pp 1167 - 1192.

[10]   TURNER, C. E. 'Further developments of a J based design curve and its relationship to other procedures ASTM STP 803 pp 80 - 102, 1983.

[11]   GARWOOD, S. J. and WILLOUGHBY, A. A. 'Fracture toughness measurements on materials exhibiting stable crack extension'. International Conference - Fracture Toughness Testing - London, June 1982, Paper 32.

[12]   ASTM E 399:72. 'Standard methods of tests for plane strain fracture toughness of metallic materials'. ASTM Book of Standards, Part 10.

[13]   BS 5447:1977. 'Methods for test for plane strain fracture toughness (K_{Ic}) of metallic materials'. British Standards Institution.

[14]   BS 5762:1979. 'Methods for crack opening displacement (COD) testing'. British Standards Institution.

[15]   BS DD19:1972. 'Methods for crack opening displacement (COD) testing'. British Standards Institution.

[16]   DAWES, M. G. 'The application of fracture mechanics to brittle fracture in steel weldments'. PhD Thesis (CNAA), The Welding Institute, December 1976.

[17]   WU SHANG-XIAN. 'Plastic rotational factor and J-COD relationship of three point bend specimen'. Engineering Fracture Mechanics, 18, No. 4, pp 83 - 95, 1983.

[18]   ASTM E813:81. 'Standard test for J_{Ic}, a measure of fracture toughness'. ASTM Book of Standards Part 10.

[19]   JSME Standard, S001-1981. 'Standard method of test for elastic plastic fracture toughness J_{Ic}'.

[20]   SLATCHER, S. and KNOTT, J. F. 'The ductile fracture of high strength steels'. ICF-5, 1. pp 201-207.

[21]   GIBSON, G. P. and DRUCE, S. G. 'A study of elastic plastic fracture parameters through the thickness of fracture toughness specimens using silicone rubber crack impressions'. UKAEA Harwell report, AERA R10815, 1983.

[22] GARWOOD, S. J., PRATT, P. L. and TURNER, C. E. 'Measurements of slow stable crack growth in structural steels and prediction of unstable fracture'. Presented at OECD CSNI specialist meeting, Daresbury, May 1978, Paper 11.

[23] GREEN, G. and KNOTT, J. F. 'On effects of thickness on ductile crack growth in mild steel'. J. Mech. Phys. Solids, 23, No. 3, pp 167 - 83, 1975.

[24] GARWOOD, S. J. 'Geometry and orientation effects on ductile crack growth resistance'. Int. J. Pres. Vessel Piping, 10, pp 297 - 319, 1982.

[25] SCHWALBE, K. H. and HELLMANN, D. 'Geometry effects on J-R and δ-R curves'. ASTM 15th National symposium on Fracture Mechanics, Maryland, July, 1982.

[26] WILLOUGHBY, A. A., PRATT, P. L. and TURNER, C. E. 'The meaning of elastic plastic fracture criteria during slow crack growth'. International Journal of Fracture, 17, No. 5, October 1981, pp 449 - 466.

[27] SUMPTER, J. D. and TURNER, C. E. 'Method for laboratory determination of Jc'. ASTM-STP 601, pp 3 - 18, 1976.

[28] RICE, J. R., PARIS, P. C. and MERKLE, J. G. 'Some further results on J-integral analysis and estimates'. ASTM STP 536, 1973, pp 231 - 245.

[29] LANDES, J. D. and BEGLEY, J. A. 'Test results from J-integral studies: and attempt to establish a J_{Ic} testing procedure'. ASTM STP 560, 1974, p 170.

[30] DOLBY, R. E. 'HAZ toughness of structural and pressure vessel steels - improvement and prediction '. Welding Journal, 1979, 59 (8), 225s - 238s.

[31] DOLBY, R. E. 'Factors controlling HAZ and weld metal toughness in C-Mn steels'. In Fracture 79, Proc. 1st National Conference on 'Fracture', Johannesburg, November 1979, South African Institution of Metallurgists.

[32] GARLAND, J. G. 'Factors controlling fracture toughness in submerged arc welds'. Welding Institute Seminar 'COD fact or fiction', Sheffield, April, 1980.

[33] JOHN, R. 'Factors affecting COD properties in MMA welds'. Ibid.

[34] DAWES, M. G., PISARSKI, H. G., TOWERS, O. L. and WILLIAMS, S. 'Fracture mechanics measurements of toughness in welded joints'. International Conference 'Fracture toughness testing', London, June, 1982, paper 33.

[35] PISARSKI, H.G. and PARGETER, P.J. 'Fracture toughness of weld heat affected zones (HAZ's) in steels used in constructing offshore platforms'. International Conference 'Welding in energy related projects', Toronto, September, 1983.

[36] PISARSKI, H. G. and DAWES, M. G. 'Measurement of crack opening displacement in weldments with particular reference to offshore structures'. Presented at Int. Conference on Analytical and Experimental Fracture Mechanics, Rome, Italy, June 1980, pp585-614. Institute Members Report 111/1980.

[37] UK Briefing Group on Dynamic Fracture Testing, IIW Commission X. 'Proposed standard test methods for measuring K_{Ic} and CTOD at high loading rates'. International conference 'Fracture toughness testing', London, June, 1982, paper 35.

[38] KAMATH, M. S. and GITTOS, M. F. 'The evidence of pop-ins in fracture toughness testing'. Welding Institute Research Bulletin, 20, No. 4, pp 114 - 122, 1979.

[39] GARWOOD, S. J., WILLOUGHBY, A. A. and RIETJENS, P. 'The application of CTOD methods for safety assessment in ductile pipeline steels'. Fitness for Purpose Validation of Welded Constructions, London, November, 1981, paper 22.

[40] MARANDET, B. and PHELIPPEAU, de ROO, P. 'Effect of specimen dimensions on J_{Ic} at initiation of crack growth by ductile tearing'. ASTM 15th National Symposium on Fracture Mechanics, Maryland, July, 1982.

[41] GIBSON, G. P. and DRUCE, S. J. 'The effects of specimen size, specimen geometry and data analysis procedures on ductile crack growth resistance curves'. UKAEA Harwell report AERE R10906, 1983.

[42] BEGLEY, J. A. and LANDES, J. D. 'The J integral as a fracture criterion'. ASTM STP 514, pp 1 - 23, 1972.

[43] CLARKE, G. A. and LANDES, J. D. 'Evaluation of J for the compact specimen'. Journal of Testing and Evaluation, 7, No. 5, pp 264 - 269, 1979.

[44] NEALE, B. K., CURRY, D. A., GREEN, G. and HAIGH, J. R. 'A procedure for the determination of the fracture resistance of ductile materials'. CEGB draft report, February, 1983.

[45] INGHAM, T., BLAND, J. T. and WARDLE, G. 'Influence of specimen size on the upper shelf toughness of SA A533B-1 steel'. SMIRT 7, Chicago, August, 1983.

[46] GARWOOD, S. J. 'The effect of temperature orientation and constraint on the toughness of A 533B plate'. A-F M-MS, Freiburg, June, 1983. pp 939-950

[47] TOWERS, O. L. and GARWOOD, S. J. 'Maximum load toughness'. International Journal Fracture, 16, April, 1980, pp R85-R90.

[48] TOWERS, O. L. and GARWOOD, S. J. 'The geometry dependence and significance of maximum load toughness values'. Proc. of ECF-3, London, September 1980, pp 57 - 68.

[49] TOWERS, O. L. and GARWOOD, S. J. 'Fracture assessment in ductile tearing situations'. Proc. of ICF-5, Cannes, March 1981, Vol. 4, pp 1731 - 1740.

[50] ASTM E 561:81. 'Tentative recommended practice for R-curve determination'. Annual Book of ASTM Standards Part 10.

[51] WHEELER, C. et. al. 'Recommendation for the measurement of R curves using centre cracked panels'. Royal Aircraft Establishment Technical Report 81142, November 1981.

[52] ALBRECHT, P. et. al. 'Tentative test procedure for determining the plane strain J_I R-curve'. ASTM Journal of Testing and Evaluation. 10, pp 245 - 251, 1982.

[53] WILLOUGHBY, A. A. and GARWOOD, S. J. 'The application of maximum load toughness to defect assessment in a ductile pipeline steel'. ASTM 16th National Symposium on Fracture Mechanics, Columbus, August, 1983.

DUCTILE FRACTURE MATERIAL CHARACTERIZATION BY J-R CURVES

J.G. Blauel
Fraunhofer-Institut für Werkstoffmechanik
Wöhlerstr. 11
7800 Freiburg FRG

ABSTRACT. Crack growth resistance J-R or δ-R curves are considered as the most promising candidates for the characterization of the fracture resistance of materials in the upper shelf toughness regime. Their use for the prediction of ductile tearing instability of components has already been successfully demonstrated. Mainly 3 methods to generate J-R curves are used and are described in detail: multiple specimen interrupted loading, potential drop, and single specimen partial unloading compliance.

The advantages and deficiencies of each of these methods are discussed and results are compared. Additional methods for crack growth monitoring and using also crack tip opening displacement as a loading parameter are mentioned. Finally the parameters are shown which influence the crack growth resistance curves and deduced material values, such as temperature, specimen-geometry and size, side grooving, notch acuity or loading rates; validity criteria are discussed.

1. INTRODUCTION

For ductile materials as used in pressure vessels and piping growth of cracks begins in a stable manner and only after macroscopic plastic deformation and crack tip blunting (see Fig.1). As shown in Fig. 2 the fracture mode is then characterized by dimples indicating a mechanism of void nucleation, growth, and coalescence taking place in front of the primary crack. But a transition into fast unstable fracture with quasi cleavage characteristics or into ductile instability may well occur depending on the respective material and mechanical conditions.

A uniform description of the different stages and processes of ductile failure is possible by crack growth resistance curves. As shown in Fig. 3 they deliver a quantitative measure of the actual fracture resistance of the material in a specimen or in a component as a function of the stable crack extension in terms of an elastic-plastic loading and crack driving parameter. The energy contour integral J and the crack tip opening displacement δ are considered as the most promising parameters. Their use for material characterization and prediction of

117

L. H. Larsson (ed.), Elastic-Plastic Fracture Mechanics, 117–143.
© *1985 ECSC, EEC, EAEC, Brussels and Luxembourg.*

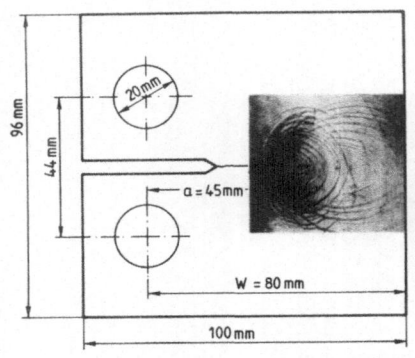 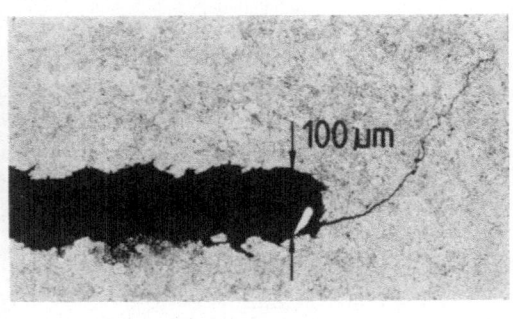

Fig. 1: Plastic zone formation and crack tip blunting during ductile
fracturing; material: reactor pressure vessel (RPV) steel
22 Ni Mo Cr 37, RT

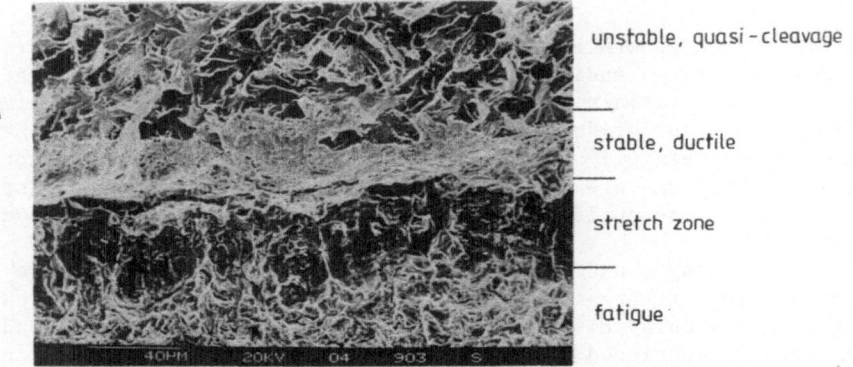

Fig. 2: Stages of ductile failure: SEM-fractography of WOL25X-specimen
of RPV material 22 Ni Mo Cr 37 tested at RT

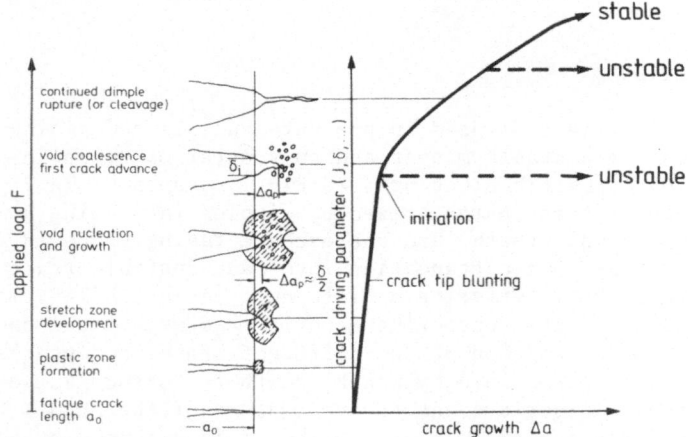

Fig. 3: Successive stages of ductile failure; microscopic processes at
crack tip and description by crack growth resistance curve

component behaviour has already been successfully demonstrated [1,2,3], see also the two lectures by BAKKER in this seminar; they will be utilized in the following.

2. CONCEPTS AND METHODS

2.1 Crack Tip Field parameters

For a certain structural situation, i.e. a loaded specimen or component with cracklike defect, the crack tip field can be completely characterized by the J-integral. J - originally introduced by CHEREPANOV and RICE as a path independent contour integral - describes the stress/strain singularity and controls the energy flow into the crack tip region when the crack extends. The theoretical background of such a description and the influence of material properties and geometry conditions have been discussed in BAKKER'S first lecture. From that working limits have been derived for a meaningful evaluation of J-levels at constant crack size and for a crack growth situation to be "J-controlled" [4,5]:

$$\Delta a \lesssim 0.1 \cdot b$$

$$\omega = \frac{b}{J} \frac{dJ}{da} > \begin{cases} 2.5 & \text{bend specimen} \\ 20 & \text{tension specimen} \end{cases}$$

$$b \geqslant 15...25 \ \frac{J}{\sigma_y} \qquad \qquad B > b$$

with Δa = stable crack growth, B = specimen thickness, and b = ligament width. The final two relationships are meant to guarantee a limiting state of plane strain.

Relying on the energy interpretation experimental techniques like the multiple specimen compliance method of BEGLEY and LANDES [6] and the approximate single specimen method of RICE, PARIS, MERKLE [7] are now used to determine J for different laboratory specimens. Numerics are still necessary to verify the applied formulas by tying the crack tip to the farfield parameters load, crack length and load line opening displacement in each specific situation [8]. For the most frequently used fracture mechanics compact specimen it is found [6]:

$$J = \frac{\eta \cdot U}{B(W-a)} \quad \text{with } U = \int_0^V F \ dV$$

$$\text{and} \quad \eta = 2 \ \frac{1+\alpha}{1+\alpha^2} \qquad \alpha = \left[(\frac{2a}{b})^2 + 2 \ (\frac{2a}{b}) + 2 \right]^{1/2} - (\frac{2a}{b} + 1)$$

$$\approx 2.2 \text{ for deep cracks}$$

An equivalent description of the crack tip field is possible through the crack tip opening displacement CTOD $= \delta$. The physical meaning of CTOD is immediately obvious from a direct observation on a metallographic cut as in Fig. 1. But the CTOD is a local parameter which is not easily accessible in real structural situations and which depends strongly on microstructural details as well as on the material deformation behaviour along the crack front. Therefore techniques and formulas had to be developed to derive its respective value from measurable farfield parameters like force and displacement. The actual status of the discussion is described in Fig. 4 for a compact specimen following the British Standard 5762 [9].

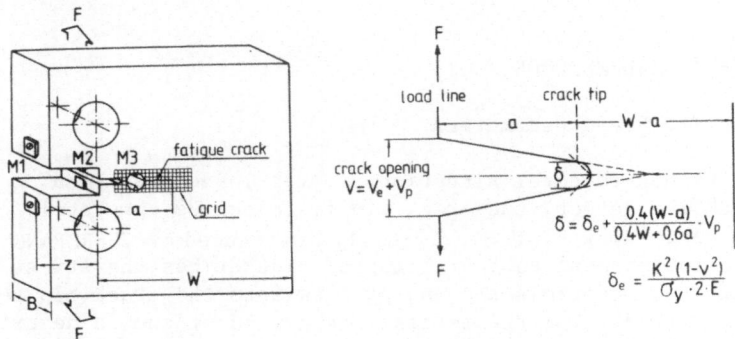

Fig. 4: Evaluation of crack tip opening displacement δ according to BS:5762 [9]

From experimental and analytical/numerical results a linear relationship is found between the J-integral and the crack tip opening displacement

$$J = \beta \cdot \sigma_y \cdot \delta$$

with $0.5 \lesssim \beta \lesssim 3$ depending on state of stress and material strain hardening exponent. This shows the concepts of J and CTOD to be equivalent, as discussed in more detail in BAKKER's first lecture.

2.2 Crack Growth Resistance Curves

Under conditions of ductile material behaviour the strongly non-linear load displacement curves for cracked specimens or components do not allow for a determination of the onset and of the extent of crack growth. To evaluate corresponding material values of J and δ from critical values of load and displacement independent methods therefore had to be developed. They rely on a direct observation of fracture phenomena or on the measurement of crack length dependent properties of the specimen as for instance compliance, electrical resistance, ultrasonic transmissivity or reflectivity, or acoustic emission. Some of the most frequently used methods for the evaluation of complete crack growth resistance curves are described in the following sections.

2.2.1 Multiple Specimen Unloading Method (MSU)

This method - first applied by SMITH and KNOTT [10] and schematically shown in Fig. 5 - allows for the determination of crack initiation and of a full crack growth resistance curve. It is used in terms of J as the basis for the ASTM-Standard E 813-81 [5] and in terms of δ it is mentioned in the British standard BS:5762 [9]. The method is expensive because of the number of specimens required, but it is easy to perform

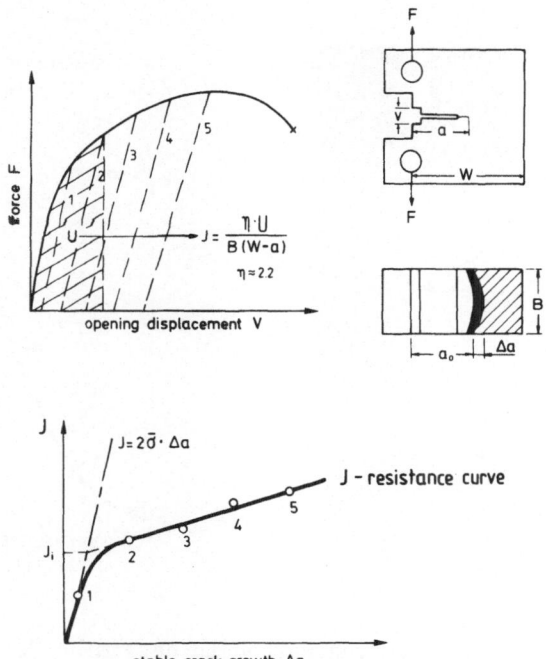

Fig. 5: Principle of multiple specimen unloading (MSU) method

and reliable and is therefore widely used for calibration purposes and to check the validity of other methods based on single specimen techniques.

As shown in Fig. 5 a set of 4 to 6 identical specimens (compact toughness (CT) or single edge notched bend (SENB) specimens fatigue precracked and instrumented for COD or J determination) are loaded to different levels of displacement below that required for complete failure and then they are unloaded. The actual crack advance is marked by either staining, heat tinting (typically 15 min at 300 to 400°C), or an additional fatigue loading. When the test pieces are subsequently broken open the increase in fractured area is clearly visible. Sometimes it is sufficient to break the specimens after unloading at the temperature of liquid nitrogen to get a good distinction between the dimple type ductile rupture of stable crack growth and the crystalline quasi cleavage fracture. The total increase of fracture area or the maximum value of stable growth in the middle of the specimen or an average value over a certain number of points along the crack front is measured under a microscope. A scanning electron microscope is necessary to be able to differentiate between stretch zone and real material separation by ductile crack extension (see Fig. 2). The Δa-values then are plotted together with the load/displacement curves and are extrapolated to give critical values F_i and V_i for the onset of stable crack growth - or they are plotted together with the respective J-values to give a complete J-R curve.

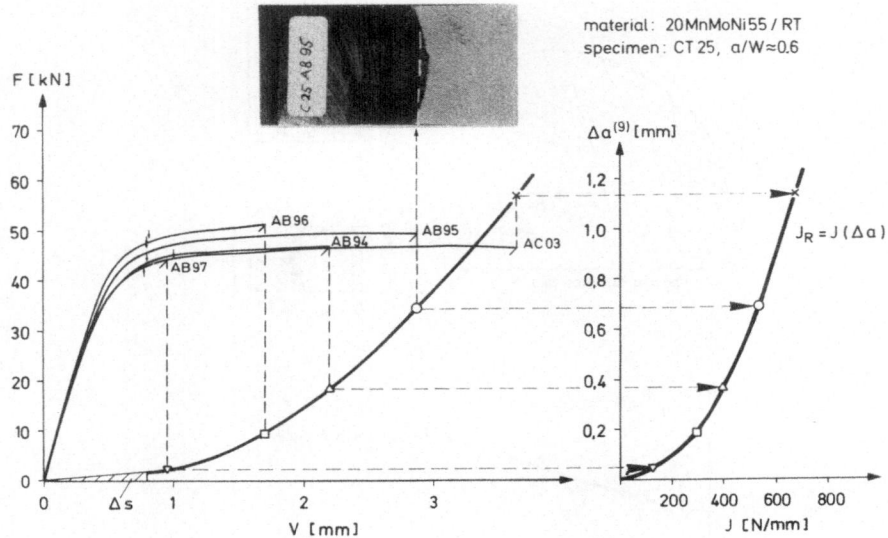

Fig. 6: Multiple specimen unloading method to determine "crack initiation" and "crack growth resistance curve" (stretch zone width
Δs from SEM-measurements; $\Delta a^{(9)}$= 9 point average of stable
crack growth)

Figure 6 shows as an example results from a test series with 25 mm
thick compact specimens (CT25) of the reactor pressure vessel steel
20 MnMoNi 55 (corresponding to the American steel A 533 B) tested at
room temperature. Some more results for different high strength structural steels are shown in Fig. 7 in terms of the crack tip opening displacement δ. A large number of tests is necessary to get a reliable
extrapolation to onset of physical crack growth.

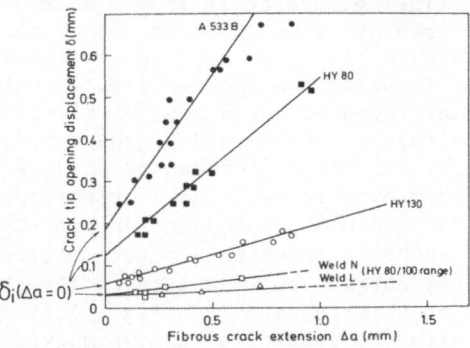

Fig. 7: Plots of crack opening displacement δ at the original fatigue
crack tip versus fibrous crack extension from interrupted loading tests for various high strength structural steels at room
temperature [12] (Δa without stretch zone!)

2.2.2 Single Specimen Partial Unloading Compliance Method (SSPUC)

To overcome the disadvantages in time and costs of the multiple specimen method different single specimen procedures have been investigated. Nowadays the so called "partial unloading compliance method" as first proposed by CLARKE et al. [12] is favoured to monitor crack growth for determining an initiation toughness as well as developing a material specific crack growth resistance curve.

The method is illustrated schematically in Fig. 8. Whereas the crack driving parameter J is again evaluated from the work done on the cracked specimen (i.e. from the area under the load displacement diagram) the increasing crack length is assessed from successive measurements of the elastic compliance C of the specimen which is derived from relatively small superimposed unloadings and known functions a = f (C,E, specimen geometry). Step by step a crack growth resistance curve $J_R(\Delta a)$ is developed. Meeting the standard conditions of [6] for the number of data points (unloadings) and appropriate data grouping a J_{Ic}-value is determined by extrapolation which - in a technical sense - characterizes initiation for each specimen; additional statistical information is then gained by testing several specimens.

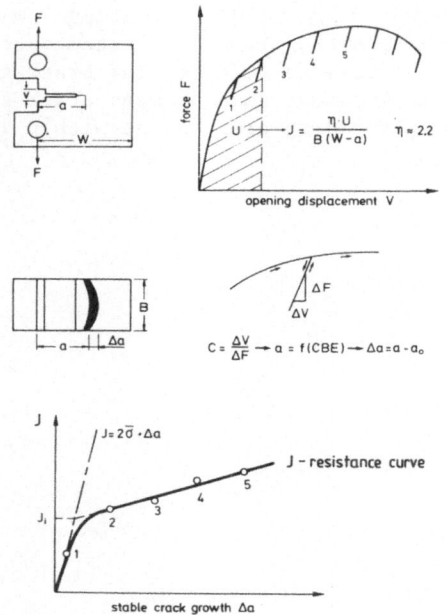

Fig. 8: Principle of single specimen partial unloading compliance (SSPUC) method

The major problems with the partial unloading compliance method are the requirements of a very high accuracy, stability, and linearity of all components of the measurement and data aquisition system and of a low friction loading device. Fig. 9 shows schematically the IWM-system [13].

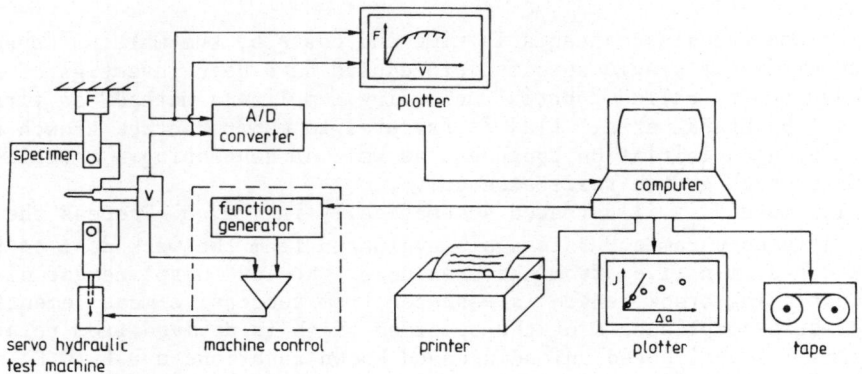

Fig. 9: Schematic of IWM test- and data acquisition system for SSPUC

The test is controlled and evaluated on line by a desk top computer, thus delivering necessary information for optimum control of test parameters while the test is running. All important primary data are stored for later reevaluation. Figs. 10 and 11 show results of two tests run at temperatures of 300 and 285°C, demonstrating low scatter of the first J(Δa)-points along the blunting line and system performance at service temperature of reactor pressure vessels. The crack growth prediction from the compliance measurement and the mean crack growth measured on the surface of the fractured specimen agree within .1 mm.

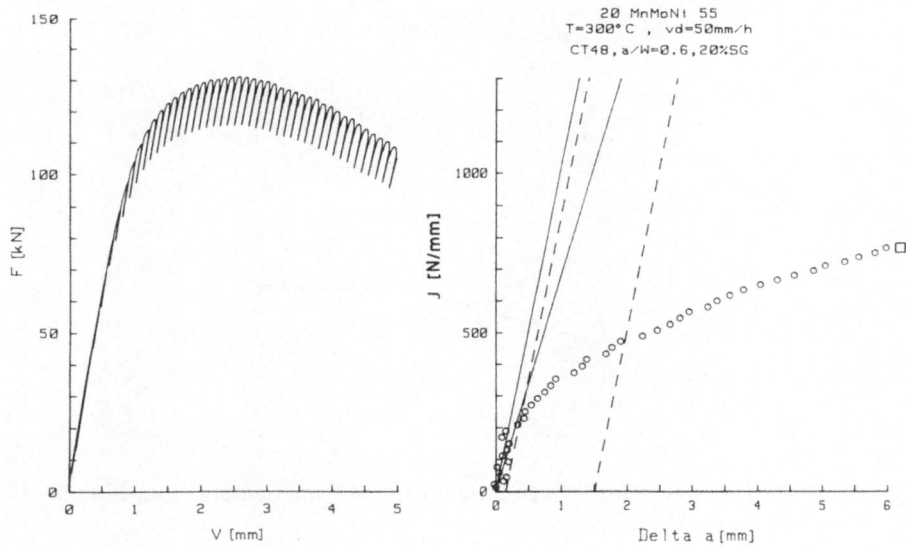

Fig. 10: Force (F) vs. opening displacement (V) diagram and corresponding J-R curve determined by SSPUC-method (square mark = Δa measured on fracture surface)

Note that the a-steps marked by arrows are related to load drops and are not due to experimental scatter (Fig. 11). The load drops occurred beyond the limit of validity for Δa assumed by [5] (Δa_{max} 4 mm for this specimen).

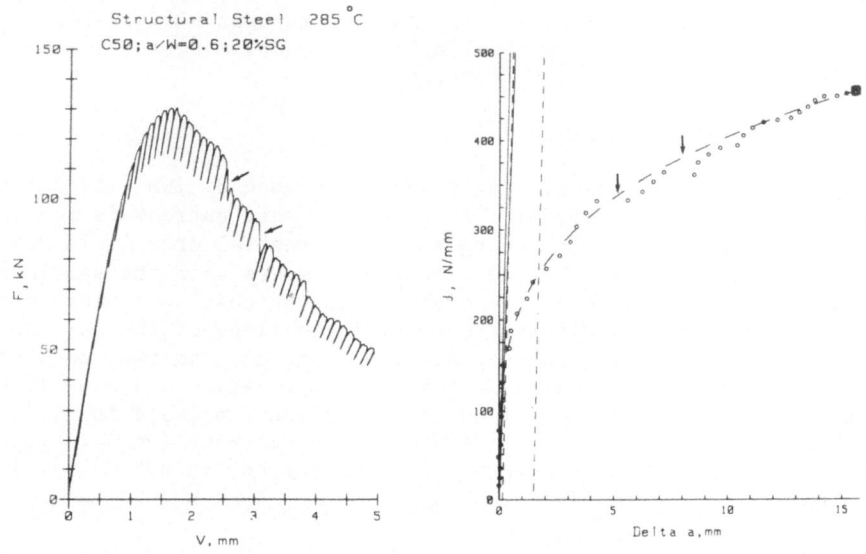

Fig. 11: F vs. V diagram and corresponding J-R curve for a CT 50 specimen tested at 285°C (543°F)

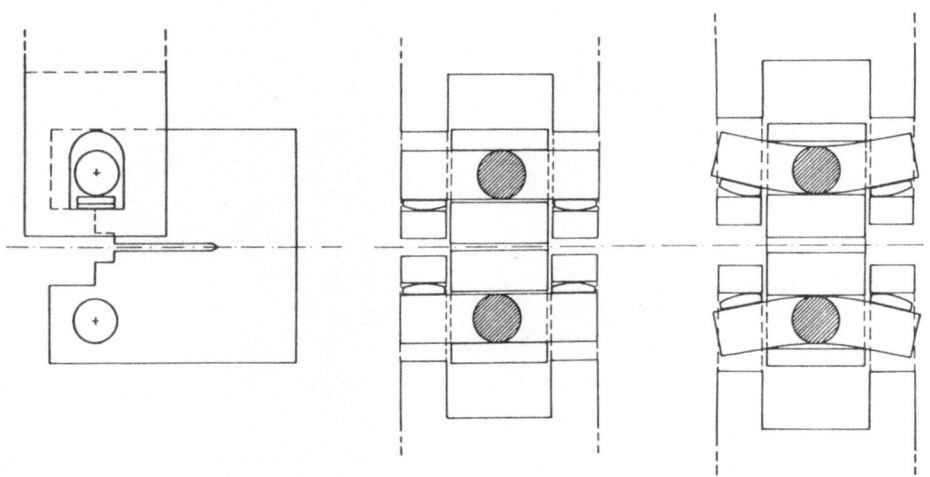

Fig. 12: Alternative clevis design [14] for SSPUC-test: side view and front view with CT specimen; curved hardened plates allow for bending of the loading bolts and maintaining the loading geometry

In the beginning of a SSPUC-experiment local plastic **deformations** and friction in the region of the loading bolts can influence the specimen compliance in a way as to induce an apparent negative crack growth. This can be fully avoided by using special grips ensuring free rolling of the loading bolts on flat surfaces. The system developed by VOSS [14] is shown in Fig. 12; it has been successfully tested at temperatures up to 800°C in partial unloading compliance tests with low scatter of crack length measurements.

2.2.3 Potential Drop Methods

The principle of a potential drop method as used by IWM [15] for compact specimens is shown in Fig. 13. A constant direct current is fed into the specimen in the plane of loading and the potential drop $\Delta\varphi$ is measured at two contact pins across the crack. $\Delta\varphi$ changes when the specimen is loaded and especially when the crack grows. At onset of stable crack growth a more or less distinct change of the slope of the $\Delta\varphi$-V curve is found and the further change of $\Delta\varphi$ is proportional to the crack growth. In this way a critical value J_i for crack initiation and – by linear interpolation between the values at initiation ($\Delta\varphi_i$, J_i) for initial crack length a_0 and at termination of the experiment ($\Delta\varphi_{max}$, J_{max}) for final crack length – a complete J-R curve may be evaluated [15, 16].

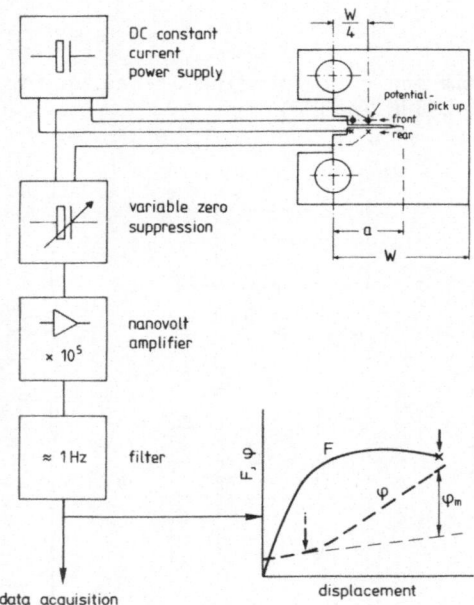

Fig. 13: Principle of the IWM setup for direct current potential drop (DCPD) measurement [15]

Figure 14 shows as an example a force/potential-displacement diagram together with the fracture surface of a 300°C-test of a 50 mm thick compact specimen of the steel 20 MnMoNi 5 5.

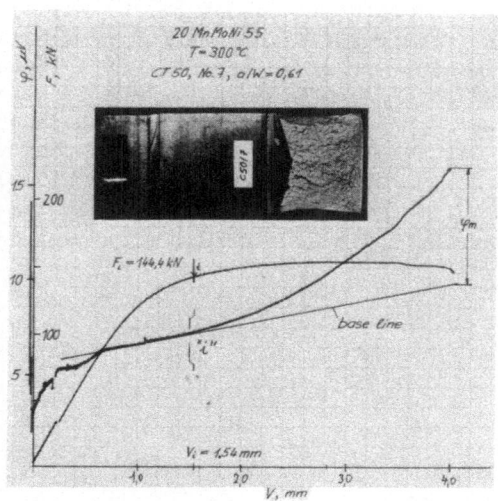

Fig. 14: Force/potential-displacement diagram for test No. 7; CT 50-specimen of steel 20 MnMoNi 55 at 300°C (572°F)

In Fig. 15 the corresponding continuous J(Δa)-curve is shown together with additional PD-results evaluated in the same way. The final points of the interrupted loading tests (filled symbols) confirm the primary J-R curve. The crack extension by crack tip blunting has not been taken into account here. Some material scatter is indicated by the different tests.

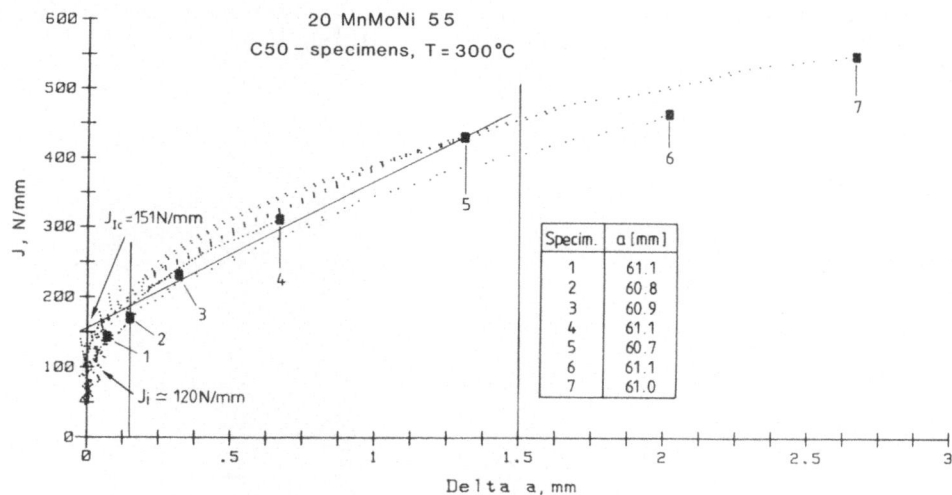

Fig. 15: J-R curve data from an interrupted loading test series (including test No.7 from Fig.14) together with a potential drop evaluation; CT50 specimens of steel 20 MnMoNi 55 tested at 300°C

The J_{Ic}-value defined according to ASTM E 813-81 [5] by an extrapolation to $\Delta a = 0$ of the regression line through the four data points between $0.15 \leqslant a \leqslant 1.5$ mm (exclusion lines, see GARWOOD's lecture) is about 25% higher than the J_i-value calculated for the displacement at "i" in Fig. 14 which is well confirmed by the first interrupted loading point. Using a power law fit according to LOSS [17] to the "valid" interrupted loading data and - in order to add crack tip blunting - an initial crack tip at $\Delta a = -0.15$ mm one finds $J_{Ic}^{LOSS} = 174$ N/mm and $J_i = 125$ N/mm. (See Fig. 16 for the definition of different critical values.)

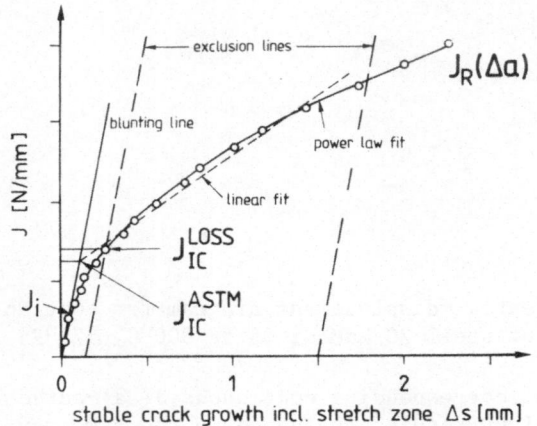

Fig. 16: Different methods to determine critical values of J at "initiation" of stable crack growth

Figure 17 gives some additional insight into the influence of experimentation and evaluation: In this example initiation is uncertain between $0.5 \lesssim V_i \lesssim 0.7$ from the $\Delta\varphi/V$ curve resulting in a variation of the

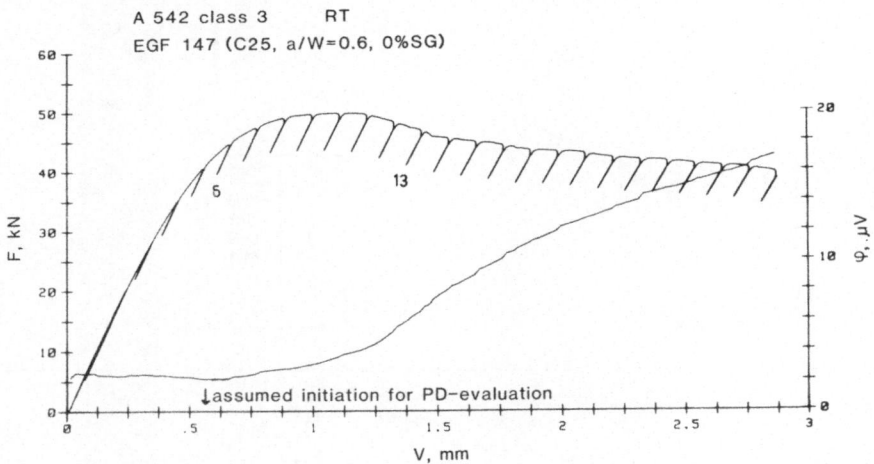

Fig. 17: Force (F)/DC potential drop ($\Delta\varphi$) vs. displacement (V) for CT25 specimen ECN 147 of the structural steel A 542 C13 tested at RT

PD critical value $43 \lesssim J_i \lesssim 81$ N/mm; for a more precise determination of J_i testing of an additional specimen to very little crack growth is therefore recommended. But the uncertainty of the point of initiation has only little influence on the resulting J-R curve and the ASTM-J_{Ic} evaluated from it; in this example only a negligible variation of $143 \lesssim J_{Ic} \lesssim 150$ N/mm results.

Besides the magnitude of the current the observable voltage change with crack extension is dependent on the position of the input and the signal pick-up leads. By an analogue technique RITCHIE et al. [18] have found highest sensitivity for current input normal to the crack plane of compact toughness specimens, but the potential gradients at the crack tip then are very steep and positional variations of the probes will cause large errors. Therefore a current input on the edge at the open end of the crack with a moderate sensitivity is preferred and the most suitable position for the probes is at the same surface across the notch and as near to the crack as possible.

The DC methods seem to be widely used although some disadvantages are obvious which are most completely discussed in a paper by MARANDET et al. [19]. These are the problems of using heavy constant currents of some 10 A, which must not heat the test piece, and of measuring very small changes of voltage which may be disturbed by thermal effects (variation of resistivity, contact voltages at the connections) and electrical effects (drift of the high gain amplifiers). Some of these difficulties may be overcome by using an AC technique. Fig. 18 shows the working principle of the equipment developed at IRSID to detect the initiation and growth of cracks in steel specimens: A 50 Hz current of about 20 A is supplied to the test specimen which is in series with an identical reference specimen. The signals picked up from the specimen across the crack have an inverted phase, they are amplified, (factor 1000), filtered and rectified to give a signal which is recorded together with the load and the loadpoint displacement on an XYY recorder.

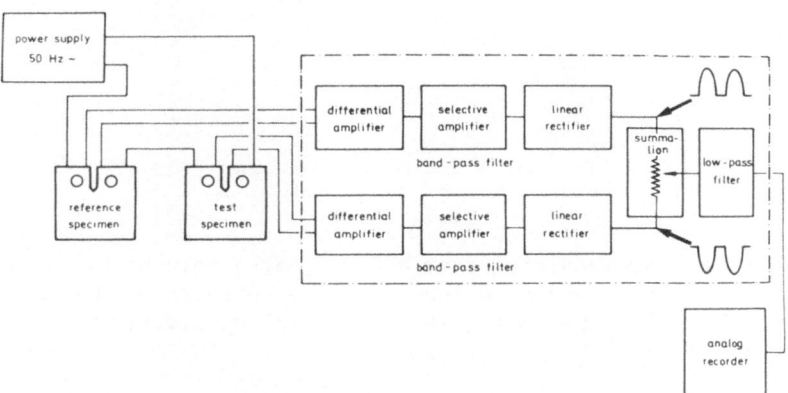

Fig. 18: IRSID equipment for an AC potential drop measurement of crack initiation after Marandet et al. [19]

2.2.4 Acoustic Methods

Active irradiation of ultrasonics during a fracture toughness experiment
and registration of the response or passive pick-up of acoustic emis-
sions generated from the crack tip failure process zone itself can be
used to detect early crack advance. The methods applied are all developed from experience in locating and sizing defects in non-destructive
testing; they have gained same importance for crack initiation detection
[11,20-23]; an on-line crack growth measurement seems still not to be
possible.

3. MATERIAL CHARACTERIZATION
3.1 Comparison of Different J-R Methods

Interrupted loading tests have been compared in Fig. 15 to potential
drop measurements. In Fig. 19 the nearly continuous sequence of data
points from one SSPUC-test at 150°C is compared to the mean curve constructed through 8 interrupted loading results for nominally equal conditions of material, geometry, and loading. Essentially the results
agree but the gain in information and in accuracy of the J-R curve from
the SSPUC-test is obvious. For other material conditions than those discussed here it should again be carefully checked whether an influence of
the high number of unloadings on the stable crack growth exists (see
[16]).

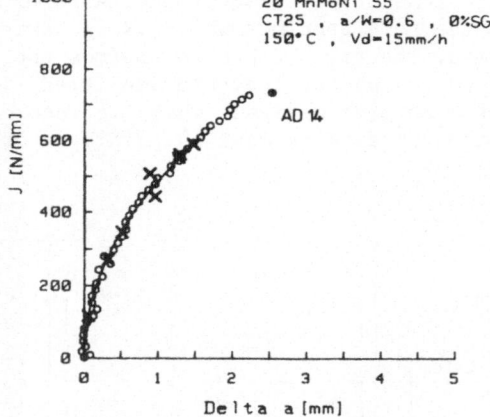

Fig. 19: J-R curves resulting from MSU and SSPUC method: mean curve
through 8 interrupted loading tests compared to SSPUC-test
AD 14; CT25 specimen (a/W = 0.6; no side grooves; $\dot{V}$ = 15 mm/h;
tested at 150°C)

Figure 20 shows J-R curves resulting from the partial unloading experiment with simultaneous measurement of potential drop in Fig. 17 (for
clarity the $\Delta\varphi$-data during unloadings have been omitted). The calibration of potential was adjusted to the crack lengths measured by the 6th
and 13th unloading which are marked. For this special choice (though

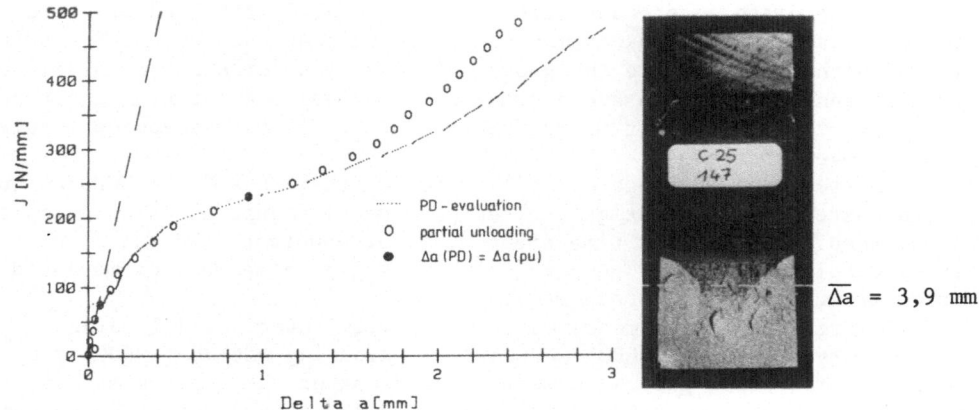

$\overline{\Delta a}$ = 3,9 mm

Fig. 20: J-R curve evaluation from Fig. 17; Δa for DCPD adjusted at the two marked partial unloadings; fracture surface of non-side grooved specimen shows pronounced tunneling

somewhat arbitrary) both resulting J(Δa)-curves are in good agreement for small amounts of crack growth. But for Δa > 1.5 mm the slopes of both curves increase and they deviate from each other. Both predictions of final crack growth underestimate the measurement $\overline{\Delta a}$ = 3.9 mm on the fracture surface by different amounts. It seems reasonable to relate the increase in slope with the onset of pronounced tunnelling which is visible on the fractured specimen (Fig. 20) having no side grooves. This is supported by analytical and experimental results of PRIJ [24] and PRANTL [25] showing that for tunnelled cracks the crack length will in general be underestimated by the compliance function of [4]. Specifically in [25] for a similar geometry as for the specimen in Fig. 20 an error of about 40% is found for the crack length and this is almost

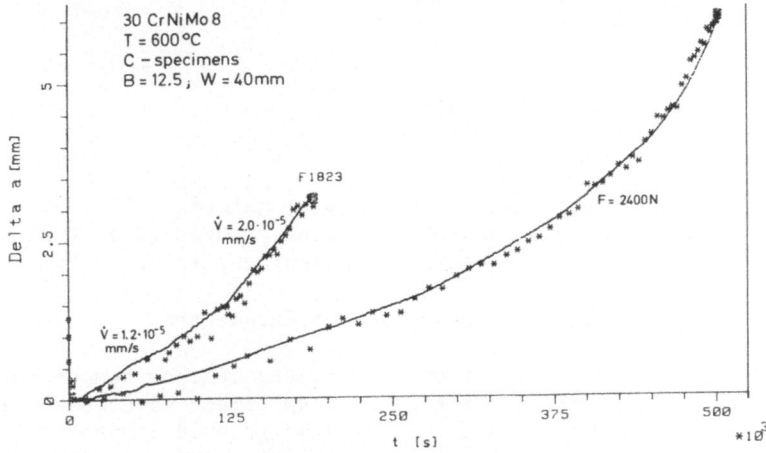

Fig. 21: Short time creep crack growth test at 600°C (1112°F) using direct current potential drop (DCPD) and partial unloading (PUC)

exactly the difference found here between PUC prediction and fracture
surface measurement. Thus it may be concluded that the PUC and the DCPD
method both underestimate the growth of tunnelling cracks, but by slightly different amounts. To avoid problems like these in defining a correct
effective crack length it is recommended as in [4] to use appropriately
side grooved specimens.

The temperature range of application of the PUC and the DCPD method
can be extended to investigate creep phenomena. Figure 21 as an example
shows results of 2 short time creep tests at constant load and at constant displacement rate, respectively, with good agreement of measured
crack growth rates from both methods [26].

In Fig. 22 results measured at 300°C with interrupted loading
(MSU), potential drop (DCPD) and partial unloading method (SSPUC) are
compared. They show good agreement independent of the testing method
and specimen thickness. Even 20% side-grooving seems to have little influence on the crack growth resistance for this test temperature and
for the thickness tested, although the side-grooved specimens showed
homogeneous crack growth over the full thickness instead of pronounced
tunelling in the smooth specimens. The lower one of the PUC-curves is
also shown in Fig. 10.

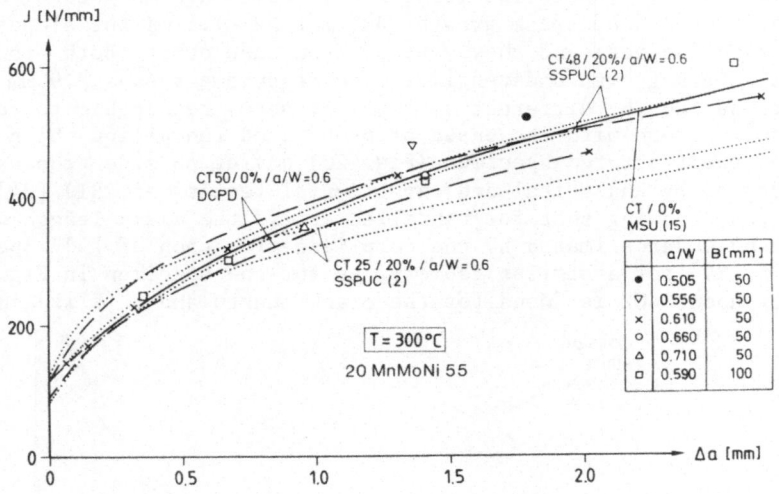

Fig. 22: J-R curves from different methods; material: 20 MnMoNi 55;
CT-specimens with and without side grooves (SG); test temperature 300°C (572°F); number of specimens in brackets

3.2 Different Materials and Influences of Parameters

Figure 23 summarizes results from SSPUC-tests at room temperature for
different materials. 25 mm thick compact specimens have been used. In
Fig. 24 J-R curves are shown for 25 mm side grooved compact specimens
taken from different positions of a thick wall weld in a low temperature
Al-alloy tank [27]. The SSPUC-tests have been conducted at a temperature
of -190°C.

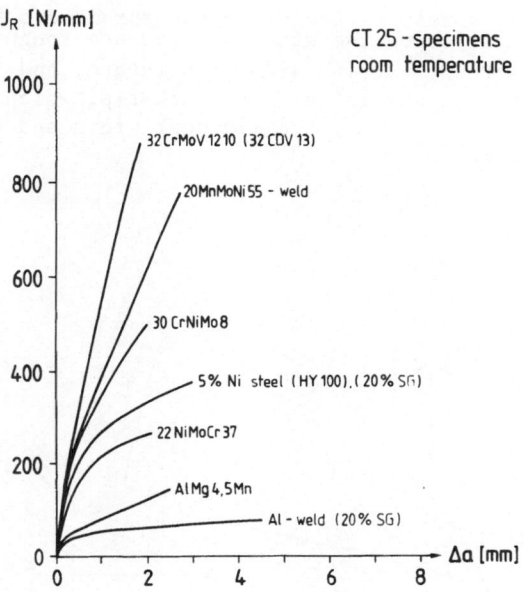

Fig. 23: Room temperature J-R curves for different materials from SSPUC-tests with 25 mm compact specimens

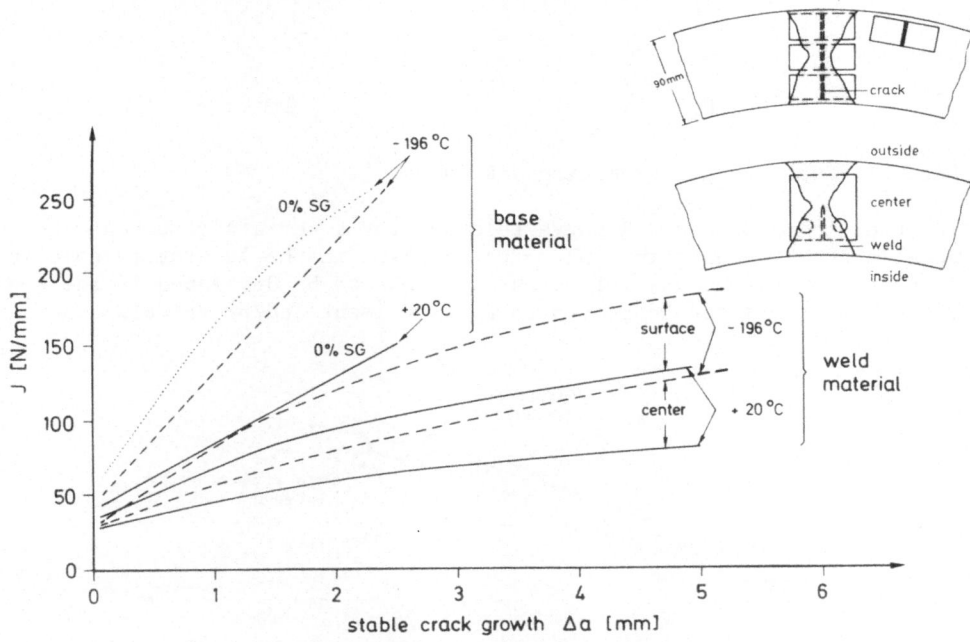

Fig. 24: J-R curves of weld- and base material from a thick wall weld joint AlMg4.5Mn [27] SSPUC tests with side grooved CT25 specimen at RT and -190°C

In comparing such results the different parameters have to be taken into account which influence the material fracture toughness. They are mainly: the yield stress, varying with temperature, and - as sketched in Fig. 25 - the local constraint at the crack tip, which is influenced by specimen/component thickness, the ligament width and distribution of stresses in the ligament.

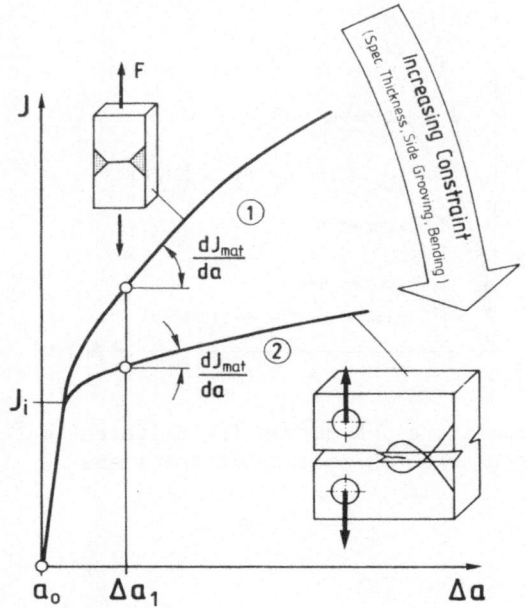

Fig. 25: Parameters influencing J-R curves [28]

In order to measure a J-R curve relevant for a certain component the same degree of constraint has to be realized in the laboratory specimen test - or a maximum possible constraint has to be developed in the specimen - and this has to be proven by experiment and/or calculation -

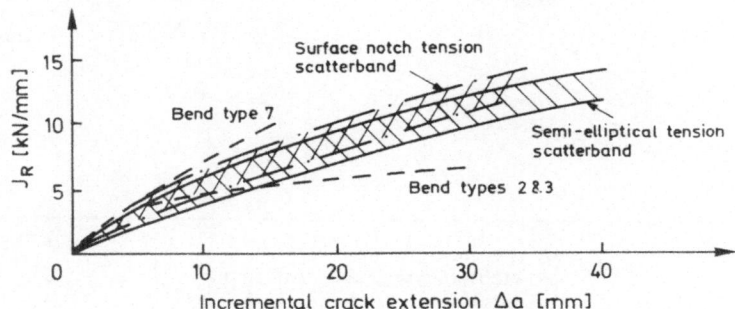

Fig. 26: J-R curve trends from different laboratory and structural specimens after [29]; material: A533B1, room temperature

to allow for a conservative assessment of the component. It is only in this sense that J-R curves can be expected to be material characteristics. The examples in Figs. 26 and 27 show that the evaluation of a lower bound material J-R curve seems possible.

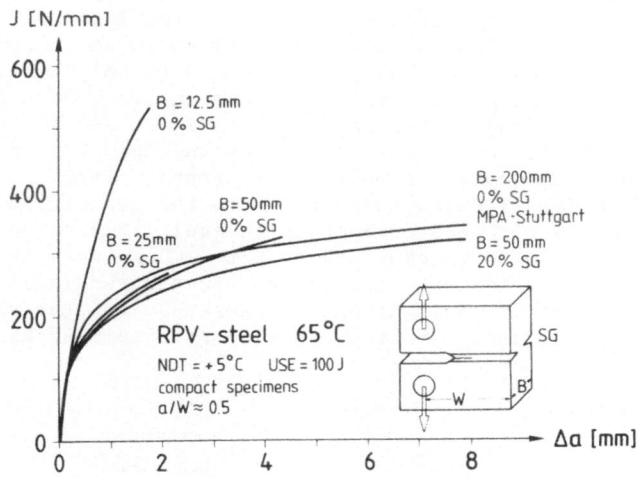

Fig. 27: J-R curves for specimens of different size indicating a lower bound value from CT50 with 20% side grooves; material: 22 Ni Mo Cr 37, 65°C

The influence of the parameters as discussed in Fig. 25 is demonstrated for the RPV material 20 MnMoNi 55 in the following [30]. Figure

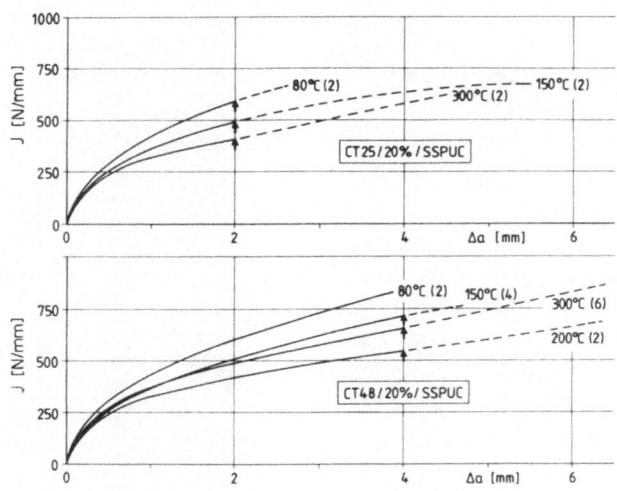

Fig. 28: J-R curves as a function of temperature for CT25- and CT48-specimens; number in brackets giving number of SSPUC-tests used for curve fit; crack growth Δa=10% ligament width marked by arrow ⬆ [30]

28 shows J-R generated by the SSPUC-method for test temperatures in the upper shelf regime. Two sizes of compact specimens, CT25 and CT48, are used having crack lengths a/W 0.6 and 20% side grooves: Mean curves through 2 to 6 data-sets (numbers in parenthesis indicate the number of specimens) available from replicate tests at each condition are drawn for the full range that has been measured; but the limit of "J-controlled crack growth" according to ASTM E 813 [5] at 10% of the remaining ligament width is marked. Both sets of curves with only small differences in the absolute values show the same temperature trends for the crack resistance: J-R(Δa) decreases from 80 to 200°C and increases again for T = 300°C. At room temperature the CT48- and the CT25-specimens failed in an unstable way after only little stable crack growth.

　　Results of a detailed investigation into the geometry dependence are given in Fig. 29 for three temperature levels. Smooth and side grooved compact specimens with thickness B = 10 to 200 mm have been used for MSU- and SSPUC-test series and mean curves are plotted (again the number of tests for each condition is given in brackets). For the selected temperatures at 80, 150, and 300°C the 20% side grooved CT25-specimens

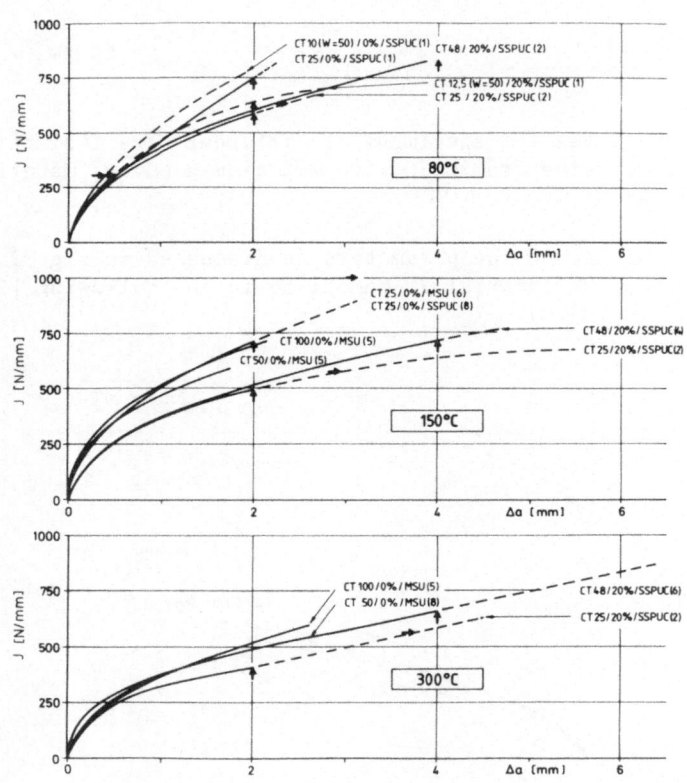

Fig. 29: Geometry dependence of J-R curves at different temperatures [30]; ➝ = minimum thickness condition; ⬆ = 10% ligament limit of J-controlled growth

deliver the lowest J-R curves - bounding also results from specimens as thin as 10 mm and as thick as 48 and 100 mm. But the J-R curves in Fig. 29 should only be compared in their regimes of validity (full continuous lines) which are given by the 10% ligament crack growth condition (for CT25 specimens with a/W = 0.6:Δa $\leqslant$ 0.1 · (W-a) = 2 mm !) and the minimum size criterion of ASTM E 813 [5] (for the CT 10-specimen B $\geqslant$ 15 J/σ_y delivers J $\lesssim$ 310 N/mm !). Considering these limits and accepting a certain experimental error from procedural and material variations no systematic size effect but the influence of side grooving remains. The side grooving effect is temperature dependent and vanishes for 300°C as shown by the more detailed plot in Fig. 22. In addition the results in Fig. 22 indicate that for the regime investigated there is no effect of relative crack length a/W.

The observations made about the general behaviour of the J-R curves are reflected in the specific values derived from them. Fig. 30 shows the temperature dependence of the initiation toughness as characterized by J_i, J_{Ic}^{ASTM}, J_{Ic}^{LOSS} (see Fig. 16). The values of J_i extrapolated from MSU-series or estimated as the point of first deviation on SSPUC-results from the formal blunting line prove to be independent of temperature and geometry parameters. A mean value of about 180 N/mm is found and the scatterband also includes the instability failures at room temperature. The technical initiation values J_{Ic}^{ASTM}- evaluated from a linear fit to the J-R curve according to ASTM E 813 [5] - and J_{Ic}^{LOSS}- evaluated from a power law fit following LOSS et al. [17] - are 30 to 50% higher and show the same temperature dependence which corresponds to that of the J-R

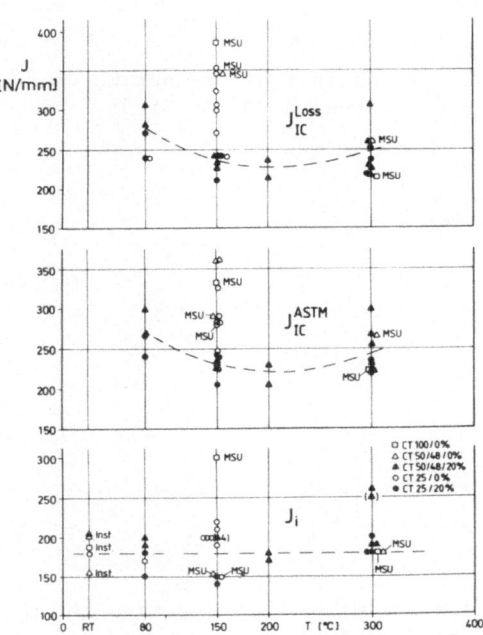

Fig. 30: Initiation toughness as function of temperature [30]

curves. The increase of J_{Ic} over J_i is a consequence of J_{Ic} not being evaluated for real onset but for a finite amount of stable crack growth in the range of 0.3 to 0.4 mm. J_{Ic}^{ASTM} is found to scatter more than J_{Ic}^{LOSS} and to be a few percent smaller. No significant influence of specimen size or side grooving could be separated out of the data available. The tendency of several of the MSU-series delivering much higher J_{Ic}-values than the SSPUC-tests has to be further analyzed.

Values of the tearing modulus T evaluated from the ASTM-linear fit through the valid $J/\Delta a$-points between the exclusion lines are given in Fig. 31 as a function of the test temperature. In addition Fig. 31 includes values of the parameter J_{50} according to JOHNSON and PARIS [31] and Fig. 32 shows an example how these have been evaluated from the underlying $J/\Delta a$ data. Again minimum values are indicated for about $200^\circ C$.

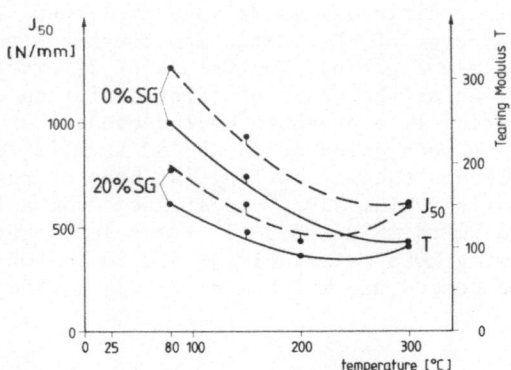

Fig. 31: Tearing modulus T and instability parameter J_{50} as a function of temperature for steel 20 Mn Mo Ni 55

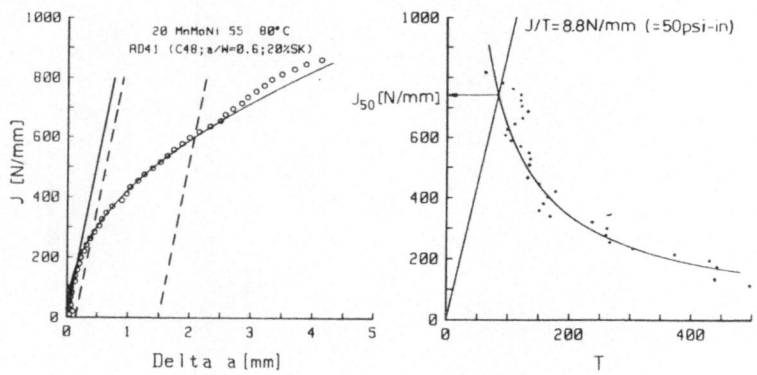

Fig. 32: Example for evaluation of J_{50}, to derive the material J/T-curve sequences of 5 neighbouring $J/\Delta a$-values and a power law fit $J = C\Delta a^n$ through the valid data points between the exclusion lines were used

The influence of net thickness/ligament ratio on the **J-R curves for** a similar steel ASTM A 533 B at room temperature is shown by the results of DAVIES et al. [32] in Fig. 33. The initiation toughness seems to be relatively independent of geometry, again if validity criteria for thickness and remaining ligament according to [5] are met.

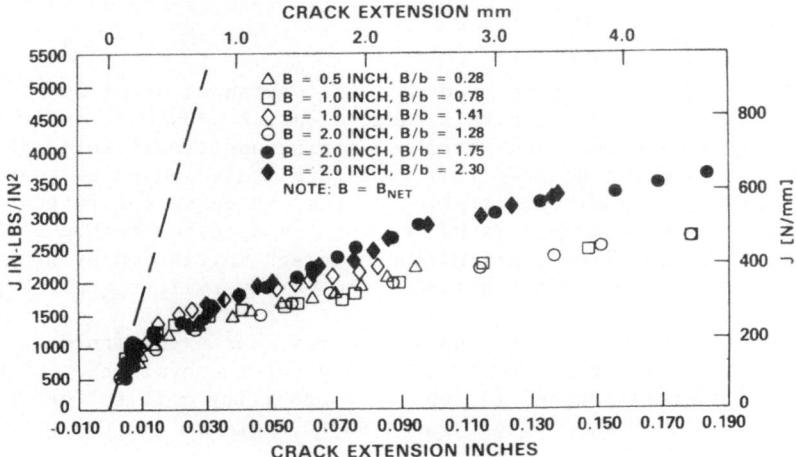

Fig. 33: J versus crack extension data for CT50 specimens of ASTM 533 B steel (ASTM E 813 [5] criteria fulfilled) [32]

The influence of testing speed is demonstrated in Fig. 34. In this example the effect is small; larger effects can be expected for other materials and when going into more dynamic conditions (see for instance [34]).

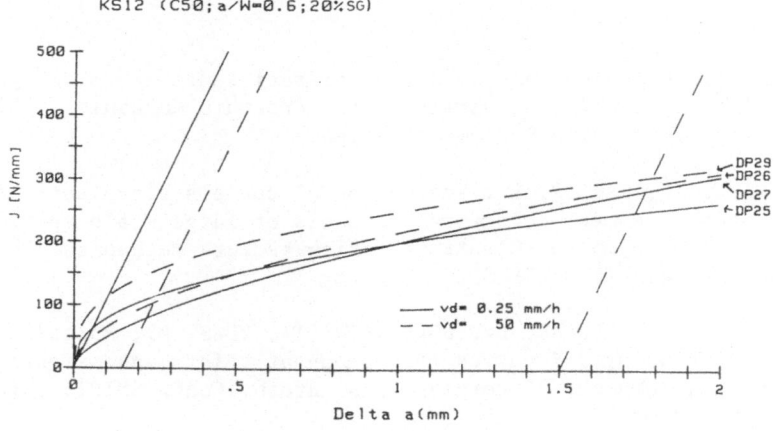

Fig. 34: Influence of different testing speeds on the J-R curves for RPV-steel 20 Mn Mo Ni 55 at 200°C [33]

4. CONCLUSIONS AND DISCUSSION

- The multiple specimen unloading (MSU) method, the direct current potential drop (DCPD) method, and the single specimen partial unloading compliance (SSPUC) method can be used to generate crack growth resistance J-R curves of different materials in the temperature regime -190° up to 800°C. Comparable results are derived from all three methods.

- The DCPD- and the SSPUC-method deliver continuous or at least closely covered curves for the full range of crack growth of interest from each single specimen. This permits gaining additional information about the material scatter from replicate tests and still enables a complete and reliable description of the J-R curve for further evaluation. The DCPD method is no real single specimen method because after all it requires confirmation of crack initiation by breaking and evaluating one additional specimen with very little crack growth.

- The MSU-method on the other hand requires less expenditures on testing but more material and - especially for the evaluation of deduced parameters for initiation and instability - it suffers from the small number and the pertinent scatter of the $(J, \Delta a)$ points.

- The J-R curves of 20 Mn Mo Ni 55 are temperature dependent showing a minimum in slope and absolute value at about 200°C. The critical J-values for onset of stable crack growth in the scatter of results available seem to be independent of temperature: $J_i \approx 180$ N/mm for 25 to 300°C. The values of the technical initiation toughness J_{Ic} according to ASTM 813 [5] or LOSS [17], the tearing modulus T, and the instability parameter J_{50} show temperature trends similar to the J-R curves with minima at about 200°C.

REFERENCES

[1] MILNE, E., Experimental validation of resistance cruve analysis, 2nd. Intern. Symp. on Elastic-Plastic Fracture Mechanics, Philadelphia, (1981) ASTM STP 803 to appear

[2] KUSSMAUL, K., ISSLER, L., Evaluation of the elastic-plastic fracture mechanics methodology on the basis of large scale specimens, 2nd. Intern. Symp. on Elastic-Plastic-Fracture Mechanics, Philadelphia (1981) ASTM STP 803 to appear

[3] BLAUEL, J.G., HODULAK, L., HOLLSTEIN, T., VOSS, B., Material characterization by J-R curves for component safety assessment. - A Study on a 20MnMoNi 55 forging, 7th Intern. Conf. SMIRT, Chicago 1983, paper G2/2

[4] ALBRECHT, P., ANDREWS, W.R., GUDAS, J.P., JOYCE, A.J., LOSS, F.J., McCABE, D.E., SCHMIDT, D.W., VanDer SLUYS, W.A., Tentative test procedure for determining the plane strain J_I-R-curve, Journal of Testing and Evaluation, JTEVA, Vol. 10, No. 6 (1982), 245-251

[5] ASTM E 813-81: Standard test for J_{Ic}, a measure of fracture toughness, Annual Book of ASTM Standards, Part, 10, Philadelphia (1981)

[6] BEGLEY, J.A., LANDES, J.D., The J-integral as a failure criterion. The effect of specimen geometry on J_{Ic}, ASTM STP 514 (1972), 1-20, 24-3

[7] RICE, J.R., PARIS, P.C., MERKLE, J.G., Some further results of J-integral analysis and estimates, ASTM STP 536 (1973), 231-245

[8] SCHMITT, W., Numerical simulation of post yield fracture mechanics experiments as a basis for the transferability to components, Proc. 8th. MPA Seminar, Stuttgart, Oct. 1982

[9] British Standard, BS 5762: 1979, Methods for crack opening displacement (COD) testing, British Standards Institution (1979)

[10] SMITH, R.F., KNOTT, J.F., Crack opening displacement and fibrous fracture in mild steel, Proc. Conf. Practical Application of Fracture Mechanics to Pressure Vessel Technology, Paper 9, pp. 65-75, Inst. of Mech. Engineering, London (1971)

[11] CLARK, G., KNOTT, J.F., Acoustic emission and ductile crack growth in pressure vessel steels, Metal Science (1977), 531-36

[12] CLARKE, G.A., ANDREWS, W.R., PARIS, P.C., SCHMIDT, D.W., Single specimen tests for J_{Ic}-determination, in "Mechanics of Crack Growth", ASTM STP 590 (1976), 27-42

[13] VOSS, B., MAYVILLE, R.A., On the use of partial unloading compliance method for the determination of J-R curves and J_{Ic}, ASTM Symposium on User's Experience with Elastic-Plastic Fracture Toughness Test Methods, Louisville, Kent., April 1983

[14] VOSS, B., On the problem of "Negative Crack Growth" and "Load Relaxation in single specimen partial unloading compliance tests, Proceedings of CSNI Workshop on Test Methods for Ductile Fracture, Paris, OECD-NEA (1983)
patent for alternative clevis design

[15] KLOSS, G., Rechnergestützte J-R Kurvenbestimmung nach dem Potentialverfahren, Vorträge der 13. Sitzung des AK Bruchvorgänge, Hannover 1981, DVM Berlin (1982)

[16] VOSS, B., BLAUEL, J.G., Experimental determination of crack growth resistance curves, Proceedings of the 4th ECF-Conference, Leoben 1982, Vol. 1 edit. K.L. Maurer, F.E. Matzer, EMAS, Cradley Heath, UK (1982)

[17] LOSS, F.J., MENKE, B.H., GRAY, R.A.Jr., HAWTHORNE, J.R., J-R curve characterization of irradiated nuclear pressure vessel steels, Proc. US NRC/CSNI Specialist's Meeting on Plastic Tearing Instability, Wash. Univ., St. Louis, Missouri, Sept. 1979

[18] RITCHIE, R.O., GARRET, G.G., KNOTT, J.F., Crack growth monitoring optimization of the electrical potential technique using an analogue method, Int. J. Fract. Mech. 7 (1971), 462-67

[19] MARANDET, B., LABBE, G., PINARD, J., TRUCHON, M.: Détéction de l'amorçage et suivi de la propagation d'une fissure par variation du potentiel électrique en régime alternatif, IRSID Rep. RE 549 June (1978)

[20] KLIMA, S.J., FISCHER, D.M. BUZZARD, R.J., Monitoring crack extensions in fracture tests by ultrasonics, J. Testing and Evaluation 4, 6 (1976) 397-404

[21] UNDERWOOD, J.H., WINTERS, D.C., KENDALL, D.P., End-on ultrasonic crack measurements in steel fracture toughness specimens and thick wall cylinders, Proc. Conf. Detection and Measurement of Cracks, The Welding Institute, Abington, (1976), 31-39

[22] MIRABILE, W.W., ATTERIDGE, J.F., LESSOR, J.F., An acoustic emission investigation of microscopic ductile fracture, Metallurgical Trans. 6A, April (1975), 797-801

[23] LOTTERMOSER, J., WASCHKIES, E., ZENNER, P., VOSS, B., GÖTZ, J., Laboruntersuchungen zur Erarbeitung von Interpretationsmodellen für die Bewertung von Fehlstellen bei der Schallemission am Kernreaktor, Bericht Nr. 780236-TW zum Forschungsvorhaben BMFT RS 196, Fraunhofer-Institut für zerstörungsfreie Prüfverfahren, Saarbrücken (1978)

[24] PRIJ, J., Some finite element results of CTS Specimen, Proc. of CSNI Workshop on Test Methods for Ductile Fracture, Paris (1982), OECD-NEA (1983)

[25] PRANTL, G., Assessment of Crack Extension by different methods, Proc. of CSNI Workshop on Test Methods for Ductile Fracture, Paris (1982), OECD-NEA (1983)

[26] HOLLSTEIN, T., VOSS, B., Hochtemperatur-Rißwiderstand und Kriech-
      rißwachstum - Untersuchungen mit dem Gleichstrompotential - und
      dem Teilentlastungsverfahren, Vorträge der 15. Sitzung des Arbeits-
      kreises Bruchvorgänge, Darmstadt, Deutscher Verband für Material-
      prüfung, Berlin (1983)

[27] BLAUEL, J.G., MAYVILLE, R., LENZ, H.W., Toughness evaluation for
      a thick weld joint of AlMg 45 Mn, Ber. II.9. Sec. Intern. Conf. on
      Alum. Weldm. München (1982)

[28] ISSLER, L., MPA-Seminar, Jan. 1981

[29] GARWOOD, S., Crack propagation toughness of structural steels at
      temperatures above brittle-ductile transition temperature,
      Rep. 3545/8/79, The Welding Institute, Cambridge UK (1979)

[30] BLAUEL, J.G., HODULAK, L., HOLLSTEIN, T., VOSS, B., Material
      characterization by J-R-curves for component safety assessment.
      - A study on a 20 MnMoNi 55 forging.
      7th. Intern. Conf. SMiRT, Chicago 1983, Paper G2/2

[31] JOHNSON, R., PARIS, P.C., Resolution of the Task A 11: Reactor
      Vessel Material Toughness Safety Issue, Rep. NUREG 0744 Vol. 1
      (1982)

[32] DAVIES, D.A., VASSILAROS, M.G., GUDAS, J.P., Specimen geometry
      and extended crack growth effects on J_I-R curve characteristics
      for HY 130 and ASTM A 533 B Steels, Rep. NUREG/CR-3089 (1983)

[33] VOSS, B., Zeitabhängigkeit von Rißwiderstandskurven bei 200 und
      285°C. IWM-Bericht V 43/82 (1982)

[34] JOYCE, J.A., Static and dynamic J-R curve testing of A 533 B
      steel using the key curve analysis technique, Rep. NUREG/CR-2274,
      USNRC Washington, July 1981

EXPERIMENTAL DETERMINATION OF FRACTURE MECHANICS PARAMETERS OF
IRRADIATED MATERIALS

J. Bernard and G. Verzeletti
Commission of the European Communities
Joint Research Centre - Ispra Establishment
21020 Ispra (Varese) - Italy

ABSTRACT.   This lecture starts with a short description of the effects
of radiation on materials. Problems arise when irradiating and subsequently testing FM specimens: the irradiation facility has its space
limitations; the gamma heating generated thermal gradients and the specimen post-irradiation activity (necessitating handling in hot cells)
increase with increasing size. Therefore, attention must be given to
the development of measurement techniques that are suitable also for
subsized specimens. The lecture describes the JRC research programme
set up for elaborating J-R curves of Liquid Metal Fast Breeder Reactors
(LMFBR) materials (specially the AISI 316H) submitted to neutron damage from 0.1 dpa to 2 dpa. The adopted method uses key curves yielding
one R-curve from each specimen. Results are given for 0.1 dpa and
0.3 dpa damage for base material, weld and heat affected zone. The method is quite efficacious and has wider fields of application than the
one related to irradiated materials.

1.   INTRODUCTION

In a material submitted to a neutron flux, the atoms are displaced from
their original positions with the generation of vacancies; depending on
the impinging neutron energy, the displaced atoms can produce further
collisions and thus further displacements. The displaced atoms can oc-
cupy an existing vacancy or a position between the lattice (intersti-
tial). The radiation damage is expressed in dpa (displacement per atom)
and represents the mean number of displacements that each atom has
suffered during the irradiation. The radiation damage depends on the
neutron flux integrated in the time (fluence) and on the neutron spectrum. The damage is provoked by neutrons having energies higher than
0.1 MeV.
      The neutron irradiation induces a series of phenomena:
- diffusion and segregation of some elements in the material;
- recrystallisation of metastable phases;
- void nucleation;
- helium bubble formation.

145

L. H. Larsson (ed.), Elastic-Plastic Fracture Mechanics, 145–164.

These phenomena depend on, amongst other factors, the alloy composition and initial microstructure; for example, it has been established that the magnitude of the embrittlement induced by neutron irradiation in the steels used in the construction of pressure vessels for Light Water Reactors (LWR) is increased by the presence of copper.

For the evaluation of safety margins of a nuclear structure having suffered various amounts of neutron damage, fracture mechanics play an important role, both for the fatigue crack propagation and for the stability of the structure. The behaviour of cracks in a structure submitted to fatigue is evaluated using the two parameter Paris law applying the concepts of LEFM. The stability of a structure submitted to a transient overload is studied applying the concepts of EPFM. Several approaches are used for which material parameters are needed like J_{IC}, T, δ_i, etc. These parameters, for a given material, are functions of temperature, ageing, environment and irradiation.

This lecture deals with the problems connected with the determination of EPFM parameters as functions of radiation damage in AISI 316H austenitic stainless steel used in the construction of LMFBRs and pipings of LWRs. The results are evaluated in terms of J resistance curves on three point bend (3 PB) specimens having dimensions of 15 x 20 x 110 mm.

2. MATERIALS AND METHODS

2.1 Definition of Specimen Geometry

The conditions for a J-controlled crack growth were discussed in Bakker's first lecture. The J-Δa curve of a material is valid as long as the minimum size requirements are satisfied: the thickness B > 20 J/$\bar{\sigma}$ and the ligament length b > 20 J/$\bar{\sigma}$ for bent specimens, $\bar{\sigma}$ being the material flow stress taken as the average of the yield stress and the ultimate tensile stress. In addition the maximum crack growth Δa $\leq$ 0.1 b and (b/J)(dJ/da) $\geq$ 5. The need for large specimens when studying the FM material properties of tough materials is incompatible with the relatively restricted available space of the irradiation facility. In order to obtain significant radiation damage in short time the irradiations are performed in the High Flux Reactor (HFR) of the JRC Petten (Netherlands) Establishment. In that case the space is limited to a diameter of 29 mm (Fig.1a) and a length of 450 mm. On the basis of this, the specimen size was set at 15 mm thickness by 20 mm width and 110 mm length. The fact that the heating of the specimens is done by absorption of γ radiation emanating from the reactor leads to the generation of thermal gradients in the specimen. This problem is more severe for large-sized specimens. In the case of our specimens having a cross-section of 15 mm x 20 mm, the γ-heating that did not exceed 2.5 W/g led to an acceptable temperature gradient of about 30°C between centre and surface of the specimens. Finally, the size is limited since the specimen remnant post-irradiation activity must be compatible with the hot cells limited shielding efficiency. Our specimen size is acceptable as far as this point is concerned.

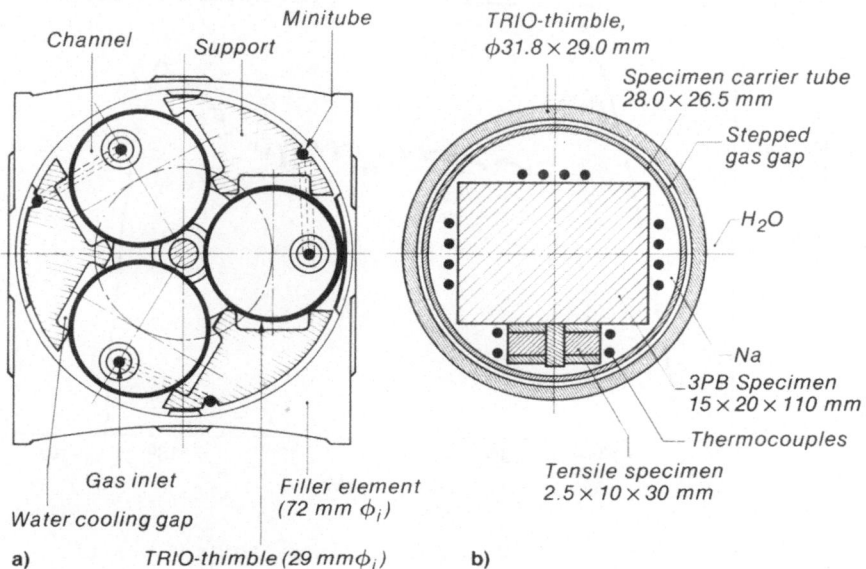

a) TRIO-thimble (29 mm ϕ_i)

b)

Labels in figure (a): Channel, Support, Minitube, Gas inlet, Water cooling gap, Filler element (72 mm ϕ_i)

Labels in figure (b): TRIO-thimble, $\phi31.8 \times 29.0$ mm; Specimen carrier tube 28.0 × 26.5 mm; Stepped gas gap; H_2O; Na; 3PB Specimen 15 × 20 × 110 mm; Thermocouples; Tensile specimen 2.5 × 10 × 30 mm

Figure 1. Horizontal cross-section of the TRIO-capsule (a) and TRIO-thimble with specimen arrangement (b).

2.2 Irradiation Facility

HFR is a light water cooled and moderated material testing reactor. The nominal power is 45 MW and the coolant operates at about 50°C. The irradiation is done in sodium filled specimen carriers (Fig.1b) containing one column of four 3 PB specimens and four tensile specimens each. These carriers are inserted in TRIO irradiation channels. There are three water-cooled TRIO channels in the irradiation capsule. The TRIO channels provide a gas gap to the inner sample carriers that are used for temperature control by gas mixtures. The temperatures of the samples are detected inside the specimen carriers by means of chromel alumel thermocouples, which are arranged in the liquid close to the samples at nine different levels of the columns.

The nominal temperatures are set at 350°C and 550°C corresponding to the cold and hot leg of the LMFBR. The neutron flux densities range between $5 \cdot 10^{13}$ and $5 \cdot 10^{14}$ n.cm^{-2}.s^{-1}. In Fig.2 the neutron spectrum is given for our irradiation positions called H_2 and H_8.

The nominal radiation damages in the first campaign were 0.1 dpa and 0.3 dpa. For the H_2 and H_8 positions in HFR this requires 26 and 78 days, respectively.

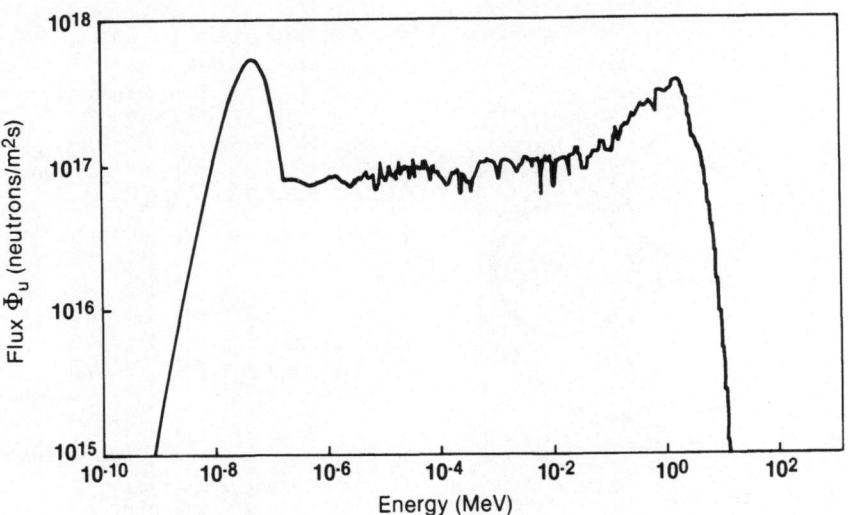

Figure 2. Neutron spectrum in the irradiated positions.

2.3 Material Specification

The material tested is an AISI 316H stainless steel. In order to have a stock of material for further irradiations and investigations, a plate 1200 x 6500 x 50 mm has been supplied. The chemical composition is given in Table I.

TABLE I
Chemical composition of the investigated AISI 316H steel in weight percent

Base material

C	Si	Mn	P	S	Cr	Ni	Mo
0.056	0.34	1.61	≤0.02	0.009	15.2	12.1	2.44

Co	Ti	Nb	Cu	B		V	N
0.06	≤0.005	≤0.005	0.22	0.00097		0.087	0.082

Weld material

C	Si	Mn	P	S	Cr	Ni	Mo
0.051	0.43	2.25	≤0.02	0.009	16.0	8.9	2.36

Co	Ti	Nb	Cu	B		V	N
0.06	≤0.005	≤0.005	0.17	0.00088		0.08	0.082

The mechanical properties are given in Table II at different temperatures. The material has been tested as a base material (B), weld deposit (W) and heat affected zone (HAZ).

TABLE II
Mechanical properties of the non-irradiated steel

Material	Temperature °C	Yield strength MPa	Ultimate tensile strength MPa	Uniform elongation percent
Base (B)	20	240[a]	570[a]	65[a]
Base (B)	350	133	483	36
Base (B)	550	105	440	35
Weld (W)	350	300	445	16
Weld (W)	550	250	382	14

[a] Result obtained in one test on a ⌀ 4 mm tensile specimen, gauge length was 20 mm.

2.4 Specimen Preparation

The 15 x 20 x 110 3 PB specimens (Fig.3) are cut with the long dimension parallel to the rolling direction, and the width 20 mm in the thickness direction (standard L-S type). The samples are provided with a machined notch, as shown in Fig.3, propagated by fatigue up to a/W = 0.55, before irradiation.

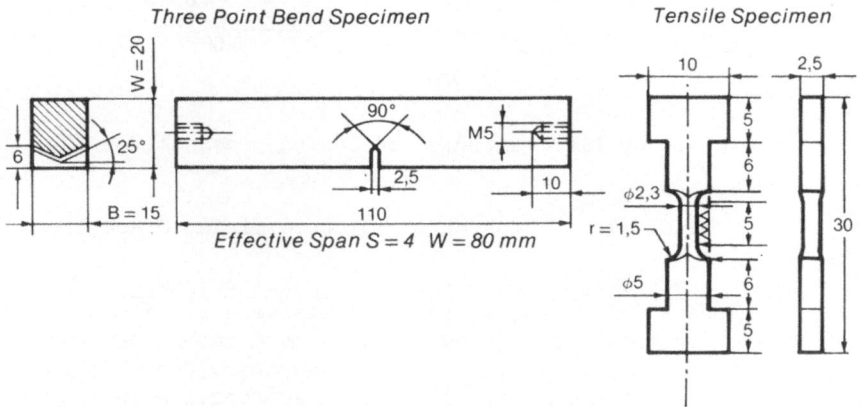

Figure 3. Dimensions of 3 PB and tensile specimen.

A method whereby the displacement cycle (20 Hz sinusoidal wave) is kept constant has been used. The system utilises an electronic device which controls the load amplitude $\Delta P = P_{max} - P_{min}$ where P_{min} is set at 0.5 kN. In order that P_{max} would remain within the limits set by the British Standard Institutions in their DD19:1972 document, the maximum allowable fatigue stress intensity at the crack tip K_f is limited to a $0.63 \bar{\sigma} B^{0.5}$ value where B is the thickness of the specimen and $\bar{\sigma}$ is taken as the average of the yield and the ultimate tensile stress. For the base material with a 405 MPa $\bar{\sigma}$ value, K_f equals 31.25 MPa√m. Following this method it can be seen in Fig.4 that the DD 19 criterion is satisfied from a/W = 0.48 onward.

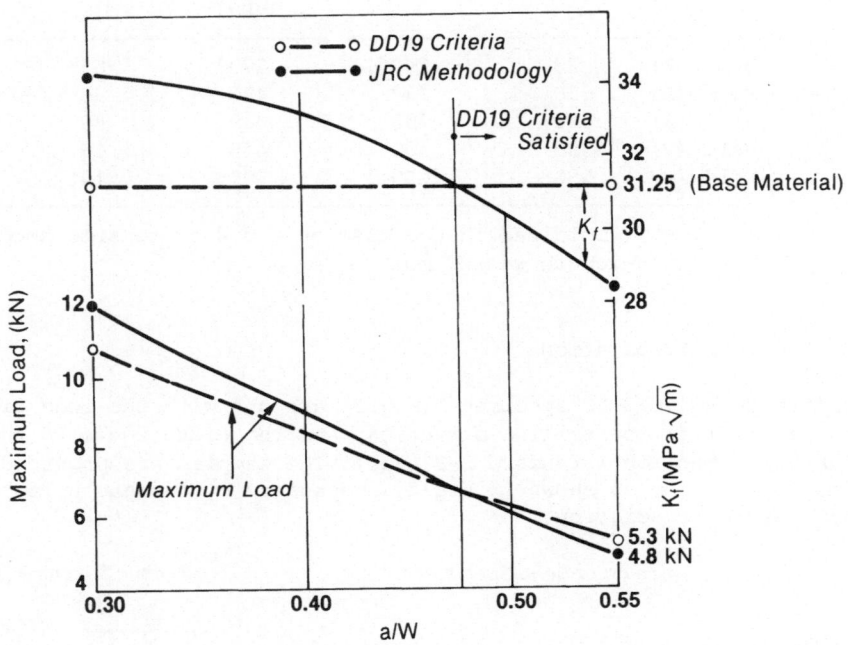

Figure 4. Applied maximum load during precrack.

2.5 Review of Candidate Methods Aiming at Determining Crack Growth

2.5.1. <u>Single specimen partial unloading compliance, SSPUC</u>. When this specific research was tackled, screening tests were run aiming at defining the most suitable crack growth measurement technique. Various techniques were discussed by Blauel in this seminar. The widely used SSPUC approach was not deemed to be the most appropriate in our case due to the following circumstances:
- the method uses highly complex and precise instrumentation that cannot be readily installed in hot cells;
- in the case of the stainless steel tested having very high toughness and large amounts of angular deformations (bent specimens), in-plane

deformations of the ligament area occur so that the measured change
in compliance during unloading yields underestimated crack growth values.
This problem is even more acute when the base material is tested at low
temperatures.

2.5.2 Potential drop direct current technique, DCPD. This method was
initially considered as most suitable for our purposes. In particular,
it presented obvious advantages over the first-mentioned one as far as
telemanipulation in hot cell is concerned. Participating in a round-
robin exercise where crack growth resistance curves were estimated for
a ferritic steel at room teperature, consistent μV/μm crack growth ratios were found for this material in 75% of the tests and therefore the
technique was considered to be acceptable. In fact, the resistance curves appearing in Fig.5 convey the idea.

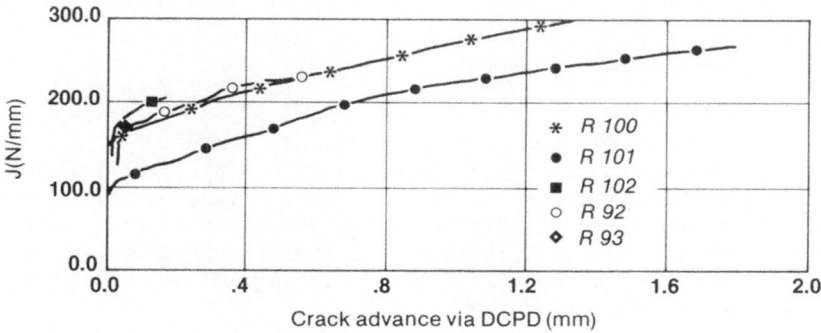

Figure 5. J-R curves by DCPD on ferritic steel.

Unfortunately, this very promising technique did not fare well in
the case of the stainless steel specimens. Indeed, the ratio μV/μm
crack growth was found to be highly variable and no systematic use
could thus be made of this method. The fact that the specimens bend at
least 10 degrees in the case of the base material at $350^{\circ}C$ and for the
maximum crack growth of roundabout 800 μm, and since in addition the
ligament section is strongly deformed, may explain why such difficulties arose (see Appendix).

2.5.3 Crack extension calculation by dimensional analysis. This approach presents great advantages in this specific research since it
uses exclusively the load-load line displacement measurements. It is
based on dimensional analysis for deeply cracked specimens in bending
as described in Bakker's second lecture.
From dimensional analysis applied to the case of pure bending of a
small remaining ligament, it has been shown /1/ that the relationship
between angular displacement θ_c and bending moment per unit thickness M

is

$$\theta_c = f\left(\frac{M}{b^2}\right) \tag{1}$$

Calling $\theta = \theta_{nc} + \theta_c$ the specimen load point rotation at any given M, then θ_{nc} is the rotation of an uncracked specimen under the same M, θ_c is the contribution due to the presence of the crack, and b is the current ligament size.

The function f depends only on the material monotonic stress-strain properties. It can be observed that the only significant geometrical parameter entering the form (1) is the ligament size; i.e. the ratio b/W is not relevant in the case of pure bending.

Inverting the form (1) leads to

$$\frac{M}{b^2} = F(\theta_c) \tag{2}$$

or

$$M = b^2 F(\theta_c) \tag{3}$$

where $F(\theta_c)$ is referred to as the material "key curve".

It is possible, in theory, to determine the $F(\theta_c)$ function using specimens having ligaments of relatively smaller lengths than the one encountered in the frame of this research. The specific difficulties inherent to irradiated materials are such that the problem had to be addressed in a different way /2/. Considering eq.(3) and in the case of 3 PB specimens, one may write:

$$0.25 \ PS = b^2 F(\theta_c) \tag{4}$$

where S is the specimen span set at 80 mm.

It is convenient to carry the analysis further considering only the significant parameters from eq.(4). A function

$$C(\theta_c) = \frac{4}{S} F(\theta_c) \tag{5}$$

is introduced. S being constant throughout the tests, $C(\theta_c)$ differs from $F(\theta_c)$ by a scaling factor 4/S that is constant in all our tests and has consequently the same characteristic of being, in theory, unique for a series of tests characterized by a material, the temperature and the fluence. From (4) and (5) follows

$$P = b^2 C(\theta_c) \tag{6}$$

Following an analogous rationale, a function $C'(\theta_c)$ is defined as follows:

$$P = b_o^2 C'(\theta_c) \tag{7}$$

b_o is the initial ligament length.

The $C'(\theta_c)$ function is known from the experimental load-load line displacement measurements, the angular displacement θ_c being proportional to the load-line displacement Δ_c (where the subscript c again refers to the displacement contributed exclusively by the cracked specimen)

$$\theta_c = \frac{4\Delta_c}{S} \tag{8}$$

It is evident, from (8), that in the rationale that follows the parameter θ_c may always be directly substituted by the parameter Δ_c. Some data reduction steps are needed in order to obtain Δ_c.

Indeed, the data acquisition system monitors the total load-load line displacement Δ at an applied load P where

$$\Delta = \Delta_{nc} + \Delta_c + \Delta_r \quad \text{and}$$

Δ_{nc} = uncracked specimen load line displacement at applied load P;
Δ_c = cracked specimen load line displacement at applied load P;
Δ_r = additional deformations of rollers, roller bearing plates and specimen indentations.

By loading geometrically similar uncracked specimens in the test rig, calibration curves at 350°C and 550°C are obtained that relate the non-relevant load line displacements $\Delta_{nc} + \Delta_r$ to the currently applied load P. At any load attained during a test carried out on the cracked specimens the data reduction procedure automatically subtracts the value of the non-relevant deformations from the measured one, thus yielding the Δ_c value.

Henceforth, the parameters P, θ_c (or Δ_c) and b will be used with subscripts that refer to specific phases of the bending tests. Thus $(P_f)_k$ is the applied load (final load) applied at the end of test k. In this research, a maximum of four tests per case characterized by material (B, W or HAZ), temperature, ageing period in Na and fluence, were available.
Subscript $k = 1,\ldots,4$ corresponds to one of the four planned tests. $k = 4$ is arbitrarily ascribed to the parameters characterizing the test corresponding to the major crack extension;
$(\theta_{cf})_k$ is the angular displacement at the end of test k;
$(a_0)_k$ is the initial crack length of the specimen used in test k;
$(b_0)_k$ is the initial ligament length of the same specimen;
b_k is the current ligament length of the same specimen;
$(b_f)_k$ is the final ligament length at the end of test k;
$(b_f)_k$ is obtained by heat-tinting the specimen at 550°C, separating the specimen halves by fatigue cracking and sawing the remaining ligament. The crack extension $(\Delta a_f)_k$ is measured using the nine points average technique. Then

$$(b_f)_k = (b_0)_k - (\Delta a_f)_k$$

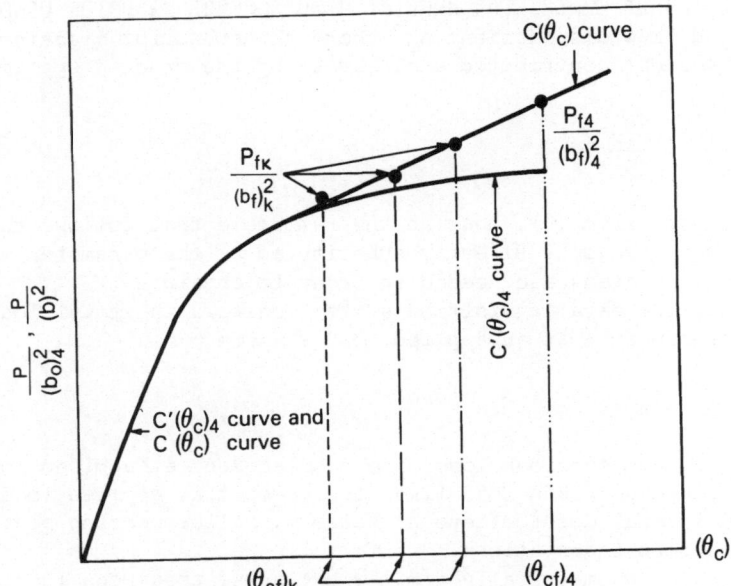

Figure 6. Normalized load-angular displacement record using all results.

In Fig.6, a $C'(\theta_c)_4$ curve is represented schematically as well as the $(P_f)_k/(b_f)_k^2$; $(\theta_{cf})_k$ points (k = 1,...,4) corresponding to the four (at the most) end of test conditions of the tests run in a specific case. As already observed, these points lie on the $C(\theta_c)$ curve defined by eq.(6). It has been verified for all cases concerning the base material and weld material and that are described in Table III, that the four (at the most) $(P_f)_k/(b_f)_k^2$; $(\theta_{cf})_k$ points line up in a most satisfactory manner on a tangent to the $C'(\theta_c)_4$ curve. Thus the $C(\theta_c)$ function is adequately represented by a straight line (elastic deformation), a curve and a straight line tangent to the curved part, see Fig.6.

By straightforward data reduction procedure, the tangent, in any test, to the $C'(\theta_c)$ curve and that passes through an imposed $P_f/(b_f)^2$; θ_{cf} end of test point, is defined. With this procedure, the single specimen method (allowing one J-R curve per specimen) is applied as follows. From (6) and (7):

$$b_i = b_o \sqrt{\frac{C'_i}{C_i}} \qquad (9)$$

The current crack growth $\Delta a_i = a_i - a_o$ is derived from (9):

$$\Delta a_i = b_o \left(1 - \sqrt{\frac{C'_i}{C_i}}\right) \qquad (10)$$

where the subscript refers to the specific point where b_i and Δa_i are evaluated, The HP 1000 data acquisition system evaluates the applied force and load line displacement signals at equi-spaced time intervals of roundabout $T/1500$ s duration where T is the time needed to carry out the bending test. T is of the order of 300 s at the most.

An incremental extension of the crack length $a_{i+1} - a_i$ follows directly from (10):

$$a_{i+1} - a_i = b_o \left[\sqrt{\frac{C_i'}{C_i}} - \sqrt{\frac{C_{i+1}'}{C_{i+1}}} \right] \tag{11}$$

3. J CALCULATION

By virtue of the deformation theory definition /3/

$$J = \int_O^P \left(\frac{\partial \Delta}{\partial a} \right)_P dP \tag{12}$$

where Δ is the load point displacement through which the applied load P works. In the case of bending (12) equates to

$$J(a, \theta c) = - \int_O^M \left(\frac{\partial \theta_c}{\partial b} \right)_M dM \tag{13}$$

where $b = W-a$ and W is the width of the 3 PB specimen.

From (13) and (1) and on the assumption that the material is history independent, the following expression for J (referred to as J_D) is derived /1/

$$J_D = 2 \int_O^{\theta_c} \frac{M}{b} d\theta_c - \int_{a_O}^a \frac{J_D}{b} da \tag{14}$$

The term $Md\theta_c$ following the first integral sign of (14) is by virtual work concept, equivalent to $Pd\Delta_c$ since all deformations are, by hypothesis and in the case of deeply cracked specimens, confined to the ligament area. Substituting M and θ_c by P and Δ_c has the advantage that the latter couple of parameters is, in fact, the one monitored by the data acquisition system.

The J value was calculated from the relationship applicable to plane-sided three point bend specimens and expressed in incremental form /4/

$$J_{i+1} = \left[J_i + \frac{2}{b_i} \cdot \frac{A_{i,i+1}}{B} \right] \left[1 - \frac{a_{i+1} - a_i}{b_i} \right] \tag{15}$$

where $A_{i,i+1}$ = area under the load versus load-line displacement record between lines of constant displacement at points i and i+1. This area does not include the uncracked body energy. The values to be introduced in eq.(15) are

$\qquad$ b_i $\quad$ from Eq.(9)

$\qquad$ $a_{i+1} - a_i$ $\quad$ from Eq.(11)

and

$\qquad$ B = 15 mm

Eq.(15) used at the point where incipient crack growth is detected by the potential drop technique yields a J_{I_c} value referred to as $(J_{I_c})_{PD}$ in Table III.

TABLE III
(J_{I_c}) from dimensional analysis versus J_{I_c} from DCPD technique

Material and no. of specimens	Temp. °C	Fast fluence dpa min. & max.	$(J_{I_c})_{PD}$ N/mm	$(J_{I_c})_{DA}$ N/mm	Δa offset µm
B-7	350	0	230-280	230-280[a]	80
B-6	550	0	170-250	180-240[a]	80
B-4	350	0.084-0.105	240-300	260-300[a]	120
B-3	550	0.076-0.109	100-140	100-130	50
B-4	350	0.19 - 0.29	210-300	250-280[a]	120
B-4	550	0.24 - 0.33	240-280	230-280[a]	150
W-8	350	0	38 - 62	38 - 62	0
W-7	550	0	30 - 50	45 - 55	0
W-4	350	0.146-0.187	41 - 52	44 - 55	0
W-4	550	0.13 -0.179	43 - 45	54 - 58	0
W-4	350	0.19 - 0.27	44 - 75	40 - 60	0
W-4	550	0.21 - 0.28	20 - 54	44 - 52	0

[a] The J-integral capacity /4/ of 139 N/mm for the base material at 350°C and of 123 N/mm for the base material at 550°C is exceeded in these cases.

4. RESULTS

4.1 J-resistance Curves for B and W material

Twelve cases, characterized by temperature, ageing in Na and fluence and concerning the B and the W material were studied. They are described in the three first columns of Table III.

Fifty nine specimens were tested. Since no effect of ageing in Na (either 26 days or 78 days) on the J-resistance curves was found for the non-irradiated B and W materials, the results cover a number of up to eight specimens in this case, the only significant parameters being the material and the temperature. Typical results appear in Figs.7 to 11.

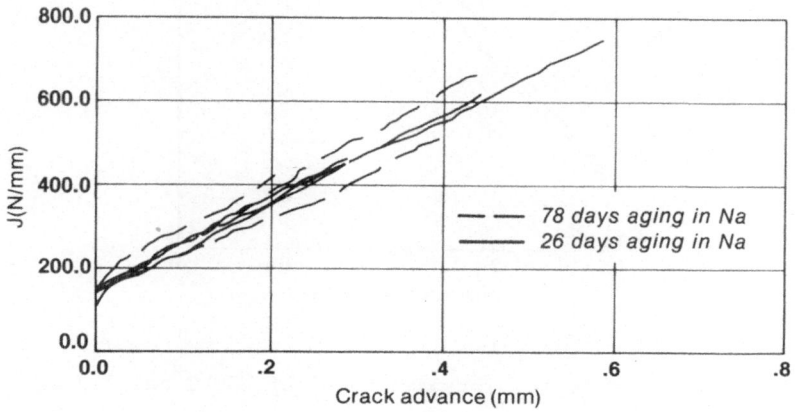

Figure 7. J-R curves for the base material, non irradiated at 350°C, after 26 days ageing in Na (full lines) and after 78 days ageing in Na (dashed lines).

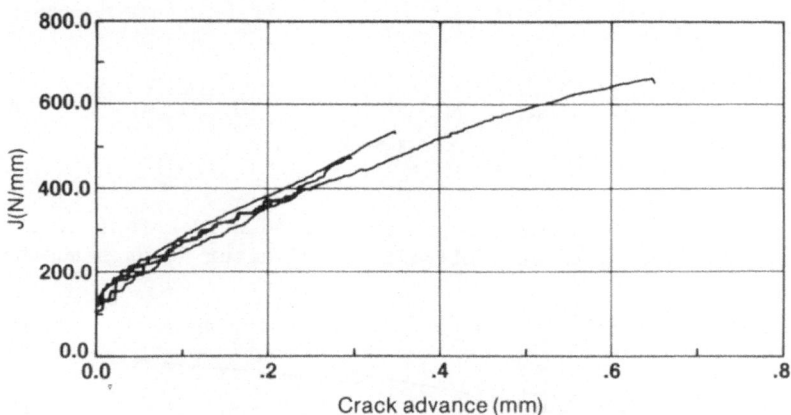

Figure 8. J-R curves for the base material, at 350°C and 0.1 dpa.

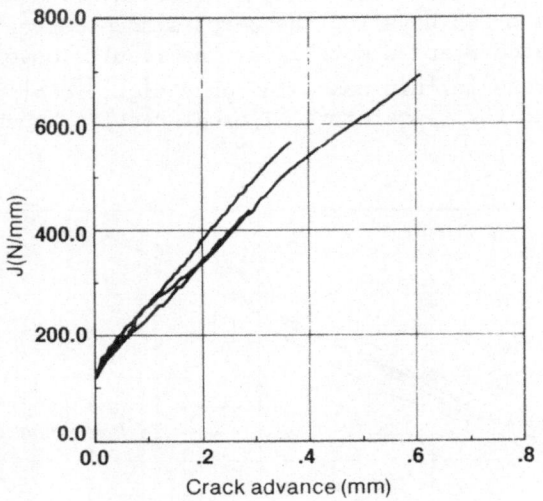

Figure 9.   J-R curves for base metal at 350°C and 0.3 dpa.

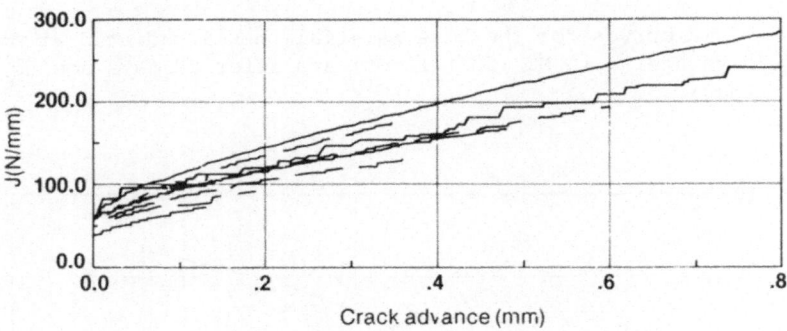

Figure 10.   J-R curves for the non irradiated weld material, at 350°C, after 26 days ageing in Na (full lines) and after 78 days ageing in Na (dashed lines).

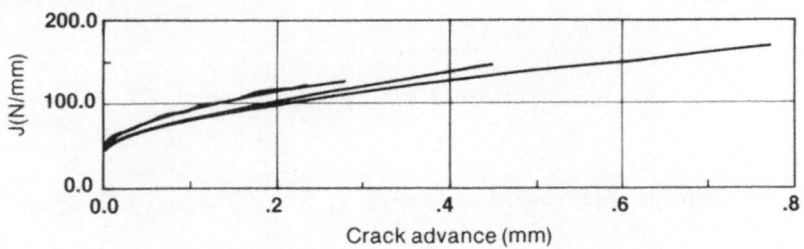

Fig.11.   J-R curves for the weld material, at 550°C and 0.3 dpa.

4.2 J_{I_c} and dJ/da Values for the Base and the Weld Material

The average J_{I_c} values yielded by the potential drop method and referred to as $(J_{I_c})_{PD}$ are found to correspond to averaged values of J resistance curves ordinates at $\Delta a = 0$ for the weld metal and at some offset Δa ranging from 50 to 150 μm for the base metal (see Table III). In addition, the J_{I_c} values taken on the J resistance curves at the tabulated Δa and referred to as $(J_{I_c})_{DA}$ in Table III have a spread of ± 6% to ± 24% for all results, inferior to the ± 8% to ± 46% (this maximum value corresponding to the last case of Table III, discarding it, the maximum is ± 26%) of the $(J_{I_c})_{PD}$ values, suggesting a better reproducibility of the results using the key curve approach. Since, with this approach, growing Δa values are found onward from the point of tangency between the $C(\theta_c)$ and the $C'(\theta_c)$ curves, it may be inferred that the case of the weld metal corresponds to a negligibly small stretch zone width, whereas the Δa values for all base metal cases suggest some amount of stretching before actual stable crack growth.

Finally, the variation in dJ/da in each case tabulated in Table III was found to be practically nil and reaching a maximum of ± 20% in the case of the non irradiated weld material at 350°C.

4.3 Test Results of the HAZ Material

From the results concerning the B and W materials it has been checked that the widely spread HAZ results lie practically within the base material at 350°C results and the weld material at 550°C for the non irradiated and for the irradiated case. Typical results appear on Fig.12.

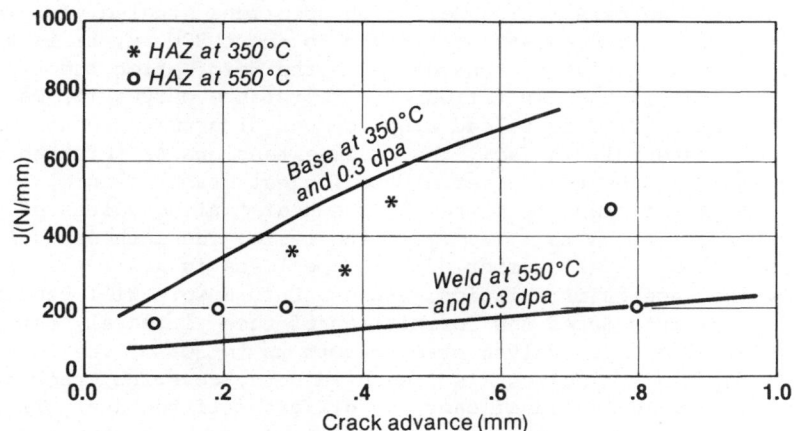

Figure 12. HAZ, results at 0.3 dpa nominal fluence.

These are not unexpected results considering that in the case of the HAZ material, the crack front actually "cuts" through all three materials, B, W and HAZ, in some random way due to the fact that the HAZ is, in fact, extremely narrow and the B-W interface is somewhat warped. On the basis of these observations it has been decided to exclude the HAZ material from future irradiation tests.

4.4 Validity of the Results

The criteria set on the validity of the results were mentioned in 2.1. The requirement $b > 20 \, J/\bar{\sigma}$ (more severe than $B > 20 \, J/\bar{\sigma}$) is not attained in the cases indicated by the superscript a in Table III. Therefore, the J_{I_c} values corresponding to these cases are very probably non conservative upper bound values. This requirement may definitely be somewhat relaxed /5/ by a factor of about four in the case of elaboration of crack-growth resistance curves.

The criterion $(b/J) \, (dJ/da) \geq 5$ /6/ has been found to be fully met in these tests. It can be rewritten as follows

$$\frac{b}{5} \geq J/\frac{dJ}{da}$$

or for $b = 9 \cdot 10^{-3}$ m nominal value

$$J/\frac{dJ}{da} \leq 1.8 \cdot 10^{-3} \text{ m}$$

In these tests $\Delta a \leq 0.1b$, always.

5. CONCLUSIONS

In the case of the relatively small 3 PB specimens studied, for which the J controlled crack growth is limited to about 800 μm, it is shown how a dimensional analysis (together with the deformation theory of plasticity) enables the derivation of J resistance curves for materials having distinctly different load displacement diagrams such as the base and the weld material. In fact, the crack extensions Δa inferred using this procedure on the test covering the longest crack advance, are corroborated in a satisfactory manner by the heat tinting values of Δa found in three (at the most) accompanying tests. The method should thus be suitable when a small number of 3 PB specimens is available, as is obviously the case in irradiation studies. Future work will benefit from a recently /7/ introduced new formulation of the J integral, referred to as J_M, which allows to analyse crack growth up to 30% of the ligament under bending loads. The results presented here, covering crack growths of about 8%, are in no significant way different if the analysis follows either the J_D or the J_M route. It is our intention to study crack-growths up to 15% of the ligament in future tests and to elaborate the test data using the J_M concept. As far as the J_{I_c} values are concerned, they correlate reasonably well with the potential drop defined ones when taken on the J resistance curve at some Δa offset value of about 100 μm for all base material tests and at zero offset for the welded material. Recently, and following difficulties encountered when applying the ASTM E-813 standard test for the J_{I_c} measure of toughness, particularly with respect to the definition of the stretch zone width in high toughness material such as the AISI 316H steel, a defect assessment procedure aiming at the determination of J_{I_c} on the crack growth resistance curves at some offset Δa, say 100 μm, is evolving. One of the conclusions of this work is that such a procedure is valid, even in the case of the weld material since the relatively low dJ/da value found for this material will lead to a reasonably small overestimation of J_{I_c} at roundabout 100 μm of Δa with respect to the tabulated values.

Finally, it may be observed that the width of 20 mm of the L-S type specimen is relevant to the wall thickness of most of the LMFBR's primary circuit components such as the vessel wall.

6.   ACKNOWLEDGEMENTS

The authors wish to thank the following persons: Messrs. Manzotti and Tognoli, Applied Mechanics Division, JRC Ispra, who performed the calibrations and the measurements; Dr. Lölgen and co-workers from HFR, JRC Petten for their work concerning the irradiation of the test specimens.

7.   REFERENCES

1.   Hutchinson J.W. and Paris P.C., 'Stability analysis of J controlled crack growth', Elastic-Plastic Fracture, ASTM STP 668, J.D. Lander, J.A. Begley and G.A. Clarke, Eds., American Soc. for Testing and Materials, 1979, pp.37-64.
2.   Bernard J.A. and Verzeletti G., 'Elasto-plastic fracture toughness characteristics of irradiated 316 grade stainless steel', paper presented at the Symposium on User's Experience with Elastic-Plastic Fracture Toughness Test Methods, April 20-22, 1983, Louisville, USA, to be published in an ASTM STP.
3.   Rice J.R., 'Mathematical analysis in the mechanics of fracture', Fracture, An Advanced Treatise, H. Liebowitz, Ed., Vol.2, pp.791-311, Academic Press, 1968.
4.   Albrecht P. et al., 'Tentative test procedure for determining the plane strain J-R curve', J. of Testing and Evaluation, JTEVA, Vol.10, No.6, Nov. 1982, pp.245-251.
5.   Turner E.C., 'Application of J-R curves to slow stable crack growth and unstable tearing in the plastic regime', Advances in Elasto-Plastic Fracture Mechanics, L.H. Larsson, Ed., pp.139-164, Applied Science Publishers Ltd., 1980.

6.   Paris P. and Johnson R., 'A method of application of elastic-plas-
      tic fracture mechanics to nuclear vessel analysis', Seminar on
      EPFM, sponsored by IWM, Freiburg (Germany), June 1981.
7.   Ernst H.A., 'Material resistance and instability beyond J-control-
      led crack growth', Elasto-Plastic Fracture: Second Symposium,
      ASTM STP 803, C.F. Shih and J.P. Gudas, Eds., ASTM, 1983, to be
      published.

APPENDIX - Direct Current Potential Drop (DCPD) Technique

In Fig.A1 a typical arrangement of the DCPD method is presented.

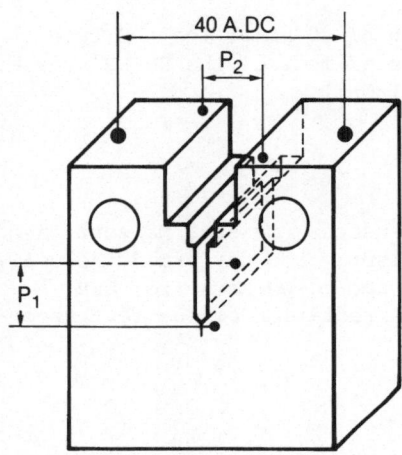

Figure A1. Location of PD pick-ups
P_1 and P_2 for a CT specimen.

In our case, two positions of the pick-up points have been used. The
amount φ of the increase in potential drop is a function of the deformation of the specimen and integrates the following phenomena:
1) φ_G which is related to the change in geometry;
2) φ_P related to the increasing amount of plasticity;
3) φ_ρ related to the variation of the material resistivity;
4) $\varphi_{\Delta a}$ related to the crack advance.

     If the integrated effect of the first three phenomena is small com-
pared to the fourth one, it becomes possible logically to measure the
crack advance during deformation. In Figs.A2 and A3 the potential drop
versus load-line displacement is presented in the case of a ferritic
steel.
     In the case of the P_2 measurements only a limited amount of varia-
tion in potential drop occurs before initiation.
     It is widely accepted that the crack growth can be determined by
measuring the potential $\varphi_{\Delta a}$ using as datum line the extrapolation of
the phenomena φ_G, φ_P, φ_ρ. Assuming a linear relationship between $\varphi_{\Delta a}$

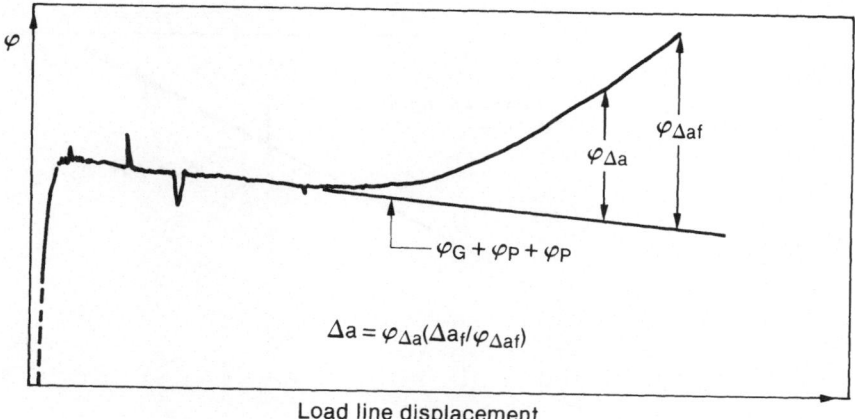

Figure A2. Potential drop P_2 in the case of a ferritic steel.

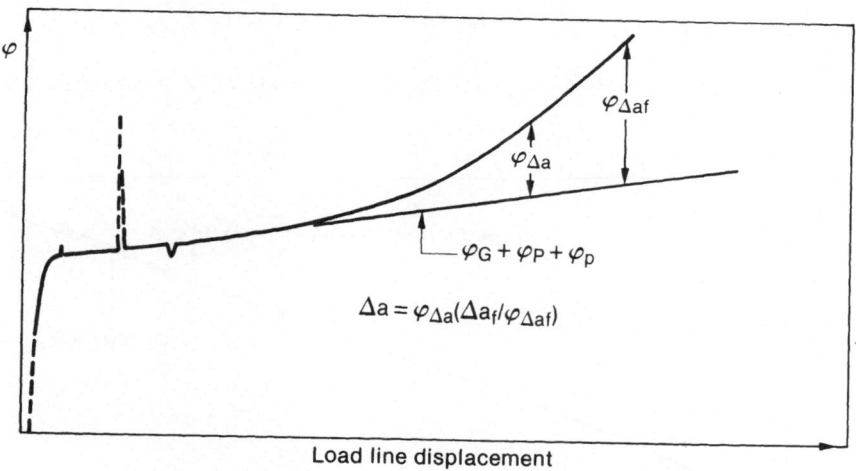

Figure A3. Potential drop P_1 in the case of a ferritic steel.

and the crack growth, and using as a calibration point the final amount of crack growth measured by heat-tinting, the current crack length is defined. In the case of the stainless steel (Figs.A4 and A5) the phenomena φ_G, φ_P, φ_ρ are preponderant excluding thus the possibility of measuring the crack advance. Anyway, the crack initiation can still be identified.

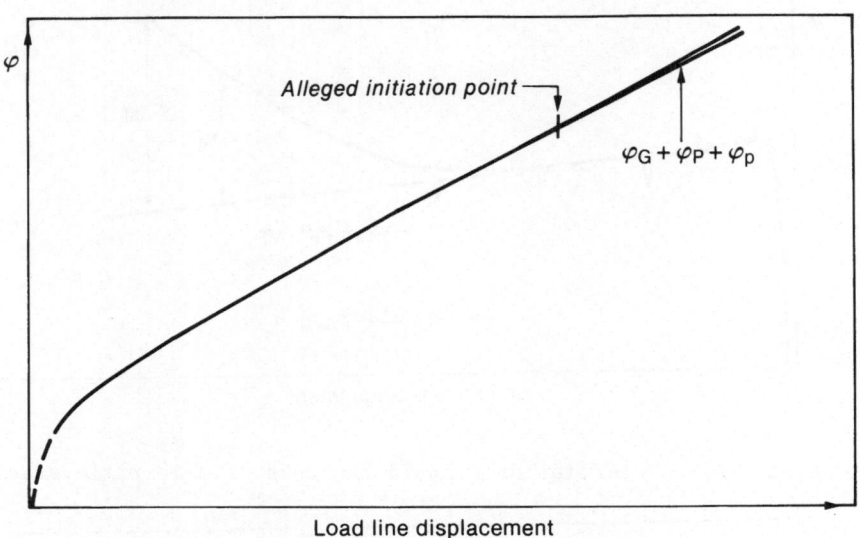

Figure A4. Potential drop P_1 in the case of a stainless steel.

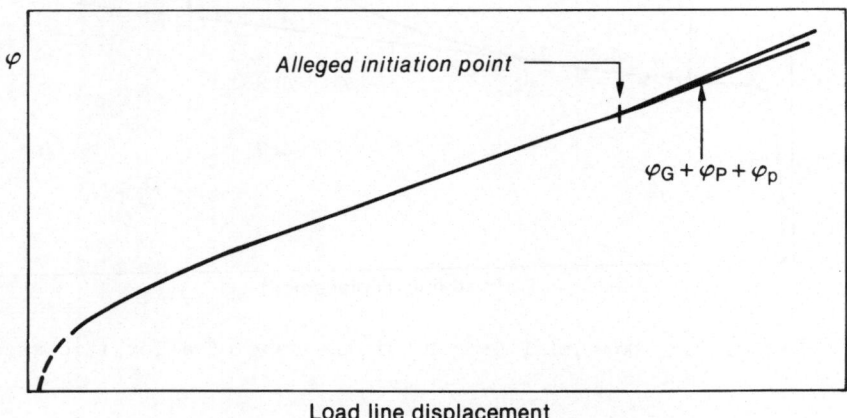

Figure A5. Potential drop P_2 in the case of a stainless steel.

SURFACE CRACKS

D. Munz
University of Karlsruhe, Institute for reliability and
failure analysis
Postfach 6380
D-7500 Karlsruhe
FR Germany

ABSTRACT. For the evaluation of the behaviour of cracked components,
plane specimens with a straight crack, characterized by its length a,
are considered. Real cracks usually are two-dimensional surface cracks
with a curved crack front. The aim of fracture mechanics is the prediction of critical loads for components with these cracks from material
data obtained with the plane specimens such as compact or bend specimens.
The main problem is the varying stress and strain field along the crack
front. For linear-elastic behaviour the determination of the stress intensity factor in the case of stress gradients and the fatigue crack propagation are discussed. Then the methods of plastic zone correction,
crack opening displacement, J-Integral and limit load criteria are
described. Finally some experimental results are presented.

1. DIFFERENCE BETWEEN FRACTURE MECHANICAL TEST SPECIMENS AND
   REAL CRACKS IN COMPONENTS

The crack growth behaviour of a material is characterized by material
parameters such as K_{Ic}, K_{Ia}, J_{Ic}, δ_c or by material-specific relations
such as K_I-Δa-, J-Δa- or da/dN-ΔK-curves. These parameters and relations
are obtained with simple specimens under simple loading conditions. The
specimens mostly are plates with through-the-thickness cracks, e.g. compact, bend or center cracked specimens and it is aimed to perform the
test either under plane-stress (thin plates) or plane-strain (sufficiently thick plates) conditions. Thus, we have a two-dimensional (2D) problem. It is well known that very often a 2D-situation is difficult to
maintain during crack extension. Proper side grooving is a convenient
method of assuring 2D-behaviour.
    Real components very often are not plates, and the cracks mostly
are surface cracks or embedded cracks. Thus, instead of a two-dimensional, we have a three-dimensional (3D) problem. Due to the curved crack
shape the stresses and strains near the crack tip vary along the crack
front. In addition, the stress distribution in a plane test specimen
(without a crack) usually is uniaxial and either constant or with a
moderate gradient, whereas in a real component the external loads may

165

L. H. Larsson (ed.), Elastic-Plastic Fracture Mechanics, 165–201.
© 1985 ECSC, EEC, EAEC, Brussels and Luxembourg.

lead to multiaxial stresses and to steep gradients (for notched components or for thermal transients).

The aim of the theory of fracture mechanics is to predict the behaviour of real cracks from the results of tests involving the simple specimens. For this purpose, it is necessary to idealize first the real cracks. The shapes of the real cracks are conveniently described by semi-elliptical or rectangular surface cracks, elliptical embedded cracks or quarter-elliptical corner cracks. Rules for crack idealization are available if the real shape is known[1]. However, very often it is difficult to get reliable information from non-destructive testing results. Then assumptions must be made about the size and shape of the most dangerous crack.

In the following sections only surface cracks will be considered. These cracks are more important than embedded cracks because very often a crack starts from the surface of a component.

2. GENERAL BEHAVIOUR OF SURFACE CRACKS

If we consider a given crack, which may be introduced during fabrication, e.g. a welding crack, or initiated by the alternating loads during operation of the component, we can distinguish between different steps in further crack extension (fig. 1). First the crack may extend further in depth and length as a fatigue crack (fig. 1b). If the crack size has sufficiently enlarged and a sufficiently large external load is applied, the crack penetrates the component and gives rise to a leak in case the component is pressurized (fig. 1c). This event is called local failure or local instability, if it involves some fast unstable crack extension. After wall penetration the crack may get arrested and further increase by fatigue until unstable crack extension and complete (global) failure of the component occur (fig. 1d). This behaviour is called leak-before-break for pressurized components. It is also possible that the crack does not become arrested after wall penetration and in this case global failure follows immediately local failure.

For the wall penetration (local instability) under rising load different behaviours are possible, dependent on the toughness of the material, the loading conditions, and the size and shape of the crack and the component (fig. 2):

a) For brittle materials such as glasses, ceramics or ferritic steels at low temperatures the crack extends in an unstable mode after reaching a critical load. The load-displacement curve is linear up to the critical load.

b) The crack extends in a stable mode, leading to a non-linear load-displacement curve and then suddenly extends through the wall and possibly through the whole component. This behaviour can be observed for ferritic steels in the ductile-brittle transition region, where stable crack extension takes place in a ductile, dimple mode and unstable crack extension in a brittle, cleavage mode. This mechanism was discussed in the lecture by Dahl and Dormagen. An example in the case of a surface crack in a plate under tension loading is shown in fig. 3.

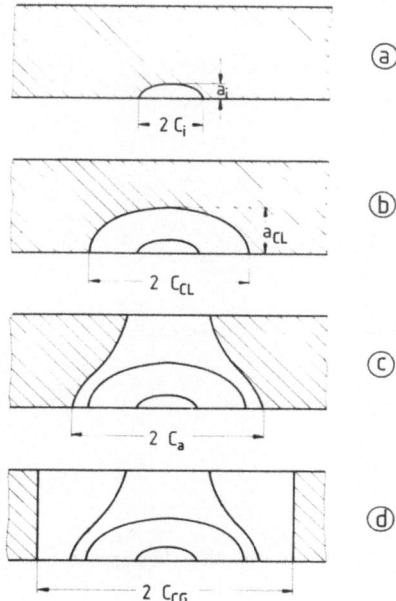

Figure 1. Different steps in crack extension: a) initial crack,
b) fatigue crack extension, c) wall penetration, d) fatigue extension
until global instability

c) The crack extends through the wall in a stable mode. In a displacement-controlled deformation the load for wall penetration (L in
fig. 2) may occur before, at or beyond the maximum load. If it occurs
beyond the maximum load, a behaviour according to b) is observed under
load-controlled deformation.

d) For tough materials necking similar to that in a tensile specimen
may occur after plastic deformation of the ligament. Again, depending
on the strain hardening capacity of the material, wall penetration may
occur before, at or beyond maximum load.

In case a) the plastic zone at the crack tip is normally small and
linear-elastic fracture mechanics can be applied. In cases b) and c) the
plastic zone sizes may also be small and the crack extension can be described by a K_R-Δa crack-growth resistance curve. However, it is also
possible that the ligament has fully yielded before the onset of crack
extension. Then elasto-plastic fracture-mechanics methods have to be
applied. In case d) the net-section collapse or limit load criteria are
used to obtain the critical load.

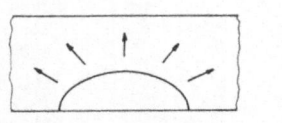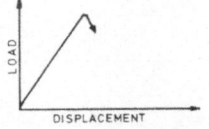

a) brittle (unstable) crack extension

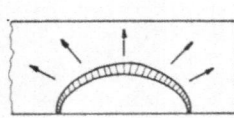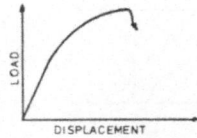

b) stable crack extension before unstable
   crack extension

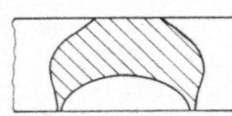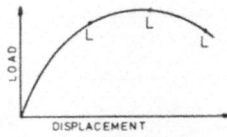

c) stable crack extension through the ligament

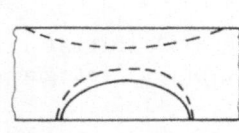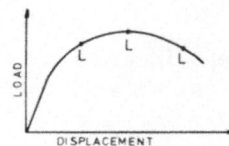

d) ligament yielding and necking

Figure 2.   Different fracture modes under increasing load

3.   LINEAR-ELASTIC FRACTURE MECHANICS

3.1 Local and Averaged K

In the range of applicability of linear-elastic fracture mechanics
(LEFM) fracture occurs if the applied stress intensity factor K_{Iappl}
exceeds the material resistance K_{Ic}. For surface cracks K_{Iappl}
varies along the crack front. Due to the rapid development of numerical
methods in recent years results for a great variety of crack geometries
and stress distributions are now available. Methods used include finite
element[2-4], boundary-integral equation[5], line-spring[6,7] and alternating
method[8].
   For semi-elliptical surface cracks in plates under load in tension
or bending and in pipes under internal pressure Newman and Raju[4,9] developed closed-form equations for $K(a,c,\varphi)$ by curve fitting finite element results. These results are restricted to $0.2 < a/c < 1$ and $0 < a/t$

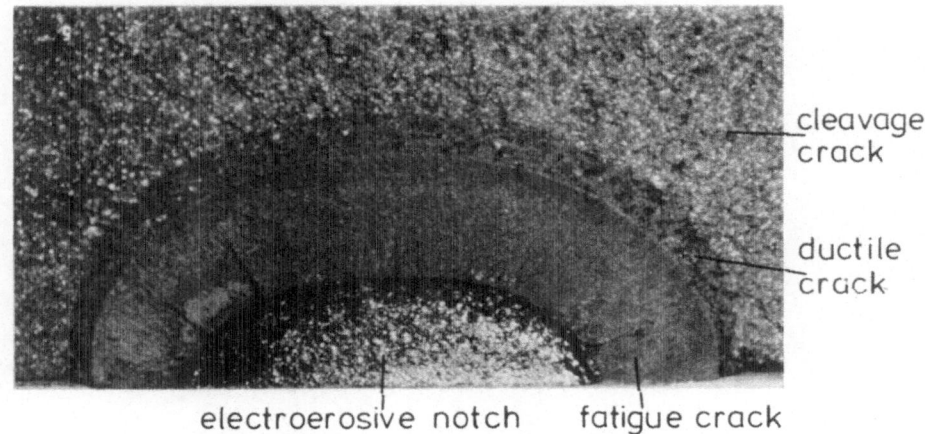

cleavage
crack

ductile
crack

electroerosive notch   fatigue crack

Figure 3. Ductile and brittle fracture under increasing loads, starting
from a fatigue crack (ferritic steel)

< 0.8, Huget et al.[10] recently extended these calculations to a/t up to
0.95 and for a/c < 0.2 using the line spring method. The results usually
are presented in the form

$$K = \frac{\sigma \sqrt{\pi\ a}}{\phi}\ M(a/c,\ a/t,\ \varphi) \tag{1}$$

The definition of the angle φ is given in fig. 6. ϕ is a function of
a/c in form of an elliptical integral, which can be approximated by

$$\phi = \left[1 + 1.464\ (a/c)^{1.65} \right]^{1/2} \tag{2}$$

for a/c ≤ 1. Some results are shown in figs. 4 and 5. Depending on the
size and shape of the crack, the maximum in K occurs at the apex or at
the surface of the crack. For very shallow cracks Huget et al.[10] found
a maximum in K between $\varphi = 0$ and $\varphi = \pi/2$.

For brittle fracture the question arises whether unstable fracture
occurs if the maximum K-value applied along the crack front reaches K_{Ic}.
It is also conceivable that a K-value averaged in some way along the
crack front has to be used as K_{Iappl} taking into account the smaller
K-values near the point of maximum K. Cruse and Besuner[11] proposed the
use of weighted averaged K-values in fatigue-crack growth evaluation.
For the deepest point A and the surface point B (fig. 6) the averaged
K-values are calculated according to

$$\overline{K}_A^2 = \frac{1}{\Delta S_A} \int_{\Delta S_A} K^2(\varphi)\ d\ \left| \Delta S_A(\varphi) \right| \tag{3}$$

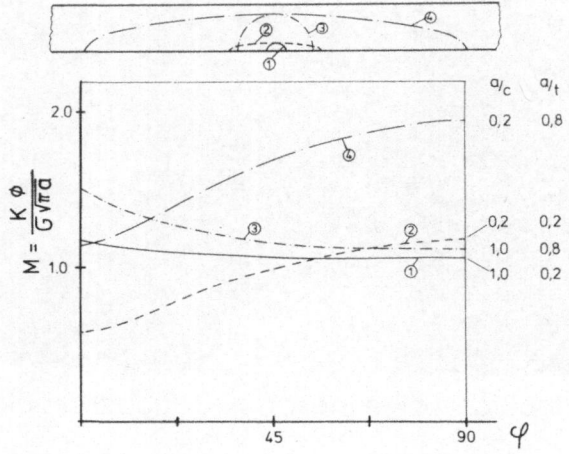

Figure 4. Stress intensity factor along crack front for surface cracks in tension (Newman, Raju[3])

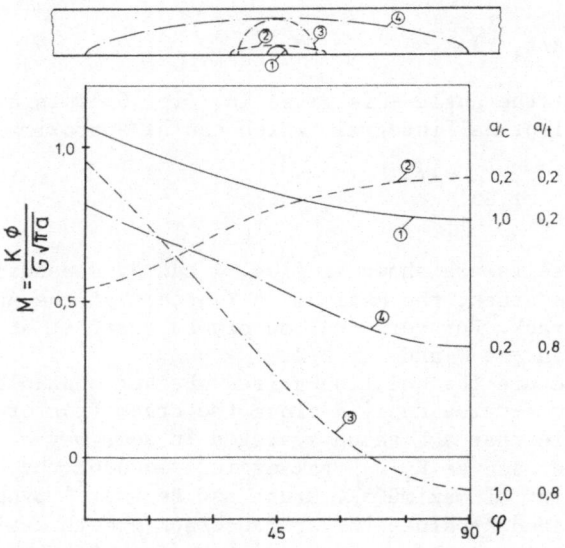

Figure 5. Stress intensity factor along crack front for surface cracks in bending (Newman, Raju[3])

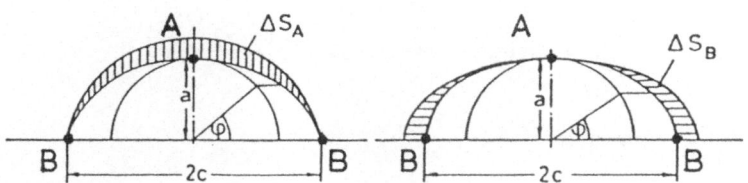

Figure 6. Incremental changes in crack area for the calculation of averaged $\overline{K}_A$ and $\overline{K}_B$

$$\overline{K}_B^2 = \frac{1}{\Delta S_B} \int_{\Delta S_B} K^2(\varphi) \, d \, |\Delta S_B(\varphi)| \tag{4}$$

The integration area is obtained by incremental changes in the crack area growing only in depth and length, respectively (fig. 6).

3.2 K for Stress Gradients

With respect to the distribution of stress in the x-direction results obtained with finite-element and boundary-integral equation methods are available for discrete values of a/c and a/t and polynomial expressions of stress distribution up to the 4th power[4,5,12]. For arbitrary stress distributions the weight-function method is a very useful method of calculating stress-intensity factors. The principle of this method was developed by Bueckner[13] and Rice[14]. The application to three-dimensional problems, already outlined by Bueckner and Rice, is described by Labbens et al.[15].

A rather inexpensive calculation procedure based on weight functions was developed by Mattheck et al.[16,17]. This method will be briefly described.

Calculation of the stress-intensity factor for an arbitrary stress distribution $\sigma(x)$ in the uncracked component along the prospective crack plane calls for the knowledge of the stress-intensity factor K_r and the crack-opening displacement $u_r(x,a)$ of one loading case, for instance a constant stress. Then in case of the stress distribution $\sigma(x)$ for a 2D-problem K is given by

$$K = \frac{H}{K_r} \int_0^a \sigma(x) \frac{\partial u_r(a,x)}{\partial a} \, dx \tag{5}$$

with $H = E/(1-\nu^2)$. Whilst for many configurations at least for one load case a $K_r(a)$ relation can be found, the corresponding crack-opening displacement $u_r(a,x)$ is very often not available. Petroski and Achenbach[18] proposed an approximate relation for $u_r(a,x)$ taking the form

$$u_r(a,x) = \frac{\sigma_o}{H\sqrt{2}} \left[4F(a) \sqrt{a} \ \sqrt{a-x} + G(a) \frac{(a-x)^{3/2}}{\sqrt{a}} \right] \tag{6}$$

where K_r is given by

$$K_r = \sigma_o \sqrt{\pi a} \ F(a) \tag{7}$$

σ_o is a stress parameter of the stress field $\sigma_r(x)$ for the reference problem. Very often a constant stress $\sigma_r(x) = \sigma_o$ is used as a reference stress field. The first term in eq. (6) is the near-crack-tip opening displacement, which is not dependent on the special configuration. The function $G(a)$ can be determined by setting $K = K_r$ and $\sigma(x) = \sigma_r(x)$ in eq. (5). This leads to

$$G(a) = \left[I_1(a) - 4 \ F(a) \ \sqrt{a} \ I_2(a) \right] \sqrt{a}/(I_3(a) \tag{8}$$

with

$$I_1(a) = \pi \ \sqrt{2} \ \sigma_o \int_o^a F(a)^2 a da \tag{9}$$

$$I_2(a) = \int_o^a \sigma_r(x) \sqrt{a-x} \ dx \tag{10}$$

$$I_3(a) = \int_o^a \sigma_r(x) \ (a-x)^{3/2} \ dx \tag{11}$$

If $\sigma_r(x) = \sigma_o$ = constant is chosen for the reference stress field eq. (8) reduces to

$$G(a) = \frac{2 \ \sqrt{2\pi}}{5 \ a^2} \int_o^a F(a) a da - \frac{20}{3} F(a) \tag{12}$$

For the 3D-problem of semi-elliptical surface cracks averaged K-values at the points A and B according to Cruse and Besuner[11] are given by

$$\overline{K_A} = \frac{H}{\overline{K}_{rA}} \int \sigma(x,y) \frac{\partial u_r(x,y,a,c)}{\partial S_A} \ dS \tag{13a}$$

$$\overline{K}_B = \frac{H}{\overline{K}_{rB}} \int \sigma(x,y) \frac{\partial u_r(x,y,a,c)}{\partial S_B} \, dS \qquad (13b)$$

where $\overline{K}_{rA}$ and $\overline{K}_{rB}$ are the averaged K-values of the reference load and $u_r(x,y,a,c)$ is the two-dimensional crack opening displacement field of the reference load. The integration has to be performed over the crack area S.

For a stress distribution symmetric with respect to the x-axis, $\sigma(x,y) = \sigma(x,-y)$, eqs. (13a) and (13b) can be written as

$$\overline{K}_A = \frac{H}{\overline{K}_{rA}} \frac{4}{\pi a} \int_{y=0}^{c} \int_{x=0}^{x(y)} \sigma(x,y) \frac{\partial u_r}{\partial a} \, dx \, dy \qquad (14a)$$

$$\overline{K}_B = \frac{H}{\overline{K}_{rB}} \frac{4}{\pi c} \int_{y=0}^{c} \int_{x=0}^{x(y)} \sigma(x,y) \frac{\partial u_r}{\partial c} \, dx \, dy \qquad (14b)$$

The reference crack-opening displacement field for the constant stress σ_o is calculated in the following steps (s. fig. 7):

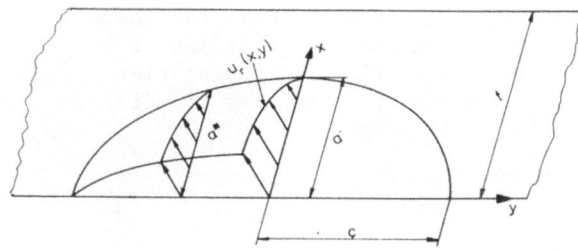

Figure 7. Crack opening displacement of a surface crack

a) First the crack-opening displacement $u(x,y = 0)$ along the axis $y = 0$ is calculated. For this purpose, the semi-elliptical crack is assumed to behave at its deepest point like an edge crack. However, for this edge crack the stress intensity factor at the deepest point of the semi-elliptical crack is used as a reference K_r. Then the method of Petroski and Achenbach (eq. 6) is applied to get $u(x,y = 0)$.

b) For crack-opening displacement at the surface $u(y,x = 0)$ a relation similar to that for a through-the wall crack is assumed to be:

$$u(y,x=0) = u(x=0, y=0) \sqrt{1- (\tfrac{y}{c})^2} \qquad (15)$$

$u(x = 0, y = 0)$ is obtained from the first step.

c) In the third step the crack-opening displacement $u(x)$ is calculated for a fixed value of y, with $0 < y < c$. Again eq. (6) is used with a crack depth

$$a^x = a \sqrt{1- (\tfrac{y}{c})^2} \qquad (16)$$

For $F(a)$ in eq. (6) the value of the corresponding y and Γ, respectively has to be used. The unknown $G(a)$ is not determined by eq. (8). $G(a)$ is defined by the "surface condition", eq. (15). Combined with eq. (6) it gives

$$G(a) = \frac{u(x=0, \; y=0) \; H\sqrt{2}}{a\sigma_o} - 4F(a) \qquad (17)$$

This procedure allows to calculate the complete crack-opening displacement $u_r(x,y)$ of the reference load. Then eqs. (14) can be used to calculate $\overline{K}_A^r$ and $\overline{K}_B$ for any symmetrical stress distribution $\sigma(x,y) = \sigma(x,-y)$.

The quality of the results was checked by comparison with finite-element results of Yagawa et al.[12] and Heliot et al.[5].

This method was applied for calculating stress-intensity factors under thermal shock conditions[19,20]. For a semi-elliptical crack in a plate shocked on one side $\overline{K}_A$ and $\overline{K}_B$ for a crack on the shocked side are dependent on a/c, a/t, the time after the beginning of the shock, and on some thermo-physical properties. With increasing time K passes through a maximum. In fig. 8 these maxima are shown for two different values of a/c as a function of the crack depth a/t.

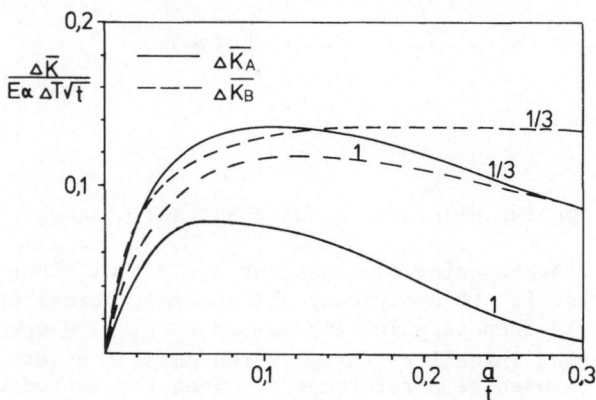

Figure 8. Maximum change in stress intensity factors $\overline{K}_A$ and $\overline{K}_B$ for thermal shock of a plate as a function of crack depth for a/c = 1 and a/c = 1/3

3.3 Plane Stress - Plane Strain

The crack extension under rising or alternating load is not only dependent on the stress-intensity factor but also on the stress state. In
sufficiently thick plane specimens with a through-the-plate crack the
center region is under plane strain, with the stress component σ_{zz} along
the crack front given by

$$\sigma_{zz} = \nu(\sigma_{xx} + \sigma_{yy}),$$

whereas at the surface there is a plane stress state with σ_{zz} = 0. With
increasing distance from the crack tip there is a continuous transition
from plane strain to plane stress also in the central part of a plate.
This is shown in fig. 9 for a center crack.

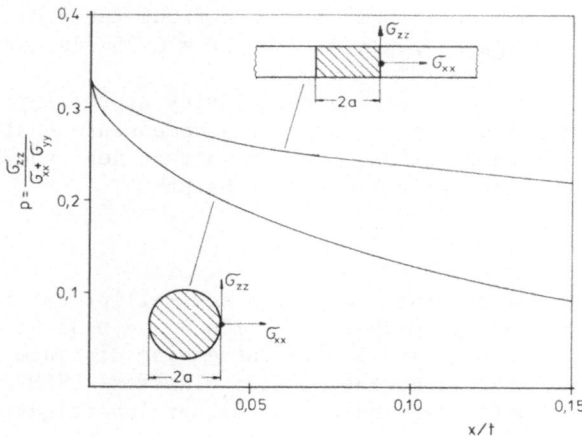

Figure 9. Change in constraint factor p with the distance from the
crack tip for a center crack in a plate and a circular crack (Sneddon[21])

For a circular (penny-shaped) crack in an infinite body there is
also plane strain at the crack tip, however, the decrease in the con-
straint $p = \sigma_{zz}/(\sigma_{xx} + \sigma_{yy})$ with increasing distance is much larger than
for the central through-the-plate crack (fig. 9). This difference may
affect the fracture behaviour; however, it is not clear to which extent.
For a semi-elliptical surface crack a similar effect is expected. In
addition, there is a transition from plane strain to plane stress at
the surface. Finite-element results of de Lorenzi[22] are shown in fig. 10
for a distance of 2% of the remaining ligament t - a ahead of the crack
front. There is a near-plane strain situation along the largest part of
the crack front and a steep decrease of σ_{zz} near the surface.
In case of deep cracks approaching the rear face it can be expected
that also for $\varphi = \frac{\pi}{2}$ a near-plane stress state exists. In this case,

plane strain can be expected in a limited region near $\varphi = \frac{\pi}{4}$ only.

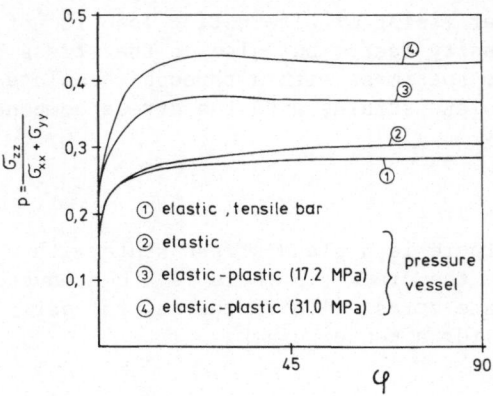

Figure 10. Constraint factor p along crack front (at a distance of 2% of the remaining ligament) a/c = 1/3, a/t = 0.25 (de Lorenzi[22])

A special problem is the stress singularity at the intersection of the crack front with the free boundary of a component. Analytical[23,24] and experimental[25] investigations have shown that near the boundary the inverse square root singularity no longer exists:

$$\sigma_{ij} \sim \frac{1}{r^{\lambda}}$$

where λ depends on Poisson ratio ν. For a semi-elliptical flaw with a/c = 0.76 and a/t = 0.525 Smith et al[25] found $\lambda = 0.33$ at the surface and a gradual increase to $\lambda = 0.5$ with increasing distance from the surface. The effect of this deviation from classical LEFM on the crack growth behaviour is not clear and needs further investigations.

4. EXTENSION OF SURFACE CRACKS BY FATIGUE

The crack extension during fatigue of a plane specimen with a crack of length a is described by the relation between the crack extension per load cycle and the range of the stress-intensity factor ΔK. Due to the varying ΔK along the crack front of a surface crack, the extension per cycle of the crack also varies along the crack front. Therefore, during continuous cycling the shape of the crack changes. To predict the crack growth rate and the number of cycles until wall penetration this change in the geometry of a surface crack has to be calculated. This can be performed in the following steps:

a) For a starting crack characterized by a_i and c_i or a_i/c_i and a_i/t the ranges of stress-intensity factors at the deepest point ΔK_A^i and at the surface ΔK_B are calculated. These can be the local values ΔK_A (a_i/c_i, a_i/t, $\varphi = \frac{\pi}{2}$) and ΔK_B(a_i/c_i, a_i/t, $\varphi = 0$) or the averaged values $\Delta \bar{K}_A$ and $\Delta \bar{K}_B$.

b) The crack extension during one cycle is calculated in the directions of length and depth

$$\Delta a = f(\Delta K_A), \quad \Delta c = f(\Delta K_B).$$

The relation $f(\Delta K)$ may for instance be the Paris-equation $f = C(\Delta K)^n$.
   c) New crack axes are calculated
      $a = a_i + \Delta a, \quad c = a_i + \Delta c$
   d) New K_A- and K_B-values are calculated using the new crack geometry and assuming that the shape of the crack stays semi-elliptical. By repetition of this procedure the continuous change of the crack front can be calculated.

Some examples are shown in figs. 11 and 12 for alternating tension and bending[26]. There is an effect of the exponent n and the use of local or averaged ΔK-values.

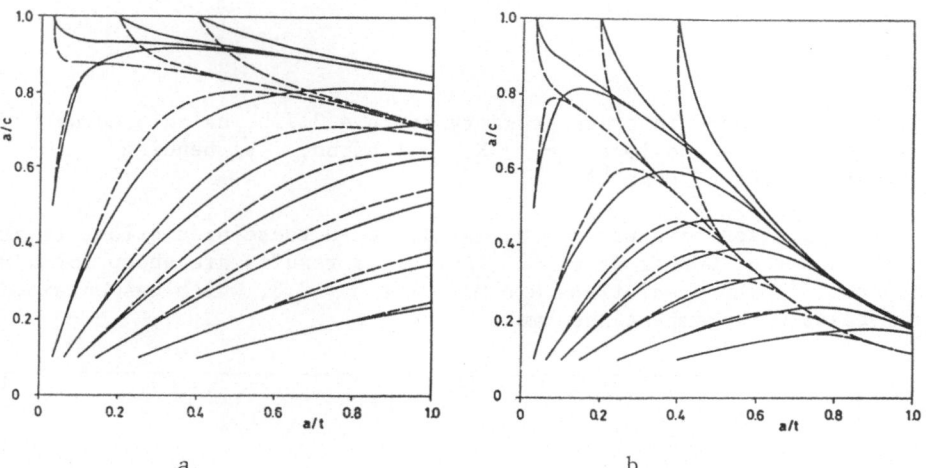

Figure 11. Change in crack geometry for exponent of Paris law n = 2 (———) and n = 4(–––) a) tension,   b) bending (Görner, Mattheck, Munz[26])

The general trend of alternating tension loading is:
-   Semi-circular cracks grow to semi-elliptical cracks reaching a limiting a/c in the range of 0.7 to 0.8.
-   Semi-elliptical cracks with small a/c values grow in the direction of semi-circular cracks, reaching a maximum possible final a/c ratio of 0.7 to 0.8. Long and deep cracks (a/c small, a/t large) do not have the chance of reaching this limiting value before penetrating the plate.

The general trend of loading in the alternating bending mode is:
-   Semi-circular cracks grow to long semi-elliptical cracks with a limiting a/c in the range of 0.1 – 0.2.
-   Semi-elliptical cracks with small a/c-values first grow in the direction of a semi-circular crack. After reaching a maximum in the a/c ratio they are growing faster in length than in depth.

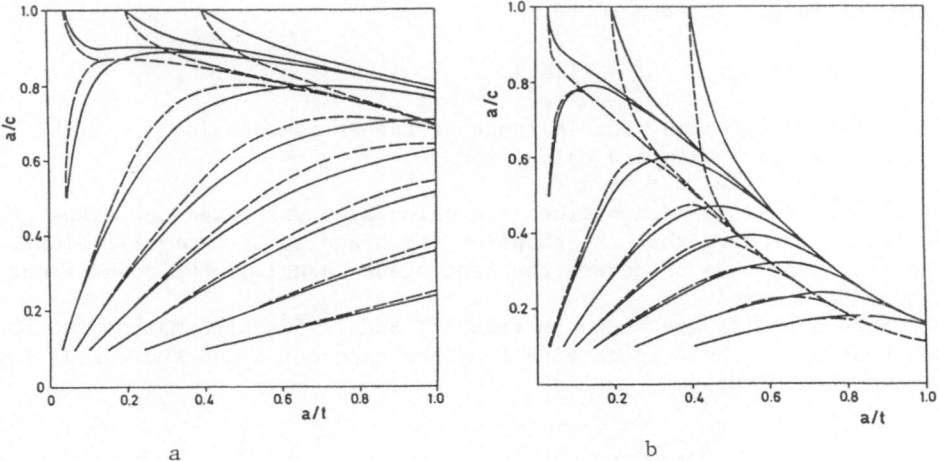

a

b

Figure 12. Change in crack geometry for n = 3.726, calculated with averaged (———) and local (---) K, a) tension, b) bending (Görner, Mattheck, Munz[26])

A comparison of experimental results with these predictions is given in the paper of Görner et al.[26]. In fig. 13 results are shown for a reactor steel[27]. The predictions are made for n = 3.3, which was obtained from tests with compact specimens.

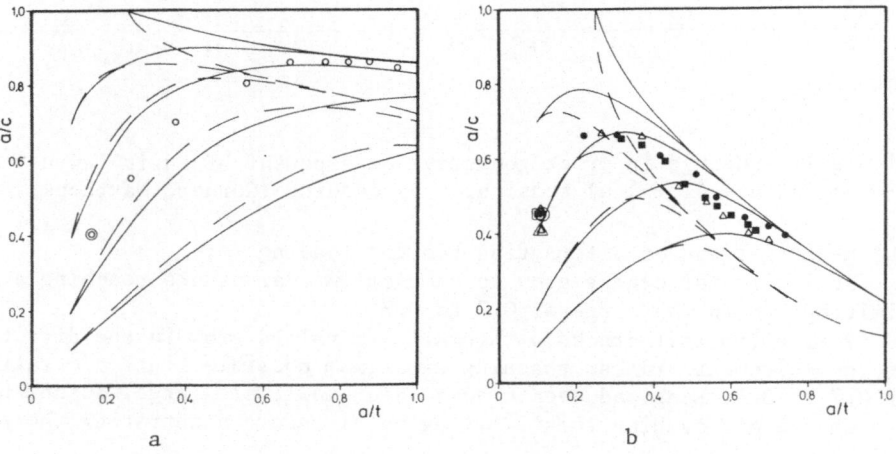

a

b

Figure 13. Comparison of experimental observed and predicted (——— averaged, --- local ΔK) change in a/c for cracks in plates for a reactor steel a) tension b) bending (Müller et al.[27])

5. PLASTIC DEFORMATION

5.1 General

Under increasing load a plastic zone develops ahead of the crack tip.
The size of this zone at fracture is dependent on the toughness of the
material, but also to some extent on the size of the crack and the component. Three stages of plastic yielding are shown in fig. 14[28]:
  a) Small-scale yielding.
  b) Ligament yielding, contained in an unyielded region. Due to the
surrounding elastic region, the displacement in the plastic region is
limited.
  c) Fully yielded cross-section with the possibility of large deformation.

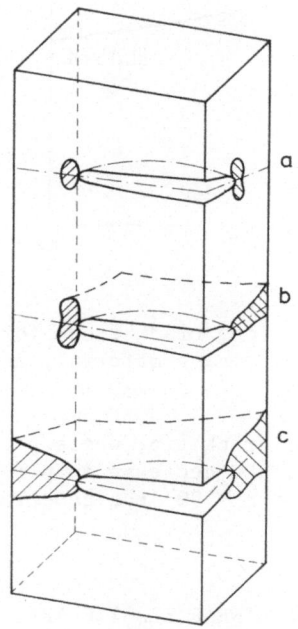

Figure 14. Different stages of plastic yielding (after Pellissier-
Tanon[28])

If fracture occurs at stage a), LEFM can be applied provided that the
plastic zone is small enough. For plastic-zone sizes approaching stage
b), the fracture has to be described using elasto-plastic fracture mechanics methods, as crack-opening displacement or J-Integral method. If
fracture occurs at stage c), elasto-plastic methods may still be applicable. For large deformation (fig. 2d) plastic collapse criteria have
to be applied.

5.2 Plane Stress – Plane Strain

Whilst for the elastic behaviour the plane-strain constraint $p = \sigma_{zz}/(\sigma_{xx} + \sigma_{yy})$ is equal to the Poisson's ratio ν, it is equal to 0.5 for a perfectly plastic material. Therefore, it is expected that for increasing load p increases as well. This is shown in fig. 10 and also in fig. 15 for a semi-circular crack[29]. The curves in fig. 15 are for a distance of

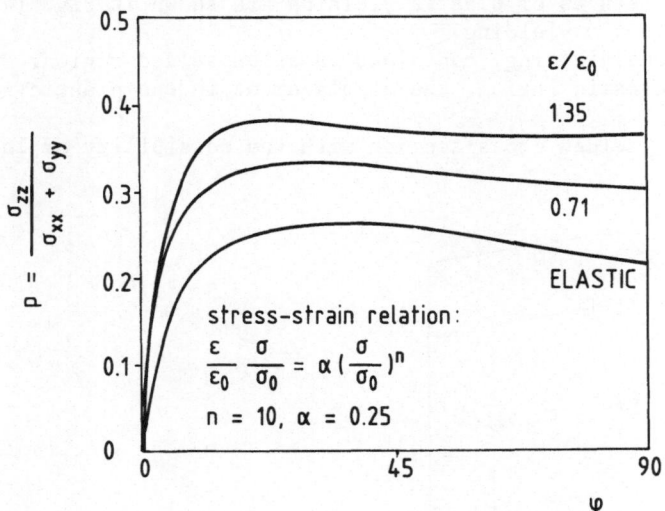

Figure 15. Change in constraint factor along crack front for a semi-circular crack in a plate for a strain hardening material, and different applied strain (Trantina et al[29])

2.5% of the crack length in front of the crack tip, where p is slightly below the value at the crack tip itself. The decrease in p to the free surface starts at about $\varphi = 15°$. For $15 < \varphi < 90°$ the constraint is nearly constant.

6. PLASTIC-ZONE CORRECTION AND LIMIT OF LINEAR ELASTIC FRACTURE MECHANICS

The application of LEFM is restricted to a plastic zone sufficiently small compared to the remaining ligament and the crack size. The range of application of LEFM can be extended by applying Irwin's plastic zone correction, with the crack depth replaced by

$$a^*_{eff} = a + r_{pl} \tag{18}$$

r_{pl} is half of the size $\omega(=$ "radius") of the plastic zone, which is given approximately by

$$\omega = \frac{1}{\alpha\pi} \cdot \left(\frac{K}{\sigma_y}\right)^2 \tag{19}$$

Normally, $\alpha = 3$ is used for plane strain and $\alpha = 1$ for plane stress.

For surface cracks the stress-intensity factor very often is written in the form of eq. (1):

$$K = \frac{\sigma \sqrt{\pi a}}{\phi} M(a/c, \quad a/t, \quad \varphi) \tag{1}$$

If the plastic-zone correction is applied only to $\sqrt{a}$ and not to M, a_{eff} is given for plane strain by

$$a_{eff} = a \left[1 + \frac{M^2}{6\phi^2} \left(\frac{\sigma}{\sigma_y}\right)^2 \right] \tag{20}$$

Then eq. (1) can be rewritten as

$$K = \frac{\sigma \sqrt{\pi a}}{\sqrt{Q}} M \tag{21}$$

with

$$Q = \phi^2 - \frac{M^2}{6} \left(\frac{\sigma}{\sigma_y}\right)^2 \tag{22}$$

Values of $M^2/6$ for the deepest point of a surface crack in a plate exposed to tension loading are given in table 1. For $0.4 < a/c < 1$ and

TABLE I $M^2/6$ for semi-elliptical surface cracks in a plate under tension

a/t \ a/c	1	0.8	0.6	0.4
0	0.180	0.187	0.193	0.199
.2	0.183	0.191	0.201	0.213
.4	0.191	0.205	0.224	0.254
.6	0.201	0.224	0.258	0.315

$0 < a/t < 0.6$ the values of $M^2/6$ are between 0.18 and 0.32. Instead of eq. (22), very often the simpler relation[30]

$$Q = \phi^2 - 0.212 \left(\frac{\sigma}{\sigma_y}\right)^2 \tag{23}$$

is used. The factor 0.212 results from $\alpha = 2\sqrt{2}/\pi$ and $M = \sqrt{1.2/\pi}$.

If one applies eq. (22) or (23), the failure stress for a material with the fracture toughness K_{Ic} is given by

$$\sigma_{fracture} = \frac{K_{Ic} \phi}{\sqrt{t} M \left[\pi a/t + \frac{1}{6t} (K_{Ic}/\sigma_y)^2 \right]^{1/2}} \tag{24}$$

$$\sigma_{fracture} = \frac{K_{Ic} \phi}{\sqrt{t} M \left[\pi \; a/t + \frac{0.212}{M^2 t}(K_{Ic}/\sigma_y)^2\right]^{1/2}} \tag{25}$$

For a given relative crack depth a/t the plastic-zone correction is dependent on the dimensionless quantity

$$A = \frac{1}{t} \left[\frac{K_{Ic}}{\sigma_y}\right]^2 \tag{26}$$

Therefore, plastic-zone correction increases for a given relative crack depth with increasing fracture toughness, decreasing yield strength and decreasing size of the component.

The limit of applicability of LEFM - including plastic-zone correction - usually is also given in terms of $(K_{Ic}/\sigma_y)^2$. The minimum ligament size $t - a$ for which LEFM can be applied is expressed as

$$t - a = \beta(K_{Ic}/\sigma_y)^2 = At\beta \tag{27}$$

In the ASTM standard for determination of fracture a value of $\beta = 2.5$ is indicated. For compact specimens a rather smaller value of about 0.4 and for center-cracked tension specimens a value of about 1 were found[31,32]. Using $\beta = 1$ the limiting crack depth for the applicability of LEFM would be

$$\frac{a}{t} = 1 - A \tag{28}$$

The minimum crack size for which LEFM can be applied is less clear. This problem which is known as the "short crack problem" will not be discussed here.

For surface cracks usually eqs. (23) and (25) are proposed up to $(\sigma/\sigma_y)^2 = 1$; at least correction-factor diagrams are plotted up to this value[1]. This is far beyond $\beta = 1$ or $\beta = 0.4$. If the application of eq. (24) is restricted to eq. (27) with $\beta = 1$, the maximum plastic-zone corrections for the failure stress within the limits of applicability are 10% for $a/t = 0.2$ and 2% for $a/t = 0.6$ in case of a semi-circular crack.

For steep stress gradients as in cracks emanating from notches or for thermal-shock loading plastic deformation is restricted to a small area near the crack. Then LEFM may be applicable beyond the relation given by eq. (28). This was pointed out by Sumpter and Turner[33].

7. THE COD CRITERION

One of the elastic-plastic fracture mechanics criteria first proposed was crack-tip opening displacement. It was assumed that independent of the relative size of the plastic zone with respect to the geometry of the component, the crack will start to extend if a critical plastic deformation at the crack tip is reached. This plastic deformation is accompanied by crack-tip blunting or crack-tip opening displacement CTOD or δ. The critical CTOD, which can be measured on suitable specimens, is for this reason a candidate material property in elasto-plastic fracture mechanics. Very often after the onset of crack extension stable crack growth is observed under rising load - or in displacement control also under decreasing load.

The different critical values of δ (δ_i, δ_m, δ_c, δ_u, δ_{tip}) are presented in the lecture of Garwood. It should be kept in mind that δ_i at the onset of stable crack extension and δ_c at the onset of cleavage should be material parameters, whereas δ_m and possibly also δ_u are dependent on the type of the specimen and on the loading systems. Therefore the predictions of the behaviour of surface cracks from results of 2D-specimens is restricted to δ_i and δ_c.

Finite-element calculations[22,29,50] have shown that δ varies along the crack front in a way similar to that of the stress intensity factor. For cracks with a/c < 1 the maximum δ occurs in the center of the crack at $\varphi = \frac{\pi}{2}$ (fig. 16). Therefore, for an assessment of surface flaws using

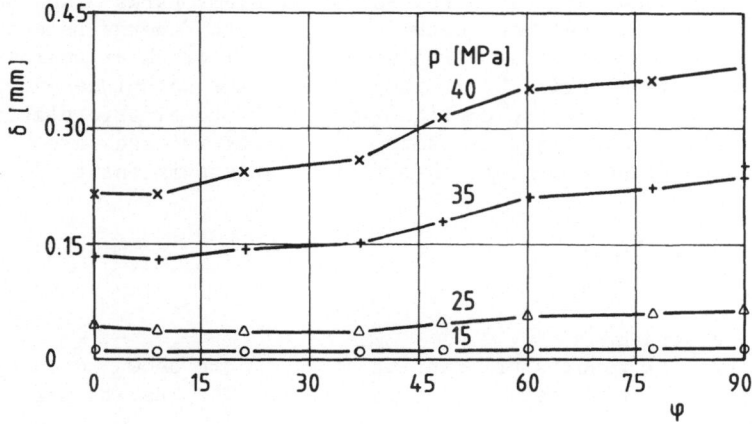

Figure 16. Crack tip opening displacement along the crack front (longitudinal surface crack in a pressure vessel, a/t = 0.59, a/c = 0.35, Brocks et al.[50])

the COD-criterion it is necessary to find relations between δ at $\varphi = \pi/2$, the crack parameters a/c and a/t, and the external load. Erdogan and co-workers[34,35] calculated these relations for rectangular cracks in plates and pipes using the principles of the Dugdale model taking into account the local bending of the ligament. The results are given as COD-stress-curves, where COD is the value at the crack tip or at the crack mouth. The crack tip results can be used to predict the onset of ductile tearing using δ_i, measured with bend or compact specimens.

For the onset of cleavage crack extension not preceded by stable crack extension δ_c-values can be used. However, it should be ensured in case of cleavage crack extension that the stress condition for the surface crack is the same as for the compact specimen. It was shown that cleavage crack extension is much more affected by a deviation from plane strain than ductile tearing[36-38].

For stable crack extension the situation is more complicated. Because of the contained yielding near the surface crack (fig. 2b) it is not possible to transorm in a simple way δ_m from a bend specimen to a surface crack. Therefore, a prediction of local failure - the penetration of the crack through the wall - is difficult with the COD-method.

The COD criterion is used for assessment of through-the-thickness and surface cracks in general form of a COD design curve. This prodedure is explained in the workshop by Turner.

8. THE J-INTEGRAL METHOD

The definitions and limits of applicability of the J-integral were discussed in the lecture by Bakker. The problems of measuring initiation values and J_R-curves were treated by Garwood and Blauel. Only one result which may be of interest for predicting the extension of surface cracks will be mentioned briefly here. For 3D-problems Parks[39] and de Lorenzi[40] developed a finite-element method to calculate "energy-release rates" G^*, for elastic-plastic deformation using virtual crack extensions. For nonlinear elastic material G^* is identical to the path-independent J-integral. The problem of the applicability of G^* to elastic-plastic deformation will not be discussed here. The energy-release rate G^* or J-integral can be transformed into an effective stress-intensity factor by

$$K^* = \left| \frac{E\ G^*}{1-\nu^2} \right|^{1/2} \tag{29}$$

for plane strain.

In fig. 17 results of Trantina et al.[29] are shown for a semi-circular surface crack in a plate loaded in tension. The results are given in terms of

$$F = \frac{K^*\ \pi}{2\sigma\sqrt{\pi a}} \tag{30}$$

The load parameter is the strain in the uncracked plate. In the elastic case the maximum in K occurs at the surface. In the elastic-plastic range a maximum in K at about $\varphi = 15^o$ is calculated. Under the assumption that G^* is the relevant parameter for crack extension, predictions about the change in geometry under increasing loads can be made based on these results, possibly including the effect of triaxiality shown in fig. 15.

These results show that the application of elastic-plastic methods for the prediction of failure loads is very complex and needs detailed information about the applied J or G^* and the material response. Similar to the COD design curve a very simple J-integral design curve was proposed, which does not require any complicated analysis[41]. This curve is described in the workshop by Turner.

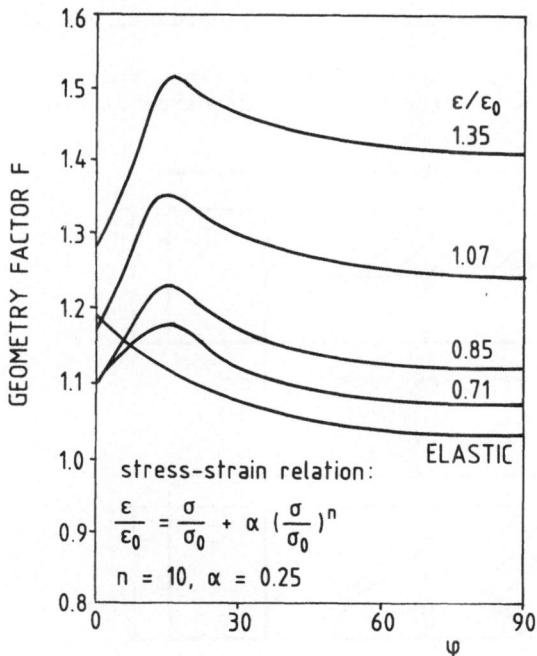

Figure 17. Effective stress intensity factor for a semi circular crack for different applied strains (Trantina et al.[29])

9. THE PLASTIC COLLAPSE

Failure by plastic collapse or net ligament instability is a very important failure mode. It can be expected for tough materials, especially for small components. The knowledge of the plastic collapse load is also important regarding the application of the failure-assessment diagram (or two-criteria method) of CEGB[42], see workshop by Dowling et al.

For surface cracks a variety of mostly simple relations and calculation procedures are available. For a better understanding of the following discussion some features of the CEGB method will be briefly summarized.

For through-the-wall cracks the net-section yield criterion was successfully applied in case of tough material, such as austenitic stainless steels[43]. For an ideal plastic material it is assumed that failure occurs if the net section has fully yielded. For superimposed tension and bending load yielding in tension and compression occur and the critical load is obtained by the requirement of equilibrium of force and moment. For strainhardening materials the same equation is used as for an ideal plastic material with a flow stress between the yield strength and the ultimate tensile strength, for instance

$$\sigma_f = \frac{\sigma_y + \sigma_{uTS}}{2} \qquad (31)$$

The approach is similar for surface cracks, however, not necessarily the whole cross section must have yielded.

In fig. 18 the approach proposed by Harrison et al.[42] is shown, where in a somewhat arbitrary way the sideward extension of the plastic region

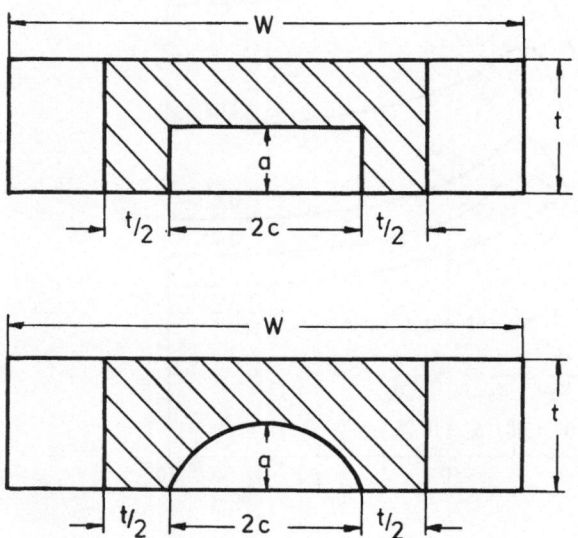

Figure 18. Plastic region at limit load for surface cracks (Harrison et al.[42])

is half of the plate thickness. It is then assumed that the flow stress multiplied by the corresponding area is equal to the applied stress multiplied by the corresponding uncracked area. This leads to

$$\sigma = \sigma_f \left[1 - \frac{\pi \ ac}{2(2c+t)t} \right] \qquad (32)$$

for a semi-elliptical crack, and to

$$\sigma = \sigma_f \left[1 - \frac{2 \ ac}{(2c+t)t} \right] \qquad (33)$$

for a rectangular crack.

In these relations the ligament failure stress is independent of the width W of the specimen.

A different relation was proposed by Chell[44] for semi-elliptical cracks

$$\sigma' = \sigma_f \left\{ 1 - \frac{a}{t} \ \frac{1 - \left[1 + 2(c/t)^2\right]^{-1/2}}{1 - \frac{a}{t} \left[1 + 2(c/t)^2\right]^{-1/2}} \right\} \qquad (34)$$

which is also not dependent on the width of the specimen.

Another relation based on a Dugdale model was proposed by Mattheck et al.[45].

For circumferential cracks in pipes under superimposed bending and tension a possible stress distribution at ligament failure is shown in fig. 19. It is assumed that the ligament has yielded and that in the remaining cross section the stress varies linearly with the distance from the pipe axis. From the equilibrium of force and moment the following re-

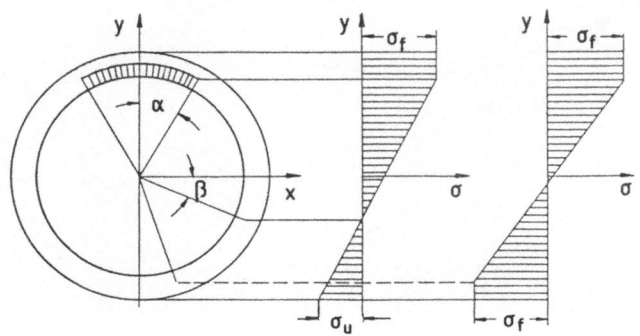

Figure 19. Assumed stress distribution for a pipe with a surface crack under combined tension and bending at local instability

lation is obtained

$$
\frac{\sigma_b}{\sigma_f} = \frac{2}{\pi} \left\{ \frac{\left[\pi(1- \frac{\sigma_m}{\sigma_f}) - \alpha\frac{a}{t} \right] \cdot \left[\sin\alpha - \cos\alpha + \frac{\pi-\alpha}{2} - \frac{\sin2\alpha}{4} \right]}{(\pi-\alpha)\cos\alpha + \sin\alpha} - \frac{a}{t}\sin\alpha \right\} \quad (35)
$$

This is a relation between the crack angle α , the crack depth a/t and the external load parameters σ_b/σ_f and σ_m/σ_f at ligament failure. σ_m is the tensile stress in the uncracked pipe, related to the external load F by

$$
\sigma_m = \frac{F}{2\pi Rt} + p \frac{R}{2t} \quad (36)
$$

σ_b is the elastic outer fiber bending stress in the uncracked pipe, related to the external moment by

$$
\sigma_b = \frac{M}{\pi R^2 t} \quad (37)
$$

In fig. 20 σ_b/σ_f is plotted versus α for σ_m/σ_f = 0.25.

Equation (35) has some drawbacks as all net section yield criteria. For small cracks plastic deformation in the ligament may not cause failure. Therefore, eq. (35) may underestimate the true failure stress. For large cracks failure may occur on the tensile side before yielding of the whole ligament. In this case eq. (35) would overestimate the failure stress.

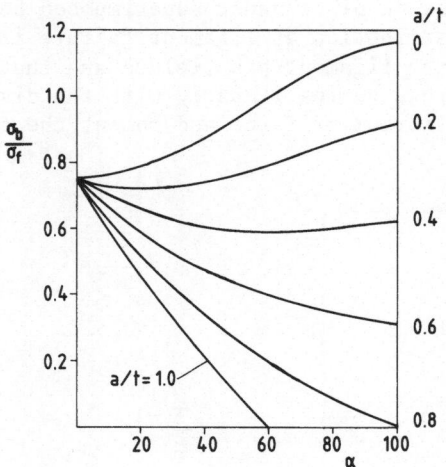

Figure 20. σ_b/σ_f versus α for a pipe with a surface crack under combined tension and bending at local instability

Generally, all relations so far mentioned are rather criteria of plastic yielding than of local failure. It is obvious that in a material with ideal plastic behaviour the ligament cannot fail if the plastic deformation is restricted to a small region surrounding the crack as shown in fig. 18, because the total deformation in the direction of loading is limited by the deformation in the elastic region. If for a strain hardening material the yield strength is replaced by the average flow stress, there will also be plastic deformation in the "elastic" region of fig. 18 or 19 at the calculated local failure stress. Then the model is no longer correct; however, the predicted failure stresses may come closer to the experimentally observed failure stresses. Only experimental results can show the quality of the different relations.

A more realistic model of local instability was developed by Hasegawa et al.[46,47]. The stress distribution assumed for local instability in a plate is shown in fig. 21. It is assumed that the stress at failure in the ligament is equal to the engineering fracture stress σ_{ef} in a tensile test. In the remaining cross-section a constant stress σ_p, dependent on the strain hardening capacity, is assumed. From experimental results

$$\sigma_p = \sigma_{uTS} - (\sigma_{uTS} - \sigma_y) \frac{a}{t} \qquad (38)$$

was found.

For pipes merely exposed to bending with circumferential cracks the stress distribution is shown in fig. 22. From this diagram the critical bending stress σ_b follows:

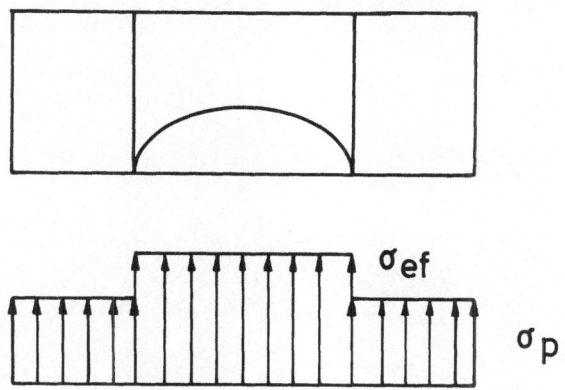

Figure 21. Assumed stress distribution at local instability in a plate under tension (Hasegawa et al[46])

$$\frac{\sigma_b}{\sigma_f} = \frac{\sigma_{ef}}{\sigma_f} \left(1 - \frac{a}{t}\right) \sin\alpha + \frac{\sigma_p}{\sigma_f} \left(2\cos\beta - \sin\alpha\right) \qquad (39)$$

with

$$\beta = \frac{\alpha}{2} - \frac{\alpha(1 - a/t)\sigma_{ef}}{2\sigma_p} \qquad (40)$$

σ_f is the average flow stress (eq. 31).

Another model of local instability including strain hardening was developed by Mattheck and Görner[48]. It is a modified Dugdale-model, with

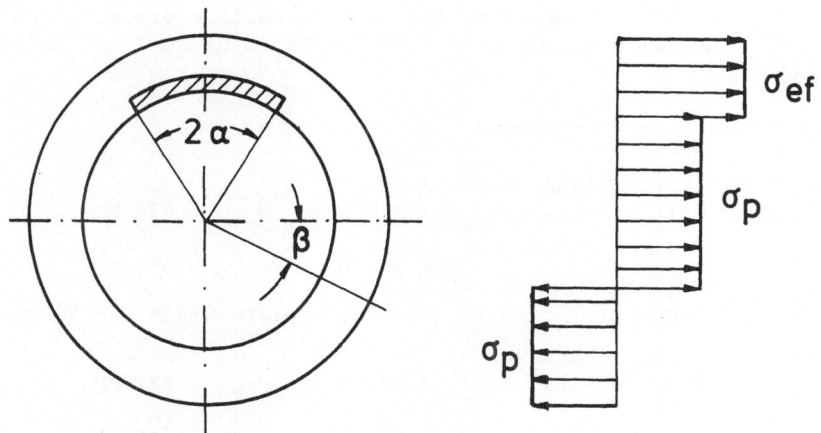

Figure 22. Assumed stress distribution at local instability for a pipe in bending (Hasegawa et al.[47])

a stress distribution at failure shown in fig. 23.

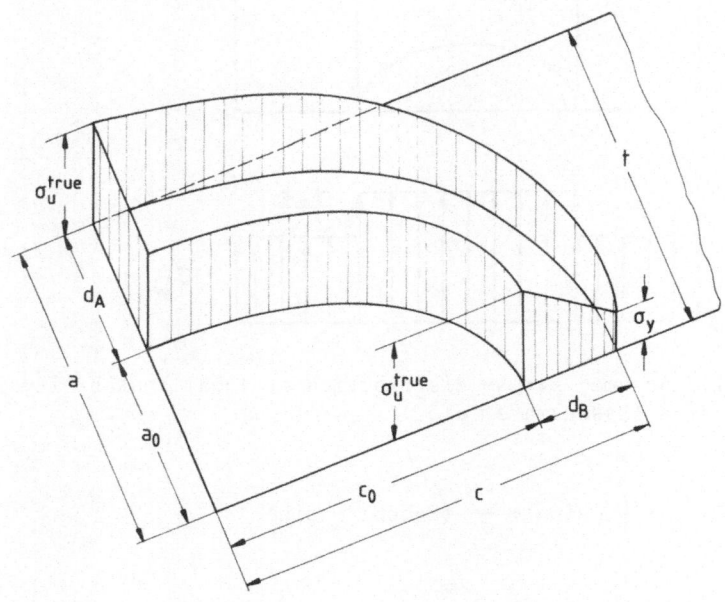

Figure 23. Assumed stress distribution at failure of a surface crack (Mattheck, Görner[48])

10.   SOME EXPERIMENTAL RESULTS

In this section some experimental results will be presented which have been obtained with rectangular plates containing surface cracks loaded by tension[49].

10.1 Materials and Specimens

a) Austenitic stainless steel (X CrNi18 8)
   Tensile test results: σ_y = 310 MPa, σ_{uTS} = 674 MPa

   σ_{ef} = 354 MPa, ψ = 81.7%.
   Plate thickness: t =   7 mm, plate width W = 50 mm.

b) Ferritic steel (90 MnV8)
   Tensile test results: σ_y = 373 MPa, σ_{uTS} = 621 MPa
   σ_{ef} = 534 MPa, ψ = 40.4%.
   Plate thickness: t =   9 mm, plate width W = 80 mm.

Cracks: semi-elliptical surface cracks introduced by fatigue starting from electroerosive notches.

10.2 Load-Displacement Curves

Load-displacement curves for the austenitic steel are shown in fig. 24.

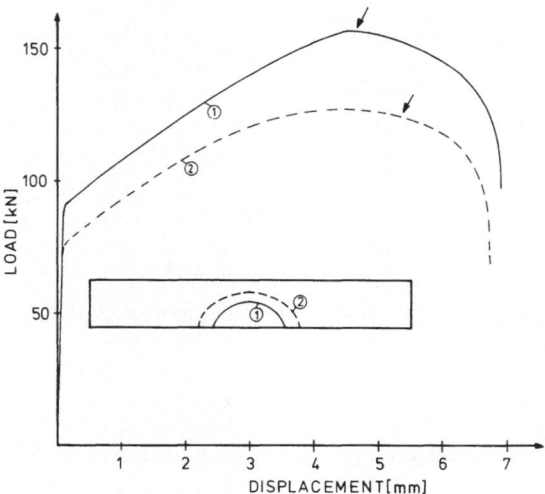

Figure 24.   Load-displacement curves for austenitic stainless steel for specimens with surface cracks

After deviating from the linear part of the curve the load still increases considerably. The crack penetrates the wall at or shortly beyond maximum load (arrows in fig. 24). In the displacement-controlled test the load decreases continuously after having passed the maximum.

For the ferritic steel three types of load displacement curves were observed (fig. 25).

Type I was found for small cracks. After a large total deformation under increasing load the crack starts to extend shortly before maximum load. The crack further extends under decreasing load, penetrates the wall and then extends sidewards.

Type II was observed for medium crack sizes (0.35 < a/t < 0.7). Crack extension starts after a small deviation from the linear part of the curve. The crack first extends in a stable manner under increasing load until suddenly instable fracture occurs.

Type III was observed for deep cracks (a/t > 0.75). As for type I, the crack extends in a stable manner through the specimen. The onset of crack extension starts as in type II after a small deviation from the linear section of the curve. Wall penetration was observed before, at or beyond maximum load.

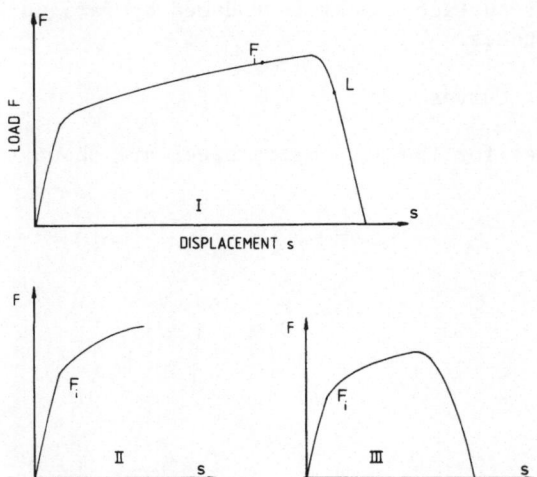

Figure 25. Three types of load displacement curves for the ferritic steel.

10.3 Fracture Appearance

The macroscopic fracture surfaces of ferritic steel are different for the three types (fig. 26).

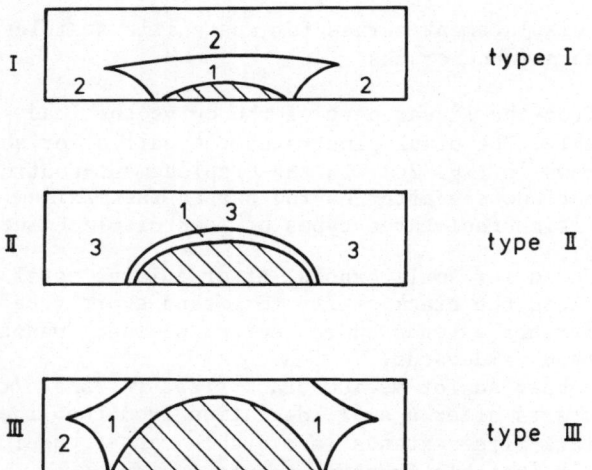

Figure 26. Fracture appearance for the test with the ferritic steel (1 flat ductile crack, 2 shear crack, 3 flat cleavage crack).

Type I: The first part of crack extension (1 in fig. 26) is perpendicular to the load applied, on the same plane as the crack. The second part is composed of shear cracks.

Type II: After some stable ductile crack extension the crack extends

by cleavage. The total fracture plane is perpendicular to the direction of load applied.

Type III: This type is similar to type I. The flat part extends through the wall and shear fracture develops after wall penetration.

10.4 Prediction of Critical Loads

Different critical loads can be used to compare experimental results with the relations given in the preceding sections:
- load at ligament yielding,
- load at the onset of stable crack extension,
- load at wall penetration,
- maximum load.

For the austenitic steel the crack penetrated the wall after maximum load was reached. In a load-controlled test the maximum load would be the failure load. Therefore, the comparison is based on the maximum stress σ_{max} (load divided by uncracked cross section). The predicted results are given in table 2 and in fig. 27. The linear-elastic prediction

TABLE II Maximum stress, stress at wall penetration and calculated stresses for austenitic steel (in MPa)

a/t	a/c	σ_{max}	σ_{Leak}	σ_1	σ_2	σ_3	σ_4	σ_5	σ_6
.771	.692	363	351	286	192	255	365	412	1431
.571	.727	449	440	357	342	372	442	586	1697
.629	.733	443	443	358	304	355	418	570	1626
.386	.321	500	477	387	376	375	457	590	1617

σ_1 Harrison[42], σ_2 Chell[44], σ_3 Mattheck[45], σ_4 Hasegawa[46], σ_5 Görner, Mattheck[48], σ_6 LEFM

is only made to show the general trend. Fracture toughness was not measured for the steel. From other investigations on stainless steel J_{Ic}-values can be found in the range between 300 and 900 N/mm. Using the lower values of J_{Ic} = 300 N/mm and the linear-elastic relation between J and K a K_{Ic} of 263 MPa$\sqrt{m}$ is obtained. Even this lower value leads to a large overestimation of maximum stress, as is always the case, if LEFM is applied to a highly ductile failure. On the other hand, the relations based on average flow stress underestimate the maximum stress. Only the predictions by Hasegawa et al. are in good agreement with the experimental results. The preductions of Mattheck and Görner overestimate the failure stress somewhat.

For the ferritic steel a comparison of measured and calculated critical loads is given in table 3 and figures 28 - 30. The maximum load can be well predicted with the relation of Hasegawa et al. and by Mattheck and Görner, whilst it is underestimated with the relations using an ave-

TABLE III

Maximum load F_{max}, load at crack initiation F_i, load at ligament yielding F_{pl} and calculated critical loads for the ferritic steel (in kN)

a/t	a/c	F_{max}	F_i	F_{pl}	F_1	F_2	F_3	F_4	F_5	F_6	F_7
0.28	0.63	414	400		308	338		430	353	671	513
0.29	1.01	433	432		328	348		433	358	771	594
0.36	0.24	333	315	212	281	270	262	378	297	432	347
0.37	0.40	340			290	291		394	320	485	391
0.40	0.54	365			292	298		393	326	510	417
0.47	1.03	410	400	278	296	314	260	387	336	587	492
0.50	0.51	290	289	218	264	253		362	294	426	360
0.50	0.78	378		220	279	284	253	374	318	521	441
0.56	0.65	334	282	230	258	244		353	292	442	380
0.60	1.01	333		238	268	266		353	308	498	431
0.71	0.83	318		202	235	189		318	261	421	372
0.71	0.27	228	208		191	128	134	247	156	246	217
0.72	0.28	222	220	162	185	125	132	246	155	244	216
0.74	0.59	276	225		210	162		293	214		
0.78	0.62	271	243	176	201	127	163	285	200	339	303
0.81	0.90	300	236	200	211	135			228	384	347
0.83	1.07	286	284		221	137			230		

F_1: Harrison[42], F_2: Chell[44], F_3: Mattheck[45], F_4: Hasegawa[46], F_5: Görner, Mattheck[48], F_6: LEFM, F_7: LEFM with plastic zone correction.

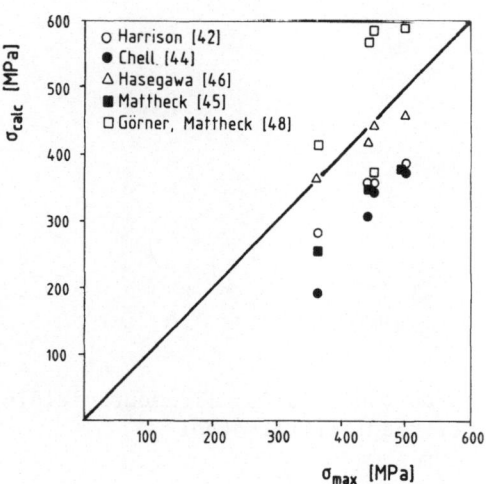

Figure 27. Comparison between calculated critical stresses and maximum stress for the austenitic steel.

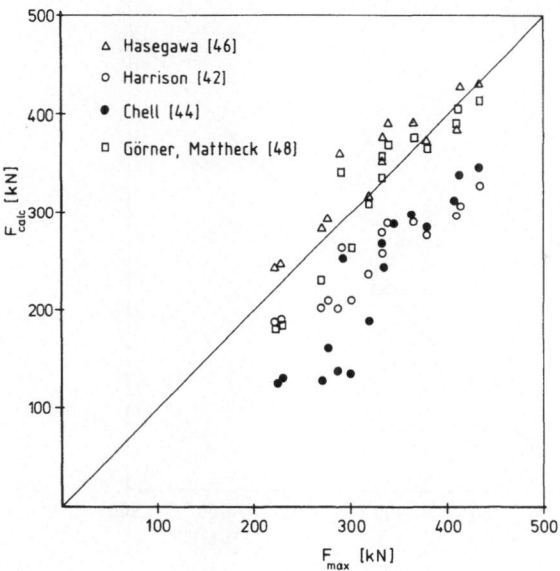

Figure 28. Comparison between maximum load and calculated critical load for the ferritic steel.

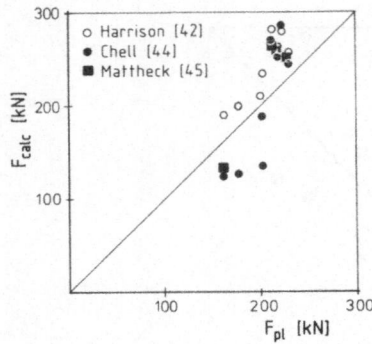

Figure 29. Comparison between the load for ligament yielding F_{pl} and calculated critical load for the ferritic steel

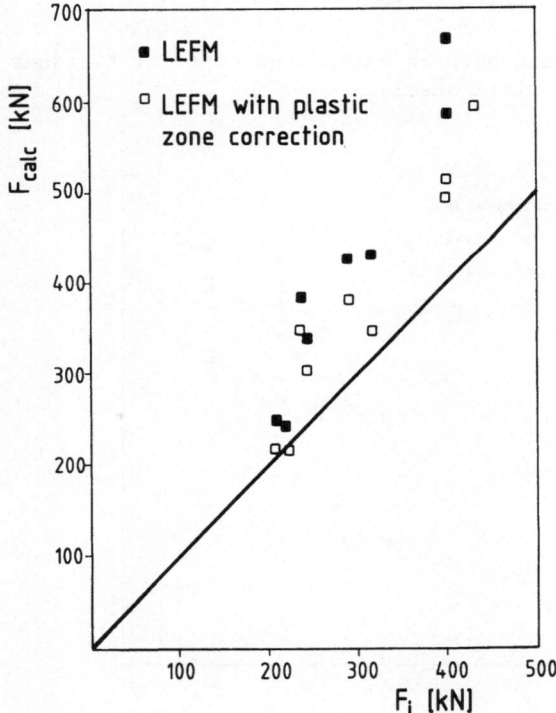

Figure 30. Comparison between load at crack initiation and linear-elastic calculated load

rage flow stress and overestimated using the linear-elastic relation. The flow-stress relations are in better agreement with the critical load for ligament yielding. In fig. 30 the linear-elastic prediction with and without plastic zone correction is compared with the critical load for the onset of stable crack extension. Only for the lowest loads corresponding to large cracks the prediction is in agreement with the measured values.

REFERENCES

1. ASME, Boiler and Pressure Vessel Code, Section XI

2. Mc Gowan, J.J. and Raymund, M., 'Stress-intensity factor solutions for internal longitudinal semi-elliptical surface flaws in a cylinder under arbitrary loading', ASTM STP 677, 1979, 365-380

3. Newman, J.C. and Raju, I.S., 'An empirical stress-intensity factor equation for the surface crack', Engng. Fract. Mech. 15, 1981, 185-192

4. Raju, I.S. and Newman, J.C., 'Stress-intensity factors for internal and external surface cracks in cylindrical vessels', Trans. ASME, J. Press. Vess. Techn. 104, 1982, 293-298

5. Heliot, J., Labbens, R.C. and Pellissier-Tanon, A., 'Semi-elliptical cracks in a cylinder subjected to stress gradients', ASTM STP 677, 1979, 341-364

6. Rice, J.R. and Levy, N., 'The part-through surface crack in an elastic plate', Trans. ASME, J. Appl. Mech. 29, 1972, 185-194

7. Parks, D.M., Lockett, R.R. and Brockenbrough, J.R., 'Stress-intensity factors for surface-cracked plates and cylindrical shells using line-spring finite element', 1981 Advances in Aerospace Structures and Materials ASME, 1981, pp. 279-285

8. Nishioka, T. and Atluri, S.N., 'Analysis of surface flaw in pressure vessels by a new 3-dimensional alternating method', Trans. ASME, J. Press. Vess. Techn. 104, 1982, 299-307

9. Newman, J.C. and Raju, I.S., 'Stress-intensity factors for internal surface cracks in cylindrical pressure vessels', Trans. ASME, J. Press Vess. Techn. 102, 1980, 342-346

10. Huget, W., Esser, K., Grüter, L., 'Stress-intensity factors for slender surface cracks', Transaction of the 7th International Conference on Structural Mechanics in Reactor Technology, 1983, paper G/F 3/4

11. Cruse, T.A., Besuner, P.M., 'Residual life prediction for surface cracks in complex structural details', Journal of Aircraft 12, 1975 369-375

12. Yagawa, G., Ichimiya, M. and Ando, Y., 'Theoretical and experimental analysis of semi-elliptical surface cracks, subject to thermal shock', ASTM STP 677, 1979, 381-398

13. Bueckner, H.F., 'A novel principle for the computation of stress intensity factors', Z. Angew. Math. Mech. 50, 1970, 529-546

14. Rice, J.R., 'Some remarks on elastic crack-tip stress fields', Int. J. Sol. and Struct. 8, 1972, 751-758

15. Labbens, R.C., Heliot, J., Pellissier-Tanon, A., 'Weight function for three-dimensional symmetrical crack problems', ASTM STP 601, 1976, 448-470

16. Mattheck, C., Munz, D. and Stamm, H., 'Stress intensity factor for semi-elliptical surface cracks loaded by stress gradients', Engng. Fract. Mech. 18, 1983, 633-641

17. Mattheck, C., Morawietz, P., Munz, D., 'Stress-intensity factor at the surface and at the deepest point of a semi-elliptical surface crack in plates under stress gradients', Int. J. Fract., 23, 1983, 201-212

18. Petroski, J.J., Achenbach, J.D., 'Computation of the weight function from a stress intensity factor', Engng. Fract. Mech. 10, 1978, 257-266

19. Stamm, H., Mattheck, C. and Munz, D., 'Berechnung von Spannungsintensitätsfaktoren im Halbraum und in der Platte unter Thermoschockbelastung', Berichtsband der 15. Sitzung des Arbeitskreises Bruchvorgänge 1983, Deutscher Verband für Materialprüfung, pp. 321-331

20. Stamm, H., Mattheck, C. and Munz, D., 'Stress-intensity factors for surface cracks under thermal fatigue conditions', Application of fracture mechanics to materials and structures, Proceedings of an International Conference in Freiburg, Martinus Nijhoff Publishers 1984, pp. 855-865

21. Sneddon, I.N., 'The distribution of stress in the neighbourhood of a crack in an elastic solid', Proceedings of the Physical Society of London 187, 1946, 229-260

22. de Lorenzi, H.G., 'Elastic-plastic analysis of the maximum postulated flaw in the beltline region of a reactor vessel', Trans. ASME, J. Press. Vess. Techn. 104, 1982, 278-286

23. Benthem, J.P., 'The quarter-infinite crack in a half space; alter-
    native and additional solutions', Intern. J. Solids and Struct. 16,
    1980, 119-130

24. Burton, W.S., Sinclair, G.B., Solecki, J.S. and Swedlow, J.L.,
    'On the implications for LEFM of the three-dimensional aspects in
    some crack/surface intersection problems', Int. J. Fract. 25, 1984,
    3-32

25. Smith, C.W., Epstein, J.S., and Olaosebikan, O., 'Boundary layer
    effects in cracked bodies: an engineering assessment', 17th National
    Symposium on Fracture Mechanics, Albany, N.Y., USA, 1984

26. Görner, F., Mattheck, C. and Munz, D., 'Change in geometry of surface
    cracks during alternating tension and bending, Z. Werkstofftech.
    14, 1983, 11-18

27. Müller, H.M., Müller, S., Munz, D., Neumann, J., 'Extension of sur-
    face cracks during cyclic loading', 17th National Symposium on Frac-
    ture Mechanics, Albany, N.Y., USA, 1984

28. Pellissier-Tanon, A., 'Fracture mechanics in structure design',
    Application of Fracture Mechanics in Materials and Structures, Pro-
    ceedings of an International Conference in Freiburg, 1983, Martinus
    Nijhoff Publishers, 1984, pp. 213-234

29. Trantina, G.G., de Lorenzi, H.G. and Wilkensing, W.W., 'Three-dimen-
    sional elastic-plastic finite element analysis of small surface
    cracks, Engng. Fract. Mech. 18, 1983, 925-938

30. Irwin, G.R., 'Crack-extension force for a part-through crack in a
    plate', Trans. ASME, J. Appl. Mech. 29, 1962, 651-654

31. Munz, D., 'Minimum specimen size for the application of linear-elas-
    tic fracture mechanics', ASTM STP 668, 1979, 406-425

32. Munz, D., 'Linear-elastische und elastisch-plastische Bruchmechanik
    von Konstruktionswerkstoffen', DFVLR-Forschungsbericht 79-31, 1979

33. Sumpter, J.D.G. and Turner, C.E., 'Design using elastic-plastic frac-
    ture mechanics', Int. J. Fract. 12, 1976, 861-871

34. Civelek, M.B. and Erdogan, F., 'Elastic-plastic problems for a plate
    with a part-through crack under extension and bending', Int. J.
    Fract. 20, 1982, 33-46

35. Ezzat, H., Erdogan, F., 'Elastic-plastic fracture of cylindrical
    shells containing a part-through circumferential crack', Trans. ASME,
    J. Press. Vess. Techn. 104, 1982

36. Sumpter, J.D.G., 'The prediction of K_{Ic} using J and COD from small specimen tests', Metal Science 10, 1976, 354-356

37. Milne, I. and Chell, G.G., 'Effect of size on the J fracture criterion', ASTM STP 668, 1979, 358-377

38. Munz, D. and Keller, H.P., 'Effect of specimen size on fracture toughness in the ductile brittle transition region of steel', Fracture and Fatigue, Proceeding of the 3. European Conference on Fracture, London 1980, pp. 105-117

39. Parks, D.M. 'A stiffness derivative finite element technique for determination of crack tip stress intensity factors', Int. J. Fracture 10, 1974, 487-502

40. de Lorenzi, H.G., 'On the energy release rate and the J-Integral for 3D crack configurations', Int. J. Fract. 19, 1982, 183-193

41. Begley, J.A., Landes. J.D. and Wilshaw, W.K., 'An estimation model for the application of the J-integral', ASTM STP 560, 1974, 155-169

42. Harrison, R.P., Loosemore, K., Milne, I. and Dowling. A.R., 'Assessment of the integrity of structures containing defects', Central Electricity Generating Board, U.K. Report R/H/R6-Rev. 2, 1980

43. Kanninen, M.F., 'Towards an elastic-plastic fracture mechanics predictive capability for reactor piping', Nuclear Engineering and Design, 48, 1978, 117-134

44. Chell, G.G., 'Elastic-plastic fracture mechanics', in Development in Fracture Mechanics - 1, Applied Science Publishers, 1979, pp. 67-105

45. Mattheck, C., Morawietz, P. Munz, D. and Wolf, B., 'Ligament yielding of a plate with semi-elliptical cracks under uniform tension', Int. J. Press. Vess. & Piping 16, 1984, 131-143

46. Hasegawa, K., Shimizu, T., Sakata, S. and Shida, S., 'Stable crack growth and leak predictions of stainless steel pipes with circumferential cracks', Transactions of the 6th Conference on Structural Mechanics in Reactor Technology', Paper L 5/3

47. Hasegawa, K., Sakata, S., Shimizu, T. and Shida, S., 'Prediction of fracture tolerances for stainless steel pipes with circumferential cracks', 4th ASME Pressure Vessel and Pipe Conference, Portland 1983

48. Mattheck, C., Görner, F., 'Leak prediction by use of a generalized Dugdale model for semi-elliptical surface flaws in plates under tension loading', Proceedings of the 5th European Conference on Fracture, 1984, Vol. II, pp. 631-642

49. Göring, J., 'Versagensverhalten von Flachzugproben mit einem halb-elliptischen Oberflächenriß aus einem Werkzeugstahl', Diplomarbeit, Institut für Zuverlässigkeit und Schadenskunde im Maschinenbau, Universität Karlsruhe (TH), September 1983

50. Brocks, W., Noack, H.D., Veith, H., 'Three dimensional elastic-plastic analysis of a semi-elliptical inner surface flaw by finite element method', Transactions of the 7th International Conference on Structural Mechanics in Reactor Technology, 1983, paper G/F 3/1

MICROMECHANISMS OF CRACK INITIATION AND CRACK PROPAGATION

W. Dahl a. D. Dormagen
Institute of Ferrous Metallurgy
Technical University of Aachen
Intzestr. 1
5100 Aachen
FRG

ABSTRACT. After summarizing fracture mechanics parameters the paper introduces micromechanical models of cleavage fracture. It is shown that cleavage can be interpreted in terms of the local tensile stress required to propagate micro-cracks. Methods to determine the microscopic cleavage stress are presented. Influencing parameters like grain size, temperature, prestraining and strain rate are discussed and the correlation between fracture toughness and cleavage stress is demonstrated, using the Ritchie, Knott and Rice-Model. Ductile fracture occurs either by nucleation, growth and coalescence of microvoids or by shear band linking. It is shown that the strains required for void nucleation and growth strongly depend on the state of stress. The micro-mechanical models are schematically used to explain the effect of temperature, strain rate, state of stress and structural parameters, such as grain size, with special regards to the cleavage/ductile transition.

1. INTRODUCTION

The necessity to establish the safety of structures has led to an interest in the micromechanisms of cleavage and ductile fracture and to their relationships to fracture mechanics parameters such as fracture toughness K_{Ic}, J-Integral and crack-opening displacement CTOD. In the following micromechanical models are presented to describe crack initiation and crack propagation under monotonic loading conditions.

2. FRACTURE MECHANICS PARAMETERS

The resistance of metallic alloys to mechanical modes of crack extension is characterized by the value of fracture toughness /1/. Under linear-elastic and quasi-linear elastic conditions the fracture toughness is expressed as a critical value of the stress intensity factor K_c. For extensive or general yielding the crack tip region is no longer characterized by the K-dominant stress field and use can be made of critical values of the J-Integral, J_c, or the crack tip opening

203

L. H. Larsson (ed.), Elastic-Plastic Fracture Mechanics, 203–225.
© 1985 ECSC, EEC, EAEC, Brussels and Luxembourg.

displacement (CTOD) δ_c. Under the conditions of small scale yielding the parameters K, J, δ are related by Eqs (1):

$$K^2 = E' \cdot J = ME' \sigma_y \cdot \delta \qquad (1)$$

where E' is Young's modulus, E, in plane stress and $E/(1-\nu^2)$ in plane strain; ν is Poisson's ratio, σ_y is the uniaxial yield stress or flow stress and M a geometrical factor, equal to unity for plane stress and approximately 2 for plane strain.

For larger amounts of plasticity the relationships in Eqs (1) are no longer valid. For extensive yielding, prior to general yield, in plane stress it is possible to relate the COD to the applied stress by the equation

$$\delta = (8 \cdot \sigma_y / \pi \cdot E) a \quad \ln\{\sec(\pi \sigma_{app} / 2\sigma_y)\} \qquad (2)$$

where a is the half-crack-length of a through-thickness crack in an infinite body, lying normal to an applied stress, σ_{app} /2,3/. Beyond general yield, it is still possible to measure J and δ, but finite element calculations indicate, that the geometrical factor M changes and depends on testpiece geometry: for deeply cracked bend specimens a value of 3 is predicted in plane strain /4/. In experimental work values in the range 1 - 3 have been measured /5,6/. At temperatures well above the ductile/brittle transition, the initiation of ductile crack extension does not immediately lead to an unstable crack propagation or to a plastic collapse of a testpiece. In this temperature regime a material's crack growth resistance curve (R-curve) is defined, when the actual values of J or δ (measured at the original crack tip) are plotted as a function of increase in crack length, $a-a_o$. The back-extrapolation to the "blunting-line" (J_R-curve) or to a_o (δ_R-curve) enables initiation-values, J_i or δ_i to be determined. The crack-growth resistance of the material is given by the slope of the curve (dJ/da) or (dδ/da). Under certain restrictions /4,7/ the stress and strain field ahead of the crack tip can be described by the single parameter J or δ, even when stable crack growth and general yielding occurs.

3. CLEAVAGE FRACTURE

3.1. Transgranular Cleavage

Cleavage fracture causes new free surfaces by the rupture of atomic bonds across well defined, low index crystallographic planes. From the macroscopic point of view it is often a brittle type of fracture without significant plastic deformation but providing microplasticity. Crack propagation occurs with high velocity in a transcrystalline manner. In polycrystalline materials cleavage fracture is characterized by facets corresponding to the individual grains or regions of cristallographic continuity. The cleavage facets often display river lines - irregular markings or steps in the fracture surface (see Fig. 1).

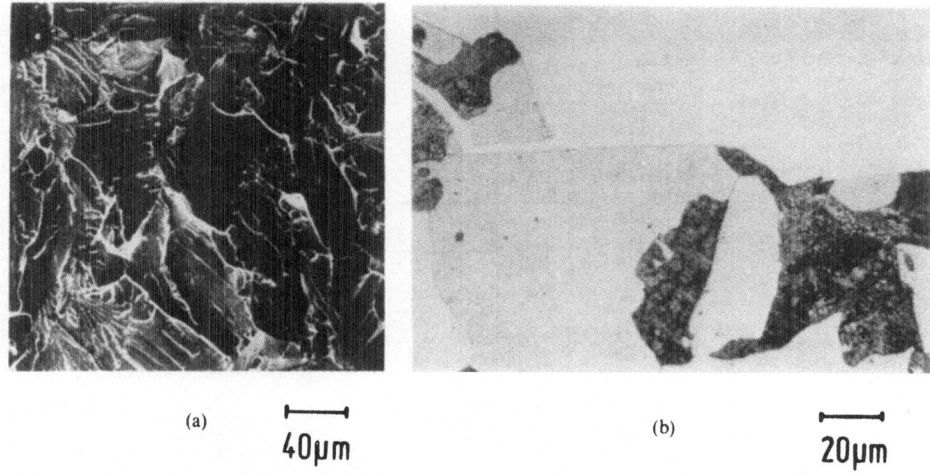

(a) ├———┤
     40µm

(b) ├———┤
     20µm

FIG. 1: Cleavage facets in mild steel:
     a) fracture surface
     b) section through nickel plated fracture surface

In ferritic steels fracture occurs along {100} planes.

Atomistic crack tip analysis suggest that in body centred cubic
iron yielding will most probably occur at a loaded crack tip before the
crack extends by cleavage /8,9/. The local balance between fracture
and yielding seems to favour yielding. Thus some form of stress
elevation mechanism must operate for cleavage to occur in steels.

From experimental work carried out by Low /10/ it is apparent that
plastic flow is a necessary prerequisite to cleavage and that yielding
is involved in the nucleation of cleavage fracture.

3.1.1 <u>Nucleation controlled cleavage.</u> It was suggested by Zener /11/
that the local stress concentration produced at the head of a dislocation pile-up could lead to cleavage fracture when the leading
dislocations were squeezed together to generate a crack nucleus.
Stroh /12/ presented an analysis of the propagation of a crack nucleus
formed by this mechanism (see Fig. 2)

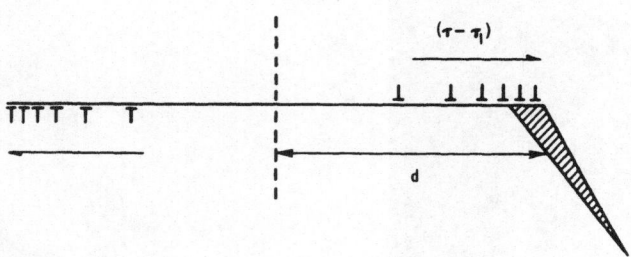

FIG. 2: Cleavage crack nucleation at the head of a dislocation
pile-up. After Zener /11/ and Stroh /12/

$$\tau_{eff} = \tau_y - \tau_i \geq \{\frac{12\gamma \cdot G}{\pi \cdot (1-\nu)d}\}^{1/2} \tag{3}$$

where τ_{eff} is the effective shear stress applied to the slip band, τ_y
is the shear yield stress, τ_i is the lattice friction stress, γ is the
effective surface energy, G the shear modulus, and d is the slip
band half length.

A significant result of this model is that crack nucleation is
predicted to be the most difficult stage in cleavage fracture. Therefore
this model does not allow stable crack nuclei, which is in contrast to
experimental observations /13/. In his analysis Stroh neglected the
effect of local shear stresses in propagating the crack and assumed that
crack formation in a non-uniform stress field was equivalent to that in
a uniform stress field. Numerical analysis made by Smith and Barnby /14/
taking into account these effects confirm this model and there is also
experimental evidence showing the cleavage fracture of zinc in both
mono- /15/ and polycrystals /16/, to be nucleation controlled as
predicted by this dislocation model of cleavage.

3.1.2 Propagation controlled cleavage. In steels cleavage fracture is
promoted by factors producing locally elevated tensile stress levels.
Cleavage of mild steels occurs at a critical tensile stress which seems
to be independent of the state of hydrostatic tension /17/. This
implies a propagation controlled mechanism. Stroh's analysis indicates
that cleavage fracture can only be propagation controlled if some
brittle phase provides a low surface energy region. A crack nucleus can
initiate and propagate through this region, but arrest at the surface
energy discontinuity provided by the phase boundary.

Cottrell /18/ proposed an alternative dislocation mechanism providing for an easy nucleation in bcc metals (see Fig. 3).

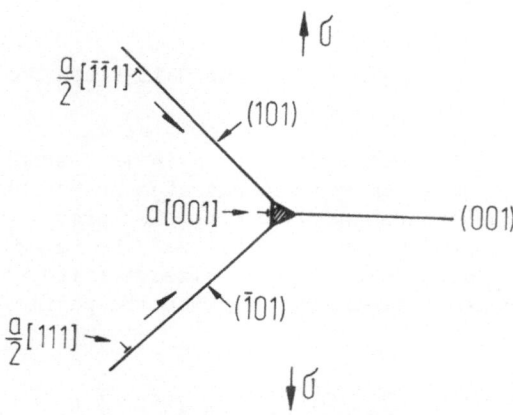

FIG. 3: Cottrell's model /18/ of cleavage crack nucleation in bcc metals

Two dislocations of Burgers vector a/2 $<\bar{1}\bar{1}1>$ slipping on intersecting {101} slip planes interact to form a sessile dislocation with its Burgers vector, a {001}, normal to the {001} cleavage plane.

$$\frac{a}{2}\underset{(101)}{\overset{}{<\bar{1}\bar{1}1>}} + \frac{a}{2}\underset{(\bar{1}01)}{<111>} \rightarrow a\underset{(001)}{<001>} \tag{4}$$

As this reaction is accompanied by a decrease in dislocation energy, crack nucleation will be easier than by the Zener/Stroh mechanism. Cleavage fracture is now propagation controlled with the fracture stress σ_f^*, given by

$$\sigma_f^*(\sigma_f^*-\sigma_i) = \frac{8 \cdot \gamma \cdot G}{d} \tag{5}$$

and setting

$$\sigma_f^* = \sigma_y = \sigma_i + k_y \cdot d^{-1/2} \tag{6}$$

gives

$$\sigma_f^* \geq \frac{8 \cdot \gamma \cdot G}{k_y} d^{-1/2} \tag{7}$$

where d is the grain size and k_y is the Hall-Petch yielding constant. This model explains the effects of grain size and yielding parameters on cleavage fracture but it neglects the possible influence of micro-structural variables other than grain size.

Similar to Cottrell's proceeding Reiff /19/ presented a model where σ_f^* depends on the grain size and the carbide thickness. The influence of carbide thickness is given by:

$$\sigma_f^* = - \frac{k_x \cdot d^{-1/2}}{D \cdot \pi (1-\nu) \cdot 4,4} + (\frac{8 \cdot \gamma \cdot G}{D \cdot \pi (1-\nu)})^{1/2} \tag{8}$$

where D is the carbide thickness and k_x the Hall-Petch-constant for a given plastic strain.

3.1.3 Smith's and Riedel's model of cleavage in mild steel.
The importance of Cottrell's omission was demonstrated by McMahon and Cohen /20/, showing that coarse carbide particles promoted cleavage fracture whereas fine carbide particles allowed the material to behave in a ductile manner. Fig. 4 shows a grain-sized ferrite microcrack associated with a cracked grain boundary cementite particle in a mild

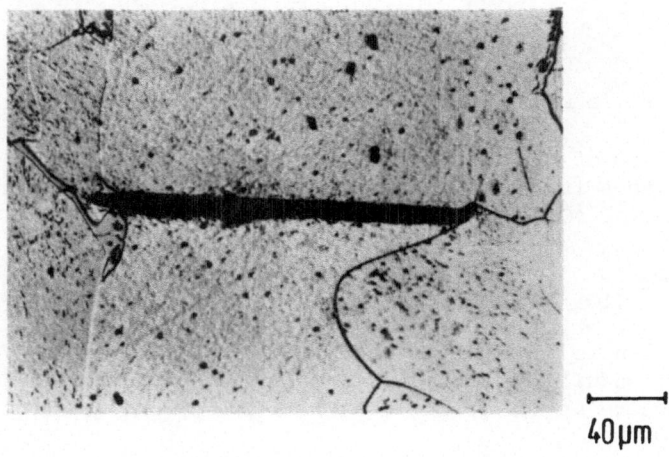

40µm

FIG. 4: Cleavage crack in mild steel associated with cracked grain boundary carbide particle

steel. A theoretical model has been proposed by Smith /21/ in which a carbide particle blocks an impinging slip band (see Fig. 5). By examining the change in energy with crack length, the cleavage fracture stress was calculated by Eq. (9) where D is the thickness of grain boundary

$$(\frac{D}{d})\{\sigma_f^*\}^2 + \tau_{eff}^2 \{1 + \frac{4}{\pi}(\frac{D}{d})^{1/2} \frac{\tau_i}{\tau_{eff}}\}^2 \geq \frac{4 \cdot E \cdot \gamma_f}{\pi (1-\nu^2) d} \tag{9}$$

carbide and γ_f is the effective surface energy of ferrite.
If the effective shear stress τ_{eff} is written as $k_y^s \cdot d^{-1/2}$ Eq. (9) reduces to:

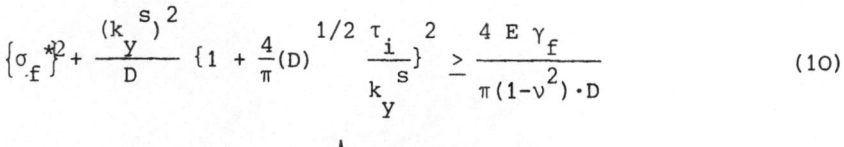

$$\{\sigma_f^{*}\}^2 + \frac{(k_y^s)^2}{D}\ \{1 + \frac{4}{\pi}(D)^{1/2}\ \frac{\tau_i}{k_y^s}\}^2\ \geq\ \frac{4\,E\,\gamma_f}{\pi(1-\nu^2)\cdot D} \qquad (10)$$

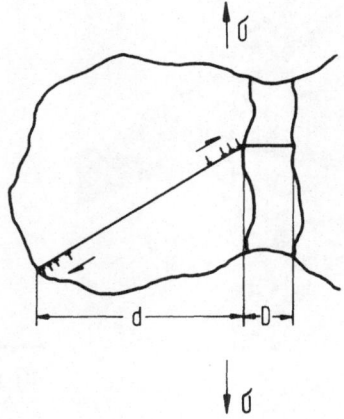

FIG. 5: Smith's model /21/ of cleavage fracture in mild steel

This predicts that the only microstructural parameter affecting the fracture stress is the carbide thickness. Riedel /22/ improves Smith's model regarding slip bands intersecting at a 90 degree angle, and the fracture stress is given by:

$$\sigma_f^{*} = \frac{1.85\ m\left(\dfrac{\gamma \cdot E}{\pi(1-\nu^2)} \cdot \dfrac{1}{d}\right)^{1/2} \cdot \left(1 + 0.37\left(\dfrac{D}{d}\right)\right)^{1/2} - 2k_y d^{-1/2}\left(1-1.46\left(\dfrac{D}{d}\right)\right)^{1/2}}{(1-\chi)\ (m+2)\ +\ 1.46\left(\dfrac{D}{d}\right)^{1/2}\left(2+\chi(m-2)\right)} \qquad (11)$$

where χ is a factor considering the stress state and is given by

$$\chi = \frac{\sigma_{xx}}{\sigma_{yy}} = \frac{\pi \cdot \omega}{1+\pi-\omega} \qquad (12)$$

and m is the Taylor factor.

3.1.4 <u>Other cleavage micromechanisms.</u> Cleavage fracture can also be initiated in steels by deformation twins produced at high strain rates and low temperatures. Knott and Cottrell /23/ observed crack nucleation in polycrystalline mild steels with coarse grain sizes at service temperatures below -150 °C <u>(see Fig. 6).</u>

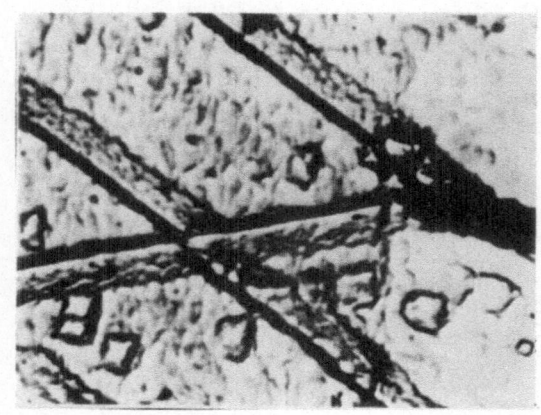

10 µm

FIG. 6: The initiation of a cleavage crack by intersecting deformation
twins in mild steel at -196 °C

3.2. Determination of Microscopic Cleavage Fracture Stress

Several methods have been used to measure the microscopic cleavage
fracture stress σ_f^*. Some of the common methods are briefly mentioned
in the following.

According to Orowan /24/ cleavage fracture is initiated when the
fracture and the yield strength of an unnotched tensile specimen
coincide:

$$\sigma_f^* = \sigma_f = \sigma_y \tag{13}$$

If fracture and yield stress do not coincide Aurich /25/ suggest   the
minimum true fracture strength as a function of temperature to be equal
to σ_f^*:

$$\sigma_f^* = \sigma_f^{min} \tag{14}$$

In some cases fracture and yield strength increase again with decreasing
temperature. Consequently Aurich and Wobst /26/ suggest   to set the
yield strength extrapolated to 0 K equal to σ_f^*:

$$\sigma_f^* = \sigma_{y(T=0\ K)} \tag{15}$$

In order to avoid tests at extremely low temperatures notched bend
or tensile specimens can be used when the stress distribution ahead of
the notch is known. Then the maximum tensile stress is considered to be
equal to σ_f^*:

$$\sigma_f^* = \sigma_{yy}^{max} \tag{16}$$

The maximum local tensile stress can be calculated by Hill's slip line field theory:

$$\sigma_{yy} = \sigma_y(1 + \ln(1 + \frac{x}{\rho})) \quad , \quad x \leq x_\beta \tag{17}$$

$$\sigma_{yy} = \sigma_y(1 + \frac{\pi-\omega}{2}) \quad\quad , \quad x \geq x_\beta \tag{18}$$

where x is the notch root distance, ρ is the notch radius, ω is the notch angle and x_β the notch root distance where the plastic zone has reached a critical size.

The solutions assume plane strain behaviour in a perfectly plastic material, the lines of maximum shear stress spreading in form of logarithmic spirals. Finite element calculations of the stress distribution can lead to better results, because the material work hardening behaviour can be taken into account.

Ritchie, Knott and Rice /27/ have modified Eq. (16) by introducing a critical distance x_c across which σ_f^* must be exceeded:

$$\sigma_f^* = \sigma_{yy}(x_c) \tag{19}$$

This fracture criterion is based on the idea that cleavage fracture is propagation controlled.

3.2.1 <u>Influence of testing parameters.</u> The influence of work hardening on the stress distribution ahead of a notch is shown in <u>Fig. 7</u>. For the mild steel C 75 the stress ratio σ_{yy}/σ_y is plotted versus the distance

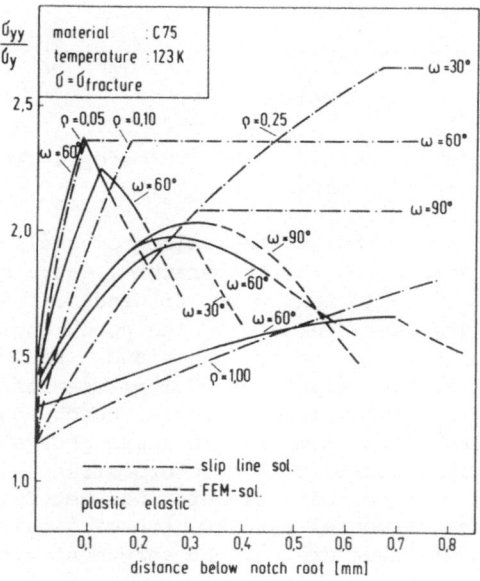

FIG. 7: σ_{yy}-stress distribution in front of the notch root at fracture load for different notch geometries /28/

from notch root for different notch angles and notch radii. Elastic-plastic finite element solutions (plane strain) are compared with slip line field theory /23/. There is quite a good agreement between the two methods between notch root and that point, where the σ_{yy}^{max}-value is reached. The greatest deviation occurs at the notch root for high stresses, due to work hardening and in the region of the maximum. The differences in the maximum value of stress concentration of FE- and SL-results may be explained by the spread of the plastic zone not only in form of logarithmic spirals in front of the notch - as predicted by SL-theory - but also in form of wings somewhat above the ligament. In Fig. 8 the fracture criteria given by Eqs (16) and (19) are compared among each other. In both cases a FE-analysis was used to determine σ_f^* for the steel C 10. For the maximum tensile stress σ_{yy}^{max} there is

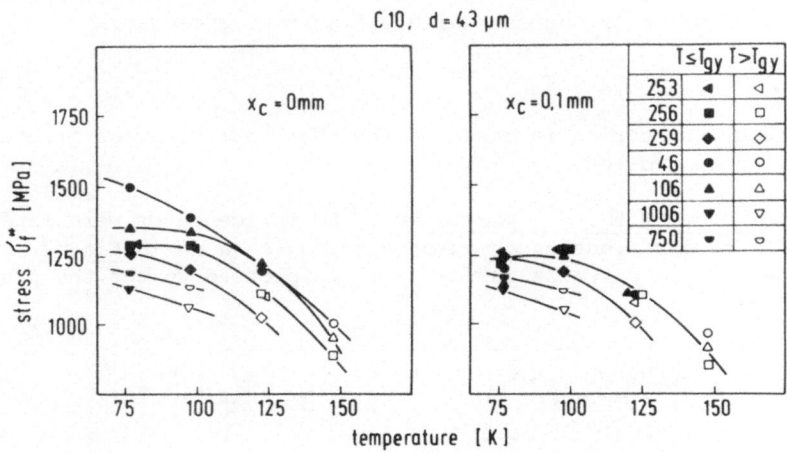

FIG. 8: Influence of notch acuity on the microscopic cleavage stress as a function of temperature /29/

a strong influence of notch radius, the values of smaller radii being significantly higher, and below the temperature, where general yielding occurs (T_{gy}) there is nearly no influence of temperature. The different series are characterized by the notch angle and the notch radius e.g. 253 stands for a 30° notch angle and a radius of 0.25 mm. Assuming a critical distance value x_c of 0.1 mm results in a microscopic cleavage stress σ_f^* being independent of notch acuity and temperature. Above the transition temperature where general yielding is observed there is still a decrease in fracture stress with increasing temperature. The mean ferrite grain diameter was determined to $d_m = 43$ µm. Thus the experimental results, the critical distance x_c being twice the grain diameter are in good agreement with theoretical predictions /27/.

The influence of loading rate on the microscopic cleavage stress is shown in Fig. 9. For the mild steel C 10 with $d_m = 90$ µm cleavage fracture stress is plotted as a function of temperature /29/.

Assuming a critical distance x_c of 0.15 mm there is no influence of strain rate by changing the cross head speeds from q = 1 mm/min to 100 mm/min.

C 10 , d = 90 µm

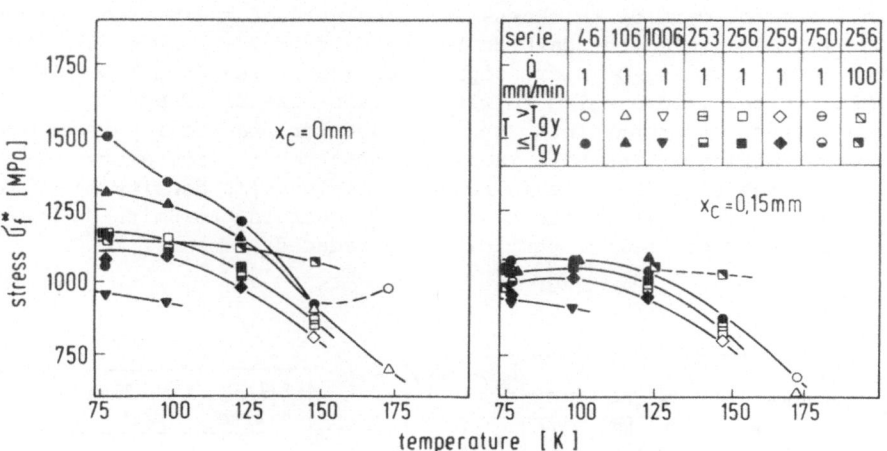

FIG. 9: Influence of loading rate on the microscopic cleavage stress as a function of temperature /29/

3.2.2 <u>Influence of metallurgical parameters</u>. It is found experimentally that σ_f^* increases with a decrease in grain size /29,30,31/ (see Fig. 10). This is in good agreement with the models

C 10, T = 77 K

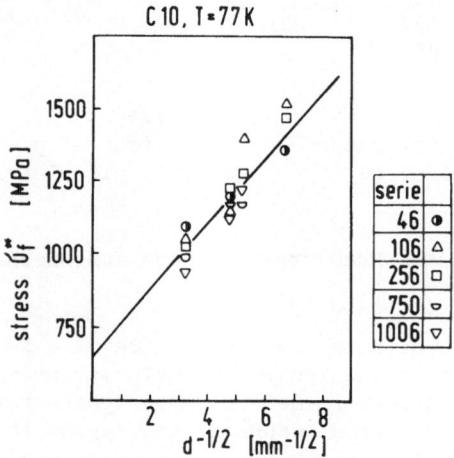

FIG. 10: Grain size dependence of the microscopic cleavage stress

presented by Cottrell, Reiff and Riedel but not predicted by Smith's model of microcrack propagation, which suggests that σ_f^* should depend only on carbide thickness. From this model the increase of σ_f^* with decreasing grain size seems to occur because carbide thickness and grain size are interrelated in steels which have been simply cooled from austenite, so that fine grains are associated with fine carbides. The influence of carbide thickness independently of grain size is shown in **Fig. 11.** Decreasing the carbide thickness from 2 - 3 μm to 0.5 μm results in an increase of microscopic cleavage stress from 1100 MPa to 1300 MPa for the mild steel C 10 with a grain size of 22 μm.

The effect of prestraining is important for practical purposes,too. After prestraining an increase in σ_f^* is observed in literature /32,33,34/ (see Fig. 12). The increase of dislocation density leads to an increase of free dislocations so that - according to Smith's theory - stresses ahead of carbide particles are reduced.

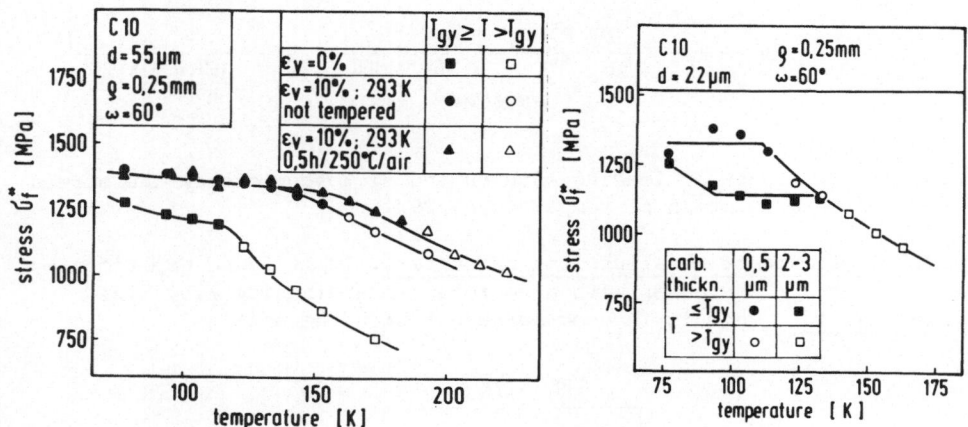

FIG. 11:
Influence of carbide thickness
on the microscopic cleavage
stress /29/

FIG. 12:
The effect of prestraining on the
microscopic cleavage stress /32/

3.3. Correlation between Fracture Toughness and Microscopic Cleavage Stress

The case to be examined is the interpretation of a K_{Ic} value at low temperatures in terms of the local crack tip stresses required to propagate small microcracks formed in second phase particles. If a testpiece is loaded, yielding occurs at the tip of the pre-crack. In plane strain this takes the form of two lobes of maximum extent, R_{IY}, at 70,5 ° to the line of crack extension, and minimum extent, r_{IY}, along the line of crack prolongation. For an infinite body,

$$R_{IY} = 0.16 \ (K/\sigma_y)^2 = 4 \ r_{IY} \tag{20}$$

Suppose that fracture occurs catastrophically at $K_{Ic} = 30$ MPa$\sqrt{m}$. If we were to freeze the loading at a K_I value of 29.9 MPa$\sqrt{m}$, it is apparent that the total area of plastic zone ($\alpha \ R_{IY}^2$) is only some 1.3 % less than the area at fracture. Virtually all the plastic work responsible for the high fracture toughness is therefore expended prior to crack extension. Plastic deformation leads to nucleation of microcracks, propagating when a critical tensile stress, σ_f^*, has been developed at the site of the microcrack nucleus. Since the stress at any given distance, r, ahead of the main crack increases with the applied stress intensity factor, K, the attainment of a critical value, σ_f^*, implies a critical value of stress intensity K_{Ic}.

A general correlation may be gained from the linear elastic expression:

$$K = \sigma_{(r)} \cdot (2\pi r)^{1/2} \tag{21}$$

where $\sigma_{(r)}$ is the stress at distance r. If a single microcrack were located at the position r*, the fracture toughness would be given by:

$$K_{Ic} = \sigma_f^*(2\pi r^*)^{1/2} \tag{22}$$

Of course, Eq. (22), is an oversimplification in two respects. First, the term σ_f^* is not single-valued, but is a function of the size of microcrack nucleus (f.i. grain boundary carbide thickness). Assuming that the microcrack behaves as a simple Griffith's crack, σ_f^* is proportional to $D_o^{-1/2}$ (D_o being the carbide thickness). Values of σ_f^* are therefore distributed, following the inverse square-root of the carbide thickness distribution. Moreover grain boundary carbides are not only distributed in terms of thickness, but also in terms of inclination to the tensile axis. Secondly the term r* is not single-valued, but is a statistical average, representing the probability of finding a carbide of representative thickness at a given distance ahead of the crack tip.

Initially the value of r* was found by Ritchie, Knott and Rice /27/ being twice the grain diameter in a mild steel, but further work by Curry and Knott /35/ showed, that this relationship is not general and emphasized the importance of the carbide thickness distribution. Thus the scatter of toughness values can be interpreted in terms of metallurgical scatter.

Using the failure criterion of Ritchie, Knott and Rice (RKR-Model), it is possible to predict fracture toughness values as a function of temperature, assuming constant values of σ_f^* and x_c, if the temperature dependence of the yield stress is known and the stress distribution for crack tip behaviour according to Tracey /36/ in small scale yielding is used.

For two different heat treatments of the steel 20 MnMoNi 55 measured and predicted K_{Ic} values are compared in Fig. 13 in a temperature range from 77 K up to 240 K. There is a good agreement within the regime of linear elastic fracture mechanics, and the results demonstrate that different types of specimen (CT- and CCP type) fail by cleavage, when the maximum tensile stress exceeds the microscopic cleavage stress σ_f^* over a critical distance x_c.

3.4. Intergranular Cleavage

It is also possible for microcracks to propagate in a cleavage-like manner along grain boundaries (see Fig. 14), if these have been segregated with trace impurity elements such as phosphorus, tin or antimony. Segregation can occur to austenite grain boundaries during

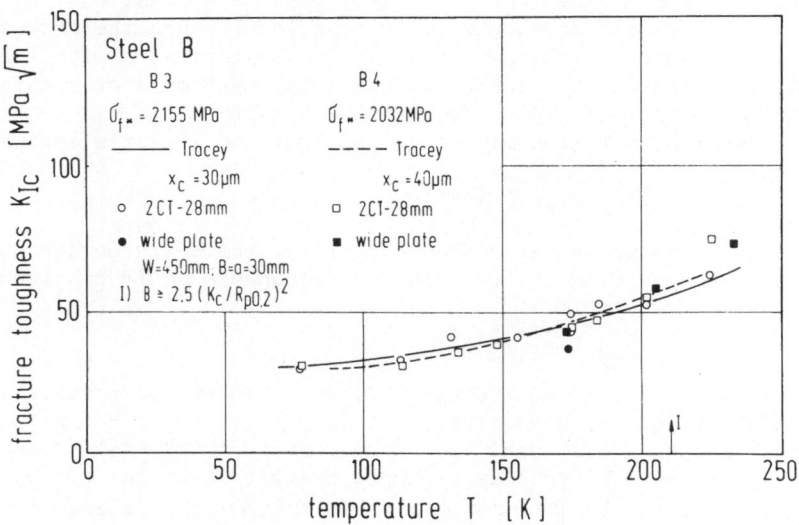

FIG. 13: The correlation between fracture toughness and microscopic cleavage stress /37/

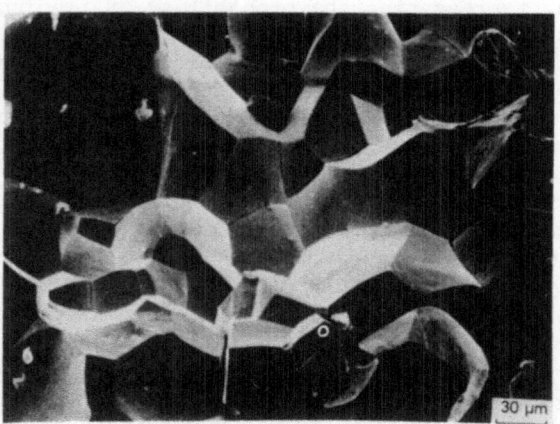

FIG. 14: Intergranular fracture at 77 K in polycrystalline iron due to segregation of sulfur to carbon depleted grain boundaries

austenisation, to prior austenite boundaries during tempering and to ferrite boundaries, coincident with prior austenite boundaries, when the steel is cooled slowly after tempering.

In the tempered condition, the steel has a microstructure of recrystallized ferrite grains and tempered carbides. The ferrite boundaries coincident with prior austenite boundaries are "high angle" or "highly disordered", because high angle boundaries may contain a high density of coincident sites. Therefore they tend to be more segregated than other ferrite boundaries. Segregation can also occur at interfaces between grain boundary carbides and the grains and at intragranular carbide/matrix interfaces.

As for transgranular cleavage, the micromechanism of grain boundary fracture is very similar: Microcracks are nucleated by plastic flow and its subsequent propagation is caused by the influence of a tensile stress /38/.

4. DUCTILE FRACTURE

4.1. Void Nucleation and Coalescence

Above the transition temperature, microcracks do not propagate by cleavage but the crack advance proceeds by the coalescence of voids centred on non-metallic inclusions or other second-phase particles /39/. At low plastic strains the inclusions decohere from the matrix and smaller microvoids can be nucleated by grain boundary carbides (of approx. 1 μm diameter). When the load applied to a testpiece is increased, voids are produced by the high plastic strains ahead of the precrack and expand under the combination of local stress field and hydrostatic stress component. Crack "initiation" is defined as the point at which the blunting precrack coalesces with a growing void. The initiation value of stable crack growth δ_i is measured at the position of the original crack tip (see Fig. 15).

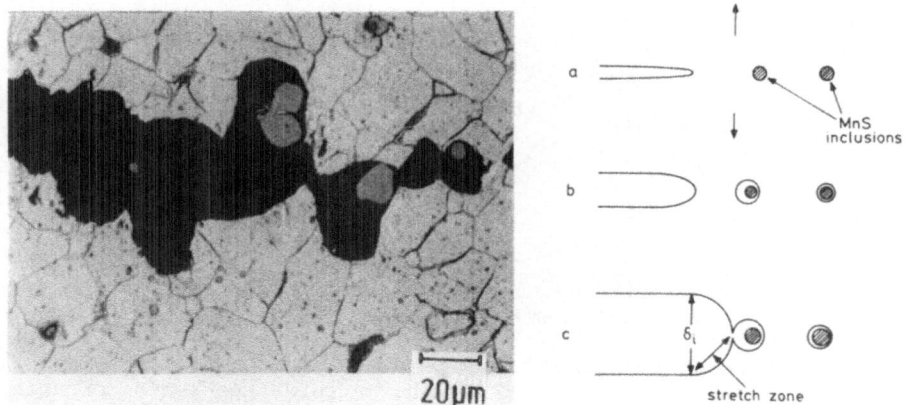

FIG. 15: Crack propagation by void coalescence

A model of void growth due to Rice and Johnson /40/ predicts that the
crack opening displacement δ_i will be in the range of 1.0 x_o to 2.7 x_o.
For materials with high volume fractions of non-metallic inclusions,
with a matrix free of second phase particles and with high work
hardening capacity the prediction is in good agreement with experimental
results. In a low carbon, free-cutting steel the value of δ_i and of
$(d\delta/da)$ is controlled almost entirely by the size, radius R_o, and the
spacing x_o, of the sulfide inclusions. Due to Rice and Tracey /41/ the
void growth depends on the state of stress in terms of $\sigma_m/\bar{\sigma}$ where σ_m
is the mean stress and $\bar{\sigma}$ is the equivalent stress:
The growth of a spherical void in a non-hardening material is given by

$$\frac{dR}{d\bar{\varepsilon}} = 0.28 \cdot R \cdot \exp (3\ \sigma_m /2\bar{\sigma}) \tag{23}$$

where R is the radius of the void and $d\bar{\varepsilon}$ the increment of equivalent
strain.

From these predictions two conclusions may be drawn. Firstly
initiation values of fracture toughness may be increased by ensuring
that inclusions are small and widely spaced. This is easily achieved in
steels with low sulfur content and sulfur shape control /42/. As weld
metal, in particular, contains a high volume fraction of closely-spaced
silicates, acting as void initiator, δ_i values are generally lower than
for the parent steel. Secondly the specimen orientation will influence
elastic-plastic fracture toughness data, due to the shape of sulfides
elongated by the rolling process /43,44/, resulting in lower toughness
values for crack initiation and crack propagation especially in the
S-L direction. Experimental results (see Fig. 16)

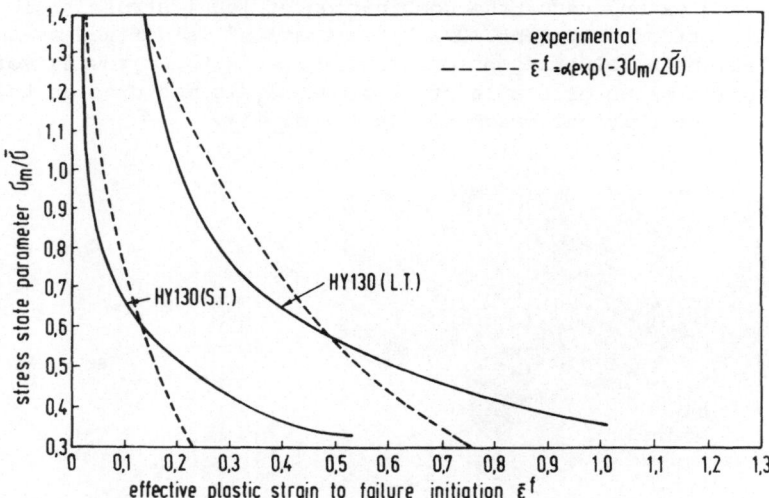

FIG. 16: Influence of stress state parameter on the effective plastic
strains to failure initiation. (After Hancock and McKenzie /46/)

have also been presented by Hancock and Cowling /45/ demonstrating,

that, in low-alloy, quenched and tempered steels, the strain required
to initiate ductile failure strongly depends on the stress state parameter, given in Eq. (23). Testing axisymmetric notched tensile specimens the stress state parameter $\sigma_m/\bar{\sigma}$ is given by

$$\frac{\sigma_m}{\bar{\sigma}} = \frac{1}{3} + \ln\left(\frac{d}{2R} + 1\right) \tag{24}$$

where R is the radius of the notch and d is the radius of the net
section and the equivalent plastic strain $\bar{\varepsilon}_p$ is given by

$$\bar{\varepsilon}_p = 2 \ln(d_o/d) \tag{25}$$

where d_o is the initial radius and d is the actual radius.
Combining these failure initiation curves with the Rice and Johnson /40/
solutions (see Fig. 17) a COD value for crack initiation δ_i = 90 μm is

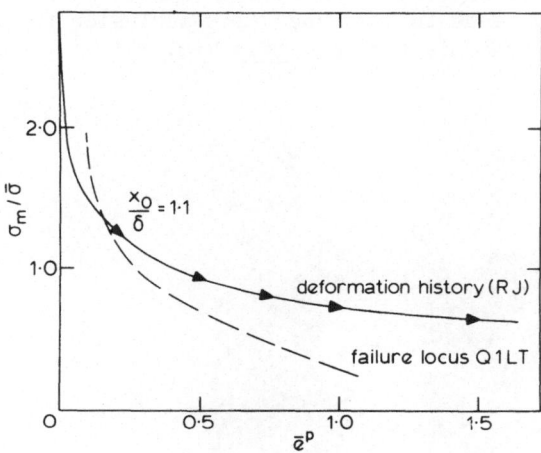

FIG. 17: Superposition of Rice-Johnson deformation history and failure
locus for HY 80. (After Hancock and Cowling /45/)

predicted for the steel HY 80. This compares with a δ_i value of 91 μm
measured in a 10:1 DEC specimen in the LT direction.
    Influences of specimen orientation on EPFM-parameters have also
been reported concerning steels with low sulfur content and sulfur shape
control /42/. Thus the influence of texture and anisotropic deformation
behaviour has also to be taken into account.
    The models presented above assume an equal distribution of non-
metallic inclusions, but this is often not the case. Inclusions are
often observed in well defined bands, and the spacing between the bands
is greater than between the inclusion colonies.

4.2 "Shear Decohesion" in Alloy Steels

When the strain in the ligament necking between two voids nucleated at inclusions becomes sufficiently large, it is possible to initiate micro-voids at other second-phase particles, such as carbides /39/. In a tempered, vacuum-melted forging steel this effect will be greatest, as non-metallic inclusions are widely-spaced and there is a high volume fraction of carbides. During deformation, dislocations tangle around the carbides stressing the carbide/matrix interface. With increasing plasticity the interface can decohere. The resistance to deformation then decreases catastrophically, so that coalescence of microvoids is rapid, once they initiate. The criterion for initiation is locally an interfacial stress, but since a critical number of dislocations in a tangle is necessary, it is convenient to represent the initiation criterion more macroscopically as a critical shear strain. The magnitude of the strain is controlled by the work-of-fracture of the carbide/matrix interface. In very clean steels, in which spacing of inclusions is extremely large microvoid nucleation on carbides occurs ahead of the crack tip and fracture tends to "Zig-zag" from inclusion to inclusion along the lobes of the plastic zone (see Fig. 18).

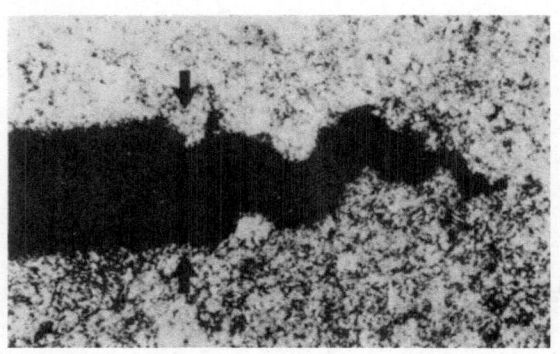

$$\approx 100 \ \mu m$$

Fig. 18: 'Zig-zag'-fracture in quenched and tempered alloy steel with position of original crack tip indicated

Ductile fracture by void coalescence can be truncated by shear deco-hesion. Due to Hancock and Mackenzie /46/ the volume fraction f of holes required for flow localisation is a function of stress state:

$$f = \{0,56 \ \exp \ (\frac{3\sigma_m}{2\bar{\sigma}})\}^{-1} \qquad (26)$$

e.g. if there is a highly tri-axial stress state, for example $\sigma_m/\bar{\sigma} = 1,4$ (eq.26) suggests, that only a 22% volume fraction of holes is required for localisation, whereas for $\sigma_m/\bar{\sigma} = 0,5$ an 85% volume fraction is necessary. As the shape of non-metallic inclusions, such as elongated Mn-sulfides, may influence the state of stress, as reported above, a tendency of shear localisation is observed especially in the S-L orientation, giving lower δ and J values /43/. The interfacial work-of-fracture does also influence fracture toughness data. Trace impurity elements such as P segregate to intragranular carbide/matrix interfaces leading to a decrease in initiation values by a factor of 2. Other factors which influence the nucleation of microvoids are the morphology and chemical composition of the second phase particles, as they may decohere, crack or shear. When a quenched steel is tempered at low temperature, carbides are precipitated not only in a fine intra-lath dispersion (ε-Carbides) but also as elongated inter-lath cementite of dimensions 2 μm x 0.2 μm. These carbides appear to be responsible for a 350°C fracture toughness minimum for a transgranular cleavage mode in Fe -0.6 C /47/, but they could also produce a low crack tip ductility by cracking at low strains to produce closely-spaced micro-void nuclei.

5. TRANSITION FROM CLEAVAGE TO DUCTILE FRACTURE

For the safety of a structure it is necessary to assure that cleavage fracture is excluded and that the component behaves in a fully ductile manner. In the following the main parameters are summarized. In Fig. 19a the influence of grain size on the transition is demonstrated. The yield strengths of smooth and notched specimens, the maximum tensile stress and the microscopic cleavage stress are plotted as functions of temperature. Decreasing the grain size results in higher yield strength and higher cleavage fracture stress, the latter being more influenced than the yield stress. This shifts the transition from cleavage to ductile fracture to lower temperatures. In Fig. 19b the effect of strain rate is shown. Increasing the strain-rate does not influence the cleavage fracture stress, but the yield stress is in-creased, resulting in a shift of transition temperature to higher values. A similar influence is observed increasing the triaxiality of stress state, see Fig. 19c. The microscopic cleavage stress is not affected by the stress state. Yielding occurs according to Tresca's yield criterion when

$$\sigma_{f1} = \sigma_1 - \sigma_3 = \sigma_y \tag{27}$$

Cleavage fracture occurs when σ_1 equals σ_f^*, thus the transition tem-perature is shifted to higher values with increasing triaxiality. The last two effects can explain cleavage fracture preceded by a certain amount of stable crack growth: the increasing load causes crack tip blunting and the plastic zone spreading into the ligament. When crack propagates by void coalescence, the notch radius of the running crack becomes smaller than at initiation. As the material is prestrained and the strain-rates are increased, the critical conditions for cleavage fracture may be locally reached. An example of this behaviour in the case of a surface crack will be shown in the lecture by Munz. When toughness parameters are evaluated as a function of temperature for bcc-alloys, a transition from cleavage to ductile fracture is observed with increasing temperature.

The microscopic failure behaviour changes from contained to general yielding. Under contained yielding conditions the Ritchie, Knott and Rice failure criterion is successfully used, but if fracture occurs after general yielding no micromechanical criterion has been established up to now. This is subject of current research work.

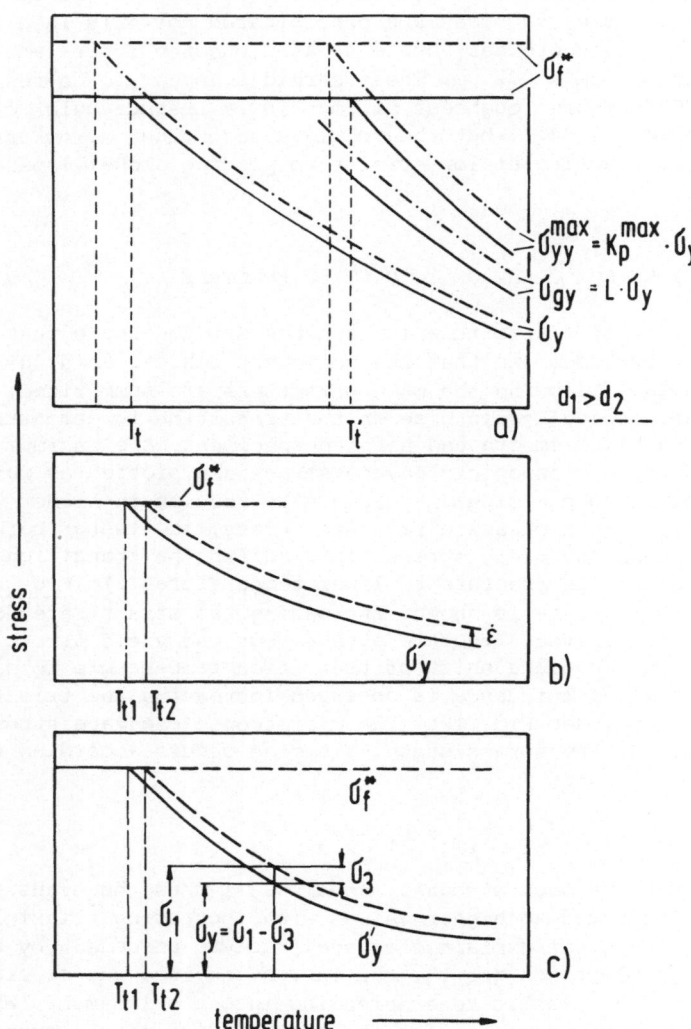

Fig. 19: Influence of grain size, strain rate and state of stress on the ductile/brittle transition

REFERENCES

/1/ Schwalbe, K.H.: in 'Bruchmechanik metallischer Werkstoffe',
K. Hauser Verlag, München, Wien, 1980

/2/ Hahn, G.T. a. Rosenfield, A.R.: 'Local Yielding and Extension of
a Crack under Plane Stress', Acta Metallurg. 13 (1965),
p. 293/306

/3/ Burdekin, F.M. a. Stone, D.E.W.: 'The Crack-Opening-Displacement
Approach to Fracture Mechanics in Yielding Materials', J. Strain
Analysis 1 (1966), p. 145/53

/4/ McMeeking, R.M. and Parks, D.M.: 'On Criteria for J-Dominance of
Crack Tip Fields in Large-Scale Yielding', in: 'Elastic-Plastic
Fracture' ed. J.D. Landes et al. ASTM STP 668, p. 175, 1979

/5/ Willoughby, A.A.: 'The Influence of Microstructure on Resistance
to Slow Crack Growth in Structural Steel', PhD thesis, Imperial
College, London, 1979

/6/ Slatcher, S.: 'The Fracture Toughness of Forging Steels', Phd
thesis, University of Cambridge, 1983

/7/ Hutchinson, J.W. and Paris, P.C.: 'Stability Analysis of J-
Controlled Crack Growth', in: 'Elastic-Plastic-Fracture' ed. J.D.
Landes et al., ASTM STP 668, 1979

/8/ Kelly, A., Tyson, A. and Cottrell A.H.: 'Ductile and Brittle
Crystals', Philos. Mag. 15 (1967), p. 567

/9/ Rice, J.R. and Thomson, R.: 'Ductile Versus Brittle Behaviour of
Crystals', ibid. 29 (1974), p. 73

/10/ Low, J.R.: 'The Relation of Microstructure to Brittle Fracture'
Trans. ASM 46 A (1954), p. 163

/11/ Zener, C.: 'The Micromechanism of Fracture', ASM 40 (1948), p. 3/31

/12/ Stroh, A.N.: 'The Formation of Cracks as a Result of Plastic Flow',
Proc. Roy. Soc. A 223 (1954), p. 404/14

/13/ Hahn, G.T., Averbach, B.L., Owen, W.S. and Cohen, M.: 'Initiation
of Cleavage Microcracks in Polycristalline Iron and Steel', in:
Proc. Int. Conf. on Atomic Mechanisms of Fracture, Swamscott
(1959), p. 91/116

/14/ Smith, E. and Barnby, J.T.: 'Crack Nucleation in Crystalline
Solids', Met. Sci. J. 1 (1967), p. 56/64

/15/ Kandar, M.H.: 'Cleavage in Zinc', Met. Trans., 2 (1971), p. 485

/16/ Curry, D.A., King, J.E. and Knott, J.F.: 'Effects of Hydrostatic
Tension on Cleavage Fracture of Pure Polycrystalline Zinc', Met.
Sci., 12 (1978), p. 247/50

/17/ Knott, J.F.: 'Some Effects of the Hydrostatic Tension on the
Fracture Behaviour of Mild Steels', J. ISI, 204 (1966), p. 104/111

/18/ Cottrell, A.H.: 'Theory of Brittle Fracture in Steel and Similar
Materials', Trans. AIME, 212 (1958), p. 192

/19/ Reiff, K.H.: 'Einfluß der Ausscheidungsdicke auf das Sprödbruchverhalten', Arch. Eisenhüttenwes., 43 (1972), p. 567/70

/20/ McMahon, C.J. and Cohen, M.: 'Initiation of Cleavage in Polycrystalline Iron', Acta Metall., 13 (1965), p. 591

/21/ Smith, E.: 'The Nucleation and Growth of Cleavage Microcracks in
Mild Steels', Proc. Conf. Physical Basis of Yield and Fracture,
Oxford, Inst. Phys. and Phys. Soc., London, 1966, p. 36

/22/ Riedel, H. and Kochendörfer, A.: 'Cleavage Fracture of Steels at Low Temperatures', Arch. Eisenhüttenwes., 50 (1979), p. 525/34

/23/ Knott, G.F. and Cottrell, A.H.: 'Notch Brittleness in Mild Steel', J. ISI, 201 (1963), p. 249

/24/ Orowan, E.: 'Fracture and Strength of Solids', Rep. Prog. Phys. 12 (1948), p. 185/232

/25/ Aurich, D.: 'Bruchvorgänge in metallischen Werkstoffen', Werkstofftechn. Verlagsgesellschaft, Karlsruhe (1978)

/26/ Aurich, D. and Wobst, K.: 'Der Einfluß plastischer Vorverformung auf den Spaltbruch von Stahl', Arch. Eisenhüttenwes., 52 (1981), p. 201/08

/27/ Ritchie, R.O., Knott, J.F. and Rice, J.R.: 'On the Relationship Between Critical Tensile Stress and Fracture Toughness in Mild Steel', J. Mech. Phys. Solids, 21 (1973), p. 395/410

/28/ Kühne, K., Redmer, J. and Dahl, W.: 'Elastic Plastic FEM Calculations Of The Stress Distribution of Notches With Different Geometry', Eng. Fract. Mech., 16, 1982, p. 845/55

/29/ Kühne, K., Dr.-Ing. Dissertation, RWTH Aachen, 1982

/30/ Dahl, W. and Uebags, M.: 'Der Einfluß der Korngröße und Korngrenzenkarbiddicke auf das Sprödbruchverhalten unlegierter Baustähle mit 0.09 - 0.14 % C bei tiefen Temperaturen', Stahl und Eisen, 97 1977, p. 486

/31/ Curry, D.A. and Knott, J.F.: 'Effects of Microstructure on Cleavage Fracture Stress in Steel', Met. Sci., 12 (1978), p. 511/14

/32/ Dahl, W., Uebags, M. and Kühne, K.: 'Der Einfluß einer Vorverformung und Alterung auf die Sprödbrucheigenschaften eines unlegierten Stahles', Arch. Eisenhüttenwes., 48 (1977), p. 541/45

/33/ Groon, J.D.G. and Knott, J.F.: 'Cleavage Fracture In Prestrained Mild Steel', Met. Sci., 9 (1975), p. 390/400

/34/ Sandström, R., Engbert, G. and Bergström, G.: 'The Influence Of Prestrain And Ageing On Cleavage Fracture In A C-Mn-Steel', Met. Sci., 15 (1981), p. 409/412

/35/ Curry, D.A. and Knott, J.F.: 'The Relationship between Fracture Toughness and Microstructure in the Cleavage Fracture of Mild Steel', Met. Sci., 10 (1976)

/36/ Tracey, D.M.: 'Finite Element Solutions for Crack-Tip Behaviour in Small Scale Yielding', J. Engng. Mat. Techn. (1976), p. 146/51

/37/ Dormagen, D., Dahl, W. and Dünnewald, H.: 'Zur Korrelation der mikroskopischen Spaltbruchspannung mit Bruchmechanik- und Großzugversuchen am Beispiel des Stahls 20 MnMoNi 55', DGM-Hauptversammlung, 1983

/38/ Ritchie, R.O., Geniets, L.C.E. and Knott, J.F.: 'Effects of Grain-Boundary Embrittlement on Fracture and Fatigue Crack Propagation in a Low Alloy Steel', in: 'Microstructure and Design of Alloys', Proc. 3rd Int. Conf. on Strength of Metals and Alloys, Cambridge 1973, Vol. 1, p. 124

/39/ Knott, J.F.: 'Micromechanisms of Fibrous Crack Extension in Engineering Alloys', Met. Sci. 14 (1980), p. 375

/40/ Rice, J.R. and Johnson, M.D.: in: Inelastic Behaviour of Solids, Kanninen et al., ed. Mc Graw-Hill (1970), p. 641/61

/41/ Rice, J.R. and Tracey, M.A.: 'On the Ductile Enlargement of Voids in Triaxial Stress Fields', J. Mech. Phys. Solids, 17 (1969), p. 201/17

/42/ Wilson, A.D.: 'Characterization of Plate Steel Quality Using Various Toughness Measurement Techniques', in 'Elastic-Plastic-Fracture', ASTM STP 668 (1979), p. 469/92

/43/ Green, G.: 'Effects of Specimen Orientation on Ductile Crack Initiation and Growth', pres. at 'Fourth European Congress on Fracture', Leoben, 1982

/44/ Dormagen, D. and Dahl, W.: 'On the Influence of Specimen Orientation on the Ductile Fracture Behaviour of the Steel Fe 350', in: 'Application of Fracture Mechanics to Materials and Structures', Martinus Nijhoff Publishers, The Hague (1984), p. 393/400

/45/ Hancock, J.W. and Cowling, M.J.: 'Role of State of Stress in Crack-Tip Failure Processes', Met. Sci., 14 (1980), p. 293/304

/46/ Hancock, J.W. and Mackenzie, A.C.: 'On the Mechanisms of Ductile Failure in High-Strength Steels Subjected To Multiaxial Stress-States', J. Mech. Phys. Sol., 24 (1976), p. 147/69

/47/ King, J.E., Smith, R.F. and Knott, J.F.: 'Toughness Variations during the Tempering of a Plain Carbon, Martensitic Steel', in Fracture 1977, Proc. 4th Int. Cong. on Fracture ed. D.M.R. Taplin, Press Pergamon, 2 (1977)

NUMERICAL ANALYSIS IN EPFM

L.G. Lamain
Commission of the European Communities
Joint Research Centre - Ispra Establishment
21020 Ispra (Va) - Italy

ABSTRACT.   Some numerical aspects of EPFM are treated in this lecture.
As well as considering the elastic-plastic overall response of the
structure, attention is given to different ways of evaluating crack
characterizing parameters. Types of elements are discussed with respect
to incompressibility constraints and crack tip modelling using the 1/4
point technique. Modified shape functions for collapsed elements are
proposed. Remarks on material modelling, taking the Bauschinger effect
into account, are made. A detailed description of the stress evaluation
process and equilibrium iteration techniques is given. Concerning the
crack characterizing parameters, different methods for calculating the
J-integral are discussed and compared and an overview of the COD definition is given. In particular the evaluation of the J-integral in
three-dimensional structures is treated in some detail. Three methods
to simulate crack growth are discussed: nodal release, stiffness reduction and node shifting. The result of a crack growth analysis of a
three point bend specimen in 3D is presented.

1.   INTRODUCTION

The finite element technique offers an almost unique possibility to
treat the most complex structural problems, because of its capability
to model structures of any shape and to take arbitrary material behaviour into account. In the analysis of cracked bodies the FE technique
has become the major technique used and much research is going on in
this field. As well as an accurate evaluation of the elastic-plastic
overall response of the structure, research is focused on crack tip modelling, simulation of crack growth and the evaluation of crack describing parameters.
    In this respect higher order isoparametric elements offer attrac-
tive possibilities: the midside nodes can be put at a quarter distance
from the crack tip to obtain a $1/\sqrt{r}$ singularity in the strain field
/1,2/, quadrilaterals can be distorted into triangles with independent
nodes in the collapsed points in order to model crack tip blunting
with a rather coarse mesh and to obtain a $1/r$ singularity in the strain

227

L. H. Larsson (ed.), Elastic-Plastic Fracture Mechanics, 227–261.

field and in three-dimensional structures curved crack fronts can be modelled. Crack growth can also be simulated by releasing the DOFs in the crack plane or by shifting the crack tip node. Often a combination of both techniques is used to obtain larger crack growth.

For the crack describing parameters, the generally accepted J-integral as defined by Rice /3/ is very attractive as it does not require a very fine modelling of the crack tip region. The numerical evaluation in FEM of the analytical expression for this line integral can be performed easily. In 1974 Parks /4/ introduced the virtual crack extension method to calculate the J-integral for the elastic case in terms of FEM. This method was later extended to elastic plastic situations /5,6/. Only a slight extension is needed to make this method applicable to 3D problems also.

Although all these techniques are widely used to analyse experiments and to predict the behaviour of structures, some precautions must still be taken. As was shown in a number of round robins /7,8,9/ the results of an elastic-plastic FE analysis can differ considerably, depending on the theoretical formulation used in the code and on the discretisation of the structure. More subjective variables such as step size of the loading and tolerances, and schemes of equilibrium iterations can also play an important role. Also the fact that most of the elements cannot describe the incompressibility in the fully plastic range can seriously degrade the accuracy of the solution.

The present lecture is intended to give an overview of the current possibilities of FEM in EPFM. Only material non-linearities are taken into account. The topics are mainly based on experience gained during the development and use of the LAMCAL code /10,11,12/. This code is focused on EPFM applications. It uses higher order elements such as the 8-node plane, the 20-noded brick and the semiloof shell elements. The iteration scheme can be specified from initial stress to an original Newton scheme. To account for the Bauschinger effects during unloading and reverse loading (this situation occurs for instance if the crack grows) a combined isotropic-kinematic hardening rule is applied. The 8-node and 20-node elements can be collapsed into a triangle or wedge, respectively, both with dependent or independent DOFs in the collapsed nodal points. If the collapsed nodes have the same DOF corrected shape functions are used to improve the behaviour of these elements.

Three routines to evaluate the J-integral are available; a contour through the nodal points, a contour through the integration points and the virtual crack extension method. Crack growth can be simulated by node release or node shifting. The latter is more accurate as it can model small amounts of growth with rather large elements. Both methods can be combined. In the present lecture a crack growth analysis in 3D, using the node shift method is discussed.

2. ELEMENT TYPES

2.1 Incompressibility Constraints

All the common element types can, in principle, be used in EPFM, but in
the literature the most frequently encountered are the higher order
elements. These elements are very attractive due to their flexibility
to model complex structures and the capability to provide good accuracy
with a rather coarse mesh. If distorted, however, the accuracy becomes
less. One of the drawbacks of most of the (higher order) elements is
that they cannot describe the incompressibility in the plastic range in
a general way. In plane strain, for instance, the incompressibility condition can be written as:

$$\frac{\partial u}{\partial x} + \frac{\partial v}{\partial y} = 0$$

where $u(x,y)$ and $v(x,y)$ are the x and y components of the displacement.
It is easy to check, using the shape functions, that this condition
cannot be satisfied. Depending on the element type, a certain number
of additional conditions, usually called incompressibility constraints,
are imposed on the interpolation functions for the displacement field.
For example, consider the special case of a 4-node quadrilaterial element with sides parallel to the x- and y-axis. The displacement field
can then be written as

$$u = a_1 + a_2 x + a_3 y + a_4 xy$$

$$v = b_1 + b_2 x + b_3 y + b_4 xy$$

The incompressibility condition requires

$$a_2 + b_3 + a_4 y + b_4 x = 0$$

in the whole element. This gives the following three incompressibility
constraints

$$a_2 + b_3 = 0$$

$$a_4 = 0$$

$$b_4 = 0$$

In /13/ the number of incompressibility constraints for the most common elements are given.

As in the crack tip region much plasticity is present, it is important that crack tip elements can describe the incompressibility to a
certain extent. As was pointed out in /13/, the least one can require

of these elements is that if the number of elements is increased, the solution converges to the right one. This is the case if the number of DOFs in a mesh increases faster than the number of incompressibility constraints. Theoretically, for the usual quadratic isoparametric 8-node element, this is not the case as the ratio DOF/constraints varies from 1 to 3/4, depending on straight or curved boundaries. But the mesh configuration also plays a role and in practice the 8-node element with straight boundaries is just able to satisfy the above mentioned requirement. If the 8-node element is collapsed into a traingle (crack tip element), the ratio reduces to 2/3. For the modified collapsed 8-node element, discussed hereafter, the ratio becomes again equal to 1. In /13/ a general way to treat the incompressibility constraints is proposed by modifying the virtual work principle.

2.2 Crack Tip Elements

In the past special crack tip elements were developed, having a displacement field proportional to $\sqrt{r}$ radiating away from the tip, giving a $1/\sqrt{r}$ singularity in the strain field, which represents the elastic case. A more attractive way of constructing these singular elements is offered by the so-called 1/4 point technique /1,2/. Here the midside nodes of the second order elements are put at a 1/4 distance from the crack tip node, to obtain the $1/\sqrt{r}$ singularity.

In EPFM, however, special elements are less needed as the singularity in the strain field is not exactly known and may even depend on material properties (HRR singularity).

For accuracy reasons, triangular and wedge type elements are more suited to model the crack tip regions than quadrilaterals or bricks. An easy way of obtaining triangular and wedge elements is to collapse isoparametric elements by putting the appropriate nodes on the same geometrical position. Two techniques can be used:
a) independent DOFs in the collapsed points, so that the nodes can move independently of each other;
b) all collapsed nodes have the same DOF.

Method a) is often used to model a kind of crack tip blunting and to make use of its $1/r$ singularity in the strain field. In /14/ it has been shown that this type of element possesses theoretically unbounded terms in the stiffness matrix. However, in practice this will not give rise to numerical problems as a bounded stiffness is obtained due to the approximate character of the Gaussian integration. If in this type of collapsed element the midside nodes are put at a 1/4 distance from the tip, a $1/r + 1/\sqrt{r}$ singularity is obtained, which should be the elasto-plastic one. However, in this case this element shows too high a stiffness in the collapsed points. In Fig.1 this is shown for a collapsed 8-node element applied in a 3PB specimen. The CTOD is here defined as the maximum displacement in the y-direction of any of the collapsed nodes. In the case of the 1/2 point elements it was the same node over the whole loading range, the node which was furthest away from the crack tip. In contrast, in the 1/4 point elements the maximum y-displacement was not always at the same node. Even negative y-displacements of some of the nodes were found.

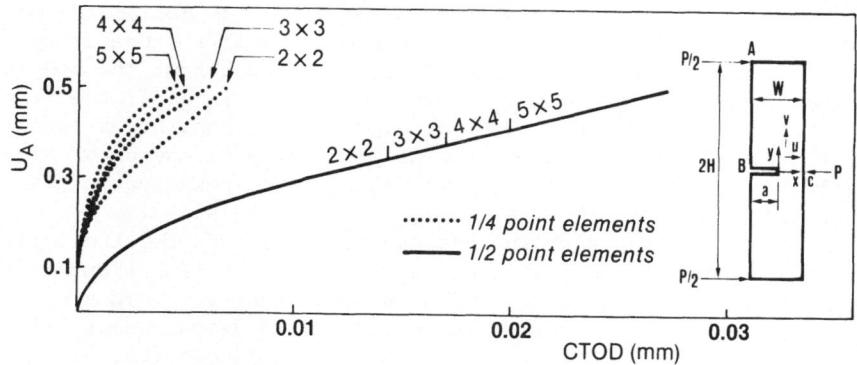

Figure 1.   CTOD as a function of the order of integration.

The conclusion that the 1/4 point elements (1/r + 1/√r singularity) behave too stiffly is based on the fact that the extrapolation method, which is discussed in chapter 6.2 and which is less sensitive to the way in which the crack tip is modelled, furnished a CTOD which was almost identical to that of the 1/2 point elements. It is interesting to note that the CTOD found with these 1/2 point elements is independent of the order of integration. This might indicate that the unbounded stiffness term of this element is not active, although the integration points are very close to the crack tip. In fact, theoretically it can be found that the unbounded terms in the 1/4 point collapsed elements are more severe, so that this might explain why here the CTOD (dotted lines in Fig.1) depend on the order of integration.

The second technique b) is to give all the collapsed points the same DOF. It is known that these types of collapsed elements can have very strange behaviour under certain circumstances, due to inconsistencies in the interpolation functions. Some of the difficulties can be avoided if the sides of these elements are kept straight with the midside nodes exactly in the middle. However, this only affects the interpolation function for the geometry and it does not remove the inconsistency in the interpolation function for the displacements. A better and more general way of improving the behaviour of the collapsed higher order isoparametric elements was proposed by Irons /15/. He modified the shape functions in such a way that inconsistencies in the interpolation function for both the geometry and the displacements are removed, even though the sides are curved.

To obtain the 1/√r singularity in the strain field by using 1/4 points, all the sides must be straight in the unmodified elements. This might be no problem for collapsed quadrilateral 8-node elements but it puts strong limitations on the use of collapsed 20-node brick elements, e.g. if curved crack fronts are to be modelled. It is found that the modified collapsed quadrilateral 8-node elements with 1/4 points possess the 1/√r singularity even if the side opposite to the collapsed points is curved.

In the same way it is possible to construct modified shape functions for the collapsed 20-node brick element. As well as better behaviour in general, all sides may be curved including the collapsed one (crack front) and if the 1/4 points are used, the $1/\sqrt{r}$ singularity is obtained all along the curved collapsed side and throughout the body of the element. Applying the limit approach presented in /16/, the stress intensity factors K_I and K_{II} can be expressed as continuous functions along the curved crack front in the displacements of the crack tip elements. If the 1/4 point technique is not used, the behaviour of the element is still improved both with straight and curved sides.

In /17/ it was shown that the modifications for the collapsed 8-node element can be obtained easily by an additional bubble function distribution over the serendipity shape functions. This distribution was found by first collapsing a Lagrangian 9-node quadrilateral element and afterwards removing the ninth (central) node (Fig.2). Analogous to this approach the modifications for the collapsed 20-node element can be found, starting from a Lagrangian 27-node brick element /18/ (Fig.3).

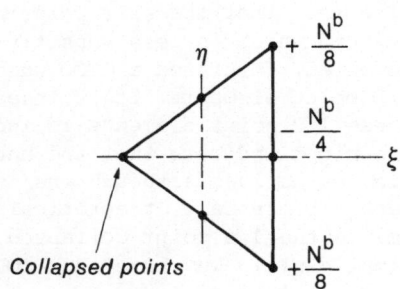

Figure 2. Modifications of the serendipity shape functions for the collapsed 8-node element by adding the bubble function.

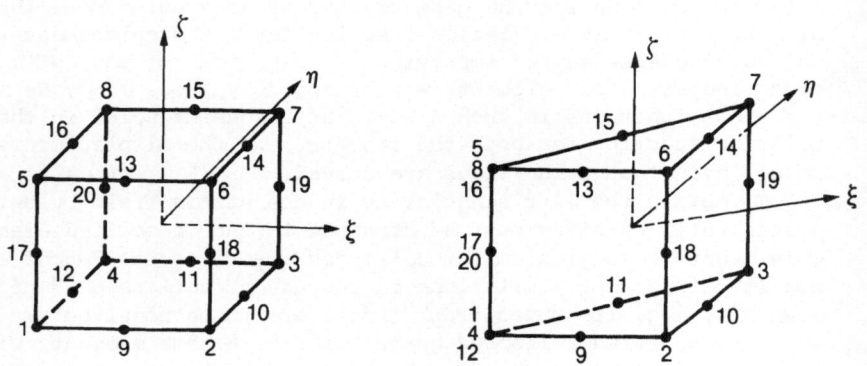

Figure 3. Collapsed 20-node brick element.

3. MESHES AND MESH GENERATION

An important and often time-consuming aspect of FEM analysis is the generation of the mesh. Many mesh generators based on several principles are currently in use, each type with its own field of optimal applicability. Examples of meshes of cracked bodies using higher order elements are given in Figs.4, 5, 6 and 7.

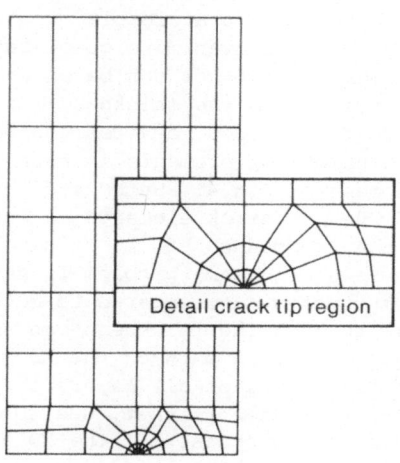

Figure 4. Mesh 1.

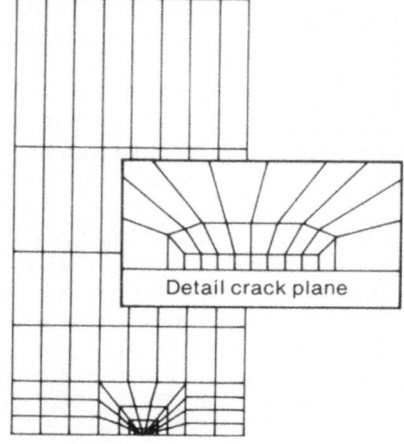

Figure 5. Mesh 2.

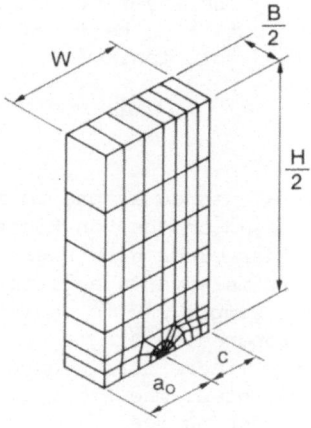
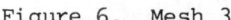

Figure 6. Mesh 3

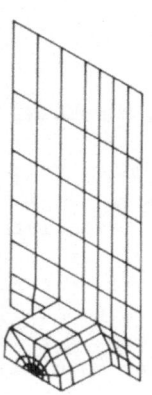

Figure 7. Mesh 4.

All these meshes are generated with the same mesh generator. This mesh generator, like many others, requires a sub-division of the structure into blocks having 4 boundaries so that it can be mapped on a square. Because of this mapping and back transformation procedure, the elements obtained can have strange shapes that can influence the accuracy of the calculation in a negative way (especially lower order elements). The usual smoothing techniques (e.g. Laplacian, equipotential) are difficult to apply to cracked bodies as the node position and the size of the crack tip elements are often fixed. A correction by hand of an automatically generated mesh is often unavoidable.

In 3D the generation of meshes becomes even more complicated. For simple structures like test specimens, 3D meshes can be obtained from the 2D ones by giving one or more layers in the thickness. In more complicated cases, ad hoc programs or graphical devices can be used. To keep the CPU time reasonable, the number of elements is rather limited. Fig.6 shows an extension of the mesh of Fig.4. Here half the thickness is modelled by only one layer of 20-node brick elements. A large difference in deflection curves of the two meshes can be found.

In Fig.7 a combined 2D/3D mesh of the same problem is shown. This one gives again a considerable improvement as compared to mesh 3 with the same CPU time. The J-integral through the thickness can also be calculated rather accurately. The connection between the 2D and 3D elements is made by tying the appropriate DOFs of all layers to those of the 2D part. A difficulty in combined meshes is that, due to different types of elements, the convergence rate decreases if the 2D elements become plastic and might even start to oscillate.

Depending on the dimensions of the structure, one has a choice between plane stress and plane strain elements for the 2D part. If the 2D part becomes plastic the overall behaviour is influenced by the choice.

From 3PB test pieces it can be seen that the plastic deformation takes place in a very restricted area around the remaining ligament, therefore a very acceptable compromise between CPU time and accuracy is obtained if the 2D elements are replaced by one layer of 3D elements. The CPU time per iteration will increase, but it is believed that fewer iterations are needed to obtain a converged solution.

4. MATERIAL MODELLING

Most of the codes to date use the Prandtl-Reuss equations to model the material behaviour. These equations are based on the Von Mises flow criterion and associated flow rule. As under crack growth situations, unloading and reverse loading can occur it is advantageous to take the Bauschinger effect into account. The most simple way to do this is by means of the kinematic hardening model proposed by Prager /19/. One of the limitations of this model is that only linear strain hardening can be used. Other models are much more complicated. The Prandtl-Reuss equations for combined isotropic-kinematic hardening read

$$d\sigma_{ij} = C_{ijkl} d\varepsilon_{kl} - \frac{2G}{S} (S_{ij} - \beta\alpha_{ij})(S_{pq} - \beta\alpha_{pq}) d\varepsilon_{pq} \qquad (1)$$

where:

C_{ijkl} = matrix of elastic material moduli

G = $E/2(1+\nu)$

S = $(2/3)\sigma_y^2(1-H'/3G)$ (H' is the hardening slope $d\sigma_v/d\varepsilon^p$)

σ_y = yield stress, σ_v = von Mises stress

E = Young's modulus

ν = Poisson's ratio

$\beta\alpha_{ij}$ = translation of yield surface in deviatoric stress space

β = combination factor isotropic-kinematic
        $\beta = 0$ isotropic
        $\beta = 1$ kinematic

$d\alpha_{ij}$ = $(2/3)H'd\varepsilon_{ij}^p$

The superscript (.P) denotes plastic components.

Eq.(1) is easy to use. However, in plane stress situations the third stress component is zero and not taken as a variable. Eliminating this component from the equations produces a non-symmetric matrix of material moduli. It can be made symmetric by introducing a modified incremental plastic strain tensor, which can be calculated as follows.
The stress strain equations are written as

$$\begin{Bmatrix} d\sigma_x \\ d\sigma_y \\ 0 \\ d\tau \end{Bmatrix} = \frac{2G}{1-2\nu} \begin{bmatrix} 1-\nu & \nu & \nu & 0 \\ \nu & 1-\nu & \nu & 0 \\ \nu & \nu & 1-\nu & 0 \\ 0 & 0 & 0 & \frac{1-2\nu}{2} \end{bmatrix} \begin{Bmatrix} d\varepsilon_x \\ d\varepsilon_y \\ d\varepsilon_z \\ d\gamma \end{Bmatrix} -2G \begin{Bmatrix} d\varepsilon_x^p \\ d\varepsilon_y^p \\ d\varepsilon_z^p \\ d\gamma^p \end{Bmatrix}$$

It should be noted that $d\gamma$ is twice the tensor component, while $d\gamma^p$ is the tensor component. Eliminating the third component gives

$$\begin{Bmatrix} d\sigma_x \\ d\sigma_y \\ d\tau \end{Bmatrix} = \frac{2G}{1-\nu} \begin{bmatrix} 1 & \nu & 0 \\ \nu & 1 & 0 \\ 0 & 0 & \frac{1-\nu}{2} \end{bmatrix} \begin{Bmatrix} d\varepsilon_x \\ d\varepsilon_y \\ d\gamma \end{Bmatrix} -2G \begin{Bmatrix} d\varepsilon_x^{*p} \\ d\varepsilon_y^{*p} \\ d\gamma^{*p} \end{Bmatrix}$$

The modified plastic strains (indicated by an asterisk) are defined as

$$d\varepsilon_x^{*p} = \frac{1}{1-\nu} d\varepsilon_x^p + \frac{\nu}{1-\nu} d\varepsilon_y^p$$

$$d\varepsilon_y^{*p} = \frac{1}{1-\nu} d\varepsilon_y^p + \frac{\nu}{1-\nu} d\varepsilon_x^p$$

$$d\gamma^{*p} = d\gamma^p$$

By a suitable back transformation the real components are obtained.

The combination factor β between the kinematic and isotropic hardening can be calculated from the cyclic uniaxial stress-strain curve
linearly, as (see Fig.8)

$$\beta = \frac{1}{2} \frac{\sigma_v - |\sigma_v^*|}{\sigma_v - \sigma_I}$$

where:

σ_I is the initial yield stress

σ_v is the Von Mises stress

σ_v^* is the Von Mises stress where reversed yielding takes place.

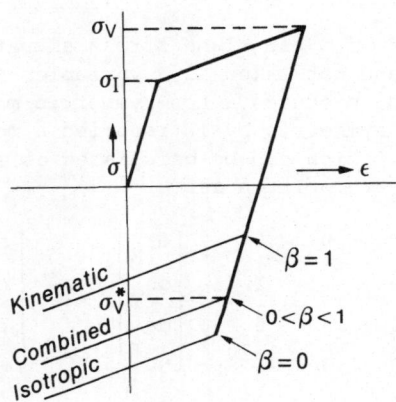

Figure 8.   Definition of β.

5. SOLUTION TECHNIQUES

In the displacement formulation of the finite element method for structural mechanics problems, the equilibrium condition is expressed by
means of the virtual work principle. For an incremental load step this
principle reads in matrix notation:

$$\int_V \delta\varepsilon'(\sigma + \Delta\sigma)\,dV = \int_S \delta u'(p + \Delta p)\,dS \qquad (2)$$

where:

V   = volume of the structure

S   = surface of the structure where the displacements are not prescribed

p   = total loading of previous increments
σ   = stress field due to p
Δp   = incremental loading
Δσ   = incremental stress due to Δp
δu   = a kinematically admissible virtual displacement field
δε   = virtual strain field due to δu
'   = transposition.

Equation (2) gives the equilibrium of the whole structure but the same equation holds if only a part of the structure is considered, e.g. the part described by one finite element. So as not to complicate the discussion too much by introducing indices and summation signs for element and node numbers, we will assume that the structure consists of one element. The equations obtained can easily be generalized by taking a summation over the number of elements. For the same reason we write the relationship between the displacement field within the element u and the nodal displacements (DOFs) $\tilde{u}$, as

$$u = N\tilde{u}$$

where N are the shape functions which depend on the element type. The strains expressed in the nodal displacements are

$$\varepsilon = B\tilde{u} \tag{3}$$

To obtain a symmetric stiffness matrix the same relation is taken for the virtual quantities

$$\delta\varepsilon = B\delta\tilde{u}$$

This relation in (2) allows us to eliminate $\delta\tilde{u}$ and the equation becomes after rearranging the known and the unknown terms:

$$\int_V B'\Delta\sigma dV = \Delta F + \left\{ F - \int_V B'\sigma dV \right\} \tag{4}$$

where F is the total nodal loading, calculated as $F = \int_S N\, p\, dS$ and ΔF is the incremental nodal loading, $\Delta F = \int_S N\, \Delta p\, dS$.

5.1 Equilibrium Iterations

The term between brackets in eq. (4) is the so-called residual load vector R and is a measure for the out-of-equilibrium state of the structure. This out-of-equilibrium is caused by material and/or geometrical non-linearities. It can be made as small as necessary by an iteration process. Several iteration processes can be applied, of which the Newton processes are the most popular.

Let us introduce the stress-strain relation in the form

$$\Delta\sigma = (C-Y)\Delta\varepsilon$$

where C are the elastic components of the material moduli and Y the plastic ones. This equation, together with eq.(3) in eq.(4), gives:

$$\left[\int_V B'(C-Y) B dV \right] \Delta \tilde{u} = \Delta F + R$$

or written in terms of the elastic K^O and plastic K_t stiffness matrices

$$(K^O + K_t) \Delta \tilde{u} = \Delta F + R$$

As this set of equations is solved in an iterative way, one can write

$$\sum_{i=1}^{p} (K^O + K_t^i) \Delta \tilde{u}^i = \Delta F + R^O$$

p is the total number of iterations. The total solution is obtained as

$$\Delta \tilde{u} = \sum_{i=1}^{p} \Delta \tilde{u}^i .$$

R^O indicates the residual load vector before the loading step ΔF. The ith contribution to the displacement vector is then

$$(K^O + K_t^i) \Delta \tilde{u}^i = \Delta F + R^O - \sum_{j=1}^{i-1} \left\{ K^O + K_t^j \right\} \Delta \tilde{u}^j$$

The last two terms of the right-hand side can be written as R^{i-1} because

$$(K^O + K_t^j) \Delta \tilde{u}^j = \int_V B' \Delta \sigma^j dV$$

This gives then the original Newton iteration scheme

$$(K^O + K_t^i) \Delta \tilde{u}^i = \Delta F + R^{i-1} \tag{5}$$

K_t^i depends on the stresses and must be assembled every iteration, K^O is constant if no geometry updating is applied and in theory it can be stored on disks so that it is assembled only once. In practice, however, it is more convenient to consider $K^O + K_t$ as one matrix and to assemble and solve it every iteration.

The initial stress method can be obtained formally from eq.(5) as

$$K^O \Delta \tilde{u}^i = \Delta F + R^{i-1} - K_t^i \Delta \tilde{u}^i$$

The last term in the right-hand side is neglected in the ith iteration, as it is not known at that moment. The equation of the initial stress method then becomes

$$K^O \Delta \tilde{u}^i = \Delta F + R^{i-1} \tag{6}$$

The advantage of this method is that the stiffness matrix is only assembled and triangulized once. This reduces the CPU time per iteration considerably, but, depending on the nature of the problem, many more iterations may be needed as compared to the Newton method. It is possible that under certain conditions no convergence can be obtained. The similarity between expressions (5) and (6) immediately offers the idea of combining both methods, by updating the stiffness matrix only from time to time. This can be written as

$$\text{updated} \quad (K^O_t + K^i_t) \Delta \tilde{u}^i = \Delta F + R^{i-1}$$

$$\text{constant} \quad (K^O_t) \Delta \tilde{u}^i = \Delta F + R^{i-1}$$

This modified Newton iteration technique can reduce the CPU time considerably, but one must have some feeling as to how the structure will behave, to optimise the iteration scheme.

Another method which does not require a reassembling and solving of the stiffness matrix is the so-called BFGS method. Here the triangularised stiffness matrix is updated by the solution of the previous iteration in a way such that the solution is approached in a secant modulus manner /29/.

The three methods can be interpreted as generalizations of the one variable cases graphically shown in Figs.9a,b,c.

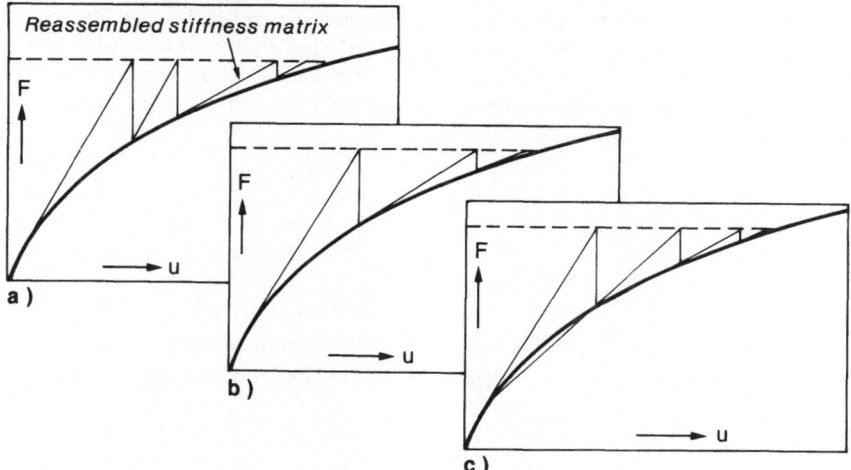

Figure 9. a) modified Newton iterations; b) original Newton iterations; c) BFGS (secant modulus) iterations.

5.2 Stress Evaluation

Due to the finite load increment the stress state will leave the yield surface if the loading in the deviatoric stress space is not radial. This can be very significant if large incremental strains are present as in the crack tip region. As this is a very local phenomenon, it is expected that the effect on the incremental displacement is less significant. On stress evaluation level some correction procedures can be applied, e.g. radial return, tangent radial return, secant method. These methods, discussed in /20/, ensure a stress state on or very close to the yield surface. To take the rotating normal into account also, they are usually applied in a substepping way, which can be seen as an incremental integration of the stress-strain relationships. The incremental strain is therefore divided into a number of smaller subincrements, and the normal plus the hardening slope are recalculated at the start of every subincrement (Fig.10). The number of subincrements can be made dependent on the angle φ between the position of the old and new stress points on the yield surface.

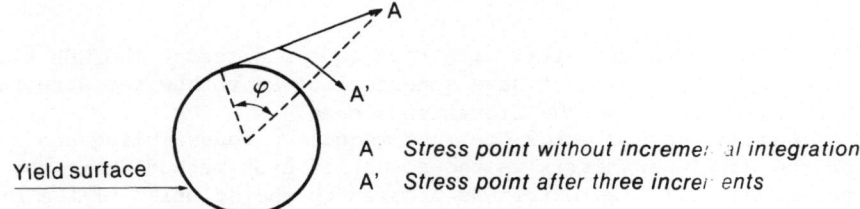

A   *Stress point without incremental integration*
A' *Stress point after three increments*

Figure 10.   Incremental integration of the stress-strain relation.

In crack growth situations it can often be noticed that an initially plastic point becomes elastic, reaches the yield surface again and undergoes reversed plasticity. It is very important that a code can detect these situations, otherwise a completely wrong stress will result (point A' instead of point B' in Fig.11).

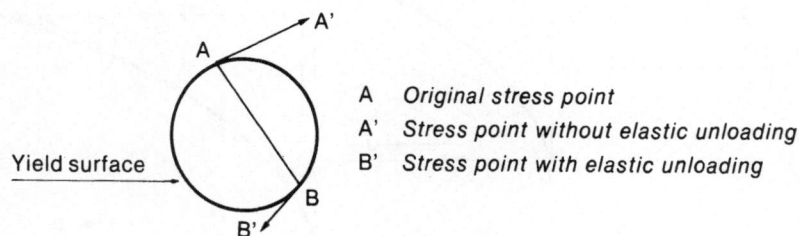

A   *Original stress point*
A' *Stress point without elastic unloading*
B' *Stress point with elastic unloading*

Figure 11.   Elastic unloading and reversed plasticity.

The complete yield criterion reads

$$f(\sigma^*) < 0 \quad \text{elastic}$$

$$f(\sigma) = 0 \quad \text{and} \quad \frac{\partial f}{\partial \sigma} \Delta\sigma^* \geq 0 \quad \text{plastic}$$

$$f(\sigma) = 0, \quad \frac{\partial f}{\partial \sigma} \Delta\sigma^* < 0 \quad \text{and} \quad f(\sigma^*) > 0 \quad \text{elastic unloading +} \\ \text{reversed plasticity}$$

$f(\sigma)$ is the yield function, σ is the stress state before the step, σ^* is the trial stress $\sigma^* = \sigma + \Delta\sigma^*$ and $\Delta\sigma^*$ is the elastic part of the stress increment.

During equilibrium iterations it may happen that, because of the path dependence of the solution, numerical unloading appears. In general it is very difficult to distinguish between physical and numerical unloading. If numerical unloading is detected, however, one should allow plastic recovery, so that the stress state remains on the yield surface.

5.3 Combination of Iteration and Substepping

In general the norm of the residual load vector R can serve as a convergence criterion. If the solution approaches the limit load, however, a small change in R can cause a large change in the displacement vector. In these cases the change in displacement during iteration is more suitable as a convergence criterion. Both criteria can be combined if the change in energy is considered. After convergence is reached, a small residual load vector will still be present. This can be added to the load vector of the next increment.

The effect of iteration and incremental integration of the stresses can be seen in Fig.12. The solutions of four combinations are given. A beam is analysed which is loaded in pure bending by prescribed displacements. Ideal elastic-plastic material behaviour and plane strain conditions are assumed. The dimensions are given in the figure. Half the beam is modelled by four isoparametric 8-node elements. It is found that in 10 plastic increments, the theoretical limit load could be obtained if original Newton iterations and 9 subintegrations for the stresses were applied. The accuracy was slightly less if subintegration was not applied but in that case many more iterations were required and the CPU time was increased by a factor 5. As can be seen the other two combinations gave completely wrong results.

6. EVALUATION OF THE CRACK DESCRIBING PARAMETERS

In this chapter the most common ways of evaluating the J-integral are discussed and a few remarks are made about COD.

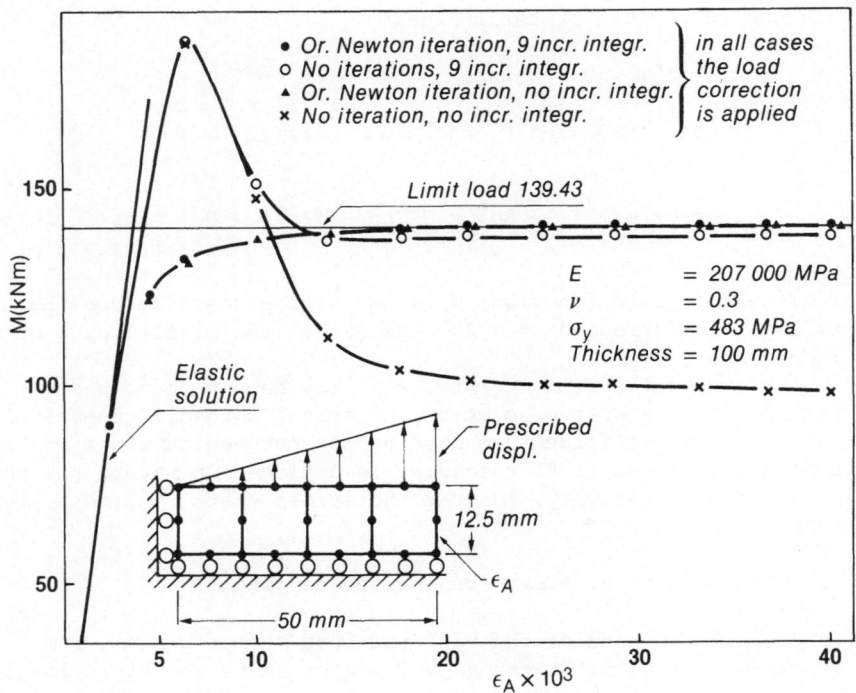

Figure 12. Outer fibre strain vs moment of beam in pure bending. Comparison of some solution techniques.

6.1 J-integral

Three different approaches to evaluate J can be distinguished:
- the line integral;
- virtual crack extension;
- from the load-line displacement curve.

6.1.1. <u>The line integral</u>. The well-known formula for the J-integral, originally introduced by Rice /3/ reads:

$$J = \int_{\Gamma} \left(W dx_2 - \bar{T} \cdot \frac{\partial \bar{u}}{\partial x_1} ds \right) \tag{7}$$

Here, Γ is a curve which surrounds the tip, $\bar{T}$ is the traction vector on Γ according to an outward unit vector $\bar{n}$ normal to the curve, W is the strain energy density.

In two-dimensional stress situations this integral is path-independent if the energy density can be written as a unique function of the stresses and the strains. In three-dimensional stress situations or if initial strains are present, a term containing a surface integral over the enclosed path must be added, as discussed by Bakker in this seminar.

The practical way in FEM of evaluating the J-integral defined by eq.(7) is along a line through the nodal points or along a line through the integration points of the elements. The first method requires an extrapolation from quantities in the integration points to the nodal points, e.g. stresses, derivatives of the displacements. The contributions made by several elements are then averaged and a kind of smoothing is obtained. It is also possible to consider the contribution from elements inside and outside the curve separately so that an impression can be obtained about the accuracy. The advantage of this method is that there are no requirements for the mesh to define a smooth path. The disadvantage is, of course, the extrapolation procedure, which requires more programming effort, especially when several integration schemes have to be taken into account. The accuracy might be less than that of the second method in which the integration path is defined by the integration points. This second method can be programmed very efficiently as it does not depend on the numerical integration scheme used to evaluate the element stiffness. In this case some requirements have to be put on the mesh to be able to define a smooth integration path (Fig.13). However, if the path is not completely closed, a reasonable answer will still be obtained. It can, of course, also be taken into account by proper programming. For some types of elements (triangles) and integration schemes it is difficult to apply this method.

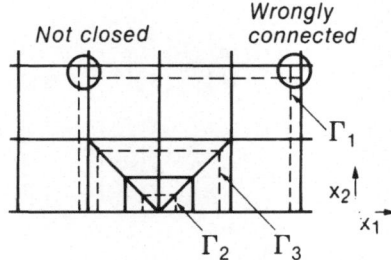

Figure 13. J-integral evaluation along a path defined by the integration points of the element.

If a 3*3 integration scheme is used to evaluate the element stiffness, three paths can be defined through one element. It has been observed that the two paths close to the boundary of the element furnish almost identical J-integrals while the one through the middle gives an about 10% higher value. It has been demonstrated by Bakker /21/ that the integrated average of the three values over the elements should be considered, which is then theoretically equal to the J found by the virtual crack extension method. The integral can be evaluated in a straightforward way. The strain energy density can be written as the sum of an elastic and a plastic part

$$W = W_e + W_p$$

in which

$$W_e = \frac{1}{2} \left\{ \frac{1+\nu}{E} s_{ij}s_{ij} - \frac{1-2\nu}{3E} \sigma^2_{hh} \right\}$$

The plastic strain energy can, in general, be evaluated incrementally as

$$W_p = \sum_{incr.} \left\{ \sigma_{ij} + \frac{\Delta\sigma_{ij}}{2} \right\} \Delta\varepsilon^p_{ij}$$

Based on

$$\sum \sigma_{ij}\Delta\varepsilon^p_{ij} = \int_{\varepsilon_p} \bar{\sigma} \, d\bar{\varepsilon}^p$$

where $\bar{\varepsilon}^p$ and $\bar{\sigma}$ are the effective plastic strain and the equivalent stress, it can be integrated analytically using the stress-strain relation, e.g.

linear hardening $\quad \bar{\sigma} = \sigma_y + H'\bar{\varepsilon}^p \rightarrow W_p = \dfrac{\bar{\sigma}^2 - \sigma^2_y}{2H'}$

power law hardening $\quad \bar{\varepsilon}^p = \left(\dfrac{\bar{\sigma}}{B_o}\right)^n \rightarrow W_p = \dfrac{n}{n+1} \bar{\sigma} \left(\dfrac{\bar{\sigma}}{B_o}\right)^n$

power law hardening $\quad \bar{\varepsilon} = \dfrac{\sigma_y}{E}\left(\dfrac{\bar{\sigma}}{\sigma_y}\right)^n \rightarrow W_p = \dfrac{n}{n+1} \dfrac{\bar{\sigma}}{E}\left(\dfrac{\bar{\sigma}}{\sigma_y}\right)^n - \dfrac{\bar{\sigma}^2}{2E}$

Here σ_y is the initial yield stress.

6.1.2. <u>Virtual crack extension method</u>. This method, originally described by Parks /4/ as the stiffness derivative method, is an energy method adapted to the finite element formulation. Extending the crack front, the change in potential energy can be calculated in the elastic case as

$$\Delta P = \frac{1}{2} u'\Delta K \, u \tag{8}$$

where:
P   = potential energy
u   = displacement vector
ΔK = change in stiffness matrix due to crack extension.

In 1977 this method was extended by Parks /5/ to non-linear material behaviour. Parks presented his work in terms of the FEM, but recently DeLorenzi /6/ gave a derivation in terms of continuum mechanics. The following derivation in terms of FEM has been used by the author for several years, writing the extended Griffith criterion in the form

$$J\delta A = \delta P + \delta U_p$$

where δA is the change in crack area, δP is the change in potential energy and δU_p is the change in plastic strain energy. The potential energy in matrix notation is

$$P = \frac{1}{2} \int_V \varepsilon_e'\sigma dV - u'F$$

where ε_e, σ, u and F is the solution of a FEM analysis, V is the volume of the structure. In the following e and p will be used as subscripts for elastic and plastic.

Keeping the loading constant, the first variation becomes

$$\delta P = \int_V \delta\varepsilon_e'\sigma dV + \frac{1}{2} \sum_{i=1}^{el.} \int_{\delta V_i^*} \varepsilon_e'\sigma dV - \delta u'F \qquad (9)$$

where δV_i^* is the change in volume of an element with a displaced node. It can be evaluated as the change in the determinant of the Jacobian in the integration points.

By applying the usual assumptions

$$\delta\varepsilon = \delta\varepsilon_e + \delta\varepsilon_p; \quad \varepsilon = Bu$$

with the first variation of the latter

$$\delta\varepsilon = B\delta u + \delta Bu$$

eq.(9) becomes

$$\delta P = \delta u'[\int_V B'\sigma dV - F] + u'\int_V \delta B'\sigma dV - \int_V \delta\varepsilon_p'\sigma dV + \frac{1}{2}\sum_{i=1}^{el.}\int_{\delta V_i^*}\varepsilon_e'\sigma dV$$

The term between brackets is zero if the structure is in a state of equilibrium, so that the unknown vector δu cancels.
The variation of the plastic strain energy can be written as

$$\delta U_p = \int_V \delta W_p dV + \int_{\delta V} W_p dV$$

where W_p is the plastic strain energy density. Hence

$$\delta P + \delta U_p \equiv JdA = u'\int_V \delta B'\sigma dV + \frac{1}{2}\sum_{i=1}^{el.}\int_{\delta V_i^*}\varepsilon_e'\sigma dV + \sum_{i=1}^{el.}\int_{\delta V_i^*}W_p dV$$

As only elements with a displaced node contribute to this expression, the summation can be restricted to these elements.
In finite terms the formula for J can be written as

$$
J = \frac{1}{\Delta A} \sum_{el.} \left\{ u' \int_{V_i^*} \Delta B' \sigma dV + \frac{1}{2} \int_{\Delta V_i^*} \varepsilon_e' \sigma dV + \int_{\Delta V_i^*} W_p dV \right\} \tag{10}
$$

V_i^*, ΔV_i^* and ΔB are the volume, the change in volume and the change in B-matrix of the considered element.
Arbitrary virtual displacements can be given as they appear only linearly in the expression. In Fig.14 a comparison of the three methods is given for a three point bend specimen.

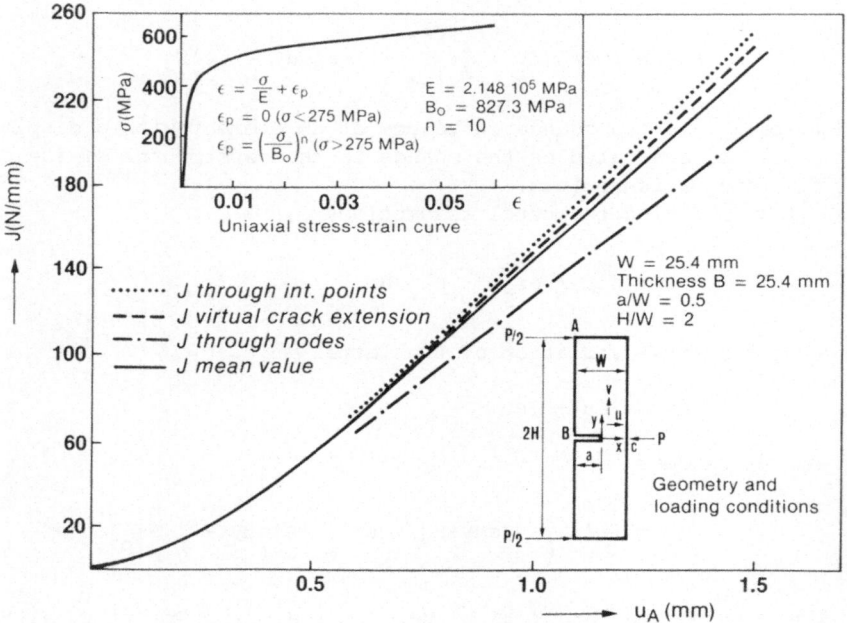

Figure 14. Comparison of several J-evaluation methods.

Equation (10) can be implemented very easily in any FEM code. As the stiffness changes only in those elements containing a displaced node, the extra calculations to be done are minimal. That this method works so well is explained by the fact that a possible error in the potential energy does not change so much with a slightly different crack tip configuration, so that the derivative of the error with respect to the crack length is almost zero. This property implies, however, that the displaced crack tip configuration must be as close to the original one as possible. For instance, if collapsed elements are used with independent node numbers in the crack tip and the crack is extended by displacing only the node that is fixed, the displaced configuration is so different from the original one that the result will be completely wrong ($J \to 0$). Good answers will be obtained if two rings of elements around the tip are displaced. This method can also be used to give an idea of the accuracy of the mesh if some arbitrary nodes are displaced. The change in energy must then be as close to zero as possible.

In plane structures this method can be applied very straightforwardly as the extension is constant over the thickness. In 3D more care is needed as the extension can change over the thickness. Here a global J^* and a local $J(s)$ can be defined. The relation is

$$J^* = \frac{1}{\Delta A} \int_{\Delta s} J(s)\delta(s)ds \qquad (11)$$

where s is the coordinate along the crack front, Δs is a certain distance along the crack front and $\delta(s)$ is the virtual displacement of the crack front. Although called global J^* it is not necessarily the J of the whole structure. It designates a mean J over a certain distance Δs along the crack front.

As crack growth always takes place over a certain distance Δs, the local $J(s)$ values must be seen in the sense of an asymptotic value for $\Delta s \to 0$. In FEM the smallest Δs is equal to the size of an element side at the crack front. The virtual crack extension method in 3D furnishes the J^* values. To know the local $J(s)$ values, additional calculations are needed. The following discussion is restricted to parabolic elements.

Consider such an element and denote the local $J(s)$ values at the nodes J_1, J_2 and J_3 (Fig.15). By displacing the nodes successively the global J_1^*, J_2^* and J_3^* can be calculated. To derive J_1, J_2 and J_3 from these values, an analytical expression must be assumed for $J(s)$ and $\delta(s)$. Say, introducing the curvilinear coordinate ξ

$$J(\xi) = J_1\xi(\xi-1)/2 + J_2(1-\xi^2) + J_3\xi(\xi+1)/2 \qquad (12a)$$

and

$$\delta(\xi) = \delta_1\xi(\xi-1)/2 + \delta_2(1-\xi^2) + \delta_3\xi(\xi+1)/2 \qquad (12b)$$

where δ_i is the virtual displacement in node i.

Note: This parabolic expression implies a linearly varying stress intensity factor K along the element side. This is in perfect agreement with the modified collapsed 20-node brick element if the shape of the element in the crack plane is rectangular.

If the midside node is in the middle

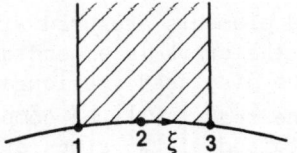

Figure 15. Crack front element.

$$s = \frac{\Delta s}{2} (1+\xi)$$

Working out eq.(11) furnishes for the change in energy

$$J^* \Delta A = \frac{\Delta s}{30} [J_1(2\delta_2 - \delta_3 + 4\delta_1) + J_2(16\delta_2 + 2\delta_3 + 2\delta_1) +$$

$$+ J_3(2\delta_2 + 4\delta_3 - \delta_1)] \qquad (13)$$

and the increase in crack area

$$\Delta A = [\delta_1 + \delta_3 + 4\delta_2] \frac{\Delta s}{6}$$

Now a choice must be made for the shape of the crack extension. From eq.(13) it can be seen that it is not possible to relate the local energy release rates J_1, J_2 and J_3 directly to the change in potential energy by a suitable choice of δ_1, δ_2 and δ_3, as this would give a negative crack increase over a part of Δs.

A practical shape for the crack increase in FEM, using quadratic elements, is shown in the three cases of Fig.16. These three shapes furnish the following relations for one element:

$$\begin{Bmatrix} J_1^* \Delta A_1 \\ J_2^* \Delta A_2 \\ J_3^* \Delta A_3 \end{Bmatrix} = \frac{\delta \Delta s}{30} \begin{bmatrix} 5 & 10 & 0 \\ 2 & 16 & 2 \\ 0 & 10 & 5 \end{bmatrix} \begin{Bmatrix} J_1 \\ J_2 \\ J_3 \end{Bmatrix}$$

with $\Delta A_1 = \Delta A_3 = \delta \Delta s/2$ and $\Delta A_2 = (2/3)\delta \Delta s$.

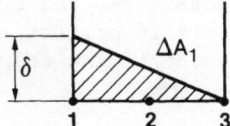 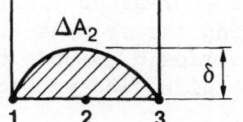 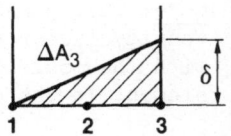

Figure 16. Possible shapes of virtual displacements.

As the displacement of a corner node may affect two neighbouring elements, this matrix should be assembled for all elements having a displaced node. For this reason the left-hand side is written in terms of energy change, as only energies can be added. Fig.17 shows an example of three elements along the crack front; this will give a 7*7 matrix.

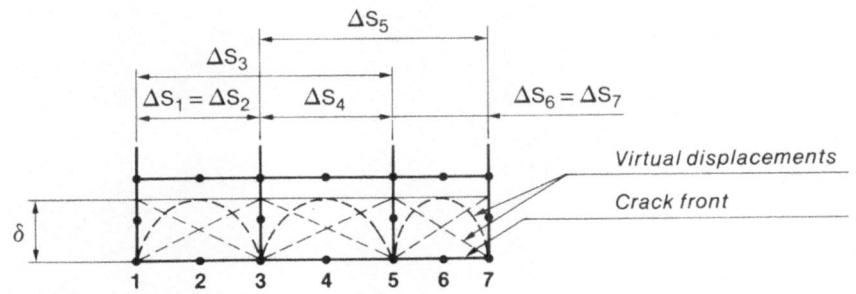

Figure 17. By displacing the 7 nodes at the crack front one after the other, the 7 global energy releases $J_i^* \Delta s_i$ are obtained.

After assembly the matrix is divided by the appropriate amount of crack increase, this cancels the unknown δ. Solving this system of equations furnishes the local J(s) values. If the crack front is curved, the length can be calculated by a numerical integration.

It should be noted here that, if a constant crack extension over the whole crack front is used, to obtain the global J^* value of the whole structure, the local J values of the midside nodes have a larger weight, e.g. for three crack front elements of equal size: 1-4-2-4-2-4-1. (Compare with the work equivalent nodal forces due to constant pressure loading.)

A disturbing fact is that the values J^* in the midside nodes are often much higher than the J^* values in the corner nodes. Especially if the crack front is curved, a completely distorted picture of J(s) can be obtained if parabolic interpolation of J(s) is used. The best way to treat this difficulty seems to be to do the interpolation of J(s) in a least square sense. Two possible least square interpolations are: a linear variation over one element side and a parabolic spline along the whole crack front. As an example the spline interpolation is explained.

The interpolation is written as

$$J(s) = a_1 + a_2 s + \sum_{i=1}^{N} a_{i+2} (s-s_i)^2 \qquad (14)$$

$$\text{if } s > s_i$$

Here N is the number of knots, the places where the parabolas are connected and s_i is the coordinate of knot i. The virtual displacements are the same as shown in Fig.16.

Eq.(14) in eq.(11) and integrated furnishes a result that can be written as

$$\tilde{J}_i^* = D_{ij}a_j \tag{15}$$

The following function G must now be minimised

$$G = \sum_{i=1}^{M} (J_i^* - \tilde{J}_i^*)^2 \tag{16}$$

M is the number of J values.

This gives the relation

$$D_{ij}J_j^* = D_{ij}D_{jk}a_k \tag{17}$$

This system of equations can be solved. Inserting the coefficients a_k obtained in eq.(14) gives the local J(s) values. Sometimes the condition (16) is too strong and unrealistic J(s) values are found. In this case one can add to eq.(16) the weighted condition that the 2nd derivative of $\tilde{J}^*$ in the knots of the two connecting parabolas must be the same in a least square sense. By varying the weight factor a certain degree of smoothing is introduced. However, if the crack front is curved, the amount of smoothing is difficult to establish.

The methods mentioned above are applied to the 3PB specimen discussed in the next chapter. In the middle of the specimen, all three methods gave approximately the same J value; closer to the free surface the differences became larger. At the free surface itself, often a difference of more than a factor two was found. Here the parabolic spline interpolation gave the smallest value and the linear interpolation the largest.

6.1.3. J calculated from the deflection curve. This method, also used in experimental practice, can only be applied to simple structures. J is calculated from the formula (see Fig.18):

$$J = \frac{\alpha U}{B(W-a)}$$

where:
α = coefficient depending on the shape of the specimen
a = crack length
B = thickness of specimen
W = height
U = area under the deflection curve.

To avoid the local deformation around the applied load point u is preferably taken at a stress-free point. U can be calculated using a

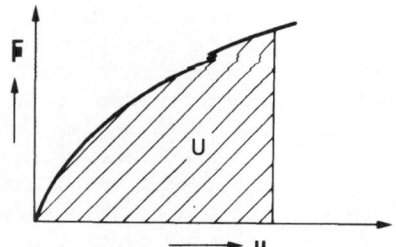

Figure 18. Load-load point displacement curve.

finite difference integration rule, e.g. the trapezoidal rule gives:

$$U = \sum_{step} \left(F_{i-1} + \frac{\Delta F_i}{2} \right) \Delta u_i$$

6.2 The Crack Tip Opening Displacement (COD)

In principle, COD can be evaluated directly from the displacements of
the crack surface found in the FEM calculation. However, besides the
fact that the crack tip region must be modelled very finely to achieve
some accuracy, it is found that the concept is very sensitive to different methods used for determining COD.
    An acceptable choice could be to measure it at the place of the
original position of the crack tip. This place is in general difficult
to find, as the crack tip does not only move due to blunting but also
due to other influences. Therefore, COD is usually defined as the opening of the crack at the intersections of lines under 45° through the
tip and the crack surfaces. In FEM, however, there is no intersection,
if the crack tip region is not modelled very finely. An alternative is
then the linear extrapolation of the crack surface to the crack tip.
This can be done by hand (graphically) or directly by the computer program by putting a straight line through a number of points in a least
square sense. Both methods are explained in Figs.19 and 20.

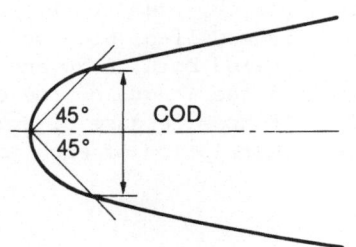

Figure 19. COD calculation
according to the 45° method.

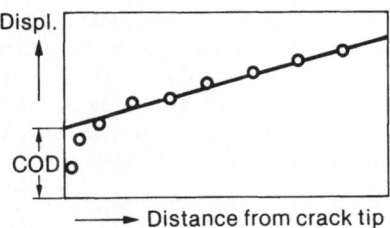

Figure 20. COD calculation according to the linear extrapolation method.

The linear extrapolation method can, of course, only be used if there
is a straight part at the crack flank. This is likely to be the case
for edge cracks. To model blunting with a coarse mesh, collapsed elements can be used with independent degrees of freedom at the collapsed
points. Because of the softening introduced in this way, the crack tip
will open more and the COD found is more in accordance with experimental values. However, it should be pointed out that the blunting obtained in this way often has an unrealistic shape. This is shown in
Fig.21, where the displacements of the collapsed crack tip nodes are
drawn thick and the crack flanks thin. The DOFs of the two methods discussed above are indicated.

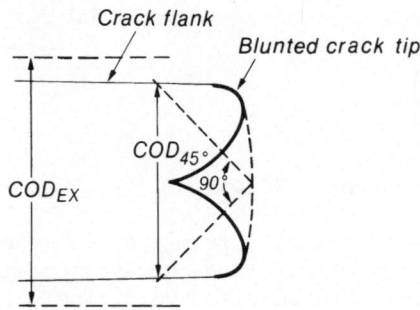

Figure 21. Shape of blunted crack tip using collapsed elements with independent DOFs in the tip.

7. SIMULATION OF CRACK GROWTH

Three methods to model crack growth are often mentioned in the literature:
- nodal release
- stiffness reduction
- node shifting.
    The first two methods are conceptually the same, the amount of
growth is determined by the length of the element side at the crack tip.
Practically these methods can only be used in 2D with very small elements in the crack plane. In 3D they cannot be used at all, if variable growth along the crack front is to be modelled. Node shifting gives
much better results and can easily be applied both in 2D and 3D. If
much growth has to be modelled (more than the length of one element
side), node shifting can be combined with node release or stiffness
reduction. In the following a detailed description of each method is
given.

7.1 Nodal Release

In this method the crack is extended by releasing the nodes in the crack
plane. The reaction forces are put with a reversed sign on the newly
created DOFs. The smallest amount of growth is equal to the side of the

element at the crack tip. This will cause a severe loading situation at the crack tip (elastic unloading and reverse plasticity), so that the load increments should not be too large. Therefore, releasing is done in a number of steps. The reaction forces can then be put proportional or non-proportional on the structure. In practice, however, only minor differences in the final situation can be observed. If higher order elements are used, the corner node and the midside node should be released at the same time to avoid overlapping of the crack faces. The ratio of the reaction forces applied to corner nodes and the midside nodes can change from step to step in the releasing process. It is essential that the elements in the crack plane are not too large compared to the crack length. Full use of the capability of the higher order elements, to give accurate results with a coarse mesh, cannot be made.

During the releasing process the external loading of the structure can be kept constant or it can be increased. If it is kept constant, a discontinuity is obtained in the deflection curve (Fig.22, u constant). If the loading is increased during the releasing process it is difficult to give a meaning to the J values calculated during the releasing process. Often the crack growth is then assumed to be proportional to the reaction forces on the released node

$$\Delta a^* = \frac{F^*}{F} \Delta a$$

where:
Δa = final amount of growth
Δa^* = current amount of growth
F = total reaction force
F^* = current reaction force on released node.

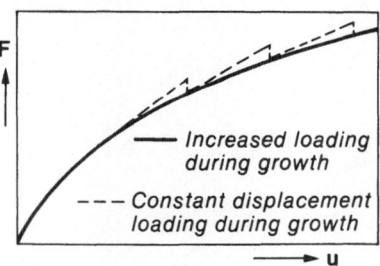

Figure 22.   Load point deflection during crack growth.

7.2 Nodal Stiffness Reduction

A variant of the method described above is the nodal stiffness reduction method. Conceptually both methods are the same, only the way of releasing nodes is different. The displacements in the crack plane are now not prescribed in the direction perpendicular to the crack plane, but an infinite (very high) stiffness is added to the diagonal term of the stiffness matrix related to these DOFs. This stiffness is so high that the nodes will not move in this direction.   A practical way to add this high stiffness is to use spring elements, where the high stiffness is the spring constant (Fig.23).

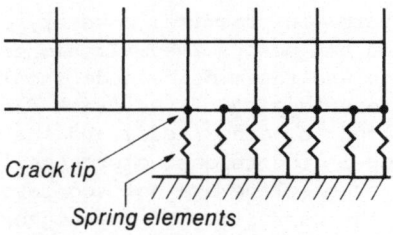

Figure 23. Crack tip plane modelled with spring elements.

Crack growth is now obtained by reducing this additional stiffness. Here also proportional or non-proportional reduction of the additional stiffness can be applied.

7.3 Node Shifting

In the previous methods the amount of crack growth is determined by the size of the crack tip elements. If higher order elements are used, this may be rather large, especially in 3D. Furthermore, as has already been said, these methods cannot be applied in 3D if local crack extension is to be modelled (Fig.24). Therefore, crack growth modelled by shifting the crack tip nodes is a more convenient way to model small amounts and local crack growth. To model large amounts of growth this method can be combined with node release.

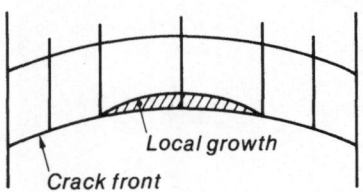

Figure 24. Local crack growth in 3D.

If the shifted crack tip node has reached the node of the next element, it can be released and the shifting can continue with the next crack tip node. The difficulty with node shifting is that the position of the integration points changes. An extrapolation of the stresses and strains from the old integration points to the new ones can be applied; however, this seems rather artificial. In any case, because of the iteration procedure to solve the system of equations, a redistribution of the stresses can be expected.

7.4 Examples of Crack Growth Analysis

7.4.1. <u>Experimental results</u>. In the laboratories of JRC-Ispra, fracture toughness experiments were performed on 3PB specimens both for irradiated and unirradiated 316H material and for several temperatures /22/. An unirradiated specimen tested at 550°C was chosen for this analysis. The dimensions of the test specimen are (see Fig.6):

- height (H)   80 mm
- width (W)   20 mm
- thickness (B)   15 mm
- initial crack length (a$_o$)   11.206 mm.

The total amount of stable crack growth (Δa) is 0.444 mm; this was calculated as a mean of local crack growth in 9 points along the crack front. The crack growth initiation point was established by the potential drop technique and took place at a displacement u = 2.7 mm. The mean crack growth rate $\Delta a/\Delta u$ = 0.126. Young's modulus and Poisson's ratio are, according to /23/: E = 1.4...1.6·10^5 MPa (depending on the state of the material), ν = 0.3.

The stress-strain curve for this material is given in /24/ in the form of the Voce relation:

$$\sigma = A \, e^{C\varepsilon_p} + B$$

At a temperature of 550°C this material shows a strain rate dependency. For a strain rate of 8.33·10^{-7} to 8.33·10^{-4} s^{-1}, the coefficients are A = -926.8...-772.2 MPa, B = 1030.4...914.0 MPa, and the exponent C = -2.633...-3.013.

The experimental load line displacement curve is given in Fig.25 and the J-integral vs load-point displacement in Fig.26. This curve is obtained using the formula

$$J = \frac{2U}{B(W-a_o)}$$

where U is the area under the load-point deflection curve.

The dJ/da value for this material shows a normal scatter. An eye fit through 14 data points gave

        dJ/da =   1000 MPa

The specimen analysed was in the lower region of the scatter band, with dJ/da = 850 MPa, J$_{Ic}$ = 170 N/mm.

7.4.2. <u>Analyses</u>. Several FEM models have been used (see Figs.4,5,6,7):
- a 2D mesh with triangular elements focused at the crack tip, used up to the initiation of crack growth;
- a 2D mesh with quadrilateral elements in the crack plane, suitable

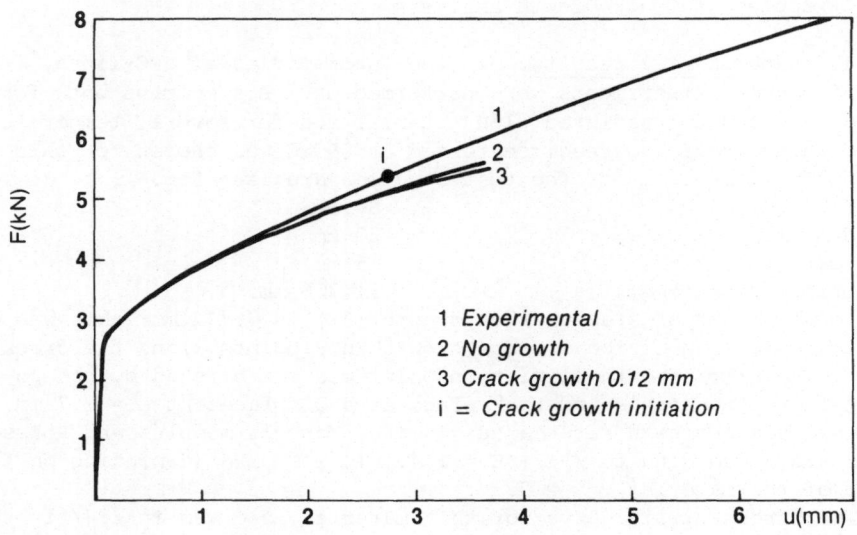

Figure 25.   Load-load point displacement 2D/3D analyses.

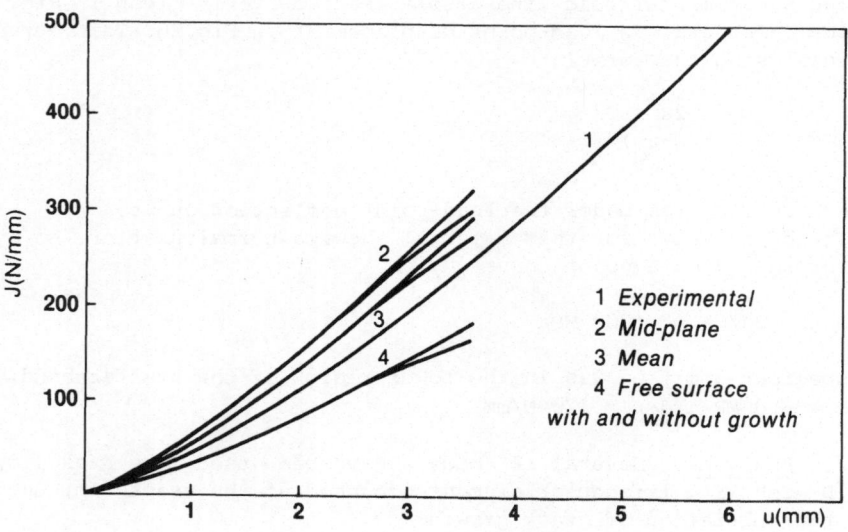

Figure 26.   J-integral vs load point displacement 2D/3D analysis.

to model crack growth by the nodal release method;
- a 3D mesh with one layer of 20-node brick elements over half the thickness;
- a combined 2D/3D mesh with three layers of 20-node brick elements over half the thickness in the crack tip region. The crack growth is obtained by a node shifting procedure.

A so-called generation phase analysis /25/ was performed. The experimental load deflection curve is the essential input to these analyses to simulate the experiment and to calculate the various fracture parameters. Especially if crack growth occurs at a high strain level, the solution is very sensitive to discretization of the model, the solution procedure of the non-linear equations, uncertainties in the material data, etc. In a generation phase calculation the effect of all these phenomena should be established before performing an application analysis. To force the calculation to follow the experimental load line displacement curve, the uni-axial stress-strain curve was modified slightly, while the non-linear equations are solved as precisely as is reasonable practically. Of course, in this iterative process, to follow the load line displacement, the true material data is taken as a basis. The LAMCAL FEM code /11/ was used for the analyses.

All calculations were performed with isotropic hardening and elastic unloading. The tangent modulus solution technique with an original Newton iteration scheme was applied. The convergence criterion was the norm of the out-of-equilibrium forces compared to the norm of the total loading (including reaction forces). The remaining residual load vector was added to the incremental load vector of the next step.

Several modified Newton iteration schemes were also tried, in order to reduce the CPU time but it was impossible to reach the required tolerance for large amounts of plasticity. Standard 9 incremental integration steps of the stress-strain relations, every load step and every iteration were performed. In the 2D case all three methods discussed in section 6.2 were applied for the evaluation of J.

In 3D only the virtual crack extension method was applied, with a least squares interpolation of the local J linear over one element. Crack growth was simulated by the nodal release method (2D) and by shifting the crack front nodes (3D). In the case of node release a predictive step was first performed to account for elastic unloading.

2D analysis. Plane strain conditions were assumed. As the test piece is a three-dimensional structure, the 2D results must be seen within the plane strain approximations. Both 2D meshes (Figs.4 and 5) consist of isoparametric 8-node elements. After scaling to first yield, the models were loaded with prescribed displacement steps of $\Delta u = 0.1$ mm. Up to three equilibrium iterations were necessary to obtain a norm of the residual load vector almost equal to zero. Up to the crack initiation point the load point deflection curve was reproduced very well with both mesh 1 and mesh 2, using the following material data: E = $1.5 \cdot 10^5$ MPa, $\nu = 0.3$, A = -916.8 MPa, B = 1031.4 MPa and C = -2.5. The J-integral vs displacement curve of mesh 1 was somewhat below and the one of mesh 2 somewhat above the experimental one. Mesh 1 was designed to serve as a comparison for mesh 2 up to the initiation point, only mesh 2 was used for crack growth analysis. The crack growth was modelled by releasing the nodes in the crack plane. To obtain a continuous deflection curve, the model was loaded at the same time. The reaction forces of the released nodes of the crack tip element were decreased in 10 proportional steps. It should be noted that the crack tip element side is released completely only at the 10th step.

Unfortunately no good results could be obtained using this method of simulating crack growth. Although initiation was assumed to be much later and the crack growth rate to be much lower than in the experiment, the experimental displacement curve after initiation could not be reproduced. Most probably the crack tip elements were too large. In /25/ and /26/ the same method was applied to constant strain elements. These elements are necessarily much smaller than higher order elements and in these cases good results were obtained. In /27/ node shifting was successfully applied with rather large higher order elements.

The 2D analysis was not repeated with the node shifting technique as the initial intention was that it should serve as a pre-calculation for the much more expensive 3D analysis.

3D analysis. Mesh 3 (Fig.6) was used with the same material data as for the 2D case. The load point deflection curve fell considerably below the experimental one and no good picture could be obtained from the local J-integral through the thickness. As the CPU time was already rather high (about 2.5 min/step on Amdahl 470/V8), a mesh with more layers over the whole thickness was considered not to be feasible. Therefore, a combined 2D plane strain 8-node/3D 20-node brick mesh (mesh 4, Fig.7) with about the same number of DOFs as mesh 3 was designed for the final analysis. Three layers of 20-node brick elements over half the thickness were used in the crack tip region with thickness ratios of 1:1:1/2 (1/2 at the free surface). At the crack front the elements were collapsed into a wedge shape (extension of mesh 1). Crack growth was modelled by shifting the position of the nodes at the crack front. This was done during a load increment. Due to the high costs, only a few iterations to follow the deflection curve could be made. With the data obtained, only the first part of the curve could be reproduced well with $E = 1.6 \cdot 10^5$ MPa, $\nu = 0.3$, $A = -877$ MPa, $B = 1000$ MPa, $C = -2.6$ (see Fig. 25).

After scaling to first yield, the structure was loaded with prescribed displacement steps of $\Delta u = 0.05$ mm. Three and sometimes four equilibrium iterations were required to obtain a norm of the residual forces of about 10% of the norm of the total forces. Because different types of elements were used in the same mesh, the convergence was much slower than in the 2D case.

The global J-integrals were obtained by the virtual crack extension method. The J(s) values reported in Fig.26 are obtained by assuming a linear variation over one element side.

Other interpolations were also performed (e.g. a least squares parabolic spline with smoothing) but the reported one was found to be the most satisfactory. With all methods the mid-plane values were about the same while the free surface value showed a larger scatter (often much lower than curve 4 of Fig.26).

The crack growth was modelled by shifting the nodes of the crack front during a loading step. A constant crack growth rate of 0.126 all along the crack front was applied and initiation was assumed at u = 2.7 mm. These values agree with the experimental ones. The result is given graphically in Figs.25 and 26. Other results are given in the following table:

	exp.	Calculated mid-plane	mean	free-surface
dJ/da (MPa)	850	695	635	200
J_{IC} (N/mm)	170	225	200	120

7.4.3. <u>Discussion</u>. For large loads the size requirement for the specimen used is not completely satisfied so that a violation of the J controlled growth can be expected in that region. /28/.

Furthermore, it seems that node release, applied on relatively large isoparametric 8-node elements, causes such a severe loading condition that the stress and strain state at the crack tip no longer represents the reality. In fact, during node release, several warnings for too large a load increment were given by the code (unloading and reverse plasticity in one step). These warnings could not be avoided by taking smaller load steps. Node shifting, applied in the 2D/3D case, gives much better results. Conceptually this procedure is also much better than node release as the crack growth rate can be modelled continuously and the structure remains in equilibrium during the extension process. Despite the fact that in 2D/3D the experimental load line displacement curve could not be reproduced exactly in the higher loading range, a total stable crack growth can be predicted which is much higher than in the 2D case. Because of the high costs involved, the calculation was stopped at a displacement of u = 3.7 mm.

In the near future the analysis will be repeated with a variable growth over the thickness of the specimen. This will be more in accordance with reality as in the experiment tunnelling effects could be noticed. The convergence is very important in this kind of analyses. The change in J-integral in one integration step can be of the same order of magnitude as the change in J due to the growing crack.

It might also be advantageous to make the thickness of the element at the free surface much smaller than in the present model. In fact, close to the free surface the J-integral drops very rapidly. Curve 4 in Fig.26 gives only a rough indication for the free surface J as the values found with the different interpolations have a very large scatter.

REFERENCES

1. Henshell, R.D. and Shaw, K.G., 'Crack tip finite elements are unnecessary', *Int. J. Numerical Methods in Engineering* 9 (1975) pp.495-509.
2. Barsoum, R.S., 'On the use of isoparametric finite elements in linear fracture mechanics', *Int. J. Numerical Methods in Engineering* 10 (1976).
3. Rice, J.R., 'Mathematical analysis in the mechanics of fracture', in: *Fracture*, H. Liebowitz, Ed., Vol.2, Academic Press, New York, 1968, pp.191-311.
4. Parks, D.M., 'A stiffness-derivative finite element technique for determination of elastic crack tip stress intensity factors', *Int. J. Frac.* 10 (1974) pp.487-502.
5. Parks, D.M., 'The virtual crack extension method for non-linear material behaviour', *Computer Methods in Applied Mechanics and Engineering* 12 (1977), pp.353-364.
6. DeLorenzi, H.G., 'On the energy release rate and J-integral for 3D crack configurations', *Int. J. of Fracture* 19 (1982) pp.183-193.
7. Wilson, U.K. and Osias, J.R., 'A comparison of finite element solutions for an elastic-plastic crack problem', *Int. J. of Fracture* 14 (1978).
8. Larsson, L.H., 'A calculational round-robin in elastic-plastic fracture mechanics', *Int. J. Pres. Ves. and Piping* 11 (1983) pp.207-228.
9. Larsson, L.H., 'EGF numerical round-robins on EPFM', *3rd Int. Conf. on Numerical Methods in Fracture Mechanics*, Eds. A.R. Luxmoore and D.R.J. Owen, Pineridge Press Ltd., Swansea, March 1984.
10. Lamain, L.G., 'Mesh generation and plotting facilities of the Lamcal program', EUR 7099 (1981) Ispra.
11. Lamain, L.G. and Blanckenburg, J.F.G., 'The finite element part of the Lamcal program - elastic-plastic fracture mechanics application', EUR 7999 (1982) Ispra.
12. Lamain, L.G. and Blanckenburg, J.F.G., 'Finite element program Lamcal - User's manual', EUR 8173 (1982) Ispra.
13. Nagtegaal, J.C., Parks, D.M. and Rice, J., 'On numerically accurate finite element solutions in the fully plastic range', *Computer Methods in Applied Mechanics and Engineering* 4 (1974).
14. Hibbit, H.D., 'Some properties of singular isoparametric elements', *Int. J. Numerical Methods in Engineering* 11 (1977).
15. Irons, B.M., 'A technique for degenerating brick-type isoparametric elements using hierarchical midside nodes', *Int. J. Numerical Methods in Engineering* 8 (1974).
16. Reynen, J., 'Surface cracks in cylinders during thermal shock', ASME 77-pvp-5, 1976.
17. Lamain, L.G., 'On the accuracy of EPFM calculations', paper M1/6, SMiRT-6, Paris (1981).
18. Lamain, L.G., 'Modified shape function for the collapsed 20-node brick element', paper G/F 3/6, SMiRT-7, Chicago (1983).
19. Prager, W. and Hodge, P.G., 'Theory of perfectly plastic solids', Wiley, New York, 1951.

20. Lamain, L.G., 'On stress evaluation in the Lamcal code', to appear as an EUR report.
21. Bakker, A., 'On the numerical evaluation of the J-integral', paper G/F 2/4, SMiRT-7, Chicago (1983).
22. Bernard, J. and Verzeletti, G., 'Elasto-plastic fracture toughness characteristics of irradiated 316H grade stainless steel', ASTM Symp., Louisville, 20-22 April 1983.
23. Aerospace structural metals handbook, Vol.2 (1981).
24. Matteazzi, S., Piatti, G. and Boerman, D., 'On relating the high-temperature flow stress of AISI 316 stainless steel to strain rate', NETS Varese, Villa Ponti, 20-22 May 1981.
25. Kanninen, M.F. et al., 'Elastic-plastic fracture mechanics for 2D stable crack growth and instability problems', in: *Elasto-Plastic Fracture*, ASTM STP 668 (1979) pp.121-150.
26. Sorensen, E.P., 'A numerical investigation of plane strain stable crack growth under small-scale yielding conditions', in: *Elastic-Plastic Fracture*, ASTM STP 668 (1979) pp.151-174.
27. Shih, C.F., DeLorenzi, H.G. and Andrews, W.R., 'Studies on crack initiation and stable crack growth', in: *Elasto-Plastic Fracture*, ASTM STP 668 (1979) pp.65-120.
28. Landes, J.D., 'Size and geometry effects on elastic-plastic fracture characterization', paper in CSNI report No.39 (1979) pp.194-225.
29. Fletcher, R., Practical methods of optimization, John Wiley and Sons, 1980, pp.43.

METHODOLOGY AND APPLICATIONS OF LOCAL CRITERIA FOR THE PREDICTION OF DUCTILE TEARING

F. Mudry
Centre des Matériaux de l'Ecole des Mines de Paris
ERA CNRS N°767 (UA 866)
B.P. 87
91003 EVRY Cédex
France

ABSTRACT.   Local criteria consist, within the field of continuum
mechanics, in a simulation of the metallurgical damage occurring at a
crack tip.   The restrictions of more usual approaches such as mode I
loading, plane strain state, isothermal situation, moderate yielding
etc..., are no longer effective at the expense of non linear finite
element calculations. Ductile damage is predicted through a simulation
of cavities growth which either influences the constitutive relations
or is a criterion for fracture assuming . a critical cavity size.
Practically, two parameters, independent on temperature are necessary.
They are measured using two different specimen geometries : one with a
crack and the other one without. These two parameters allow the
prediction of fracture of one mesh element in a finite element program.
This procedure is used to predict crack initiation and stable crack
growth in different structures.

1.   INTRODUCTION

1.1. General presentation

The subject of this lecture is "local criteria", and some explanation
of these words is required before beginning the lecture in itself. This
is why, in a first part, we will present very rapidly the different
analysis usually performed when dealing with cracks in Elastic-Plastic
Bodies. The local criteria analysis will then be introduced and the
advantages and disadvantages of each method will be discussed.
        In a second part, the physical events leading to ductile
fracture will be briefly described and different kinds of modelisation
are discussed. Afterwards, a rather simplifed equation is used to
predict ductile fracture in any configuration and a special attention
will be paid to the crack-tip. At this place, the characteristic
distance or characteristic volume will be introduced.
        In a third part, the methodology of local criteria in ductile
fracture will be presented in extenso and applications dealing with

L. H. Larsson (ed.), Elastic-Plastic Fracture Mechanics, 263–283.
© 1985 ECSC, EEC, EAEC, Brussels and Luxembourg.

specimens and industrial structures will be treated.

The local criteria methodology is, in fact, a rather old idea since Orowan |1| for cleavage fracture and more recently, McClintock |2| for ductile fracture and fatigue have used it. However, more recent developments were possible thanks to non linear finite element programs on middle size computers.

1.2. Global criteria and local criteria

The purpose of all the methods presented here is the same : is a crack, either postulated or real, potentially dangerous in faulty conditions (earthquake, loss-of-coolant accidents in nuclear reactors, etc...).

When dealing with such a problem, two different kinds of approach may successively be used.

First, simplified analysis are used. Rough, conservative estimates of the stress intensity factor (with some kind of plasticity correction), limit load, or J integral are performed. Then, following one of the available defect assessment procedures (see workshops by Dowling et al, Rousselier and Turner), the considered crack may become unstable or not.

Because many real problems involve complex geometries and loadings, some conservative assumptions must be made such as : 2D representation of the geometry, considering thermal stresses as primary stresses (or load imposed stresses), considering as mode I loading the non-symmetrical loads (mode II and III) etc...

This kind of analysis may be over-conservative in some cases. Moreover, even if the simplified analysis shows that the crack is safe, a precise knowledge of the safety margins may be desired.

In this case, more precise analysis are required. This, more than often, involves rather complex non linear finite element calcula-tions in order to compute more precisely the J integral or the COD (see lectures by Bakker and Lamain) etc. Account is taken of the true geometry and loadings. In this case, a local criterion may be used since it requires a rather precise knowledge of the stress-strain history. This kind of criterion just states that a given piece of metal (i.e. mesh element) will break for a given function of stress and strains. Most usual methods are based on energy release rates considerations (such as K or J). They will be referred here as global criteria since a precise knowledge of the stress and strain distribu-tion is not required.

On the contrary, local criteria are based on local mechanical values and try to model the metallurgical damage at the crack-tip. Let us notice that, if the stress and strain field can be described making use of a single parameter (such as J or K), then the local and global criteria will be equivalent since the critical local criterion will be reached for a critical J_{1c} or K_{1c} value.

1.3. Cost benefit analysis of the different methods

1.3.1. Simplified analysis. These methods may be of very simple use as long as problems remain within the field where they have been already employed. Handbooks of K, J and limit loads are available for different simplified geometries which can be close enough to the problem.

However, it may necessitate more complex evaluations when overconservative assumptions have been made or when an evaluation of the margins is desired. Extrapolation outside the scope of usual analysis may be difficult : mixed mode, thermal stresses, non proportional loadings, etc... Moreover, even in Mode I, the results may run outside the two theoretical bounds given for the J-design curve method (see the lectures by Bakker and Blauel for a more complete discussion). One concerns the ligament b and may be written as |3,4| :

$$\alpha \, \frac{J}{\sigma_o} < b$$

where σ_o is the yield stress and α is a numerical factor given as 25 in bending and 200 in tension. The other one deals with the tearing resistance slope dJ/da :

$$\frac{b}{J} \times \frac{dJ}{da} > \omega$$

where ω is a numerical constant, not very well known, though values around 10 have been suggested. These limitations were derived in order to be sure that the stress and strain distribution may be adequately described by the Hutchinson Rice and Rosengren (HRR) singular field |6,7|. If these conditions are not fulfilled, it may be argued that the method still remains conservative, though this assumption has never been demonstrated. One last problem is the collection of data for the critical values because, in many cases, only a limited amount of representative material is available and non-valid experiments are performed (for instance : aged welds, irradiated metal etc...). So that, more than often, a relatively safe analysis can be performed on the structure but a very uneasy interpretation of the experimental data is required.

1.3.2. More precise analysis. These methods require more complex elasto-plastic calculations which simulate as closely as possible the real geometry and loadings of the problem. The financial cost is of course of another order of magnitude though the computing cost shows a steadily decreasing trend. If such a calculation is necessary, a post processor for local criterion utilisation is of very low cost, as will be seen in a moment.

The main advantage of local criteria is that there is no theoretical limitation so that safe extrapolation to non symmetrical loadings, thermal stresses etc... can be safely made. This arises from the fact that the physical events taking place in the material are described locally. This is not dependent on what happens outside of the volume currently under consideration. Furthermore, this allows a rather

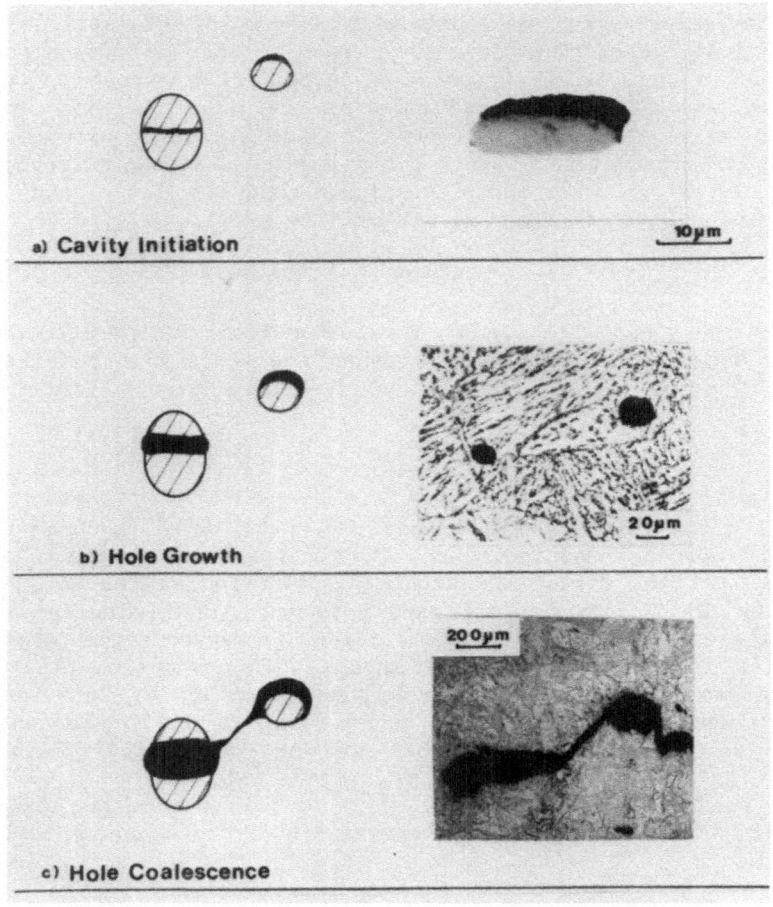

a) Cavity Initiation

10μm

b) Hole Growth

20μm

200μm

c) Hole Coalescence

Fig. 1. The three steps process of ductile fracture. a) cavity initiation, b) hole growth, c) hole coalescence

safe guide when dealing with the effect of temperature, strain rate, sulphur content, aging etc... Perhaps, another benefit could be a better understanding, through a common language, between mechanics and metallurgy.

On the other hand, this approach must be re-checked when dealing with a new class of material. For example, ferritic steels and austenitic steels may have slighly different behaviours. This implies that the tested specimens for material characterisation must be adequately modelised by finite element calculations, which means that rather small specimens may be used. In this lecture we will limit ourselves to the low alloy steels in the temperature range : -200°C, +300°C.

2. DESCRIPTION AND MODELISATION OF THE DUCTILE FRACTURE PROCESS

2.1. Description

For the ferritic steels, in the temperature range -200°C, +300°C, there are only three different modes of fracture. They are described with some details in the lecture by Dahl and Dormagen. Here, we will assume, that intergranular fracture is negligible, thanks to adequate chemistry control. The two other fracture modes are cleavage and ductile. Cleavage fracture is very dangerous because of its unstable nature and will be analysed in my other lecture. Therefore, only ductile fracture will be described here.

Briefly repeating some aspects treated in the lecture by Dahl and Dormagen, we will only define the three steps process of ductile fracture. In a first step, non metallic inclusions either rupture or disbound from the matrix. This leads to hole initiation. In a second step, these holes grow under the effect of plastic strain and stress triaxiality. At last, an unstable band links neighbouring cavities leading to the well known dimple rupture. This third step is called hole-coalescence (fig. 1).

For most low alloy steels, hole initiation is effective in the very first moments of straining. In fact, stress based criteria have been proposed for cavity formation. In most cases, the corresponding strain is small |8|. Therefore, it can be assumed that this step is negligible and that hole growth is the only operative damage.

2.2. Modelisation

Two different kinds of modelisation of the hole growth damage and coalescence have been used.

2.2.1. <u>Critical non-dimensional cavity size</u>. The first one uses an equation to predict hole growth and assumes that coalescence takes place for a critical size. In other words an incremental equation is proposed which relates the radius of an assumed cavity to the plastic strain and stress triaxiality. Many different equations have been proposed, which lead to different criteria, such as the well known

Latham and Cockroft or Oyane criteria |9,10|. However, for rather large stress triaxialities different criteria are used. McClintock, first, has given an approximate equation for hole growth which leads to a fracture criterion in ductile rupture |11|. More recently, Norris et al |12| have proposed a simple criterion for ductile rupture :

$$D = \int_o^\varepsilon \frac{d\varepsilon_{eq}^P}{(1-k\sigma_m)} = D_c$$

where D is a damage function, $d\varepsilon_{eq}$ is the incremental Von Mises equivalent plastic strain, σ_m is the hydrostatic stress, D_c and k are material constants measured on two independent specimen geometries.

MacKenzie et al |13| have proposed a very simple methodology. They show that ductility is a strong function of stress-triaxiality and that an experimental curve relating the critical strain to stress triaxiality may be measured from notched specimens (fig. 2). So that a fracture criterion could be :

$$D = \int \frac{d\varepsilon_{eq}^P}{\varepsilon_c \ (\sigma_m/\sigma_{eq})} = 1$$

where $\varepsilon_c \ (\sigma_m/\sigma_{eq})$ is the experimental critical strain for a given stress-triaxiality.

All these criteria are very similar since the following elements must be present :
i : the equation must be history dependent and, therefore, must achieve an incremental form.
ii : many experimental evidences show that hole growth is proportionnal to the exponential of strain |14,15|.
iii : the stress states must involve hydrostatic pressure and must include only stress ratios since the absolute magnitude of stresses has no effect. So that, the damage function should be of the general form :

$$\frac{dD}{D} = f \ (\frac{\sigma_m}{\sigma}) \ d\varepsilon_{eq}^P$$

where D is the damage : the radius R of an assumed cavity, for example. σ is some other invariant stress.

After a very precise mechanical evaluation of the problem, Rice and Tracey |16| have given a rather simple expression for the growth of a spherical cavity in an infinite body :

$$\frac{dR}{R} = \alpha \ x \ \exp \ (\frac{3}{2} \frac{\sigma_m}{\sigma_{eq}}) \ d\varepsilon_{eq}^P$$

where : R is the current radius of the cavity, α is a numerical constant equal to 0.283 for vanishingly small hole volume fractions. The exact form of this equation is only valid for perfect plasticity but as it stands, it gives rather good estimates of cavity growth rates in ferritic steels |16|. The very strong influence of the σ_m/σ_{eq} ratio can be noted. We assume that hole coalescence is achieved for a critical value of the non dimensional cavity size, that is :

$$\int_{R_o}^{R} \frac{dR}{R} = Ln\ (\frac{R}{R_o}) = Ln\ (\frac{R}{R_o})_c$$

where $(\frac{R}{R_o})_c$, the critical cavity size is a material parameter measuring ductility. This assumption is rather well experimentally verified |17|.

2.2.2. <u>Damage and stress-strain relationship.</u> A second kind of criteria takes into account the strain softening effect of growing cavities so that the hole volume fraction is included in the stress-strain relationship. The plastic flow potential is dependent on the hole volume fraction. This enables flow localization and rupture. Many recent works have been devoted to this subject. In particular, Gurson |18|, Tveergaard |19|, and Rousselier |20| have proposed criteria which have very similar forms. Gurson criterion is written as :

$$\frac{3}{2} \frac{S_{ij}\ S_{ij}}{Y^2(\epsilon_{eq})} + \alpha\ f\ \cosh\ (1,5\ \frac{\sigma_m}{Y(\epsilon_{eq})}) - (1+f^2) = 0$$

where $\alpha = 2$ for Gurson and 4 for Tveergaard. Rousselier criterion is written as :

$$\frac{\sqrt{3/2\ S_{ij}\ S_{ij}}}{Y(\epsilon_{eq})(1-f)} + \frac{\sigma_1\ C}{Y(\epsilon_{eq})(1-f)^2}\ f\ \exp\left[\frac{\sigma_m}{\sigma_1(1-f)}\right] - 1 = 0$$

where S_{ij} is the deviatoric part of the stress, Y is a hardening function of strain, f is the hole volume fraction, σ_1 and C are numerical constants.

In fact, all these different criteria have very similar forms and give similar results.

The Rice and Tracey formula has been retained for the applications, though interesting results have been obtained by Rousselier et al |28| and Norris et al |12|. One material parameter, the critical cavity size $(R/R_o)_c$ is necessary. It is a measure of the material ductility. It can be shown, on physical grounds, connected to Gurson's criterion that this critical value should be connected to the inclusions volume fraction |21|. Typical values are between 1.2 and

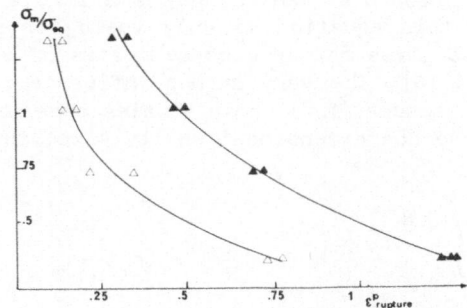

Fig. 2. Ductility as a function of stress triaxiality. After MacKenzie et al |13|. HY130 steel. Longitudinal direction (▲). Short transverse direction (Δ).

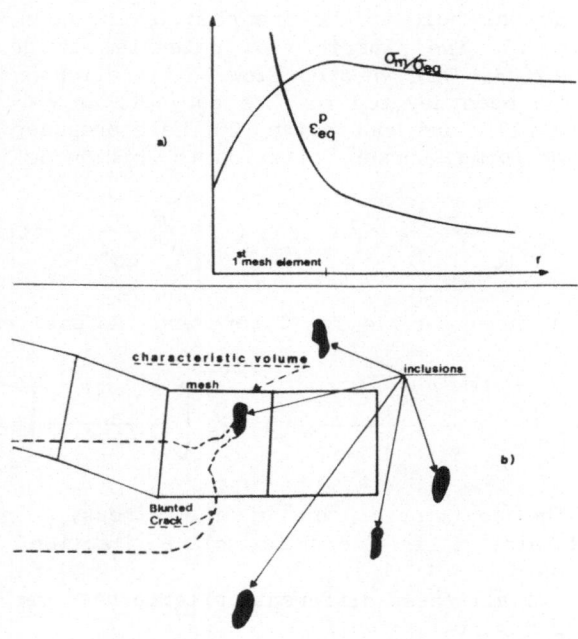

Fig. 3. a) Schematic representation of the stress triaxiality and plastic strain distribution in front of a crack tip. b) Schematic illustration of the characteristic volume concept.

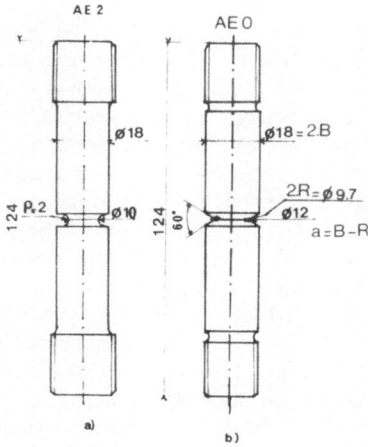

Fig. 4. Geometry of the specimens used when measuring the critical parameters. a) notched round tensile bar, b) cracked round tensile bar.

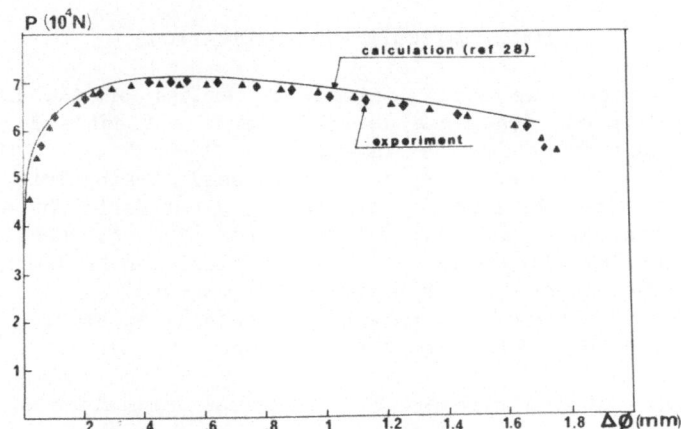

Fig. 5. Typical load versus reduction in diameter results on a notched specimen. Comparison with computed curve.

2.2. which means that the average hole radius has grown 20% to 120% of its original size.

2.3. Crack tip situation

At the crack-tip, the stress triaxiality is rather high ($\sigma_m/\sigma_{eq} \simeq 2$) and very steep strain gradients are present (fig. 3). If the cavity growth rate is computed very near the crack-tip, say 1 μm, huge values will be found. This is physically unreasonable since the probability of finding really a hole at 1μm of the crack tip is very low. Therefore, the damage function must be computed at a characteristic distance or in a characteristic volume in front of the crack tip. This parameter was first introduced by McClintock |2|. It must be connected, in some way to inclusion spacing and is, in fact, a rough way of taking into account the statistical nature of fracture.

In a finite element program, from a rigourous point of view, it should be necessary to use very fine meshes, with singular elements etc... in order to have a precise solution near the crack tip. Then, the damage function should be averaged using the statistical distribution of inclusions or a "damaged zone" modelisation |29|. This procedure would be very uneasy and expensive. That is why, a very useful, though approximate, shortcut is the use of a square mesh element of characteristic size to model the characteristic volume |22|. The damage function is computed in the first element and, when it reaches its critical value, the mesh element is relaxed either by the node release technique or similar techniques. The most important conclusion of this discussion is that the mesh size will be material dependent. It will be, in fact, a fitted parameter for a given material, though it should be connected in some way to inclusion spacing.

3. METHODOLOGY OF LOCAL CRITERIA AND APPLICATIONS

In this section, we will use mainly results of the Beremin group including "L'Ecole des Mines de Paris" and le "Centre de Calcul de la Division des fabrications de Framatome".

The Rice and Tracey formula is used, taking into account some strain hardening. As already said in the first part, the exact form of the equations may be a little different for other classes of materials. However, for low alloy steels, the methodology can be immediately used.

The methodology is a two steps process :
i : measure of the critical parameters for a given steel
ii : numerical calculation of the structure.

3.1. Measure of the parameters : numerical simulation of the experiments

As stated in the first part, the local criterion for ductile fracture requires the knowledge of two material constants. One is the critical cavity size $(R/R_o)_c$ which is a measure of material ductility. The other

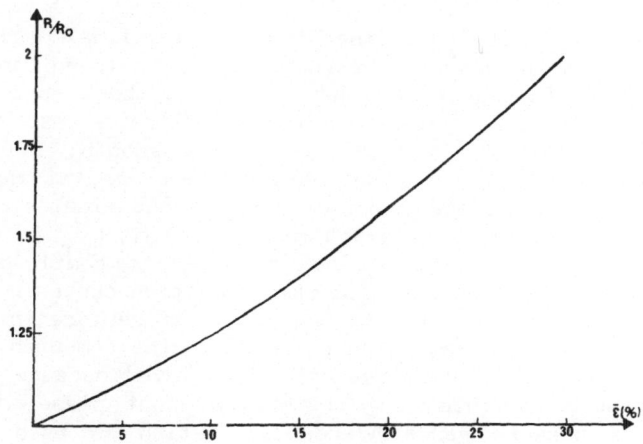

Fig. 6. Variation of the non dimensional cavity size on the specimen axis as a function of the average strain $\varepsilon = 2 \ln (\emptyset_o / \emptyset)$ where $\emptyset_o$ is the initial minimum diameter and $\emptyset$ is the current diameter.

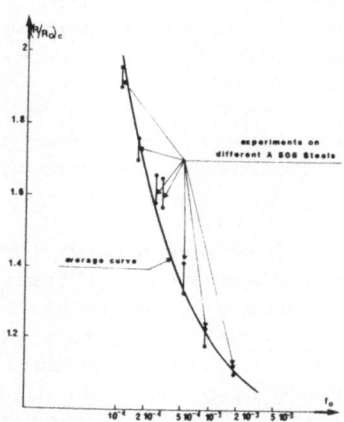

Fig. 7. Relationship between the critical cavity size and initial void volume fraction.

one is the characteristic mesh size $(\Delta a)_c$ which is a measure of the size effect. Therefore two independent experiments are necessary.

3.1.1. <u>Measure of $(R/R_o)_c$</u>. The first one involves specimen geometries with rather smooth notches. Axisymmetrically notched tensile bars such as those shown in fig. 4 have been used. In these specimens, fracture takes place on the specimen axis where the stress triaxiality is elevated, but the strain gradient is very smooth. In this case, the second parameter $(\Delta a)_c$ has no effect. Therefore, if the exact moment where rupture takes place is known and if a numerical simulation of the specimen is performed, an exact measure of $(R/R_o)_c$ is possible. When loading the specimen, the minimum diameter is continuously recorded; fig. 5 shows a typical load diameter reduction curve on this geometry. The abrupt drop on the curve is indicative of rupture initiation on the specimen axis. On the same figure, is shown the computed load-displacement curve. An excellent agreement is achieved because 2D representation is possible without any assumption about plane strain or plane stress, and because geometry reactualisation is used |23|. For any reduction of diameter, the computed load is less than 2% in error, even when the strains are larger than 30%. Fig. 6 shows a typical variation of the cavity size (R/R_o) on the specimen axis as a function of the reduction of diameter. Therefore, the knowledge of the rupture point gives a precise value for $(R/R_o)_c$.

    Furthermore, as already stated this parameter is connected to the inclusion volume fraction for a constant strain hardening exponent. The relationship is given in fig. 7. The inclusion volume fraction shown in that figure takes into account specimen orientation : long, transverse or short-transverse orientation. It is in fact a rough estimate of the initial void volume fraction which cannot be computed from chemistry alone. Quantitative metallography on three perpendicular polished sections is necessary |24|. This procedure allows a rapid estimate of (R/R_o) with a small amount of material. Furthermore, the variation of $(R/R_o)_c$ with temperature is very slow. It can be assumed that this parameter is temperature independent, at least for temperatures lower than 200°C.

3.1.2. <u>Measure of $(\Delta a)_c$</u>. This second parameter is only important when very steep strain gradients are present. Therefore it should be fitted on a cracked geometry. Because a 2D representation was desired, we have selected axisymetrically cracked round bars (fig. 4). This allows a fair agreement between computed and measured load-displacement curves |25,26| (Fig. 8). It should be pointed out that the computed curve is always within the experimental scatter band though initiation takes place far beyond limit load. CT specimens or 3 point bend specimens may be used but, for a calculation in plane strain, "adequate' side-grooves are necessary.

    The computation is performed with a square mesh size of 0.2 mm and the critical growth rate is computed in this element. Fig. 9 shows a typical variation of the non dimensional cavity size as a function of the J-integral as it is measured experimentally. If the critical point where the crack initiates is known with some accuracy

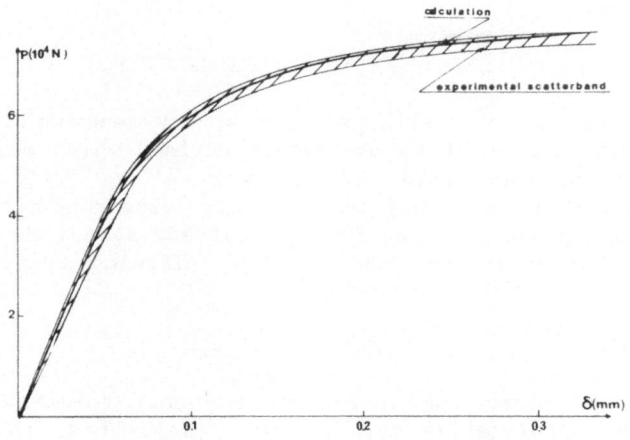

Fig. 8. Load displacement curves on cracked round tensile bars. Experiments and calculations.

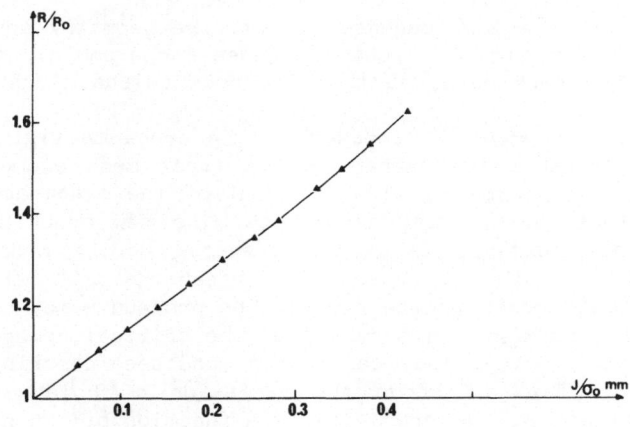

Fig. 9. Variation of the cavity size R/R_o as a function of J as it is measured experimentally.

from experiment, then a certain corresponding value of the critical $(R/R_o)_i$ is computed. $(\Delta a)_c$ is calculated from the following relation based on the HRR field :

$$(\Delta a)_c \times Ln \ (R/R_o)_c = 0.2 \times Ln \ (R/R_o)_i$$

More than often, 0.2 mm will be already a very reasonable value. Note that crack initiation must be accurately defined which means that the usual conventional definition of J_{Ic} may not be sufficient. Other experimental techniques involving scanning electron microscopy and specimen polishing must be used |30|. As already stated above, $(\Delta a)_c$ is connected to the inclusion spacing. The following relation gives a rough estimate :

$$(\Delta a)_c \simeq \frac{2}{\sqrt[3]{N_V}}$$

where N_V is the number of inclusions per unit volume which can be estimated from quantitative metallography |24|. This, in particular, implies that $(\Delta a)_c$ is, as a first approximation, independent on temperature.

3.2. Applications

3.2.1. Calculations on fracture mechanics specimens. Calculations of fracture mechanics specimens may seem a rather simple and academic exercise. It is not so, because initiation may take place largely beyond general yield. This implies that very high strain levels may be achieved. In the first calculation dealing with three point bend specimens, a prefectly plastic behaviour was assumed |22|. The local criteria parameters were : $(\Delta a)_c$ = 0.2 mm, $(R/R_o)_c$ = 1.25. These are typical values for a low toughness alloy. The geometry and meshing of the specimen is in Fig. 10. Square 8-nodes isoparametric elements are used along the crack path, so that the node release technique may be used.

A displacement is imposed at the arrow in fig. 10 when the critical cavity size is reached in the first mesh element. The two nodes lying on the symmetry line in front of the crack are realeased, keeping constant the loading conditions. The new crack tip element, already slightly damaged is further damaged during node releasing. However, further loading is necessary to reach the critical value. This means that stable crack growth occurs. The procedure may be carried on until unstable crack growth occurs i.e. the critical damage is reached without further loading. This calculation has been described in detail elsewhere |22|. Fig. 11 shows the variation of hole growth in the different elements at the crack-tip as a function of a non dimensional displacement. This example shows that stable crack growth, maximum load and instability can be predicted when using local criteria. A similar exercise has been performed by Rousselier |20| using his constitutive relations. This leads to very encouraging results. An experimental program has been undertaken and numerically modelled in order to check the predictive capability of this procedure. All results are not yet

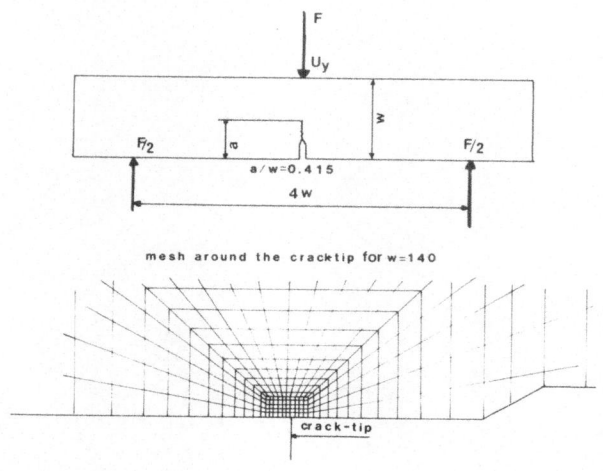

Fig. 10. Three point bend specimen geometry.

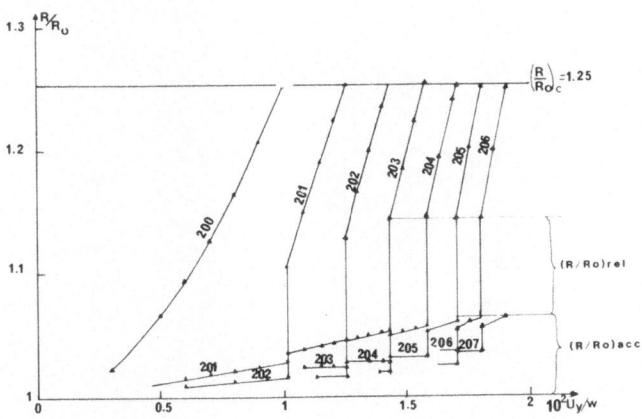

Fig. 11. Variation of the cavity size in crack tip elements. Element 200 was first crack tip element. Elements 201, 202, etc... become crack tip elements when the crack propagates. For element 206, $(R/R_o)_{acc}$ is the damage accumulated before the crack tip reaches it. $(R/R_o)_{rel}$ is the damage during node releasing of element 205. Reloading is necessary to reach the critical value.

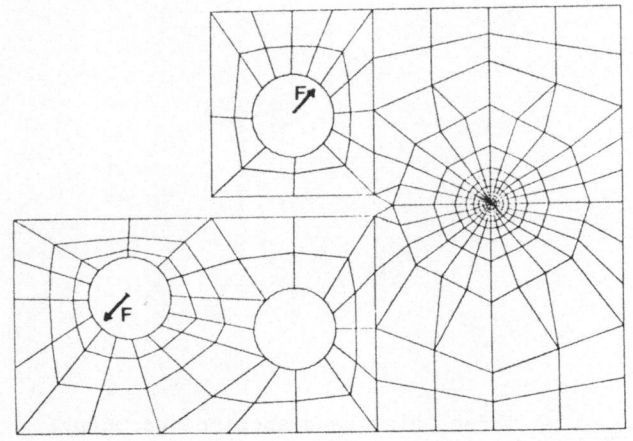

Fig. 12. Dissymetric C.T. Type specimen. (Courtesy of J.C. DEVAUX, Centre de Calcul de Framatome).

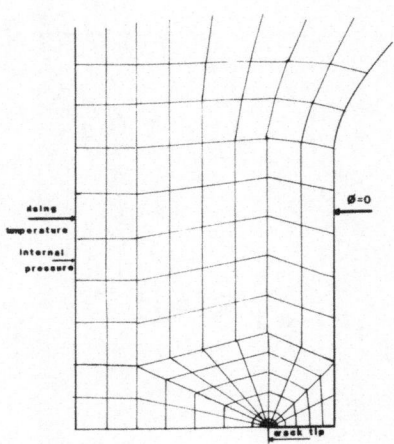

Fig. 13. Thermal shock on a thick tube. (The upper part of the drawing corresponds to the lower part of a fixation system of the tube.)

available. However, it can be stated that, with a $(R/R_o)_c$ value measured on notched specimens and a fitted value of $(\Delta a)_c$ in order to predict correctly crack initiation on one specimen geometry, the crack initiation on four different geometries is correctly predicted |26|. Moreover the stable crack growth is also predicted. It should be noticed that only two parameters, namely $(\Delta a)_c$ and $(R/R_o)_c$ are able to predict the rupture of one notched specimen and the crack resistance curves in these geometries. Fig. 12 shows another specimen geometry and meshing where some mode II loading is present. In this case also, crack initiation was correctly predicted. However, in this geometry, side grooving is impossible so that the loading curve cannot be predicted with a plane strain calculation. The direction of the crack is also correctly predicted provided that a radial meshing is used. In this case, a very small deviation angle of 7° is found, which is in accordance with experimental values. It is also found that, for lower temperature, i.e. for loads lower than the limit load the predicted angle is more important. This was also experimentally verified |27|.

3.2.2. Practical utilization of local criteria. It is difficult, except in very specific cases, to use three dimensional finite element representations of flawed structures. We may guess that it will be a current practice in the future if the decreasing trend of memory and array processor prices is confirmed. Right now, however, mostly 2D representations are tractable. This leads to some conservative assumptions. Another difficulty is the lack of information about stress-strain relationships because the cavity growth may be very sensitive to this input data. More than often, reliable information about strain hardening is not available in a wide temperature range.

As an example, we will briefly describe an assumed situation. A thick tube, initially at room temperature, is subjected to thermal stresses when a hot liquid is poured into the pipe. It is desired to know whether a quarter thickness external radial flaw may be dangerous. For symmetry reasons, the flaw is assumed to run around the circumference (fig. 13).

The calculation will have to proceed in a rather usual manner. The variation with temperature of the heat exchange coefficient, elastic constants and yield stress are input data. The temperature and pressure of the liquid are also given. Because of lack of available data and for conservative reasons, a perfectly plastic behaviour may be assumed. To check this point another calculation can be done with a strain hardening coefficient n=0.1. Heat transfer is zero on the external part of the cylinder. The characteristic mesh size is $(\Delta a)_c = 0.2$ mm.

For each time increment, the heat equation is solved with an explicit algorithm. New values for the elastic constants and yield stresses are computed and the elasto-plastic mechanical equations are iteratively solved. The cavity growth ratio is computed in the first mesh element, and compared to the critical value which is temperature independent. In the given example, no initiation took place. However, if such was the case, it was still possible to release one node keeping constant the time and afterwards, to proceed the calculation. This

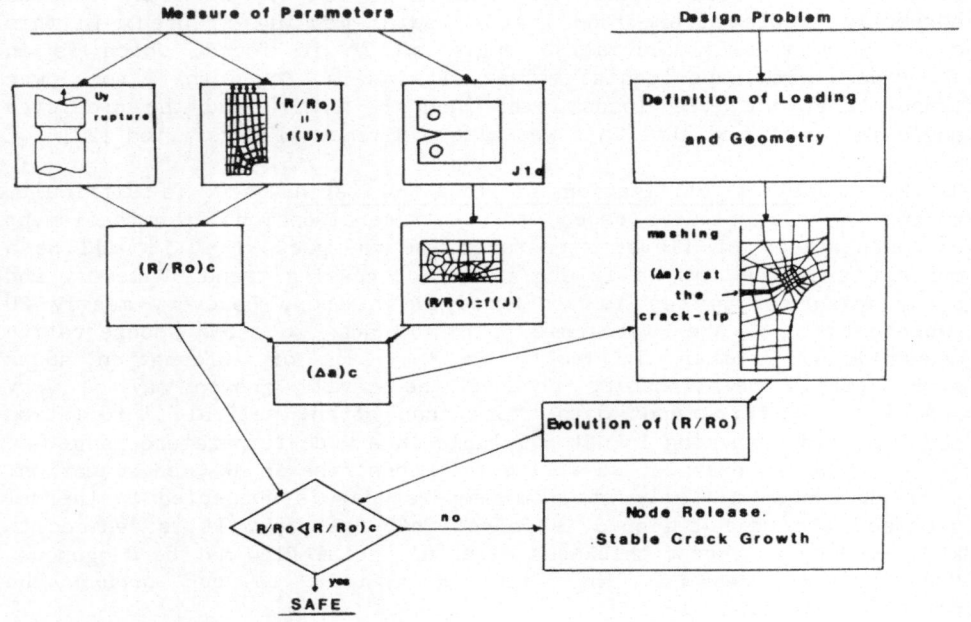

Fig. 14. Schematic representation of the local criteria methodology.

procedure with the price of a reasonably long computer calculation (5 hours on a mini PRIME computer) gives a rather precise description of the problem and shows that no initiation takes place when simplified methods cannot give firm conclusions.

4. CONCLUSIONS

The purpose of this lecture was an introduction to the local criteria methodology which was presented for pedagogic reasons, as opposite to more global methodologies as K or J. In fact, as stated above, if the given problem may be solved by simplified methods, the local criterion approach will not be very useful. Moreover, it can be demonstrated that, when the J integral approach applies, the two methodologies are equivalent. For more complex problems, however, the local criterion approach provides a methodology without theoretical limitations, which may be a very useful feature. The two major requirements are that a good thermo-elasto-plastic finite element program is available and that the rupture process of the material used is very well understood and can be adequately modelized.

In the particular case of low alloy steels, a methodology for the prediction of ductile tearing has been presented which involves the evaluation of two material parameters. Fig. 14 is a schematic representation of the methodology. The two critical parameters $(R/R_o)_c$ and $(\Delta a)_c$ are a measure of the material ductility and of the size effect. Their evaluation is possible thanks to a finite element simulation of two independent mechanical tests.

A finite element modelization of the problem under consideration allows an evaluation of the safety of a given flaw through a simple comparison of a simple damage parameter with the critical value $(R/R_o)_c$.

REFERENCES

|1| Orowan, E., 'Cleavage fracture of metals', Rep. Prog. Phys., 12, (1948), p.185.

|2| McClintock, F.A., 'Plasticity aspects of fracture', in : Fracture : an advanced treatise, H.A. Liebowitz (ed.), Academic press, 3, (1968), p.48.

|3| McMeecking, R.M., Parks, D.M., 'On criteria for J-dominance of crack tip fields in large scale yielding', in : Elastic-plastic fracture, ASTM STP 668, J.D. Landes, J.A. Begley, G.A. Clarke (eds.), ASTM, (1979), p.175-194.

|4| Shih, C.F., German, M.D., 'Requirements for a one-parameter characterisation of crack tip fields by the HRR singularity', Int. J. Fract., 17, (1981), p.27-39.

|5| Hutchinson, J.W., Paris, P.C., 'The theory of stability analysis of J-controlled crack-growth', in : Elastic plastic fracture, ASTM STP 668, J.A. Landes, J.A. Begley, G.A. Clarke (eds), ASTM, (1979), p.37-64.

|6| Hutchinson, J.W. 'Plastic stress strain fields at a crack tip', J. Mech. Phys. Sol., 16, (1968), p.13.

|7| Rice, J.R., Rosengren, G.F., 'Plane strain deformation near a crack tip in power law hardening material', J. Mech. Phys. Sol., 16, (1968), p.1.

|8| Beremin, F.M., 'Cavity formation from inclusions in ductile fracture of A508 steel', Metal. Trans., 12A, (1981), p.723-732.

|9| Latham, D.J. Cockroft, M.G., 'Ductility and workability of metals', J. Inst. Metals, 96, (1968), p.33.

|10| Oyane, M., 'Criteria of ductile fracture strain', Bulletin J.S.M.E. n° 90, (1972), p.1507.

|11| McClintock, F.A., 'Ductile rupture by the growth of holes', J. Appl. Mech., 35, (1968), p.36.

|12| Norris, D.M. Jr., Reaugh, J.E., Moran, B., Quinones, D.F., 'A plastic strain, mean stress criterion for ductile fracture', J. Eng. Mat. and Techn., 100, (1978), p.279.

|13| MacKenzie, A.C., Hancock, J.W., Brown, D.K., 'On the influence of state of stress on ductile failure initiation in high strength steels', Eng. Fract. Mech., 9, (1977), p.167.

|14| Roesch, L., Henry, G., Eudier, M., Plateau, J., 'Etude de l'influence d'inclusions d'alumines sur les propriétés mécaniques d'un fer fritté', Mém. Scient. Rev. Metall., 63, n° 11, (1966), p.927.

|15| Perra, M., Finnie, I., 'Void growth and localization of shear in plane strain tension', in : ICF4, Waterloo, Fracture 1977, D. Taplin (ed.), University of Waterloo Press, (1977), p. 415.

|16| Rice, J.R., Tracey, D.M., 'On the ductile enlargment of voids in triaxial stress fields', J. Mech. Phys. Sol., 17, (1969), p.201.

|17| Beremin, F.M., 'Study of fracture criteria for ductile rupture of A508 steel', in : Advances in fracture research (ICF 5), D. François (ed.), Pergamon press, 2, (1981), p.809.

|18| Gurson, A.L., 'Yield criteria and flow rules for porous ductile media', J. Eng. Mat. and Tech., 99, (1977), p.2.

|19| Tveergaard, V., 'Influence of voids on shear band instabilities under plane strain conditions', Technical report, n° 159, june, University of Denmark, (1979).

|20| Rousselier, G., 'Contribution à l'étude de la rupture des métaux dans le domaine de l'élasto-plasticité'. Thèse de Doctorat-d'état, Lab. de mécanique des solides, Ecole Polytechnique, (1979).

|21| Mudry, F., 'Influence of loading direction on the ductile fracture process in structural steels', in : Le comportement plastique des solides anisotropes, Colloque internation n°319 du CNRS, J. Boehler (ed.), Villard de Lans, (1981), p.407.

|22| D'Escatha, Y., Devaux, J.C., 'Numerical study of initiation, stable crack growth and maximum load with a ductile criterion based on the growth of holes', in : Elastic plastic fracture, ASTM STP 668, J.D. Landes, J.A. Begley, G.A. Clarke (eds.), ASTM, (1979), p.229.

|23| Devaux, J.C., Mottet, G., 'Simulation numérique d'éprouvettes axisymétriques entaillées', Rapport interne Framatome n° TMC/DC/80.095, (1980).

|24| Mudry, F., 'Etude de la rupture ductile et de la rupture par clivage d'aciers faiblement alliés', Thèse de Doctorat d'état, Université de Technologie de Compiègne, Ecole des Mines de Paris, (1982).

|25| Devaux, J.C., Mottet, G., 'Simulation numérique de l'éprouvette axisymétrique fissurée', note interne Framatome n°TMC/TC/80.191, (1981).

|26| Mottet, G., Devaux, J.C., 'Déchirure ductile des aciers faiblement alliés : modèles numériques' rapport interne Framatome n°TM/C DC/84.052, (1984).

|27| Devaux, J.C., Marini, B., Mudry, F., Pineau, A., 'Etude de l'amorçage et de la propagation stable des fissures en milieu tridimensionnel en utilisant un critère physique de rupture ductile', Colloque méthode de calcul, DGRST, March 3, 1983, Antibes, France, paper n° 1.

|28| Rousselier, G., Devaux, J.C., Mottet, G., 'Elastic-plastic behaviour law including ductile fracture damage', 7th International conference on structural mechanics in reactor technology, paper S/1-3, Chicago, (1983).

|29| Bui, H.D., Erlacher, A., 'Propagation of damage in elastic and plastic solids' in : Advances in fracture research, (ICF 5), D. François (ed.), Pergamon press, 2 (1981), p.533.

|30| Lautridou, J.C., Pineau, A., 'Crack initiation and stable crack growth resistance in A508 steels in relation to inclusion distribution', Fract. Mech., 15, (1981), p.55-71.

FRACTURE AND DAMAGE

D. FRANCOIS
Ecole Centrale des Arts et Manufactures
Grande Voie des Vignes
92290 - CHATENAY-MALABRY - FRANCE

ABSTRACT.   After summarising the theory of damage of Rabotnov-Katchanov and its subsequent developments by Lemaître and Chaboche, it is pointed out how this concept has been extended to ductile fracture in an overall manner, or by taking into account the physical mechanism of development of cavities. It is found that the rate of damage is proportional to the rate of plastic deformation and to the exponential of the stress triaxiality factor.   In a similar way, for heterogeneous materials such as concrete, microcracked damaged zones develop and the damage can be expressed as a function of the maximum strain. The statistical distribution of cracks, can be incorporated in the theory using Weibull's formula.   Subsequently, the way in which these concepts of damage can be introduced into fracture mechanics calculations is demonstrated. Damage can modify the model of Dugdale for the plastic zone. Bui and Ehrlacher state that the rate of available energy is dissipated in energy of plastic deformation on the one hand and energy of fracture on the other hand, the latter appearing entirely in the damaged zone, the dimensions of which are a function of the properties of the material but also of the geometry of the notched component and the method in which it is loaded.

Using local criteria of ductile fracture corresponding to a critical damage as a starting point, it is possible to calculate the stable growth of a crack by finite elements in a very satisfactory manner, even when the level of plasticity is high.

I.   INTRODUCTION

The mechanics of fracture deal with the problem of extension of a planar defect initially present in a structure or which is propagated there, for example by fatigue. It is well determined in the case of brittle fracture, if the linear elastic fracture mechanics are applicable. At the other extreme we will find entirely ductile fractures, taking place generally after necking by deformation of cavities which grow and finally coalesce. This is an example of fracture by damage which in this case affects a certain volume of the component. Another one is found in the fracture of concrete where the heterogeneities of

L. H. Larsson (ed.), Elastic-Plastic Fracture Mechanics, 285–302.
© 1985 ECSC, EEC, EAEC, Brussels and Luxembourg.

the structure stabilizes microcracks which occupy again a certain volume before they merge. These problems can be treated by mechanics of damage deriving inspiration from the works of Kachanov |1| and Rabotnov |2|. The intermediate situation of crack propagation accompanied by large plastic deformations have mainly formed the subject of investigations by extrapolation of the mechanics of fracture to the field of elastoplasticity. On the other hand the approach by mechanics of damage has been much more rarely considered. However, for several years certain investigators have made interesting attempts in this direction. We will indicate the main aspects of these treatments, confining ourselves to the case of ductile fractures and of the fracture of concrete.

We will first explain the basis of damage theory. We will then show how it can be applied to ductile fracture which occurs by the growth and coalescence of cavities, and to the fracture of a material such as concrete which occurs by the growth and coalescence of many microcracks. It will then remain to try to link damage mechanics and fracture mechanics, that is to say to study the propagation of a macrocrack in a damaging material.

2. ELEMENTS OF DAMAGE THEORY |3|

2.1. Theory was developped basically by Kachanov |1| and Rabotnov |2| to explain fracture by creep. When a load is maintained at high temperature microcracks appear which gradually reduce the useful section of the component. After damage the stress which is active is $\sigma = F/A$ where F is the load applied and A is the section. The damage is expressed by the fact that the effective stress in the section is no longer σ but :

$$\sigma_{eff} = \sigma/(1 - D) \qquad (1)$$

If D = 0 there is no damage ; D = 1 defines the complete rupture of the volume element ($\sigma_{eff} = \infty$). It is possible to associate D with the reduction of the section resulting from the multiplication and from the growth of microcracks by writing :

$$\sigma_{eff} = F/A_{eff} = \sigma/(1 - D)$$

and hence :

$$A_{eff}/A = 1 - D \qquad (2)$$

However, this assimilation is in no way indispensable, since A_{eff} is not strictly equal to the uncracked section. The fundamental hypothesis of damage mechanics is that the usual stress-strain relations remain valid under the condition that the stress is replaced by the effective stress.

The development of damage with time is a function of the effective stress :

$$\dot{D} = dD/dt = g\left[\sigma/(1 - D)\right] \qquad (3)$$

which is thus presented as a differential equation which can take different forms. The whole question is hence reduced to the suitable choice of the function g.

In the case of elasticity Hooke's law is expressed as follows :

$$\sigma_{eff} = \sigma /(1 - D) = E \varepsilon_e \qquad (4)$$

where ε_e is the elastic deformation. It is seen that D can be associated with the variation of the modulus of elasticity measured by successive unloadings during the course of a tensile test.

Thus :

$$D = 1 - (1/E) \, d\sigma /d\varepsilon_e \qquad (5)$$

and if $E_D = d\sigma /d\varepsilon_e$ is the modulus of the damaged material

$$D = 1 - E_D/E \qquad (6)$$

Figure 1 shows measurements carried out in this manner in an aluminium alloy 2024 at 20 °C [3,20] The values of the damaged moduli shown on the figure relate of course to the deformed section and not to the initial one S_0. Other definitions of the damage parameter are used. In particular :

$$D = Ln (A/A_{eff}) \qquad (7)$$

with the following effective stress in this case :

$$\sigma_{eff} = \sigma \exp D \qquad (8)$$

2.2   The internal variable D is associated with the thermodynamic potential : the specific free energy.

In the same way as $\sigma_{ij} = \rho \, (\partial \psi / \partial \varepsilon_{ij})$, the entropy $S = - \partial \psi / \partial T$, the strain energy release rate $G = \rho \, (\partial \psi / \partial a)$, a damage energy release rate Y can be defined as :

$$Y = - \rho \, (\partial \psi / \partial D) \qquad (9)$$

For an elastic damaging material at constant temperature T, the following potential can be used :

$$\psi = (1/2 \rho) \, a_{ijkl} \, \varepsilon_{ij} \varepsilon_{kl} (1 - D) - TS \qquad (10)$$

The variables associated to ε_{ij} and D are then :

$$\sigma_{ij} = \rho \, (\partial \psi / \partial \varepsilon_{ij}) = a_{ijkl} \, \varepsilon_{kl} (1 - D) \qquad (11)$$

$$Y = - \rho \, (\partial \psi / \partial D) = (1/2) \, a_{ijkl} \, \varepsilon_{ij} \varepsilon_{kl} \qquad (12)$$

As in the case of linear elastic fracture mechanics where fracture occurs when $G = G_C$, in damage mechanics the critical condition is :

$$Y = Y_C$$

At fracture, we have :

$$Y_c = (1/2) \, \mathcal{E}_{ijR} \, \sigma_{ijR}/(1 - D_R) \tag{13}$$

A pseudo dissipation potential can be introduced which is a convex function of Y, (grad T)/T and of the state variables $\mathcal{E}_{ij}$, T, D. The evolution of damage D is then given by :

$$D = (\partial \psi / \partial Y) \, (Y, \text{grad } T/T, \mathcal{E}_{ij}, D, T) \tag{14}$$

This is identical with the function g (equation 3).

3. APPLICATION TO DUCTILE FRACTURE

3.1. Mechanisms

Ductile fracture results from the initiation of cavities, their growth and finally their coalescence. It is generally accepted that this phenomenon is due to inclusions. The corresponding criteria must be a function of the plastic deformation on the one hand, but also on the other hand of the average stress and the equivalent stress and in particular of their ratio : the triaxiality factor. It is in fact clear that the development of cavities will be all the more rapid the higher this ratio whereas on the other hand the application of a hydrostatic pressure can bring about the closing of cavities, that is, a negative rate of development.

3.2. Law of Lemaitre and Chaboche |3|

From the thermodynamic considerations of the paragraph 2.2. during the course of plastic deformation the development of the damage must be in the form :

$$\dot{D} = h \, (\sigma_{eff}, \mathcal{E}_p, D) \, \dot{\lambda} \tag{15}$$

with $\dot{\lambda} = \dot{\sigma}_{eff}/E_T$
Where E_T is the tangent modulus $E_T = d\sigma_{eff}/d\mathcal{E}_p$

Following a suggestion by Broberg |5| who adopted a simple power function of σ_{eff} for h, Lemaitre and Chaboche |3| introduced in addition a threshold stress σ_D below which there was no damage and they suggested that the function g was of the following form :

$$dD = [(\sigma - \sigma_D) / S (1 - D)]^s \, (d\sigma /S) \tag{16}$$

σ_D, S and s are parameters which depend only on the material and the temperature. For a given material they must be determined by identification with an experimental curve, such as the one shown in fig. 1. For this alloy Lemaitre and Chaboche found from the evolution of the modulus of elasticity :

$$\sigma_D = 400 \text{ MPa} ; S = 580 \text{ MPa}; \text{ and } s = 0.58$$

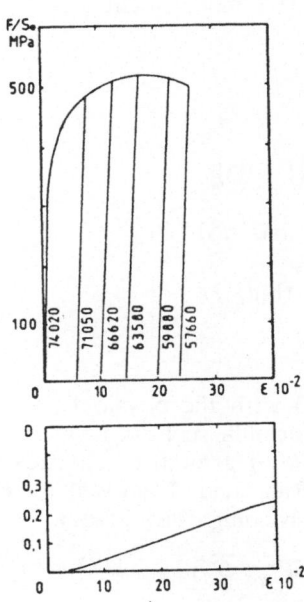

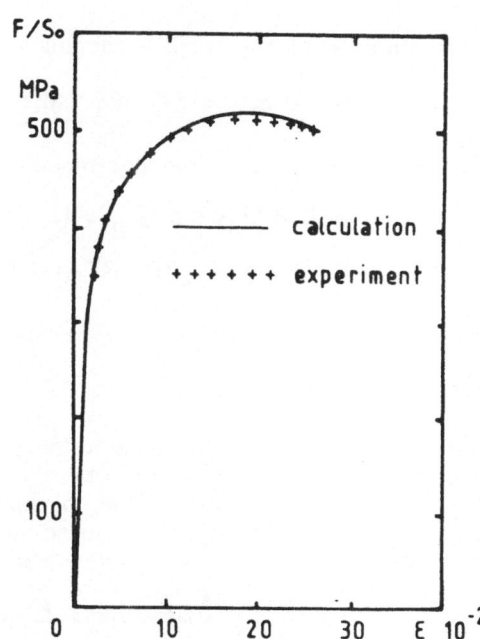

Fig. 1- Evolution of the modulus of elasticity on unloading for an alloy 2024 at 20 °C and correlative development of the damage [3,20]

Fig. 2   Strain hardening curves calculated by means of the law of damage of Lemaitre and Chaboche[3] compared with the one measured experimentally on 2024 at 20 °C

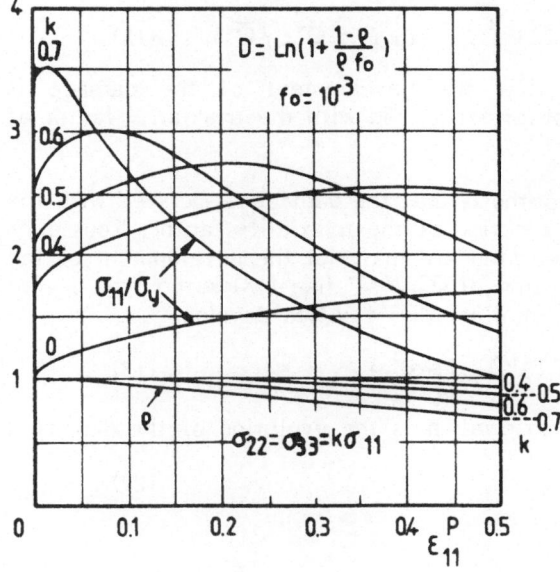

Fig. 3 - Strain hardening curve calculated by means of the law of damage of Rousselier[6] for different stress triaxiality factors and a volume fraction f_0 of inclusions of 10^{-3}

A calculation of the work hardening curve from this model with

$$\sigma < \sigma_D \; ; \; d\mathcal{E}_p = d\sigma / E_T \; ; \; dD = 0$$

E_T being the constant tangent modulus.

$$\text{and} \; \sigma \gtrsim \sigma_D \; ; \; d\mathcal{E}_p = (1 / E_T) \, d[\sigma /(1 - D)]$$

$$dD = [(\sigma - \sigma_D) / S (1 - D)]^s \, (d\sigma /S)$$

again very suitably gives the experimental curve (Fig. 2).

3.3. Law of Rousselier |6|

It is tempting to try to connect the parameter D with the physical process of damage rather than to use a purely empirical formula such as formula 16. Various authors have attempted to express the law of growth of cavities |7, 8|. One of the models most used is that of Rice and Tracey |9|. For spherical cavities in a material without strain hardening they give :

$$\dot{R}/R = 0.283 \, \dot{\mathcal{E}}_{eq} \exp (1.5 \, \sigma_m / \bar{\sigma}) \qquad (17)$$

In this formula R is the radius of the cavities,
$\dot{\mathcal{E}}_{eq}$ the equivalent plastic deformation,
σ_m the mean stress, and
$\bar{\sigma}$ the flow stress.

This suggests writing the differential equation of damage (3) in the form :

$$\dot{D} = \alpha \, \dot{R}/R = \alpha \, 0.283 \, \dot{\mathcal{E}}_{eq} \exp (1.5 \, \sigma_m / \bar{\sigma}) \qquad (18)$$

Such an expression indicates that the development of the damage is associated with that of plastic deformation and with the triaxiality factor of the stresses.

Using as a starting point the hypothesis that the damage associated with the variation of density is solely a function of the triaxiality factor, Rousselier |6| showed that for a generalised standard material the development of the damage was an exponential function of $\sigma_m / \bar{\sigma}$ like (18) without having to formulate in detail the manner in which the cavities develop.

$$\dot{D} = A \, \sqrt{3} \, \dot{\mathcal{E}}_{eq} \exp (B \, \sigma_m / \rho \bar{\sigma}) \qquad (19)$$

A and B are constants of the material, ρ is the evolution of the density d

$$\rho = d/d_o \qquad (20)$$

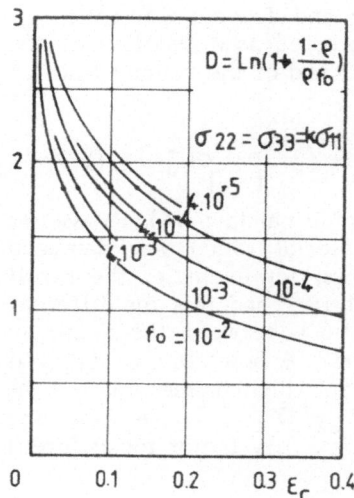

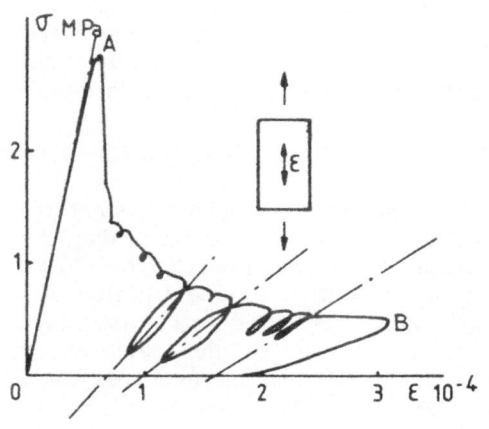

Fig. 5 Stress strain behavior of concrete in a tensile test[12]

Fig. 4 Calculated development of the deformation at instability as a function of the stress tria-xiality factor for different vo-lume fractions f_o of inclusions[6]

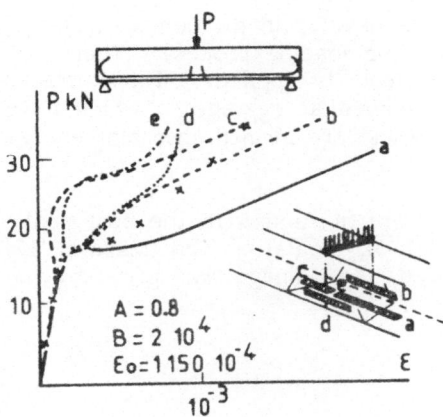

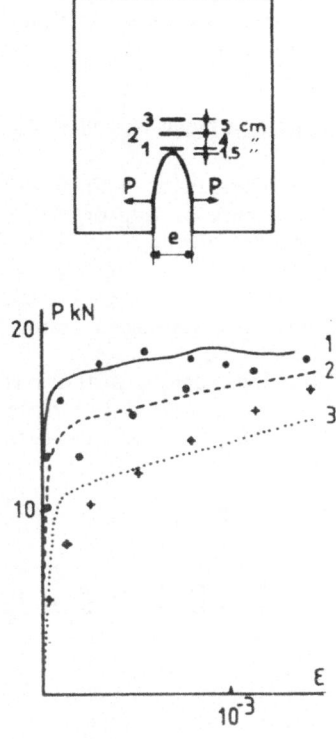

Fig. 6 Experimental (a,b,c,d,e) deter-mination and analytical calculation (stars) of the strains at various loca-tions on the lower face of a beam loaded in bending[12]

Fig. 7 Experimental determination (dots) and numerical calculations (lines) at various locations near the notch of the specimen[12]

By identification with the equation (17) of Rice and Tracey, Rousselier showed that the damage could therefore be expressed as a function of the volume fraction f of cavities (including inclusions) and of the volume fraction f_0 of the inclusions by the formula :

$$D = Ln \ [f(1 - f_0)/f_0 \ (1 - f)] = Ln \ [1 + (1 - \rho) / \rho f_0] \quad (21)$$

As Lemaitre and Chaboche, starting from the empirical law (16), Rousselier could construct work hardening curves for material which was damaged according to the law (19) or (20). Fig. 3 shows, for example, the result obtained with a volume fraction of inclusions of 10^{-3} and this for different triaxiality factors characterised here by the ratio k between the minimum principal stress and the maximum principal stress. A point of instability is seen to occur on these curves for a deformation which is the smaller the higher the value of the triaxiality factor $k = \sigma_{22}/\sigma_{11}$. Fig. 4 shows the deformation at instability as a function of the triaxiality factor for different volume fractions of inclusions.

The ductility defined in this manner is the smaller, the greater the volume fraction of inclusions. These different calculations are in good agreement with the experimental results, for example, those obtained by means of test pieces notched in such a manner as to modify the triaxiality factor.

4. APPLICATION TO THE FRACTURE OF HETEROGENEOUS MATERIALS

4.1. In a material such as concrete cracks develop in the cement paste at the boundaries of aggregates early during the loading process. There are even cracks already present before the load is applied. Those cracks propagate in the cement paste but are stopped at aggregates which have large radii. Thus a large number of microcracks are formed absorbing energy and creating a damaged zone.

4.2. Following various authors |10, 11| by identification with the post peak behavior in a uniaxial tensile test (Fig. 5), Lemaitre and Mazars |12| proposed that for the concrete they tested, the damage was given by the law :

$$\varepsilon_M < \varepsilon_0 \qquad\qquad D = 0$$

$$\varepsilon_M > \varepsilon_0 \qquad D = 1 - A/exp[B(\varepsilon_M - \varepsilon_{D_0})] - \varepsilon_{D_0}(1-A)/\varepsilon_M \qquad (22)$$

where ε_M is the largest principal strain, ε_{D_0} the damage threshold taken at the peak of the tensile curve and A and B are constants which for the concrete tested were equal to 0.8 and 2.10^4 respectively.

4.3.   Tensile tests are difficult to perform on a material such as concrete and it is much easier and more usual to use bend tests. It turns out that the threshold strain varies from one test to the next and furthermore it is systematically larger in the bend tests. Of course this is not an unexpected behavior if the statistical nature of the crack distribution is taken into account.   More recently a better treatment was developped using Weibull's theory |13|.   The damage thresholds $\mathcal{E}_{D_0}$ are supposed to be distributed according to a distribution function S ($\mathcal{E}_{D_0}$).

If P_D ($\mathcal{E}$, ΔV) is the damage probability for a volume element ΔV under a strain $\mathcal{E}$, the hypothesis of independence between adjacent volume elements ΔV, yields the damage probability of a volume V under a strain $\mathcal{E}$:

$$P_D (\mathcal{E},V) = 1 - \exp (- k \mathcal{E}^{m}V) \tag{23}$$

where m is Weibull's parameter and k a constant, both depending upon the material tested.   In the case of a non uniform strain equation (23) becomes :

$$P_D (\mathcal{E},V) = 1 - \exp (- k \int_V \mathcal{E}^{m} dV) \tag{24}$$

The local strain can be expressed in a given specimen as a function of a reference strain $\mathcal{E}_N$:

$$\int_V \mathcal{E}^m dV = \mathcal{E}_N^{m} \int_V t^m dV \tag{25}$$

t being the ratio of the strain $\mathcal{E}$ to the reference strain $\mathcal{E}_N$. It depends only upon the geometry of the test piece and the type of loading, but not upon its intensity.

It can then be shown that the mean $\overline{\mathcal{E}}_{Do}$ of the thresholds $\mathcal{E}_{Do}$ is given by :

$$\overline{\mathcal{E}}_{D_0} = \Gamma(1 + 1/m) / [k \int_V t^m dV]^{1/m} \tag{26}$$

where Γ is the Γ function.
For two specimens of volume V_1 and V_2 it is shown that :

$$\int_{V_1} \mathcal{E}^m dV = \int_{V_2} \mathcal{E}^m dV = W_0 = Const. \tag{27}$$

Using tensile tests on cylinders 36 mm in diameter, 140 mm long, the fracture strain was found to be $\mathcal{E}_T = 10^{-4}$.

Using three point bend tests on 10 x 10 x 40 mm^2 beams the fracture strain was found to be $\mathcal{E}_B = 1.4.10^{-4}$.

From the relation (27), using indexes B for bend and T for tensile tests,

$$\mathcal{E}_B = \mathcal{E}_T [2V_T(m+1)^2/V_B]^{1/m} \tag{28}$$

was deduced, giving m = 7 and $W_O = 1.43.10^{-26}$ m^{3}for the concrete tested.

In finite element calculations $W = \sum \varepsilon_i^m \Delta V_i$ can be calculated at each step. When W reaches W_O the damage threshold is reached in the most highly loaded node where $\varepsilon_M = \varepsilon_{D\sigma}$

4.4.   The damage laws just given strictly applies only to uniaxial loadings. For more complex situations a three dimensional law is needed. From the physical nature of the damage it is clear that it is anisotropic : most microcracks develop in a direction perpendicular to the maximum principal strain. Thus the damage parameter D must be tensorial. This introduces large complications which so far have seldom been considered. Lemaitre and Mazars |12| more simply made the assumption that the damage was scalar. It still follows the law expressed by equation (22), the uniaxial strain   being replaced by the equivalent strain :

$$\tilde{\varepsilon} = \left[<\varepsilon_1^2> + <\varepsilon_2^2> + <\varepsilon_3^2> \right]^{1/2} \tag{29}$$

$<\varepsilon_1>$ $<\varepsilon_2>$ and $<\varepsilon_3>$ being the positive principal strains.
This damage law can now be used to study the behavior of structures either by analytical calculations in simple cases (beams) or by finite element calculations. Figure 6 shows results obtained by Lemaitre and Mazars |12| for reinforced concrete beams in bending. The initiation of cracks is well predicted and the average behavior on further loading is well described by simple analytical calculations.

Figure 7 shows the comparison between experimental and numerical results for a notched specimen. Here again the agreement is quite good and it is further ascertained by acoustic emission results which allowed to localize the development of damage in various specimens |12, 14|.

5.   TRANSITION FROM MECHANICS OF DAMAGE TO MECHANICS OF FRACTURE

5.1. Introduction

At the tip of a crack a plastic zone develops within which appears the damage described above. Hence a relationship must exist between this damage and the fracture toughness of the material. In order to establish it, since the criterion of damage in ductile fracture is a function of the development of the deformation and the stress, it is indispensable to know the latter exactly in the plastic zone. This problem is not easily solved by analytical methods.

5.2. Damage in the Dugdale zone

Janson and Hult |4| extended the Dugdale plastic zone model to include a damage which was assumed to be a power function of the deformation

$$D = D_1 \varepsilon^n \tag{30}$$

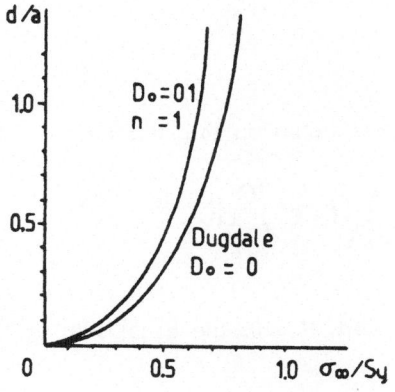

Fig. 8 Dimension of the Dugdale zone
with or without damage as a function
of the relationship between the stress and
the limit of elasticity. The damage is
associated with the deformation by the
formula $D = D_1 \varepsilon^n$

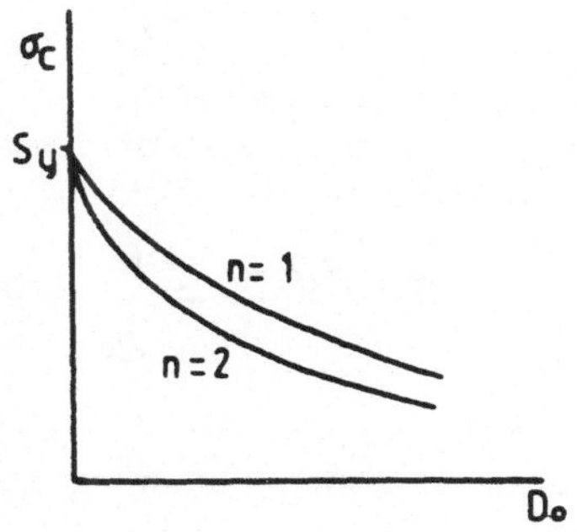

Fig. 9 Critical stress of the
Dugdale model[14]

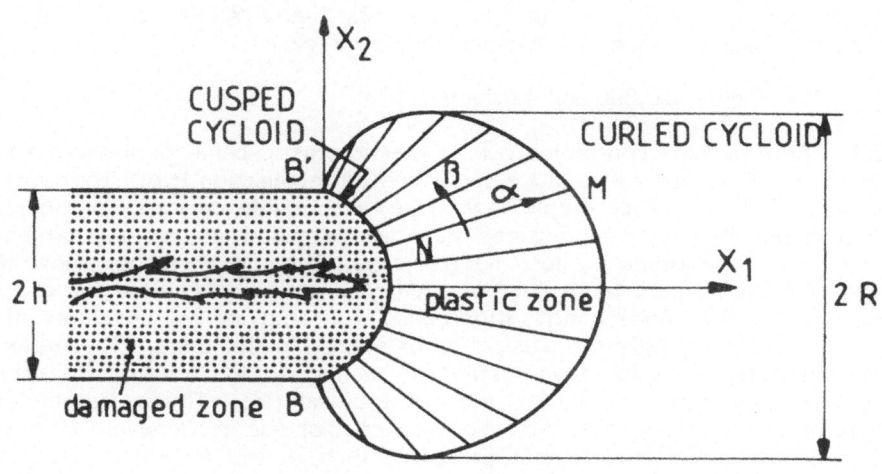

Fig. 10 Damaged zone and plastic zone in mode III[15]

and hence a function of the distance x to the tip of the crack

$$D = F(x/d) \qquad (31)$$

d being the length of the plastic zone.
Using as a definition of the effective stress equation (8), the size of the plastic zone results from :

$$\pi\sigma/2\sigma_y = \int_\alpha^1 \exp\left[-F\left(\frac{t-\alpha}{1-\alpha}\right)\right]\left(1-t^2\right)^{1/2} dt \qquad (32)$$

where $\alpha = a/(a + d)$
In case of a non damaging material, i.e. with F = 0, the usual size of the Dugdale zone results :

$$a/(a + d) = \cos\left(\pi\sigma/2\sigma_y\right) \qquad (33)$$

Assuming a x^{-1} singularity for the strain, the function F is given by

$$F(x/d) = D_0 \left(x/d\right)^{-n} \qquad (34)$$

$$\text{with } D_0 = D_1 \left(\sigma_y/E\right)^n \qquad (35)$$

Figure 8 shows the result of a numerical integration which demonstrates that the damage increases the size of the plastic zone.
The critical stress σ_c is attained when this zone size grows indefinitely. putting $\alpha = a/(a + d) = 0$ in equation (32), figure 9 results, showing the strength reduction when the damage D_0 is increased.

5.3.   The theory of Bui and Ehrlacher |15|

5.3.1. These authors consider a law of elastic plastic behavior without strain hardening. Fracture takes place for a critical elongation ε_R. The limit of elasticity is σ_y. This simple law makes it possible to carry out certain calculations, but the conclusions can be extended to strain hardening materials.   Like other authors before them, Bui and Ehrlacher show that within the plastic zone there is a damaged zone sometimes called the process zone (figure 10). Their contribution consists in studying the development of this zone and especially the dissipation of energy in this zone. The material within it being considered as completely damaged cannot carry any stress. The stress becomes zero within and on the boundaries. Thus it resembles a notch which is propagated, but whose exact shape is not known. However, unlike a notch, the mass flow is preserved through the frontier. Along the length of the, latter, there is therefore a positive entropy jump. The strain energy release rate $\mathcal{G}$ is dissipated in two ways : in the form of plastic deformation energy and in the form of fracture energy. The dissipated plastic energy is designated by Dp and the rate of fracture energy G. The balance is given as :

$$\mathcal{G} \,(da/dt) = Dp + G \,(da/dt) \qquad (36)$$

This energy G dissipated on the boundary of the damaged zone BB', in a quasi static loading

$$G = \int_{BB'} W\,(\mathcal{E}_e)\, n_1 \, ds \tag{37}$$

where $W\,(\mathcal{E}_e)$ is the density of elastic energy and n_1 the component of the normal to the boundary parallel to the x axis.

The damaged zone has a width 2h. This is only zero if the ductility $\mathcal{E}_R$ is infinite. This implies that G, the rate of fracture energy, is positive. G tends towards zero, if $\mathcal{E}_R$ tends towards infinity and if h hence tends towards zero, that is, for a crack without a damaged zone.

In the quasi-static case, in plane strain, eq. 37 yields :

$$G = (1 - \nu^2)\,\sigma_y^2\, h / E \tag{38}$$

2h, the width of the damaged zone is a function of the properties of the material (in particular of its limit of elasticity and of its ductility) but also of the geometry and the method of loading. It is a quantity which it would be important to know better but which is not easy to calculate.

5.3.2. Bui and Ehrlacher were able to do it in mode III (fig. 10). In a quasi-static situation, the boundary of the damaged zone was found to be a cusped cycloid and the boundary of the plastic zone a curled cycloid. The damaged zone has a width 2h given by

$$2h = K_{III}^2 / k^2 (2\delta_R/\delta_y - 1) \tag{39}$$

where k is the yield stress, δ_y the yield strain and δ_R the fracture strain, all in shear.

It is clearly seen that h tends towards zero if the ductility tends toward infinity. For a brittle material $\delta_R = \delta_y$ and $k = \tau_R$, then

$$2h = (K_{III}/\tau_R)^2 \tag{40}$$

The plastic zone height 2R is found to be given by

$$2R = \frac{K_{III}^2}{\pi k^2} \sin \cos^{-1}\!\left(-\frac{2hk^2}{K_{III}^2}\right) + \frac{h}{\pi} \cos^{-1}\!\left(-\frac{2hk^2}{K_{III}^2}\right) \tag{41}$$

The limits are R = h for a brittle material and $2R = (1/\pi)\,(K_{III}/k)^2$ for a perfectly plastic material ($\delta_R = \infty$ and h = 0).

G is given by

$$G = 2h\,\tau_R^2 / \mu \tag{42}$$

and Dp is given by

$$D_p = (2 K_{III}\,\dot{K}_{III}/\mu k)(K_{III}^2/2\pi k - hk/\pi) \tag{43}$$

where μ is the shear modulus.

Quite consistently with the size of the plastic zone, Dp = 0 for a brittle material and Dp tends toward $K_{III}^2 \dot{K}_{III}/\pi\mu k^2$ for a perfectly plastic material (h = 0).

5.3.3. In quasistatic mode I, solutions can be found by extrapolation of the preceding ones in mode III, or by finite element calculations. However in small scale yielding, it is known that

$$J = K_1^2(1-\gamma^2)/E \qquad (44)$$

Integrating over the damage zone boundary BB' yields.

$$J = \int_{BB'} W\,dx_2 = (1-\gamma^2)\,h\,\sigma_R^2/E \qquad (45)$$

σ_R being the fracture stress.
Hence the damaged zone thickness is

$$2h = 2\left(K_I/\sigma_R\right)^2 \qquad (46)$$

analogous to equation (40)
At fracture

$$2h_R = 2\left(K_{Ic}/\sigma_R\right)^2 \qquad (47)$$

For concrete with K_{IC} = 2 MPa$\sqrt{m}$ and σ_R = 4 MPa, the size of the damaged zone reaches $2h_R$ = 0.5 m, whereas in mild steel at low temperature if the cleavage stress is 1 500 MPa, for K_{IC} = 30 MPa$\sqrt{m}$, $2h_R$ = 800 μm, of the order of a few grain sizes.

For a stationary crack it is tempting to anticipate that a formula analogous to equation (39) holds :

$$2h = 2(K_1/\sigma_y)^2/(2\varepsilon_R/\varepsilon_y - 1) \qquad (48)$$

A finite element calculation gave the result shown on figure 11, where the profile of the damaged zone is more blunted than in mode III.

For a moving crack there would be a wake zone behind the active plastic zone. Again by analogy with mode III equation (43), the plastic dissipation rate is probably of the form

$$D_P = AK_1^3\dot{K}_1 - BK_1\dot{K}_1 h \qquad (49)$$

where A and B are some material constants. Then the balance equation (36) can be written

$$\frac{1-\gamma^2}{E}K_1^2\frac{da}{dt} = (AK_1^3 - BK_1 h)\frac{dK_1}{dt} + \frac{1-\gamma^2}{E}\sigma_y^2 h\frac{da}{dt} \qquad (50)$$

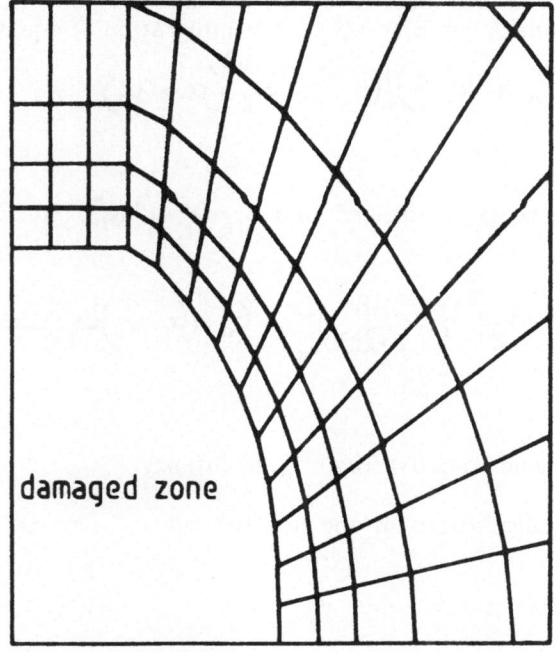

Fig. 11   Damage front in mode I by the use of a finite
element method[15]

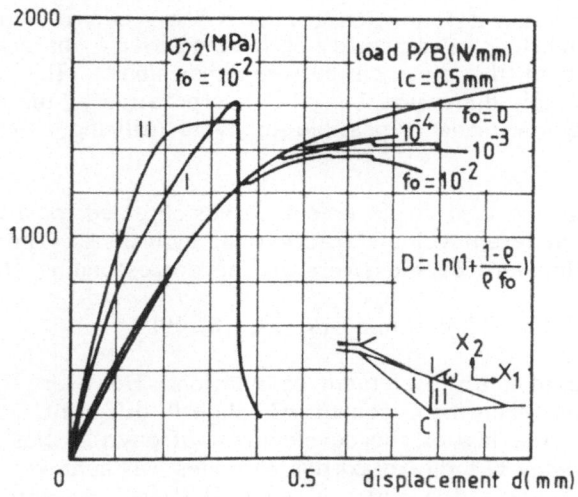

Fig. 12   Load-displacement curves of a cracked three point bend
test piece calculated by Rousselier[6] for various values of the
volume fraction of inclusions.

This differential equation describes the R curve of the material. If for example it is assumed that $B/A = \sigma_y^2$, the integration of equation (50) yields

$$G_c \approx G_o + (2/A)\left[(1-\gamma^2)/E\right]^2 (a-a_o) \qquad (51)$$

for the R curve.

If on the other hand it is assumed that $h = \alpha + \beta K_1^2$, the R curve takes the form

$$a - a_o = \frac{1}{2}\left(\frac{1-\gamma^2}{E}\right)^2\left(\frac{A - B\beta}{1 - \beta\sigma_y^2}\right)\left[G_c - G_o + (G_1 - G_o)Ln\left(\frac{G_1 - G_c}{G_1 - G_o}\right)\right] \quad (52)$$

which increases monotonically from G_o to infinity.

5.4. Numerical calculations of the development of a crack

5.4.1. Calculation of Rousselier |6|

Rousselier calculated the stable growth of a crack by the method of finite elements. The law of work-hardening used was a function of the damage as in Fig. 3 and the damage was defined by identification with the law of growth of cavities of Rice and Tracey (Equation (17). At the tip of the crack where the stress and deformation gradients are great, it was necessary that the critical conditions of instability be reached over a characteristic distance I_c connected with the distance between inclusions. The elements of the calculation had this dimension I_c. The incorporation of the damage into the work-hardening law made it possible to avoid utilising a local criterion of fracture.

Figure 12 shows the load displacement curves obtained for a three point bend test piece for different volume fractions of inclusions. The curves I and II show the development of the stress on the corresponding elements.

5.4.2. Calculation of Beremin |16, 17, 18, 19|

Beremin also used a finite element calculation. However, the way in which the damage was taken into account was slightly different from that adopted by Rousselier. The law of the development of cavities was here again that of Rice and Tracey but the relaxation of nodes was achieved when the radius of the cavities reached a critical value $(R/R_o)_c$. As with Rousselier the elements had a characteristic dimension I_c associated with the distance between inclusions. The crack advances by small steps of length l_c. $(R/R_o)_c$ can be measured on notched axisymmetrical test pieces which makes it possible to vary the triaxiality factor. Calculations show that in these test pieces the ratio between the strain and average strain and the ratio between the stress and the average stress are stabilized when the deformation is greater than a few per cent. In this case the relationship between the hydrostatic stress and the equivalent stress remains constant and the law of development of cavities of Rice and Tracey can be easily integrated. Beremin found, for example, values of $(R/R_0)_C$ of the order of 1.7 in a steel A 508 at 100 °C in the longitudinal direction. J resistance curves predicted by finite element calculations compared well with experimental results. More details on the experimental and numerical procedure are found in Mudry's lecture on local criteria.

6. CONCLUSIONS

Taking the damage into account makes it possible to calculate the propagation of cracks in the test pieces or any structures, whatever the geometry and whatever the level of loading. This can be done by using finite element calculations in elastoplasticity with releasing of the nodes when a criterion of ductile fracture is reached locally. It proves also extremely useful for a brittle material like concrete where no sound simple method is yet available to predict the behavior of a cracked structure. This is still a costly method but it makes it possible to eliminate uncertainties when the state of the stresses and deformations at the tip of the crack is no longer a function of a single parameter.

AKNOWLEDGMENTS

I am quite grateful to J. Mazars, J. Lemaitre and H. Bui, and to their collaborators, who greatly helped me in understanding these problems, provided me with most of the material for this review lecture and with whom very fruitful work was achieved.

REFERENCES

|1| KACHANOV, L.N. Kakd, Nauk. SSR. Ord. Tekh. Nauk. 8 (1958) 26-31

|2| RABOTNOV Y.M. Creep Fracture Proc. XII Int. Cong. Appl. Mech. 1968 Stanford Springer (1969)

|3| LEMAITRE J. and CHABOCHE J.L. J. de Mécanique Appliquée 2 (1978) 317

|4| JANSON J. and HULT J. J. de Mécanique Appliquée 1 (1977) 69

|5| BROBERG H. J. of Mech. 41 (1974) 809

|6| ROUSSELIER G.,"Finite deformation constitutive relations including ductile fracture damage" in "Three Dimensional Constitutive Relations and Ductile Fracture" S. Nemat-Nasser-Editor. North Holland Pub. Co (1981) (331)

|7| McCLINTOCK F.A. and ARGON A.S. Mechanical Behaviour of Materials. Addison Wesley (1966)

|8| McCLINTOCK F.A. Proc. Roy. Soc. A. 285 (1965) 58.

|9| RICE, J.R. and TRACEY D.M. J. Mech. and Phys. Solids 17 (1969) 201

|10| LOLAND K.E. Cement and Concrete Research 10 (1980) 395

|11| MARIGO J.J. Compte Rendu Acad. Sci. 294 (1981) 1309

|12| LEMAITRE J. and MAZARS J. Annales ITBTP 401 (1982) 249

|13| MAZARS J. Int. Conf. on Fracture Mechanics Technology Applied to Material Evaluation and Structural Design. Melbourne (1982)

|14| WALTER D. Thèse de Docteur Ingenieur. Université de Technologie de Compiègne (1981)

|15| BUI H.D. and ERLACHER A. in Advances in Fracture Mechanics (Fracture 1981) edited by FRANCOIS Pergamon Oxford (1981)

|16| BEREMIN F.M. J. Meca. App. 4 (1980) 307

|17| D'ESCATHA Y. and DEVAUX J.C. ASTM STP 668 (1979) 229

|18| LAUTRIDOU J.C. Engineering Doctorate Thesis - Ecole des Mines de Paris St Michel, 75006 Paris (1980)

|19| MUDRY F. State Doctorate thesis University of Technology, Compiègne B.P. 223, 60206 Compiègne (1982)

|20| LEMAITRE, J. and DUFAILLY, G.Modelling and Identification of plastic Damage in Metals. 3rd Mechanics Congress. Grenoble 1977

CLEAVAGE FRACTURE AND TRANSITION : APPLICATION TO THE WARMPRESTRESS EFFECT

F. Mudry
Centre des Matériaux de l'Ecole des Mines de Paris
ERA CNRS N°767 (UA 866)
B.P. 87
91003 EVRY Cédex
France

ABSTRACT. After a brief presentation of cleavage fracture, the attention is focussed on the transition behaviour which means sudden change in the fracture appearance from ductile to cleavage. A rough modelisation of this behaviour is proposed, based on simplified local criteria : many experimental features are explained in this way but more remains to be done for a comprehensive modelisation. Two more precise local criteria for cleavage fracture at a crack-tip are discussed. One is the Ritchie Knott and Rice model and the second one is based on Weibull statistics. Both criteria allow a good prediction of the fracture toughness transition behaviour. They are used to predict the warmprestress effect which is an apparent enhancement of toughness after deformation at temperatures above the transition temperature. Other applications of Weibull statistics are presented.

1. INTRODUCTION

Our first lecture was devoted to local criteria. The present lecture concentrates on the application of local criteria to cleavage fracture and transition behaviour. This is a very important topic since the unstable nature of cleavage is the reason of many brittle fractures. Many metallurgical tests such as Charpy-V impact notch, Pellini or DT tests have been designed to avoid brittle fracture and to make sure that operative conditions are always above the transition temperature.

However, this temperature may be shifted to elevated values when irradiation, ageing, temper embrittlement or ˉother metallurgical phenomena take place. Moreover, for example in certain faulty condi-tions in nuclear reactors, quick cooling is necessary which may lead to more embrittling temperatures. These are good reasons for a serious analysis of brittle fracture.

Fig. 1 shows two identical fracture mechanic tests at the same temperature inside the transition range. One broke in a brittle manner for rather low toughness value ($K_{1J} \simeq 70$ MPa $\sqrt{m}$), in the other one

303

L. H. Larsson (ed.), Elastic-Plastic Fracture Mechanics, 303–325.
© *1985 ECSC, EEC, EAEC, Brussels and Luxembourg.*

considerable stable crack growth took place without instability ($K_{1J} >$ 150 MPa $\sqrt{m}$). This behaviour gives rise to large scatter in that temperature region so that great care must be taken when defining a minimum K_{1J} (or K_{1R}) value for design purposes. Moreover, in certain mechanical conditions, cleavage fracture may happen far beyond the conventional transition temperature. Fig. 2 shows the fracture surface of a cracked round tensile bar at 100°C which shows 40% of unstable cleavage facets though the transition temperature (RTNDT) was -15°C |1|. However, in this case, cleavage took place after large plastic strainings.

These two examples are given only to show that cleavage fracture is still an open question.

In a first part, we will briefly describe the metallurgical events leading to cleavage and we will present the transition behaviour, using oversimplified modelisations of the cleavage and ductile fractures. A more general presentation of the micromechanisms leading to cleavage can be found in the lecture by Dahl and Dormagen. This lecture specializes on the modelling of cleavage and transition.

In a second part, we will introduce more precise modelisations of the cleavage behaviour including the Ritchie Knott and Rice model |2| and Weibull statistics |3|.

As last, these modelisations are used in some practical problems such as warm prestress effect or cold thermal shock.

2. DESCRIPTION OF CLEAVAGE FRACTURE AND TRANSITION BEHAVIOUR

2.1. Cleavage Fracture

The various events leading to cleavage fracture are not all very well understood. However, a simple two step process may be assumed.

At first, microcracks initiate in the steel from various mechanisms such as dislocation interaction with grain boundaries, carbides or twins. These microcracks are usually stable if the stress is sufficiently low. They require some inhomogeneous plastic flow to nucleate. We will simplify the analysis assuming that microcracks nucleate as soon as plasticity begins. Final rupture takes place when one of these microcracks becomes unstable for a sufficient normal stress. Energetic considerations show that the normal stress is the only operative stress inducing microcrack instability.

A very simple modelisation of this behaviour is the definition of a critical cleavage stress σ_c : it is assumed that there will be always a microcrack oriented at right angle with the maximum principal stress σ_1. So that rupture takes place for a critical maximum principal stress :

$$\sigma_1 = \sigma_c \qquad (1)$$

Moreover, it is assumed that this critical stress σ_c, is temperature independent.

We will see in a moment that the situation is a little more

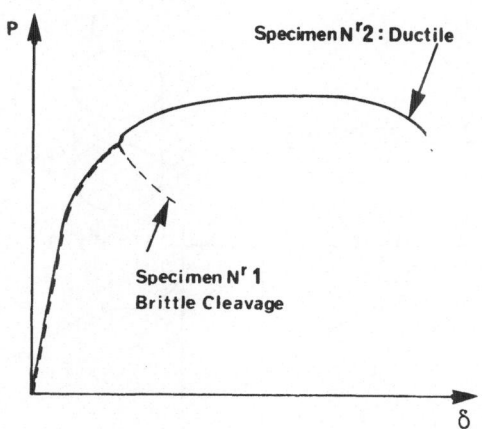

<u>Fig. 1.</u>   Two load displacement curves obtained on two specimens of the same geometry, at the same temperature.

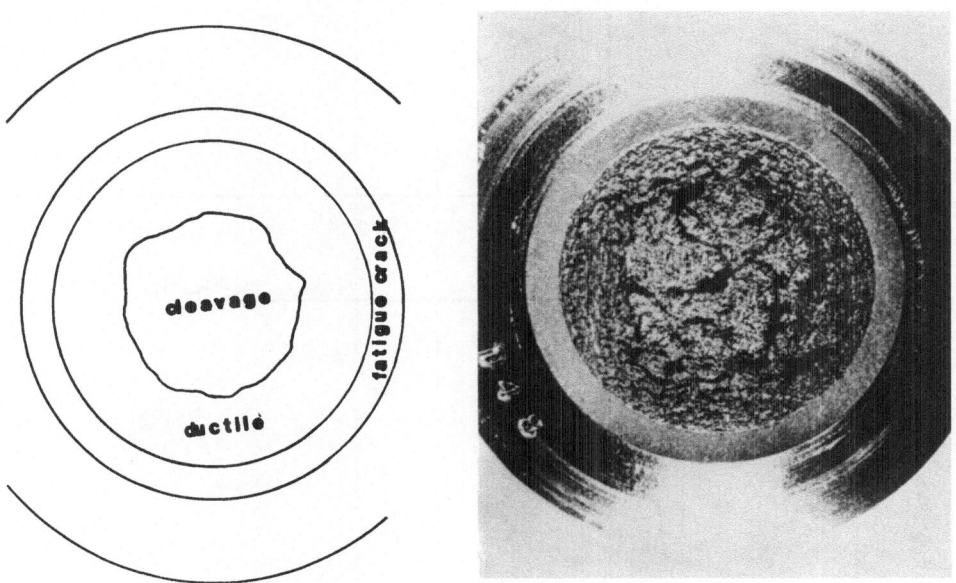

<u>Fig. 2.</u>   Fracture surface of a specimen tested at 100°C. RTNDT was -15°C in this steel (courtesy of Dr. Rousselier, Centre de Recherches E.d.F.).

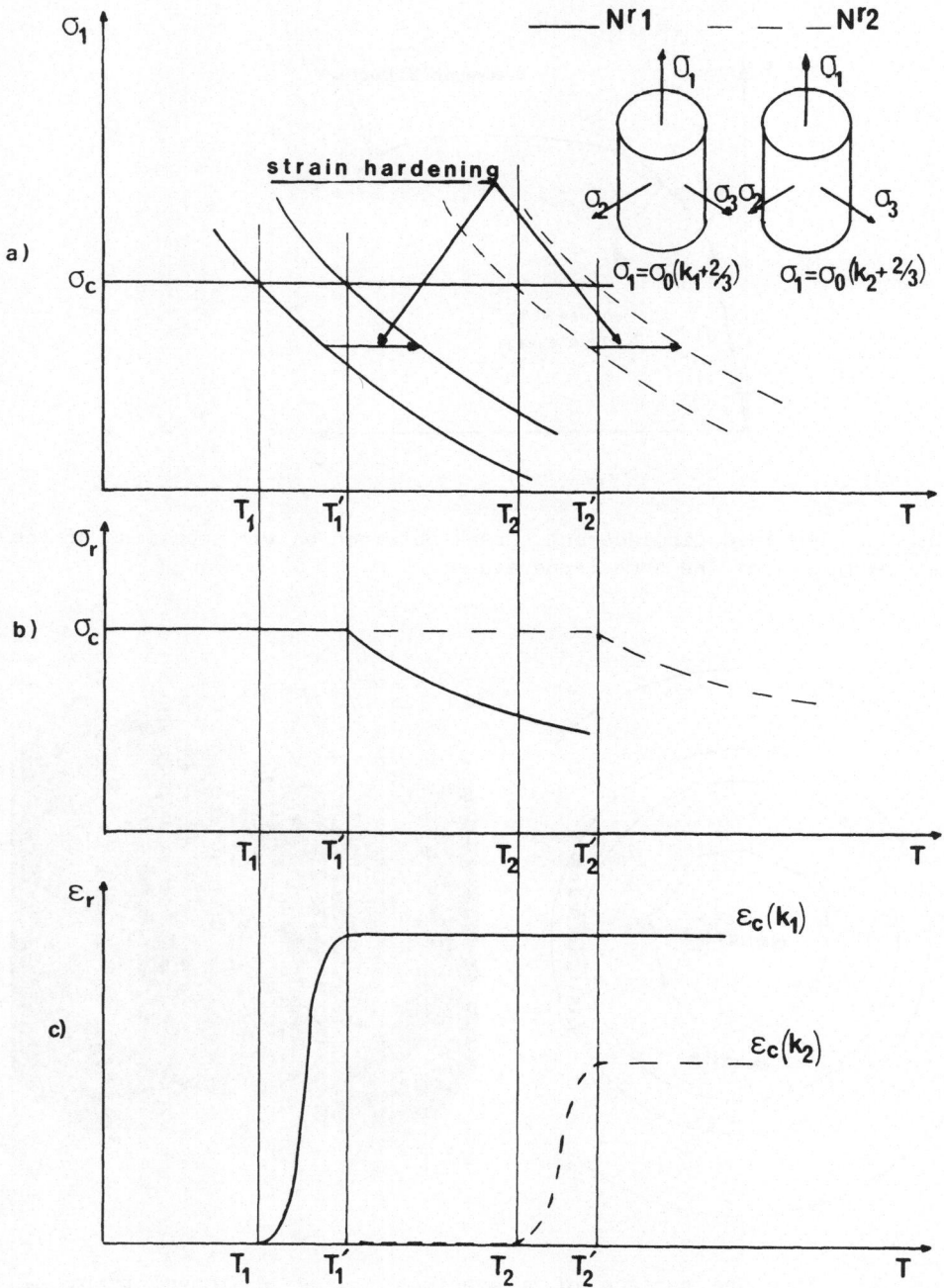

Fig. 3. Schematic representation of the transition behaviour on two hypothetical cylindrical specimens.

complex. This very simple fracture criterion, however, provides a good guide for the following discussion.

2.2. Transition behaviour

Furthermore, we will assume that ductile fracture takes place for a simplified criterion, namely a critical strain dependent on stress triaxiality (see in my first lecture, the MacKenzie et al approach |4|), but independent on temperature,

$$\varepsilon_{eq}^{p} = \varepsilon_{c}(\sigma_{m}/\sigma_{eq}) \qquad (2)$$

where σ_{m} is the hydrostatic stress and σ_{eq} is the equivalent Von-Mises stress. This simple equation does not take into account history effects. However, it is a very good guide for proportional loading.

We now assume, as a first approximation, that no interaction occurs between the two fracture micromechanisms. In other words, when loading a structure, both damage mechanisms, microcracking and hole growth, independently take place and fracture may be cleavage or ductile depending on the strain and stress state.

In this case, since cleavage is stress controlled and ductile damage is strain controlled, any hardening of the material promotes cleavage. In ferritic steels, the yield stress is a strong decreasing function of temperature, so that cleavage is reached first at low temperature whereas ductile rupture is predominant at elevated temperature. Fig. 3 schematically shows the results of two hypothetical experiments.

Two identical cylinders are tested at different temperatures. One is strained under a constant stress triaxiality $(\sigma_{m}/\sigma_{eq}) = k_{1}$ and the other one under a higher stress triaxiality $(\sigma_{m}/\sigma_{eq}) = k_{2}$. When yielding begins the maximum principal stresses are $\sigma_{1} = \sigma_{o} (2/3 + k_{1})$ and $\sigma_{1} = \sigma_{o} (2/3 + k_{2})$ respectively, where σ_{o} is the yield stress. Therefore σ_{1} is a decreasing function of temperature since it is proportional to σ_{o}. Let us define as T_{1} and T_{2} the temperatures beyond which σ_{1} is less than σ_{c} at yield. Below these temperatures, rupture takes place without noticeable plastic strain. For higher temperatures, some strain hardening takes place and the specimens still rupture by cleavage up to the point where the critical strain is achieved i.e. $\varepsilon_{c}(k_{1})$ and $\varepsilon_{c}(k_{2})$ respectively. Two other temperatures T'_{1} and T'_{2} may be defined with the equations :

$$\sigma_{c} = \sigma_{eq}(\varepsilon_{c}(k_{1}), T'_{1}) \times (2/3 + k_{1})$$

$$\text{and} \qquad\qquad\qquad\qquad\qquad\qquad\qquad\qquad\qquad (3)$$

$$\sigma_{c} = \sigma_{eq}(\varepsilon_{c}(k_{2}), T'_{2}) \times (2/3 + k_{2})$$

$\sigma_{eq}(\varepsilon,T)$ is the flow stress for strain ε and temperature T.

The corresponding rupture stresses and strains are shown as functions of temperature in Fig. 3.

This very simple representation of the transition behaviour is sufficient to explain the effect of temperature and stress triaxiality (i.e. notch effect).

Moreover, it shows the embrittling effect of an elevation of the yield stress without variation of the critical stress σ_c. Such is the case for the effect of strain rate (i.e. loading speed) and of irradiation damage. This is not so for grain refining because the microcracks are usually shorter : both the yield stress and the critical stress are enhanced.

2.3. The complex nature of the transition behaviour

This representation is however too simple to explain certain experimental features. First, it is well known that considerable experimental scatter takes place in the transition region. Some specimens rupture by cleavage and others in a ductile manner through a rather large temperature range. Moreover, a continuous variation of the fracture surface appearence from pure cleavage to pure ductile is seen. This means that both rupture modes are operative. In cracked specimens, some ductile stable crack growth is often seen on the fracture surface prior to unstable cleavage fracture. This instability may be due to an elevation of the strain rate because of ductile tearing instability or to the fortuitous presence of a low toughness point or to an elevation of stresses due to the reduction of the load sustaining area. Recent work is devoted to this particular subject of cleavage-ductile interaction |5,6|. Though significant progresses have been made, still more remains to be done.

Another difficult topic is the crack arrest test interpretation (see lectures by Brickstad and Kalthoff). In model materials, only one fracture mode is usually operative. Moreover, rather small plastic zones are present, though not always. In this case, an unstable crack achieves a speed close to the Rayleigh waves speed. Dynamic loads and wave interactions are very important. The dynamic stress intensity factor as well as the crack arrest toughness may be precisely studied.

However, for most low alloy steels, in the transition temperature region, the crack surface shows very difficult patterns with long fibrous bands remaining attached far behind the crack tip. The fracture surface is a complex mixture of cleavage and ductile. It has been shown that considerable energy is spent in joining together by shearing two unstable cleavage streams. Fig. 4 is an illustration of this phenomenon taken from Hahn et al. |7|.

There is still some debate about the true speed of such a crack and about the relevance of dynamic analysis for this complex problem |8| (see Brickstad's lecture).

3. MORE PRECISE MODELISATION OF CLEAVAGE. LOCAL CRITERIA

3.1. The Ritchie Knott and Rice (RKR) approach |2|

The R.K.R. model uses the critical stress concept to predict

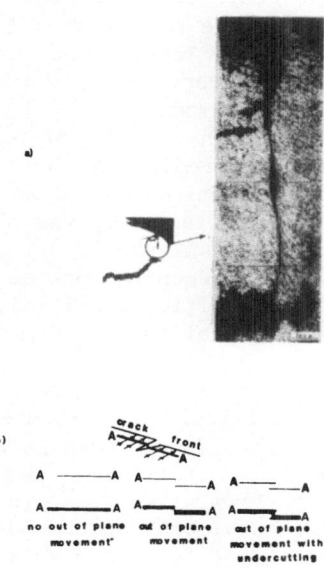

Fig. 4. a) A shear wall in a section perpendicular to the crack plane of an A533 crack arrest specimen. Test temperature was 19°C above RTNDT. The section is also perpendicular to the crack propagation direction. b) Schematic showing a possible mechanism for generating shear walls by shear of the ligament joining two non coplanar crack segments. Taken from Hahn et al |7| : (Fig. 15 and 16).

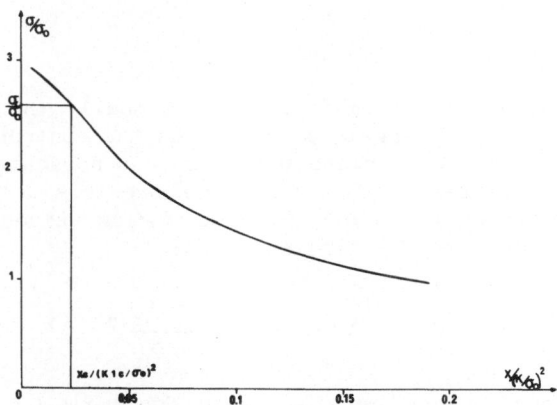

Fig. 5. Stress distribution in front of the crack-tip in small scale yielding. Relation between (K_{1C}), yield stress (σ_o), cleavage stress (σ_c) and characteristic distance (X_c) in the RKR model |2|.

the temperature dependence of fracture toughness. Using the stress distribution in front of a crack in small scale yielding (Fig. 5), they assume that fracture takes place when a critical stress σ_c is reached at a characteristic distance x_c. They showed that the lower part of the fracture-toughness transition curve was correctly predicted with a critical distance of two grain sizes.

This approach has been used by different authors |9,10| and has given very interesting predictions of the fracture toughness transition temperature, irradiation effect and loading speed effect. However, the characteristic distance x_c was shown to be only loosely connected to the grain size and should be considered as a fitted parameter. On the other hand, depending on the metallurgical situation of the matrix (martensitic, bainitic, ferritic), different relations of the critical stress with the grain size or lath packet size have been derived |11|.

However, for structures without crack, it has been shown that the critical stress was difficult to measure if different specimen geometries are used |12|. In order to make good predictions of rupture loads on notched specimens, a statistical definition of the critical stress must be introduced. This gives rise, in particular, to a rather large size effect : large specimens rupture for proportionally smaller loads because the probability of finding a weak point is higher.

3.2. Weibull statistics

This problem is well known in ceramic materials. Weibull statistics, based on a weakest link analysis is used to describe the results |13|. The material is modelled as a chain of "grains". Each grain has the same failure probability $P(\sigma)$ and the same volume V_u. The failure probability of one grain, <u>for low stresses</u>, is proportional to the applied stress to a power m, i.e. :

$$P(\sigma) = (\frac{\sigma}{\sigma_u})^m \qquad (5)$$

m is called the Weibull parameter. If m is small, the fracture stress of one grain may spread largely below σ_u. On the contrary, if m goes to infinity, no rupture takes place before σ_u and no scatter will be seen in the fracture stress. This expression predicts a very small but positive probability of failure for low stress values. Other expressions are often proposed for $P(\sigma)$ such as :

$$(\frac{\sigma - \sigma_a}{\sigma_u})^m \quad \text{or} \quad (\frac{\sigma - \sigma_a}{\sigma - \sigma_u})^m \qquad (6)$$

These expressions may be more accurate for very low stress values. However, since they introduce three parameters instead of two, Eq. 5 is preferred here.

If the volume of tested material is V ($V \gg V_u$) and if fracture takes place as soon as one grain ruptures, then the cumulative failure probability will be :

$$P(\sigma_R < \sigma) = 1 - \exp\left[-\frac{V}{V_u}\left(\frac{\sigma}{\sigma_u}\right)^m\right] \tag{7}$$

The average rupture stress will be :

$$<\sigma_R> = \left(\frac{V_u}{V}\right)^{1/m} \sigma_u \ (1/m)! \tag{8}$$

The size effect is clear in this last formulation.

If the volume V is not homogeneously stressed, the cumulative failure probability becomes :

$$P(\sigma_R < \sigma) = 1 - \exp\left[-\left(\frac{\sigma_w}{\sigma_u}\right)^m\right] \tag{9}$$

where $\sigma_w^m = \sum_i \sigma_i^m$

where σ_i is the stress in the ith grain

σ_w will be named the "Weibull stress". (For a volume V homogeneously stressed : $\sigma_w = \left(\frac{V}{V_u}\right)^{1/m} \sigma$.)

This approach can be used in a finite element program if the mesh elements are regarded as "grains" of the model. For each load increment, the maximum principal stress is computed in all mesh elements inside the plastic zone (since no microcrack is present if no plasticity has occurred yet). The Weibull stress is computed from :

$$\sigma_w^m = \sum_{\substack{\text{mesh elements}\\ \text{inside the}\\ \text{plastic zone}}} \sigma_1^m \frac{\Delta V}{V_u} \tag{10}$$

where ΔV is the volume of the mesh element considered. ΔV may be bigger than V_u if the stress variation over the element is small. The failure probability is always given by equation (9).

For most low alloy steels, the Weibull parameter is between 20 and 24. The value m = 22 is the most commonly used. This means that the fracture scatter should be rather low. Fig. 6 gives the cumulative fracture probability as a function of $\left(\frac{\sigma_w}{\sigma_u}\right)$, for m = 22 typically, a rupture stress can be determined with a $\pm$ 7% accuracy.

However, care must be taken because the relationship between the Weibull stress and the applied load may be strongly non linear in deeply notched structures.

4. METHODOLOGY OF LOCAL CLEAVAGE CRITERIA

The R.K.R. criterion and Weibull statistics are local criteria. This means that a precise description of the stress and strain field is required for their application. The methodology, as for the ductile fracture criterion, will be a two steps approach.

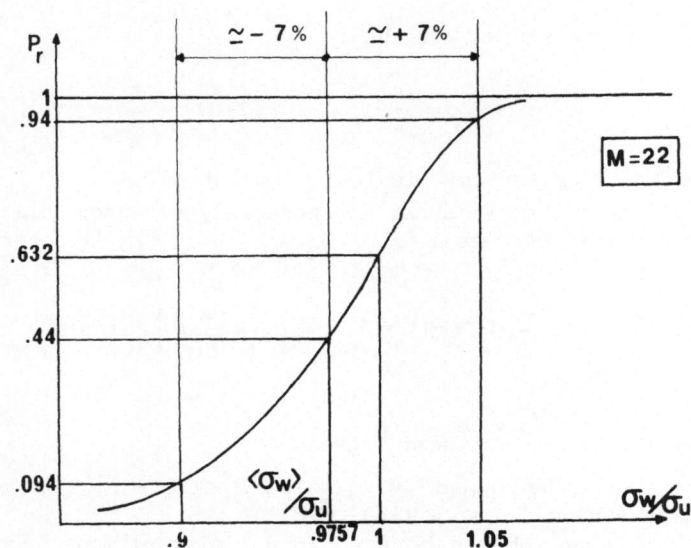

Fig. 6. Cumulative fracture probability as a function of the Weibull stress. $\langle\sigma_w\rangle$ is the mean value. A scatter band of $\pm$ 7% is given.

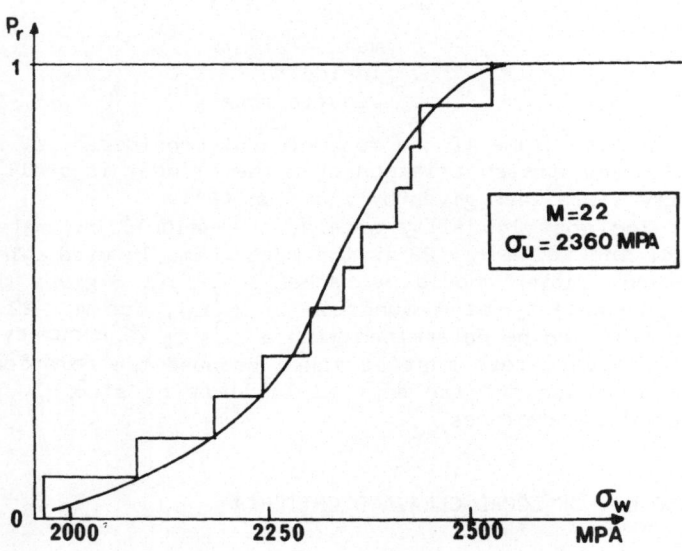

Fig. 7. Experimental and theoretical cumulative fracture probability. The theoretical curve follows equation 9. On the experimental curve, each vertical segment corresponds to one experiment.

a. Measure of the rupture parameters on selected cracked and uncracked specimens.

b. Utilisation of the critical values for the given problem which means a precise knowledge of the stress and strain distribution either through numerical modelisation or using analytical solutions if available.

4.1. Measure of the critical parameters

In order to use the R.K.R. model, only two experiments are required. One will be a measure of σ_c, the second is a determination of x_c. Usually, σ_c is measured on notched three point bend specimens. The stress distribution is known from numerical computations, or from slip line field theory. The knowledge of the rupture load, alone, is a measure of σ_c. The second manipulation is a conventional K_{IC} test. The stress distribution, shown in Fig. 5, is used. Using $\sigma_{yy} = \sigma_c$ and $K = K_{1C}$, a value of x_c is found if the yield stress σ_o is precisely known.

The procedure is very similar for the Weibull statistics. First, notched specimens similar to those used in ductile fracture are broken. A minimum of six is required to have an estimate of the scatter. An elasto plastic calculation of the specimen allows the determination of the Weibull stress |14| (Eq. 10) as a function of load. In this equation, m is taken as 22 and V_u is taken as a cubic volume with edges equal to two grain sizes. For each individual experiment, a Weibull stress is measured. Taking the average value $<\sigma_w>$, σ_u is computed from : $<\sigma_w> = \sigma_u \times (\frac{1}{22})! = \sigma_u \times 0,9757$. Then the experimental cumulative probability is compared to the theoretical expression (Eq. 10), (figure 7). A rather good correlation is usually achieved. If such is not the case, other experiments are required in order to more precisely determine the scatter. The value of V_u is determined from conventional K_{1C} tests at low temperature. For most practical applications of Weibull statistics, only the parameter σ_u should be precisely determined. This parameter is a measure of material strength. The value of m is often 22 and a precise determination of V_u is useless except for very low toughness applications.

In fact, the material strength, measured either by σ_c or σ_u, may be determined on any specimen geometry provided that the stress distribution is known. The knowledge of a toughness value K_{1C} is sufficient for the determination of the second parameter, either x_c or V_u, provided that the failure process is entirely controlled by cleavage. However, care must be taken if large strains are present when rupture takes place because large tensile straining enhances cleavage strength |15|. The reverse is true for compressive straining. This effect may be responsible of surprising fracture paths in the transition region.

4.2. Application to flawed structures

If analytical solutions (e.g. mode I, plane strain, small scale yielding solution) are not known, a finite element program must be used with a rather fine meshing of the crack tip since x_c and V_u are usually rather small ($\simeq$ 50 μm). Afterwards, the Weibull stress is computed from Equation (10) and the maximum principal stress at a distance x_c of the crack-tip is looked for. This gives a prediction of the failure behaviour. The R.K.R. model gives a precise fracture value whereas Weibull statistics also give an indication of the scatter. Moreover, for both models the direction of crack instability can be predicted if a radial meshing is used at the crack-tip (fig. 8).

5. APPLICATIONS

At this point, local criteria may seem rather useless, since linear fracture mechanics correctly predicts brittle fracture. In the following, we will show that many interesting conclusions can be drawn from their utilisation. Many of them are only refinements in the usual linear fracture mechanics field. Others are extensions in large scale yielding conditions. We will use the Weibull statistical approach in the following though many results have been obtained using the R.K.R. model. The reasons for this choice are, first that, I have all the available information on this model and second that this approach can easily be interpreted as a R.K.R. model with a description of the scatter and size effect on σ_c.

5.1. Temperature dependence of fracture toughness

One of the very interesting features of local criteria is the temperature independence of the critical parameters.

Assuming small scale yielding, the crack-tip stress distribution is a function of the stress intensity factor K alone. Fig. 5 gives this distribution in dimensionless coordinates. The solution may be written in the following form :

$$\sigma_{ij}(r, \theta) = \sigma_o \times f\left[r/(K/\sigma_o)^2\right] \times g_{ij}(\theta) \tag{11}$$

where σ_o is the yield stress, r and θ are polar coordinates around the crack tip. Using the definition of σ_w yields |3,4| :

$$\sigma_w^m = \frac{\sigma_o^{m-4} \times B \times K^4 \times C_m}{V_u} = \sigma_u^m \times Ln(\frac{1}{1-P}) \tag{12}$$

where B is the crack front length, C_m is a numerical value depending on m, slightly on the strain hardening exponent and, also, on the plastic zone size $(K/\sigma_o)^2$. Table 1 gives the corresponding variations. It is seen that C_m is almost constant except for low values of $(K/\sigma_o)^2$ because, in this case, the stress gradients are so important that averaging on V_u greatly influences C_m.

This equation is a rather good description of the fracture toughness transition (fig. 9). A very large scatter is noticed : the

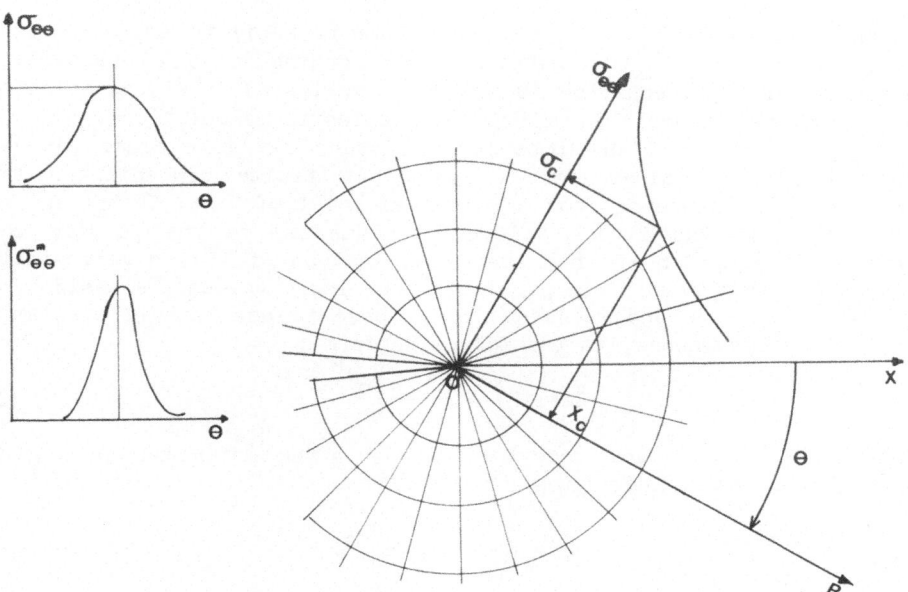

Fig. 8. Prediction of crack instability and direction in mixed-mode loading. The critical stress is compared to the maximum value of $\sigma_{\theta\theta}$ at the characteristic distance x_c. For Weibull statistics, the probability of the crack angle should be proportional to $\sigma_{\theta\theta}^m$.

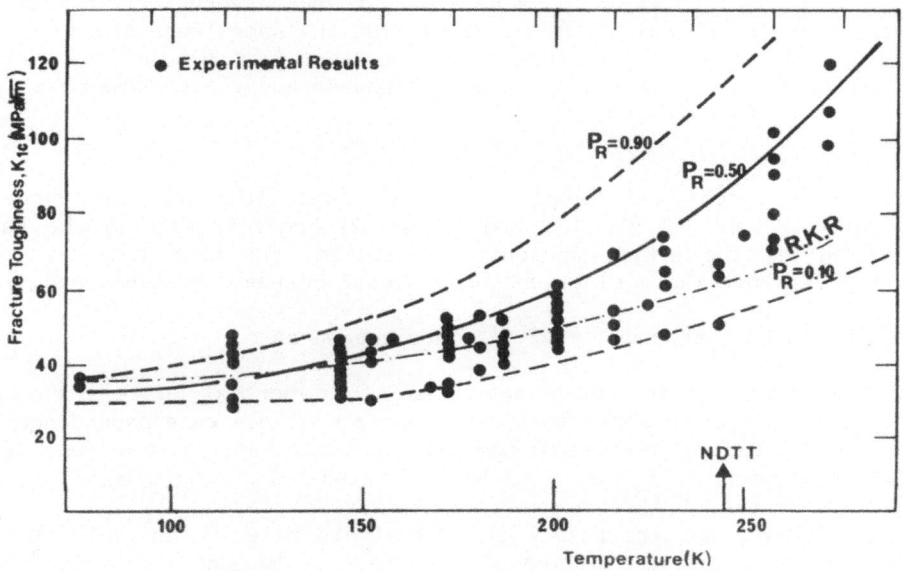

Fig. 9. Fracture toughness transition curve. Experimental results from the HSST program |Ref. 24|. Comparison with the RKR model and of the Weibull statistics model (From Ref. 3).

toughness corresponding to 90% cumulative probability should be 2.16 times bigger than the toughness corresponding to 10% probability. Remember that this equation is based on small scale yielding solutions and cannot be extrapolated without elastoplastic calculations.

Equation 12 predicts that fracture may take place for very small toughness values for a small but finite probability. This behaviour is disturbing from a practical point of view. It arises from the fact that the stress volume is connected to the plastic zone surface. If equation (6) had been used instead of (5), a very similar expression would result. In particular, no lower values are assigned to K. However, Landes and MacCabe |16| propose to use Weibull statistics for fracture toughness, in the following form :

$$- Ln(1-P) = \frac{(K - K_o)^M \times B}{H} \tag{13}$$

where M, K_o and H are constants for a given material, at a given temperature. Instead, Equation 12 predicts :

$$- Ln(1-P) = \frac{K^4 B}{H} \tag{14}$$

with $H = \dfrac{V_u \, \sigma_u^{\,m}}{\sigma_o^{\,m-4} \times C_m}$

Therefore M should be four and K_o is zero.

These two expressions have a very similar behaviour around the average, but very different probabilities are found for low values. In order to choose, it would be necessary to run a very large number of experiments ($\simeq 10^3$) on the same material at the same temperature.

Equation 12 also predicts, except for very low values of $(K/\sigma_o^2$, a relationship between fracture toughness and yield stress :

$$\sigma_o^{\,m-4} \times K^4 \simeq const. \tag{15}$$

This is a linear relation in Log-Log plot. This relation already proposed by Hahn et al |17| and Curry |9| predicts exactly the same behaviour as the R.K.R. approach. Of course, for high temperatures, ductile fracture takes place and other local criteria must be used.

5.2. Strain rate dependence of fracture toughness

When moderate strain rates are used, the inertial effects may be neglected. It can be shown that, for moderate strain rate dependence of the yield stress, representative of ferritic steel, the crack-tip solution is exactly the same as without strain rate dependence of the yield stress (Eq. 11).

Using an appropriate average strain rate in the yield stress allows a precise determination of the fracture toughness from Equation 12. Once again, similar results (without scatter) are obtained by Curry |9| using the R.K.R. model. The same approach also predicts the embrittling effect of irradiation through the variation of the yield stress |10| .

5.3. Prediction of an anisothermal behaviour : the warmprestress effect

This effect is an apparent enhancement of the fracture toughness if the flawed structure has been previously strained at a higher temperature. Fig. 10 shows schematically a fracture toughness transition curve and different stress intensity factor-temperature paths.

In case of path 1, cooling takes place under constant load conditions. Experimental results show that reloading is necessary at low temperature in order to break the specimen.

In case of path 2, the structure is unloaded, cooled down and reloaded at low temperature. Rupture is obtained for stress intensity factors a lot larger than the corresponding fracture toughness without prior thermal history. For exemple, the fracture toughness of A508 steel at 77°K is 30 MPa $\sqrt{m}$. After path 2, the measured fracture toughness was 55-65 MPa $\sqrt{m}$; and 72-77 MPa $\sqrt{m}$ after path 1. This shows that the warm prestress effect is definitely a large effect.

A safe way of interpreting the experiments is to state that no fracture is possible if the stress intensity factor is decreasing. However, we have seen that some reloading is possible without fracture and the question is raised whether path 3 is dangerous.

This behaviour can be predicted using local criteria. Curry |18| correctly predicts the warm-prestress behaviour using the R.K.R. model and small scale yielding solutions. Beremin |19| also gives good predictions in general yielding using Weibull statistics and numerical results. We will first describe briefly the small scale yielding analysis and, then, give indications for more complex finite element modelisations.

5.3.1. Small scale yielding analysis.
Equation 11 gives the stress distribution as a function of r and θ. This solution is only valid for isothermal loading. However, approximate solutions can be derived using a certain combination of such isothermal solutions |18|. We shall denote $H_{\{K,\sigma_o\}}(r,\theta)$ the solution given by equation 11.

Let us assume that the crack tip is cooled so that the yield stress is increased from σ_o to $\sigma_o + \Delta\sigma_o$, and that the stress intensity factor becomes $K + \Delta K$ where ΔK is either positive (loading) or negative (unloading). In this case the stress distribution may be approximately given by :

Loading :
$$
\begin{cases}
\text{if } \dfrac{\Delta K}{\Delta\sigma_o} < K/\sigma_o \ : \ H_{\{K,\sigma_o\}}(r,\theta) + H_{\{\Delta K, \Delta\sigma_o\}}(r,\theta) = \sigma_{ij}(r,\theta) \\[4mm]
\text{if } \dfrac{\Delta K}{\Delta\sigma_o} > K/\sigma_o \ : \ H_{\{K \,+\, \Delta K, \ \sigma_o + \Delta\sigma_o\}}(r,\theta) = \sigma_{ij}(r,\theta)
\end{cases}
\tag{16}
$$

Unloading :
$$
H_{\{K,\sigma_o\}}(r,\theta) - H_{\{-\Delta K, \ +\,2\,\sigma_o + \Delta\sigma_o\}}(r,\theta) = \sigma_{ij}(r,\theta)
$$
($-\Delta K$ is positive, in this case)

For exemple, for path number one in fig. 10, it is seen that nothing happens during cooling since K is kept constant. So that, the stress distribution at lower temperature is exactly the same as at higher temperature. Therefore, since the critical stress for cleavage is independent on temperature, no rupture is possible at lower temperature. If the initial stress intensity factor is K_i at temperature T_1 (yield stress σ_1) and the final fracture stress intensity factor is K_f at temperature T_2 (yield stress σ_2), the distribution will be approximatively [18] :

$$\sigma_{ij}(r,\theta) = H_{\{K_i,\ \sigma_1\}}(r,\theta) + H_{\{K_f-K_i,\ \sigma_2-\sigma_1\}}(r,\theta) \qquad (17)$$

if $\dfrac{K_f-K_i}{\sigma_2-\sigma_1} < \dfrac{K_i}{\sigma_1}$; otherwise no warm-prestress exists.

Using either the R.K.R. model or Weibull statistics yields almost the same value for K_f. For path number two, the final stress distribution will be :

$$\sigma_{ij}(r,\theta) = H_{\{K_i,\sigma_1\}}(r,\theta) - H_{\{K_i,2\sigma_1\}}(r,\theta) + H_{\{K_f,\sigma_2+\sigma_1\}}(r,\theta)\,(18)$$

if $\dfrac{K_f}{\sigma_2-\sigma_1} < \dfrac{K_1}{2\sigma_1}$

which also gives an approximate value for K_f which compares well with experiment. Quite surprisingly, Chell et al [20] give very similar predictions using dislocation arrays and a modified J-integral.

5.3.2. <u>Large scale yielding modelisation</u>. When the first loading K_i is no longer in small scale yielding, an elastoplastic calculation must be done if a complete modelisation is desired. The Beremin group [19] has studied path 1 and 2 on axisymmetrically cracked specimens which were loaded at 100°C around ductile initiation which means far beyond the limit load. It was shown, in case of path 2 that the warmprestress effect arose from three different origins.

a. Production of a cold worked structure at 100°C, which is metallurgically more resistant to cleavage fracture. The beneficial effet of a tensile predeformation has been demonstrated. This effect is however rather small in this case.

b. Modification of the crack-tip geometry which becomes blunted after preloading. This modifies the stress-strain distribution. It is clear that, when the characteristic distance is of the same order of magnitude as the C.O.D., reloading at lower temperature involves a notch and not a crack.

c. Introduction of residual compressive stresses during unloading. This factor is certainly the most important for moderate preloading.

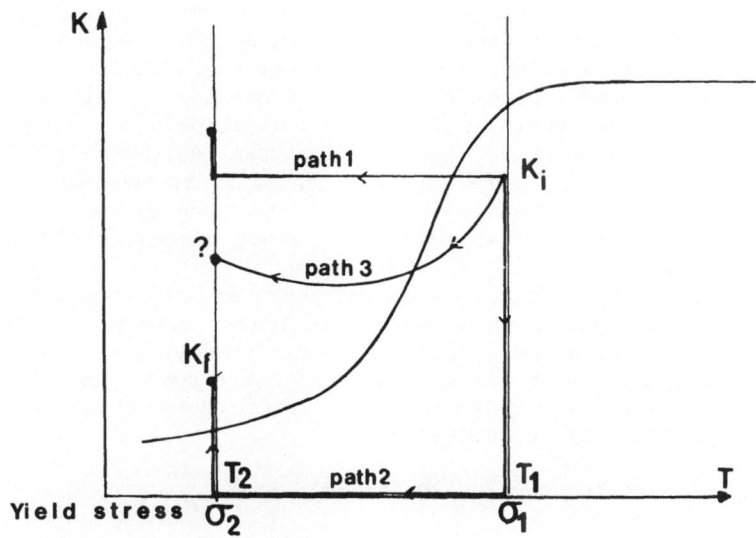

Fig. 10.   Schematic representation of the warm pre-stress effect.

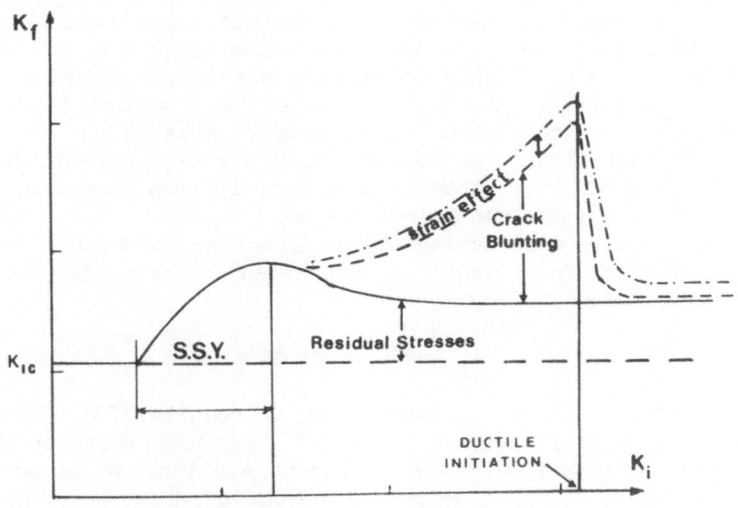

Fig. 11.   Schematic representation of the variation of final fracture toughness (K_f) as a function of the initial loading (K_i) : warm prestress effect for path 2 of Fig. 10.

Four different experiments along paths 1 and 2 have been numerically modelled. The TITUS program takes into account crack blunting and the variation of yield stress and Young's modulus with temperature. Using Weibull statistics, a rather precise prediction of fracture is achieved |19|. For path 2, the influence of residual stresses is almost constant whatever the initial loading provided that it is beyond the limit load. The blunting of the crack is almost as important as the residual stress in these experiments. Figure 11 is a schematic representation of the variation of K_f as a function of K_i in the case of path 2. In this figure, K_i is of course computed from J since general yielding is the rule.

Figure 12 gives the comparison between predicted and measured toughnesses. Therefore, the warm prestress effect is correctly predicted when using a local criterion for cleavage.

Note that, when going beyond the limit load, the influence of the residual stresses may be reduced. This shows that small scale yielding solutions may be unconservative.

5.4. Prediction of fracture in 3D cracks

In small scale yielding and mode I situations, the usual criterion for fracture of 3D cracks is that the maximum value of K along the crack front reaches K_{1C}. (For a complete discussion of 3D problems, see Munz's lecture.).

For instance, for the edge elliptical surface crack shown in figure 13, the stress intensity factor is maximum on the surface (K_b) and is minimum on the smaller axis (K_a). The theoretical criterion for fracture should be : $K_b = K_{1c}$. However, experimental programs clearly show that this equation largely underestimates the rupture load. In fact, K_{1c} is almost equal to the mean value of K_a and K_b |21|. This behaviour is easily predicted if Weibull statistics is used. Equation 12, in this particular case, predicts that $K_{1c} \simeq 0.65 (K_a + K_b)$. The product $K^4 \times B$, in this Equation, is simply replaced by : $\int K^4(s) \, ds$ where s is the curvilinear coordinate along the crack front. The effect only arises from the fact that only a very limited amount of material is interested by the maximum value of $K(s)$.

As well, considering usual fracture mechanics tests, an effect of the crack front length is predicted from equation 12 in small scale yielding :

$$K_{1c}^{4} \times B \simeq \text{const.}$$

This size effect is clear in Iwadate et al results |22|. These autors have conducted a very large quantity of toughness tests on different steels. They show that fracture toughness scatter can be analysed by Weibull statistics using a Weibull exponent of 4 for K_{IC} which is in accordance with equation 12.

5.5. Prediction of fracture in thermal shock experiments.

A large attention is devoted to the problem of cold thermal shock in

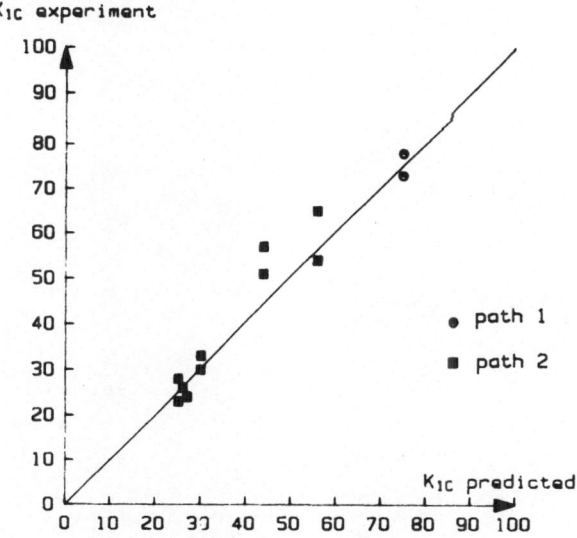

Fig. 12. Comparison of experimental and computed toughnesses.

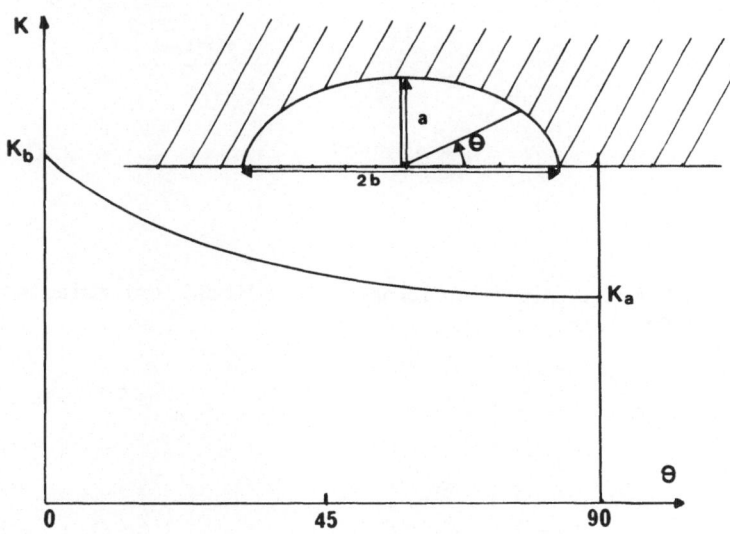

Fig. 13. Geometry of an elliptical surface crack with variation of the stress intensity factor. (Taken from reference 21).

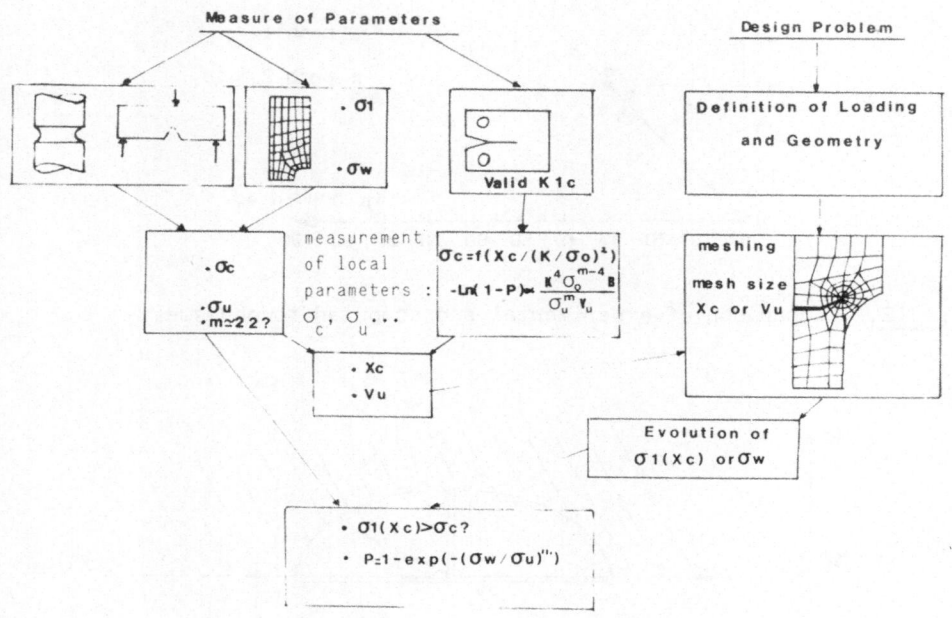

<u>Fig. 14.</u> Schematic representation of the local criteria methodology for cleavage fracture.

TABLE I : Variation of (K/σ_o) and Cm with temperature (See equations 11 and 12).

Temperature (°K)	77	173	223	273
$(K/\sigma_o)^2$ (m)	10^{-3}	$3.5 \; 10^{-3}$	10^{-2}	4.10^{-2}
Cm	2.10^5	$1.3 \; 10^6$	$1.5 \; 10^6$	$1.5 \; 10^6$

PWR reactor pressure vessels. Different experiments have been performed on large size, thick tubes in order to verify the efficiency of K_{1C} and K_{1A} concepts |23|. As a whole, these tests have demonstrated the rather good prediction capability of these methods. However, in many cases, the static and arrest toughness derived from the experiments are low either in the minimum of the toughness scatter band or, even, below this minimum. This result may be due to large material variability. However, a more simple approach is the use of a size effect predicted by equation 12. Since very long crack fronts were used in these experiments, a lower toughness is predicted.

5.6. Conclusions

a. Most trends in the transition behaviour of ferritic steels may be predicted using two local criteria, one for cleavage fracture and the other for ductile fracture. The two criteria are assumed entirely independent. This simple modelisation is sufficient for a rather good prediction of the transition behaviour. However, experimental evidences show that complex interactions in the fracture mechanisms are operative in the transition region.

b. The local criteria for cleavage fracture correctly predict a number of experimental results : temperature and strain rate dependence of fracture toughness, warmprestress effect, 3D-crack toughness and thermal shock experiments.

c. Weibull statistics provide a mean of rationalizing the very large experimental scatter in the transition. On this ground, it can be explained why some specimens fail in a ductile manner at a given temperature whereas other ones fail entirely by cleavage.

As for ductile fracture, figure 14, is a schematic representation of the local criteria approach for cleavage fracture.

REFERENCES

|1| Devaux, J.C. Mudry, F. Rousselier, G., Pineau, A. 'An experimen-
     tal programme for the validation of local ductile fracture
     criteria using axisymmetrically cracked bars and compact tension
     specimens', Engineering Fracture Mechanics, in press.
|2| Richie, R.O., Knott, J.F., Rice, J.R. 'On the relationship
     between critical tensile stress and Fracture Toughness in mild
     steel'. J.Mech. Phys. Solids, 21, (1973), p.395-410.
|3| Pineau, A. 'Review of micromechanisms and a local approach to
     predicting crack resistance in low strength steels' in Advances
     in Fracture research (ICF5), D. François (Ed.), Pergamon Press,
     2, (1981), p.553-579.
|4| MacKenzie, A.C., Hancock, J.W., Brown, D.K. 'On the influence of
     stress on Ductile Failure initiation in high strength steels'.
     Eng. Frac. Mech., 9, (1977), p.167.
|5| Kobayashi, T., Irwin, G.R., Zhang, X.J. 'Photographic Examination
     of Fracture Surfaces in Fibrous-Cleavage Transition behaviour'.
     ASTM Symposium on Fractography in Failure Analysis of Ceramics
     and Metals, Philadelphia PA, April 1982 (To be published in
     Special Technical Publications of the ASTM).
|6| Dubois, M., Buche, D, Bretin, R. 'Cleavage, Ductile Fracture
     transition study by competition between two Fracture criteria',
     7e SMIRT Conference, Chicago, (1983) : paper n° G2/8.
|7| Hahn, G.T., Hoagland, R.G., Lereim, J., Markworth, A.J.,
     Rosenfield, A.R. 'Fast Fracture Toughness and Crack Arrest
     Toughness of Reactor Pressure Vessel Steel', in : Crack Arrest
     Methodology and Applications, Hahn, G.T., Kanninen, M.F. (Eds.),
     ASTM-STP n° 711, (1980), p.289-320.
|8| Crosley, P.B., Ripling, E.J. 'Significance of Crack Arrest
     Toughness (K_{1a}) Testing' in Crack Arrest Methodology and
     Applications, Hahn, G.T., Kanninen, M.F. (Eds.), ASTM-STP n° 711,
     (1980), p.321-337.
|9| Curry, D.A. 'Predicting the Temperature and Strain Rate Depen-
     dence of the Cleavage Fracture Toughness of Ferritic Steel'.
     Materials Science and Engineering, 43, (1980), p.135-144.
|10| Parks, D.M. 'Interpretation of Irradiation Effects on the
     Fracture Toughness of a Pressure Vessel Steel in terms of Crack
     Tip Stress Analysis'. J. Eng. Mat. Techn., 98, (1973), p.30-36.
|11| Brozzo, P., Buzzichelli, G., Mascanzoni, A., Mirabile, M.
     'Microstructure and Cleavage resistance of low Carbon Steels',
     Metal Sciences, 11, (1977), p.123-129.
|12| Griffiths, J.R., Owen, D.R.J. 'An Elastic-Plastic Stress Analysis
     for a notched bar in Plane Strain Bending'. J. Mech. Phys.
     Solids, 19, (1971), p.419-431.
|13| MacClintock, F.A. 'Plasticity Aspects of Fracture' in Fracture :-
     an advanced treatise. H. Liebowitz (Ed.), Academic Press, 3,
     (1971), p.85-90.
|14| Devaux, J.C., Bouche, D. 'Développement du critère de la
     contrainte critique de clivage sur une distance caractéristique
     en vue de son application à l'analyse de la sûreté des cuves de

chaudières à eau pressurisée'. Rapport du contrat n°
78/7/3173-3174 de la Délégation Générale à la Recherche Scientifique et Technique, Paris, Juillet, (1983).

|15| Beremin, F.M. 'A local criterion for Cleavage Fracture of a
Nuclear Pressure Vessel Steel'. Metallurgical Transactions A.,
14A, (1983), pp.2277-2287.

|16| Landes, J.D., MacCabe, D.E. 'The effect of section size on the
transition temperature behavior of structural steels',
Westinghouse Internal report n° 82-1D7-Metal-P2, Pittsburg,
(1982).

|17| Hahn, G.T., Hoagland, R.G., Rosenfield, A.R. 'The variation of
K_{IC} with Temperature and Loading Rate', Metall. Trans., 2A,
(1971), p.537-541.

|18| Curry, D.A. 'A model for Predicting the Influence of Warm
Pre-stressing and strain ageing on the cleavage Fracture
Toughness of ferritic Steels'. Int. Jal. of Fracture, 22, (1983),
p.145-149.

|19| Beremin, F.M. 'Numerical modelling of warmprestress effect using
a damage function for Cleavage Fracture' in Advances in Fracture
Research (ICF5), D. François (Ed.), Pergamon Press, 2, (1981),
p.825-831.

|20| Chell, G.G., Haigh, J.R., Vitek, V. 'A theory of warm prestres-
sing : experimental validation and the implications for elastic
plastic failure criteria'. Int. Jal. of Fracture, 17, (1981),
p.61-81.

|21| Pellissier-Tanon, A., Ponsot, A., Bretin, R.N., Heliot, J.
'Critère de rupture brutale des fissures à géométrie tridimen-
sionnelle dans la partie basse de la transition des aciers
ferritiques'. Rapport contrat n° 76-7-1280. Délégation Générale à
la Recherche Scientifique et Technique, Paris, (1983).

|22| Iwadate, T., Tanaka, Y., Ono, S., Watanabe, J. 'An Analysis of
Elastic-Plastic Fracture Toughness Behavior for J_{1C} Measurement
in the transition Region'. Second Symposium on Elastic-Plastic
Fracture Mechanics, October, (1981), ASTM-STP, in Press.

|23| Cheverton, R.D., Gehlen, P.C., Hahn, G.T., Iskander, S.K.
'Application of Crack arrest theory to a thermal shock experi-
ment' in Crack arrest Methodology and Application', Hahn, G.T.,
Kanninen, M.F. (Eds.), ASTM-STP n° 711, (1980), p.392-421.

|24| Ritchie, R.O., Server, W.L., Wullaert, R.A., 'Critical Fracture
Stress and Fracture Strain Models for the Prediction of Lower and
Upper Shelf Toughness in Nuclear Pressure Vessel Steels' Metall.
Trans., 10A, (1979), p.1557-1573.

WORKSHOP 1
WORKED EXAMPLES USING THE CEGB PROCEDURE FOR THE ASSESSMENT OF THE
INTEGRITY OF STRUCTURES CONTAINING DEFECTS

A R Dowling, R P Harrison and K Loosemore
Central Electricity Generating Board
Generation, Development & Construction Division
Barnwood, Glos, U.K.

R A Ainsworth
Central Electricity Generating Board
Berkeley Nuclear Laboratories
Berkeley, Glos, U.K.

I Milne
Central Electricity Research Laboratories
Kelvin Avenue
Leatherhead, Surrey, U.K.

ABSTRACT This series of worked examples is designed to illustrate as
simply as possible the procedures in the second revision of CEGB Report
R/H/R6 "Assessment of the Integrity of Structures Containing Defects"
(1980). It is also intended to demonstrate how extra information may be
gained by performing sensitivity analyses. The examples are artificial
and the first 5 are based upon one type of problem, that of a crack in a
pressurised cylinder. The level of complication is increased from one
example to the next so that a working knowledge of the procedures can be
built up to the level of sophistication needed. In the final example,
the geometry of a crack in a flat plate is used to demonstrate how
inconsistencies arise from the conventional application of rigid safety
factors in fracture problems. Each example is structured so that it can
be read without reference to the others.
    The material properties were chosen to demonstrate certain features
in the analysis and should not be taken as typical of any particular type
of steel. For examples 1 to 4 failure is assumed when cracking is
initiated, as defined by K_{Ic} (section 5 of R/H/R6 Rev.2) while Example 5
examines the additional safety margins resulting from the inclusion of
some stable ductile crack growth following initiation.
    Details of flaw characterisation (e.g. by circumscribing a natural
defect with an ellipse or by bounding it with a rectangle) and stress
anslyses have been omitted.

NOTE: Copies of CEGB Report R/H/R6 Rev.2 may be obtained from
         Dr R A Ainsworth, Central Electricity Generating Board, Berkeley
         Nuclear Laboratories, Berkeley, Glos, U.K.

L. H. Larsson (ed.), Elastic-Plastic Fracture Mechanics, 327–371.

EXAMPLE 1

ASSESSMENT OF A FLAW IN A CYLINDRICAL VESSEL
LOADED ONLY BY INTERNAL PRESSURE

Problem

A pressure vessel contains an axial flaw in the outer surface of
its cylindrical section. The vessel is to be loaded by internal pressure
only.

Can the vessel withstand the design pressure and if so by what
margin?

Mean radius of cylinder, R	= 0.762 m
Thickness of cylinder, t	= 0.108 m
Design pressure, P	= 24 MPa
Lower bound yield stress, σ_y	= 430 MPa
Lower bound ultimate stress, σ_u	= 580 MPa
Lower bound fracture toughness, K_{Ic}	= 100 MPa√m
Flaw depth, a	= 0.007 m
Flaw length, ℓ	= 0.035 m

Solution

1.1 Flaw Characterisation

The flaw is specified as an external axial surface flaw with the
following dimensions:
        depth,   a = 0.007 m
        length, ℓ = 0.035 m
The vessel dimensions are:
        mean radius,   R = 0.762 m
        wall thickness, t = 0.108 m

1.2 Material Properties

Lower bound yield stress,	σ_y	= 430 MPa
Lower bound ultimate stress,	σ_u	= 580 MPa
Lower bound fracture toughness,	K_{Ic}	= 100 MPa√m

1.3 Stress Analysis

Applied pressure, P = 24 MPa

The hoop stress due to the applied pressure is given by:

$$\frac{PR}{t} = \frac{24 \times 0.762}{0.108}$$

$$= 169 \text{ MPa}$$

The applied pressure contributes to plastic collapse, hence
$$\sigma^P = 169 \text{ MPa}$$

1.4 Evaluation of S_r

Using A2.2.1 in Appendix 2 to R/H/R6 Rev. 2

$$Sr = \frac{PR}{\bar{\sigma}t}\left[\frac{\dfrac{t}{a} - \dfrac{1}{m}}{\dfrac{t}{a} - 1}\right]$$

where

$$m = (1 + 0.263 \frac{\ell^2}{Rt})^{1/2}$$

$$= (1 + \frac{0.263 \times (0.035)^2}{0.762 \times 0.108})^{1/2}$$

$$= 1.002$$

From paragraph 5.2 of R/H/R6 Rev.2

$$\bar{\sigma} = (\sigma_y + \sigma_u)/2$$

$$= (430 + 580)/2$$

$$= 505 \text{ MPa}$$

Substituting

$$S_r = \frac{24 \times 0.762}{505 \times 0.108}\left[\frac{\dfrac{0.108}{0.007} - \dfrac{1}{1.002}}{\dfrac{0.108}{0.007} - 1}\right]$$

$$S_r = 0.335$$

1.5 Evaluation of K_r

$$K_r = K_r^P = \frac{K_I^P}{K_{Ic}}$$

From ASME XI Article A3000,

$$K_I^P = \sigma_m M_m (\pi a/Q)^{1/2} + \sigma_b M_b (\pi a/Q)^{1/2}$$

$$a/\ell = 0.2$$

$$a/t = 0.065$$

$$\sigma_m = \sigma^P = 169 \text{ MPa}$$

$$\sigma_b = 0$$

From Figure A-3300-1 of ASME XI, Q is obtained using $(\sigma_m + \sigma_b)/\sigma_y = 0$ (no plasticity correction is used, see paragraph A3.1 of R/H/R6 Rev.2).

$$Q = 1.32$$

From Figure A-3300-3 of ASME XI

$$M_m = 1.1$$

Hence

$$K_I = 169 \times 1.1 \ (0.007\pi/1.32)^{1/2}$$

$$= 24.0 \ \text{MPa}\sqrt{m}$$

and

$$K_r = \frac{24.0}{100}$$

$$\underline{K_r = 0.240}$$

1.6 Assessment

The assessment point, A, falls within the Assessment Line on the Failure Assessment Diagram, Fig. 1.1, indicating that failure is avoided.

For this case the locus of the assessment point as a function of applied load is a straight line passing through the origin and the assessment point since both S_r and K_r are directly proportional to applied load. From Fig. 1.1 the reserve factor on applied pressure is:

$$\frac{OB}{OA} = 2.81$$

1.7 Sensitivity

The sensitivity of the assessment to variations in input data can now be examined:

1.7.1 Variations in K_{Ic}

The line CAD in Fig. 1.1 is the locus of assessment points as a function of K_{Ic}. It can be seen that the structure will tolerate a reduction in K_{Ic} by the ratio CA/CD, i.e. 4.17, to 24 MPa$\sqrt{m}$.

Thus the reserve factor on fracture toughness, F_r^K, is 4.17.

1.7.2 Variations in Flaw Size

Assuming that the ratio a/ℓ remains constant, S_r and K_r can be calculated for variations in flaw size from the following:

$$S_r = \frac{24 \times 0.762}{505 \times 0.108} \left[\frac{\dfrac{0.108}{a} - \dfrac{1}{m}}{\dfrac{0.108}{a} - 1} \right]$$

$$m = \left[1 + 0.263 \frac{\ell^2}{0.762 \times 0.108} \right]^{1/2}$$

$$K_I^P = 169 M_m (\pi a / Q)^{1/2}$$

$$K_r = \frac{K_I^P}{100}$$

Table 1.1 and Fig. 1.2 show the values of S_r and K_r derived for a range of flaw sizes. The reserve factor on pressure, F^P, was found for each flaw size and plotted on Fig. 1.3, from which the flaw size for $F^P =$ 1 is 62 mm. Thus at the pressure used in the assessment, 24 MPa the reserve factor on flaw size, F^a, is 62/7 = 8.86.

1.7.3 Other Variables

The effects of varying other parameters σ_y, σ_u, P and the formulae chosen for finding S_r and K_I could also be investigated.

1.8 Observations

The vessel containing a crack of depth 7 mm can withstand the design pressure. The crack can only cause failure at the design temperature where the lower bound fracture toughness is 100 MPa√m if the pressure is raised to 67 MPa. This pressure, which is 2.81 times the design pressure, is almost equal to the plastic collapse pressure of the vessel.

Failure under linear elastic conditions is potentially more serious because there would be no forewarning of failure by plastic distortion. Figs. 1.1 and 1.2 demonstrate two situations where this is possible. Fig. 1.1 shows that failure may occur at the design pressure with the existing size of crack if the fracture toughness is reduced to 24 MPa√m (see paragraph 1.7.1). This could only occur if the vessel were pressurised when it was much colder than the design temperature. Failure may also occur at the design pressure and temperature if the crack depth is increased to 62 mm (Fig. 1.2), for example by fatigue.

It can therefore be concluded that there is no risk of failure of this vessel provided that the design temperature and pressure are maintained and the crack does not grow significantly in service.

TABLE 1.1

a/ℓ = 0.2, Q = 1.32

a	ℓ	a/t	M_m	K_I	K_r	S_r	F^p
mm	mm			MPa√m			
0	0	0	–	0	0	0.335	2.985
7	35	0.065	1.10	24.0	0.240	0.335	2.812
20	100	0.185	1.12	41.3	0.413	0.337	2.106
40	200	0.370	1.24	64.7	0.647	0.347	1.456
60	300	0.556	1.47	93.9	0.939	0.385	1.029
70	350	0.648	1.65	113.8	1.138	0.428	0.853

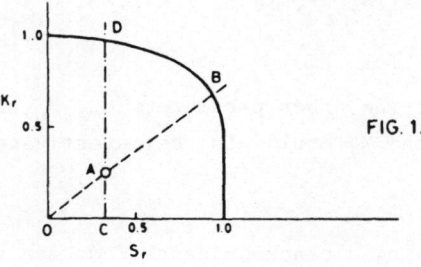

FIG. 1.1

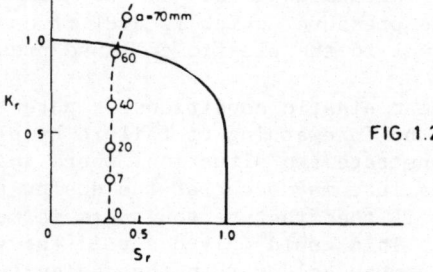

FIG. 1.2

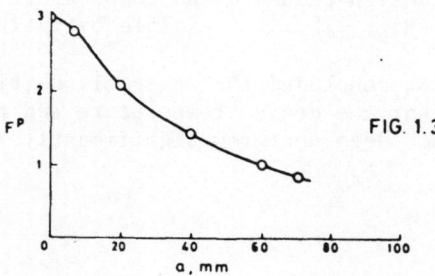

FIG. 1.3

EXAMPLE 2

ASSESSMENT OF A FLAW IN A CYLINDRICAL VESSEL LOADED BY INTERNAL PRESSURE
AND A RESIDUAL WELDING STRESS: SIMPLIFIED APPROACH, ALL STRESSES ASSUMED
TO CONTRIBUTE TO PLASTIC COLLAPSE

Problem

A pressure vessel contains an axial flaw in the outer surface of a
seam weld in its cylindrical section. Although the vessel has been post
weld heat treated, it is assumed that there is a small residual welding
stress of uncertain magnitude and profile. The vessel is loaded by
internal pressure.

Can the vessel withstand the design pressure, and if so by what
margin?

It is judged that the residual welding stress is unlikely to exceed
10% of the yield stress.

Mean radius of cylinder, R	= 0.762 m
Thickness of cylinder, t	= 0.108 m
Design pressure, P	= 24 MPa
Assumed value of residual stress, σ_{res}	= 43 MPa
Lower bound yield stress, σ_y	= 430 MPa
Lower bound ultimate stress, σ_u	= 580 MPa
Lower bound fracture toughness, K_{Ic}	= 100 MPa√m
Flaw depth, a	= 0.007 m
Flaw length, ℓ	= 0.035 m

Solution

2.1   Flaw Characterisation

The flaw is specified as an axial surface flaw with the following
dimensions:
depth,   a = 0.007 m
length, ℓ = 0.035 m
The vessel dimensions are:
mean radius,   R = 0.762 m
wall thickness, t = 0.108 m

2.2   Material Properties

Lower bound yield stress   = 430 MPa
Lower bound ultimate stress   = 580 MPa
Lower bound fracture toughness = 100 MPa√m

2.3   Stress Analysis

Applied pressure P = 24 MPa

Mode I stress due to applied pressure acting at the flaw

$$\frac{PR}{t} = \frac{24 \times 0.762}{0.108}$$

$$= 169 \text{ MPa}$$

The residual welding stress acting normal to the plane of the flaw is 43 MPa. The distribution of this stress is unknown. It will be taken to be uniform across the section containing the flaw. If this stress is self equilibrating it cannot contribute to failure by plastic collapse and should be categorised as a σ^S stress. Nevertheless a simplified and pessimistic analysis can be performed by treating it as a σ^P stress.

Assuming that both the pressure stress and the residual welding stress contribute to plastic collapse then

$$\sigma^P = 169 + 43$$
$$= 212 \text{ MPa}$$

2.4 Evaluation of S_r

Using A2.2.1 in Appendix 2 to R/H/R6 Rev.2

$$S_r = \frac{PR}{\sigma t} \left[\frac{\frac{t}{a} - \frac{1}{m}}{\frac{t}{a} - 1} \right]$$

Assuming that the pressure stresses and residual welding stresses behave in an identical manner, use

$$P = \frac{\sigma^P t}{R}$$

$$= \frac{212 \times 0.108}{0.762}$$

$$= 30 \text{ MPa}$$

$$m = (1 + 0.263 \frac{\ell^2}{Rt})^{1/2}$$

$$m = (1 + \frac{0.263 \times (0.035)^2}{0.762 \times 0.108})^{1/2}$$

$$= 1.002$$

From paragraph 5.2 of R/H/R6 Rev.2

$$\bar{\sigma} = (\sigma_y + \sigma_u)/2$$

$$= (430 + 580)/2$$

$$= 505 \text{ MPa}$$

Substituting

$$S_r = \frac{30 \times 0.762}{505 \times 0.108} \left[\frac{\dfrac{0.108}{0.007} - \dfrac{1}{1.002}}{\dfrac{0.108}{0.007} - 1} \right]$$

$$\underline{S_r = 0.420}$$

2.5 Evaluation of K_r

$$K_r = K_r^p = \frac{K_I^p}{K_{Ic}}$$

From ASME XI Article A3000,

$$K_I^p = \sigma_m M_m (\pi a/Q)^{1/2} + \sigma_b M_b (\pi a/Q)^{1/2}$$

$$a/\ell = 0.2$$

$$a/t = 0.065$$

$$\sigma_m = \sigma^p = 212 \text{ MPa}$$

$$\sigma_b = 0$$

From Figure A-3300-1 of ASME XI, Q is obtained using $(\sigma_m + \sigma_b)/\sigma_y = 0$ (no plasticity correction is used, see paragraph A3.1 of R/H/R6 Rev.2),

$$Q = 1.32$$

From Figure A-3300-3 of ASME XI

$$M_m = 1.1$$

Hence

$$K_I = 212 \times 1.1 \ (0.007\pi/1.32)^{1/2}$$

$$= 30.3 \text{ MPa}\sqrt{m}$$

and

$$K_r = \frac{30.3}{100}$$

$$\underline{K_r = 0.303}$$

2.6 Assessment

The assessment point, A, falls within the Assessment Line on the Failure Assessment Diagram, Fig. 2.1, indicating that failure is avoided.

For this example, both S_r and K_r are proportional to the total applied stress, i.e. the algebraic sum of the residual stress and the pressure stress. The locus of assessment points as a function of applied stress, or pressure, is therefore a straight line passing from the origin through the assessment point and meeting the Assessment Line at F, Fig. 2.1. From Fig. 2.1 the reserve factor on total applied stress is

$$\frac{OF}{OA} = 2.2$$

2.7 Sensitivity

The sensitivity of the assessment to variations in the input data can be examined as in Example 1. In addition the sensitivity to the assumed level of residual stress can be determined, as follows.

At zero applied pressure the coordinates of the assessment point are reduced by a factor of $43/212$ from the values calculated in paragraphs 2.4 and 2.5.

$$S_r = \frac{43}{212} \times 0.420 = 0.085$$

$$K_r = \frac{43}{212} \times 0.303 = 0.061$$

This point, labelled B in Fig. 2.1, defines the false origin with respect to pressure. The failure pressure, P_f is given by

$$P_f = \frac{OF - OB}{OA - OB} P$$

and the reserve factor on pressure is given by $(OF-OB)/(OA-OB)$. Similarly, the reserve factor on residual stress is given by $(OF-BA)/OB$. These reserve factors can be plotted as a function of the other variables to indicate the confidence obtained in the analysis.

In Fig. 2.2 the reserve factor on residual stress, F^R, is plotted as a function of applied pressure at the given crack size and in Fig. 2.3 the reserve factor on applied pressure, F^P is plotted as a function of residual stress at the given crack size.

2.8 Observations

From Figs. 2.2 and 2.3 it is clear that both F^P and F^R are large at the considered applied pressure and residual stress and it is concluded that the structure is safe.

Other observations can be made which are qualitatively similar to those discussed in Example 1, but these need to be qualified because of the assumptions made about the residual stress. It has been assumed that the residual stress contributes to bulging of the vessel in the same manner as the pressure stress. At the small levels of residual stress considered neither of these effects will be significant especially for very small cracks. At larger residual stress levels, where the residual stress exceeds about half the pressure stress, the effects become significant and the predictions will be severely pessimistic.

A common feature of the reserve factor curves of both Example 1 and Example 2 is that the reserve factor does not reduce rapidly at any stage between the assessment condition and the 'failure' condition. Such a gradual approach to F = 1 gives increased confidence in the conclusion that the structure is safe at the assessed condition.

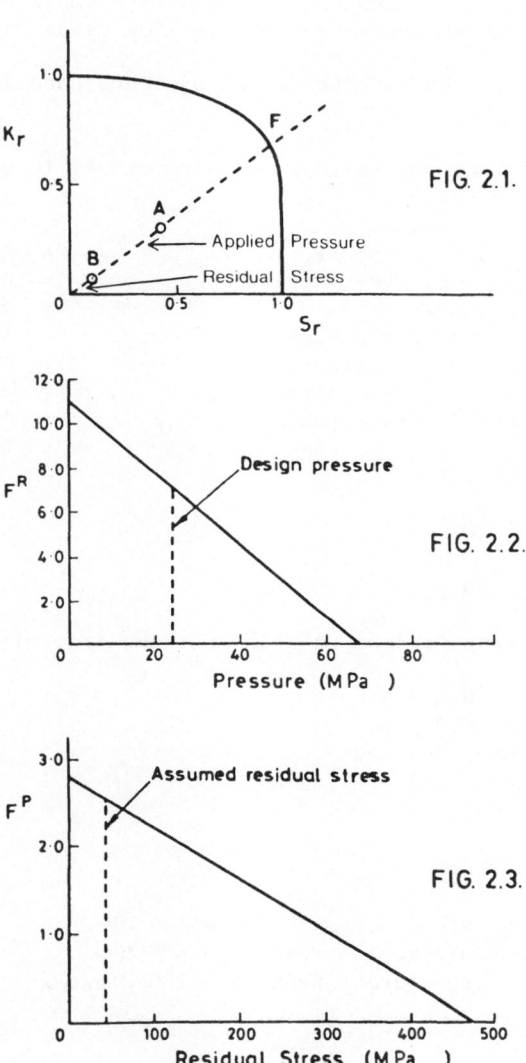

FIG. 2.1.

FIG. 2.2.

FIG. 2.3.

EXAMPLE 3

ASSESSMENT OF A FLAW IN A CYLINDRICAL VESSEL
LOADED BY INTERNAL PRESSURE AND A RESIDUAL WELDING STRESS
ASSUMED NOT TO CONTRIBUTE TO PLASTIC COLLAPSE

Problem

A pressure vessel contains an axial flaw in the outer surface of a
seam weld in its cylindrical section. Although the vessel has been post
weld heat treated, it is assumed that there is a small residual welding
stress of uncertain magnitude and profile. The vessel is loaded by an
internal pressure.

Can the vessel withstand the design pressure, and if so by what
margin?

It is judged that the residual welding stress is unlikely to exceed
10% of the yield stress.

Mean radius of cylinder, R	= 0.762 m
Thickness of cylinder, t	= 0.108 m
Design pressure, P	= 24 MPa
Assumed value of residual stress, σ_{res}	= 43 MPa
Lower bound yield stress, σ_y	= 430 MPa
Lower bound ultimate stress, σ_u	= 580 MPa
Lower bound fracture toughness, K_{Ic}	= 100 MPa$\sqrt{}$m
Flaw depth, a	= 0.007 m
Flaw length, ℓ	= 0.035 m

Solution

3.1 Flaw Characterisation

The flaw is specified as an axial surface flaw with the following
dimensions:
        depth, a = 0.007 m
        length, ℓ = 0.035 m
The vessel dimensions are:
    mean radius,   R = 0.762 m
    wall thickness, t = 0.108 m

3.2 Material Properties

        Lower bound yield stress   = 430 MPa
        Lower bound ultimate stress   = 580 MPa
        Lower bound fracture toughness = 100 MPa$\sqrt{}$m

3.3 Stress Analysis

        Applied pressure P = 24 MPa

Mode I stress due to applied pressure acting at the flaw

$$\frac{PR}{t} = \frac{24 \times 0.762}{0.108}$$

$$= 169 \text{ MPa}$$

The residual welding stress acting normal to the plane of the flaw is 43 MPa. The distribution of this stress is unknown; it will be taken to be uniform across the section containing the flaw.

$$\sigma^P = 169 \text{ MPa}$$

and

$$\sigma^S = 43 \text{ MPa}$$

3.4 Evaluation of S_r

Using A2.2.1 in Appendix 2 to R/H/R6 Rev.2

where

$$S_r = \frac{PR}{\bar{\sigma}t} \left[\frac{\frac{t}{a} - \frac{1}{m}}{\frac{t}{a} - 1} \right]$$

$$m = (1 + 0.263 \frac{\ell^2}{Rt})^{1/2}$$

$$m = (1 + \frac{0.263 \times (0.035)^2}{0.762 \times 0.108})^{1/2}$$

$$= 1.002$$

From paragraph 5.2 of R/H/R6 Rev.2

$$\bar{\sigma} = (\sigma_y + \sigma_u)/2$$

$$= (430 + 580)/2$$

$$= 505 \text{ MPa}$$

Substituting

$$S_r = \frac{24 \times 0.762}{505 \times 0.108} \left[\frac{\frac{0.108}{0.007} - \frac{1}{1.002}}{\frac{0.108}{0.007} - 1} \right]$$

$$\underline{S_r = 0.335}$$

3.5 Evaluation of K_r

$$K_r = K_r^p + K_r^s$$

3.5.1 Finding K_r^p

$$K_r^p = \frac{K_I^p}{K_{Ic}}$$

From ASME XI Article A3000,

$$K_I^p = \sigma_m M_m (\pi a/Q)^{1/2} + \sigma_b M_b (\pi a/Q)^{1/2}$$

$$a/\ell = 0.2$$

$$a/t = 0.065$$

$$\sigma_m = \sigma^p = 169 \text{ MPa}$$

$$\sigma_b = 0$$

From Figure A-3300-1 of ASME XI, Q is obtained using $(\sigma_m + \sigma_b)/\sigma_y = ($
(no plasticity correction is used, see paragraph A3.1 of R/H/R6 Rev.2).

$$Q = 1.32$$

From Figure A-3300-3 of ASME XI

$$M_m = 1.1$$

Hence

$$K_I^p = 169 \times 1.1 \ (0.007\pi/1.32)^{1/2}$$

and

$$K_r^p = \frac{24.0}{100}$$

$$K_r^p = 0.240$$

3.5.2 Finding K_r^s

From Appendix 4 of R/H/R6 Rev.2

$$K_r^s = \frac{K_I^s(a_1)}{K_{Ic}} + \rho$$

From ASME XI Article A3000,

$$K_I^s = \sigma_m M_m (\pi a/Q)^{1/2} + \sigma_b M_b (\pi a/Q)^{1/2}$$

$$a/\ell = 0.2$$

$$a/t = 0.065$$

$$\sigma_m = \sigma^s = 43 \text{ MPa}$$

$$\sigma_b = 0$$

From Figure A-3300-1 of ASME XI, Q is obtained using $(\sigma_m+\sigma_b)/\sigma_y=0$ (no plasticity correction is used, see paragraph A3.1 of R/H/R6 Rev.2).

$$Q = 1.32$$

From Figure A-3300-3 of ASME XI

$$M_m = 1.1, \text{ for } a/t < 0.1$$

Hence

$$K_I^s = 43 \times 1.1 \ (\pi a/1.32)^{1/2}, \text{ for } a/t < 0.1$$

Figure 3.1 shows K_I^s as a function of flaw size. The value of K_I^s at the flaw size of interest (a = 0.007 m) is 6.1 MPa√m.

From Appendix 4 of R/H/R6 Rev.2 the plastic zone size η is given by:

$$\eta = \frac{1}{2\pi} \left(\frac{K_I^s(a_1)}{\sigma_y} \right)^2$$

$$= \frac{1}{2\pi} \left[\frac{6.1}{430} \right]^2 = 3.2 \times 10^{-5} \text{ m}$$

The corrected crack length, a_2, is then
$$a_2 = a_1 + \eta$$

From Fig. 3.1, $K_I^s(a_2)$ remains at 6.1 MPa√m.

Thus

$$\frac{K_I^s(a_1)}{K_I^s(a_2)} = 1$$

From Table A4.1 of R/H/R6 Rev.2, $\rho = 0$

Hence $\quad K_r^s = \dfrac{6.1}{100} = 0.061$

3.5.3 Finding K_r

$$K_r = K_r^p + K_r^s$$

$$= 0.24 + 0.061$$

$$= 0.301$$

3.6 Assessment

The assessment point, A, falls within the Assessment Line on the Failure Assessment Diagram, Fig. 3.2, indicating that failure is avoided.

For this example both S_r and K_r^p are directly proportional to applied load. Thus the locus of the assessment points as a function of applied pressure is a straight line passing through the K_r^s intercept on the K_r axis, and the assessment point.

From Fig. 3.2 the reserve factor on applied load is:

$$\frac{BC}{BA} = 2.70$$

3.7 Sensitivity

The sensitivity of the assessment to variations in input data can now be examined.

3.7.1 Sensitivity to level of residual stress

For any level of residual stress only K_r needs to be recalculated as S_r is independent of the residual stress.

Using the method in paragraph 3.5 above K_r^p is constant and Table 3.1 shows the values of K_r^s and K_r obtained for various values of σ^s.

From these values, and using $S_r = 0.335$, Fig 3.3 shows the locus of assessment points for the various values of σ^s assumed. From this the maximum tolerable value of σ^s is estimated as 400 MPa.

3.7.2 Sensitivity to the plasticity correction model

Previously the plastic zone size correction factor η has been calculated assuming plane stress conditions. The effect of assuming plane strain conditions can now be considered.

For plane strain conditions

$$\eta = \frac{1}{6\pi} \left[\frac{K_I^s \ (a_1)}{\sigma_y} \right]^2$$

Values for η using this expression are listed in Table 3.2, as are the resultant values for ρ and K_r, and the assessment points are plotted in Fig. 3.4. It is clear from this that for this particular example the plastic zone size correction of K_r^s makes no difference at low levels of residual stress and only a small difference at high levels.

3.8 Observations

The vessel, containing a crack of depth 7mm and in the presence of an assumed residual welding stress of 10% of the lower bound yield stress, can withstand the design pressure.

The crack can only cause failure at the design temperature where the lower bound fracture toughness is 100 MPa√m if the pressure is raised to 64.8 MPa. This pressure is 2.7 times the design pressure.

The main assumptions are that the residual stress was considered to be acting uniformly across the section and did not contribute to failure by plastic collapse. Using these assumptions it is demonstrated that a residual welding stress of almost yield magnitude could be tolerated at the design pressure.

It is also demonstrated that for this particular example the analysis is relatively insensitive to variations in the plasticity correction model chosen.

TABLE 3.1

σ^s	100	200	300	400	432
a_1	0.007	0.007	0.007	0.007	0.007
$K_I^s(a_1)$	14	28.5	42.5	57	61.5
η	1.69×10^{-4}	7.0×10^{-4}	$1.55.10^{-3}$	2.8×10^{-3}	3.26×10^{-3}
a_2	0.007169	0.0077	0.00855	0.0098	0.01026
$K_I^s(a_2)$	14	30	47	67	73
$\dfrac{K_I^s(a_1)}{K_I^s(a_2)}$	1	0.95	0.9	0.85	0.84
ρ	0	0.086	0.14	0.186	0.195
$\dfrac{K_I^s(a_1)}{K_{Ic}}$	0.14	0.285	0.425	0.57	0.615
K_r^s	0.14	0.371	0.565	0.756	0.810
K_r	0.38	0.611	0.805	0.996	1.05
S_r	0.335	0.335	0.335	0.335	0.335

TABLE 3 2

σ^s	100	200	300	400	432
a_1	0.007	0.007	0.007	0.007	0.007
$K_I^s(a_1)$	14	28.5	42.5	57	61.5
η	5.6×10^{-5}	2.3×10^{-4}	5.2×10^{-4}	9.3×10^{-4}	1.1×10^{-3}
a_2	0.00706	0.0072	0.0075	0.0079	0.0081
$K_I^s(a_2)$	14	29	44	60.5	66
$\dfrac{K_I^s(a_1)}{K_I^s(a_2)}$	1	0.98	0.965	0.94	0.93
ρ	0.0	0.048	0.070	0.098	0.109
$\dfrac{K_I^s(a_1)}{K_{Ic}}$	0.14	0.285	0.425	0.570	0.615
K_r^s	0.140	0.333	0.495	0.668	0.724
K_r	0.380	0.573	0.735	0.908	0.964
S_r	0.335	0.335	0.335	0.335	0.335

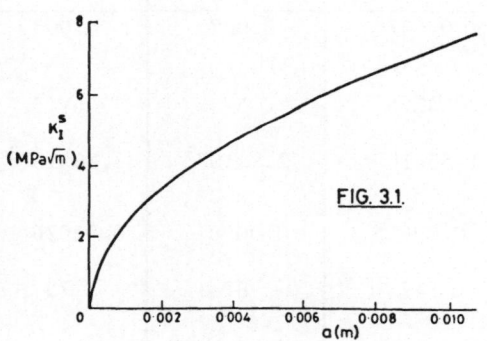

FIG. 3.1.

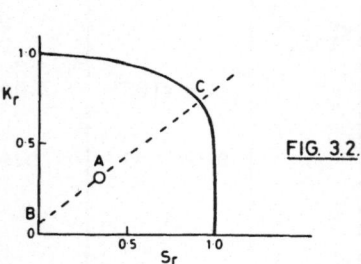

FIG. 3.2.

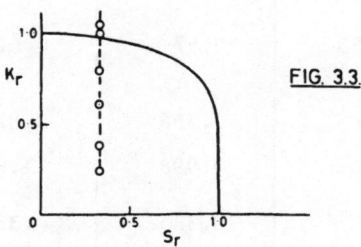

FIG. 3.3.

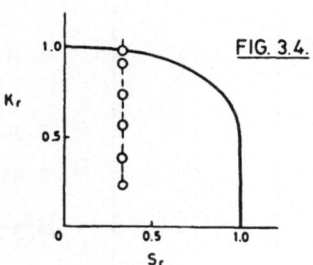

FIG. 3.4.

EXAMPLE 4

DETERMINATION OF A TOLERABLE FLAW SIZE IN A CYLINDRICAL VESSEL
LOADED BY INTERNAL PRESSURE AND A RESIDUAL WELDING STRESS
ASSUMED NOT TO CONTRIBUTE TO PLASTIC COLLAPSE

Problem

A pressure vessel contains an axial flaw in the outer surface of a
seam weld in its cylindrical section. Although the vessel has been post
weld heat treated, it is assumed that there is a small residual welding
stress of uncertain magnitude and profile. The vessel is loaded by
internal pressure.

What size of flaw can the vessel withstand assuming a constant flaw
depth to length ratio of 0.2?

It is judged that the residual welding stress is unlikely to exceed
10% of the yield stress.

Mean radius of cylinder, R	=	0.762 m
Thickness of cylinder, t	=	0.108 m
Design pressure, P	=	24 MPa
Assumed value of residual stress, σ_{res}	=	43 MPa
Lower bound yield stress, σ_y	=	430 MPa
Lower bound ultimate stress, σ_u	=	580 MPa
Lower bound fracture toughness, K_{Ic}	=	100 MPa√m

Solution

4.1 Flaw Characterisation

The flaw is specified as an axial surface flaw with $a/\ell = 0.2$
The vessel dimensions are:
mean radius,   R = 0.762 m
wall thickness, t = 0.108 m

4.2 Material Properties

Lower bound yield stress   = 430 MPa
Lower bound ultimate stress   = 580 MPa
Lower bound fracture toughness = 100 MPa√m

4.3 Stress Analysis

Applied pressure P = 24 MPa

Mode I stress due to applied pressure acting at the flaw

$$\frac{PR}{t} = \frac{24 \times 0.762}{0.108}$$

$$= 169 \text{ MPa}$$

The residual welding stress acting normal to the plane of the flaw is 43 MPa. The distribution of this stress is unknown; it will be taken to be uniform across the section containing the flaw.

$$\sigma^P = 169 \text{ MPa}$$

and

$$\sigma^S = 43 \text{ MPa}$$

4.4   Evaluation of S_r

Using A2.2.1 in Appendix 2 to R/H/R6 Rev.2

$$S_r = \frac{PR}{\bar{\sigma}t} \left[\frac{\frac{t}{a} - \frac{1}{m}}{\frac{t}{a} - 1} \right]$$

where

$$m = (1 + 0.263 \frac{\ell^2}{Rt})^{1/2}$$

From paragraph 5.2 of R/H/R6 Rev.2

$$\bar{\sigma} = (\sigma_y + \sigma_u)/2$$

$$= (430 + 580)/2$$

$$= 505 \text{ MPa}$$

S_r has been calculated for a range of flaw sizes, Table 4.1.

4.5   Evaluation of K_r

$$K_r = K_r^P + K_r^S$$

4.5.1 Finding K_r^P

$$K_r^P = \frac{K_I^P}{K_{Ic}}$$

From ASME XI Article A3000,

$$K_I^P = \sigma_m M_m (\pi a/Q)^{1/2} + \sigma_b M_b (\pi a/Q)^{1/2}$$

$$a/\ell = 0.2$$

$$\sigma_m = \sigma^P = 169 \text{ MPa}$$

$$\sigma_b = 0$$

From Figure A-3300-1 of ASME XI, the value Q = 1.32 was obtained using$(\sigma_m + \sigma_b)/\sigma_y = 0$ (no plasticity correction is used, see paragraph A3.1 of R/H/R6 Rev.2).

From Figure A-3300-3 of ASME XI, M_m was found for a range of flaw sizes, Table 4.1.

Hence K_r^p is found for a range of flaw sizes, Table 4.1.

4.5.2 Finding K_r^s

From Appendix 4 of R/H/R6 Rev.2

$$K_r^s = \frac{K_I^s}{K_{Ic}} + \rho$$

From ASME XI Article A3000,

$$K_I^s = \sigma_m M_m (\pi a/Q)^{1/2} + \sigma_b M_b (\pi a/Q)^{1/2}$$

$$a/\ell = 0.2$$

$$\sigma_m = \sigma^s = 43 \text{ MPa}$$

$$\sigma_b = 0$$

From Figure A-3300-1 of ASME XI, the value Q = 1.32 was obtained for a range of flaw sizes using $(\sigma_m + \sigma_b)/\sigma_y = 0$ (no plasticity correction is used, see paragraph A3.1 of R/H/R6 Rev.2).

From Figure A-3300-3 of ASME XI, M_m was found for a range of flaw sizes, Table 4.1.

Hence K_I^s is found for a range of flaw sizes, Table 4.1, and is plotted in Fig. 4.1.

From Appendix 4 of R/H/R6 the plastic zone size η is given by

$$\eta = \frac{1}{2\pi} \left[\frac{K_I^s (a_1)}{\sigma_y} \right]^2$$

Then, as $a_2 = a_1 + \eta$, $K_I^s (a_2)$ is found from Fig. 4.1 for a range of flaw sizes, Table 4.1.

Values of K_r^s are found for a range of flaw sizes, Table 4.1, using values of ρ from Table A4.1 of R/H/R6 Rev.2.

4.5.3 Finding K_r

$$K_I = K_I^p + K_I^s$$

K_r is found for a range of flaw sizes, Table 4.1.

4.6 Assessment

The locus of derived values of assessment points as a function of flaw size is shown in Fig. 4.2. It intersects the Assessment Line at a = 0.05 m and this is the maximum tolerable flaw size.

TABLE 4.1

$a/\ell = 0.2 \qquad Q = 1.32$

a_1 (m)	0.01	0.02	0.03	0.04	0.06	0.08*
m	1.004	1.016	1.035	1.062	1.135	1.229
$\dfrac{t/a - 1/m}{t/a - 1}$	1.0004	1.0036	1.013	1.034	1.149	1.532
S_r	0.335	0.336	0.339	0.346	0.385	0.513
M_m	1.1	1.12	1.14	1.23	1.47	1.90
K_I^p	28.7	41.3	51.5	64.1	93.9	140.1
K_r^p	0.287	0.413	0.515	0.641	0.939	1.401
$K_I^s (a_1)$	7.3	10.5	13.1	16.3	23.9	35.6
η	5×10^{-5}	1.0×10^{-4}	1.5×10^{-4}	2.3×10^{-4}	4.9×10^{-4}	11×10^{-4}
a_2	0.01005	0.0201	0.0301	0.0402	0.0605	0.081
$K_I^s (a_2)$	7.3	10.5	13.1	16.35	24.0	35.8
ρ	0.0	0.0	0.0	0.01	0.015	0.025
K_r^s	0.073	0.105	0.131	0.173	0.254	0.381
K_r	0.360	0.518	0.646	0.814	1.193	1.782

* This crack depth is outside the range covered by Fig. A-3300-3 of ASMEXI.

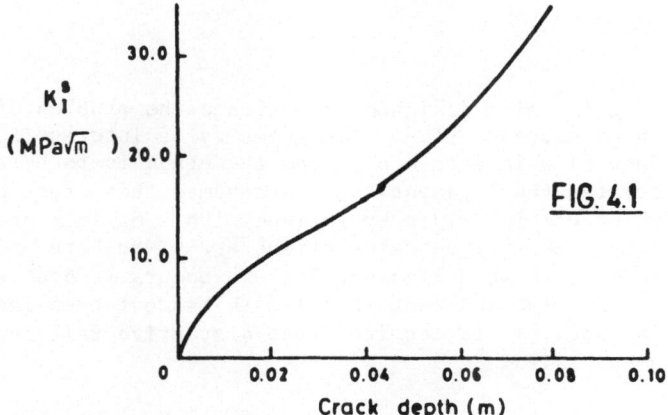

FIG. 4.1

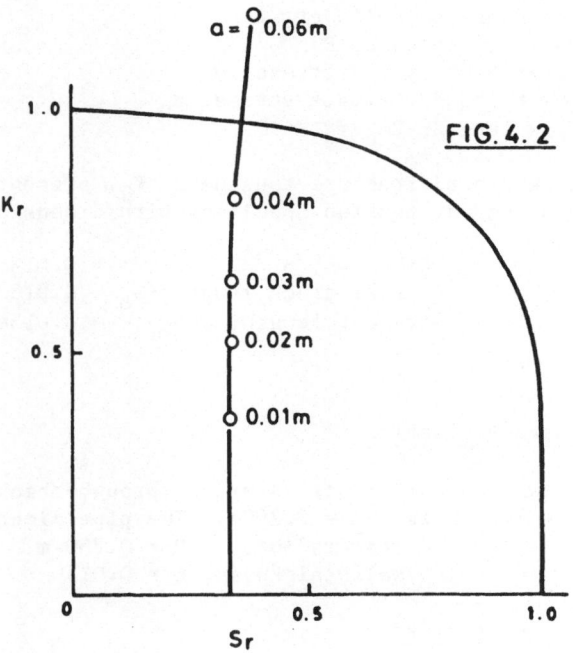

FIG. 4.2

EXAMPLE 5

ASSESSMENT OF A FLAW IN A CYLINDRICAL PIPE LOADED ONLY BY INTERNAL
PRESSURE AND WHERE FAILURE FOLLOWS DUCTILE CRACK GROWTH

Problem

A test rig is being designed to evaluate the problem of crack
extension in a pressurised pipe. The geometry of interest is an axial
external surface flaw in a thin pipe and the crack is to be extended
until it penetrates the ligament. It is assumed that crack extension
through the thickness is driven by fatigue with a maximum pressure in the
cycle of 5.7 MPa. No fatigue calculations are given here and instead it
is simply assumed that when ligament failure occurs it does so over the
full length of the flaw and that this length has not been increased by
the fatigue loading. It is required that disruptive failure is avoided
when the ligament fails.

Can the pipe withstand the pressure of 5.7 MPa when the ligament
fails and if so by what margin? What is the maximum length of through-
thickness crack which can be tolerated at a pressure of 5.7 MPa?

Mean radius of cylinder, R	= 0.380 m
Thickness of cylinder, t	= 0.013 m
Maximum pressure, P	= 5.7 MPa
Lower bound yield stress, σ_y	= 500 MPa
Lower bound ultimate stress, σ_u	= 700 MPa
Flaw length, $2a_o$	= 0.200 m

The lower bound fracture toughness, K_Q, presented in Fig. 5.1 was
obtained from compact tension specimens with dimensions

thickness, B		= 0.013m
initial crack length, a_o	=	0.013m
ligament length, $w-a_o$	=	0.013m

Solution

5.1 Flaw Characterisation

The flaw is specified as an axial through-thickness defect. The
original flaw length is $2a_o$ = 0.200m. The pipe dimensions are:-

mean radius,   R = 0.380 m
wall thickness, t = 0.013 m

5.2 Material Properties

Lower bound yield stress,	σ_y	= 500 MPa
Lower bound ultimate stress,	σ_u	= 700 MPa

The lower bound fracture resistance, K_Q, obtained for crack
extensions up to 5mm is shown in Fig. 5.1.

5.2.1 Limitations on Fracture Resistance Data

To ensure that the stress state in the structure is similar to that in the specimen, the restrictions of paragraph A5.4(b) of R/H/R6 Rev. 2 apply. These are evaluated as follows:

(i) crack growth limit

(a) Based on specimen dimensions

$$\Delta a < 0.06 \ (w-a_0)$$
$$< 0.06 \times 0.013m$$
$$\Delta a < 0.8mm$$

(b) Based on structure dimensions

$$\Delta a < 0.06 a_0$$
and $\Delta a < 0.06 \ (t-a_0)$
$(t-a_0)$ is not known but is large compared with a_0, hence $\Delta a < 6mm$
The specimen provides the limit to crack extension, $\Delta a < 0.8mm$

(ii) Specimen size limit

$$w-a_0 > \frac{25 \ K_Q^2 \ (1-\nu^2)}{\sigma_y E}$$

$$B > w-a_0$$

At $\Delta a = 0.8mm$, $K_Q = 250 \ MPa\sqrt{m}$

Taking $\nu = 0.3$ and $E = 210 \ GN \ m^{-2}$

$$\frac{25 \ K_Q^2 \ (1-\nu^2)}{\sigma_y E} = 0.013m$$

Thus $B > w-a_0 > 0.013m$
From the initial data $B = w-a_0 = 0.013m$. Therefore the specimen size condition is satisfied for $\Delta a < 0.8mm$.

(iii) Resistance curve slope limit
$$\frac{dK_Q}{da} > \frac{5K_Q}{b}$$
where b is the smaller of w-a in the specimen or t-a in the structure and a is the instantaneous flaw length. Since the specimen is smaller than the structure, the specimen dimensions apply.

$$w-a = w-(a_0 + \Delta a_0)$$
$$= 0.026 - 0.0138$$
$$= 0.0122m$$

Hence $\dfrac{5K_Q}{b} = \dfrac{5 \times 250}{0.0122}$

$$= 102 \times 10^3 \text{ MPa}\sqrt{m}/m$$

In Fig. 5.1 the slope of the resistance curve at $\Delta a = 0.8$mm is given as

$$\frac{dK_Q}{da} = 108 \times 10^3 \text{ MPa}\sqrt{m}/m$$

So the resistance curve slope limit condition is satisfied.

From (i) (ii) and (iii) above Δa_{lim} is defined as

$$\Delta a_{lim} = 0.8\text{mm}$$

Although Δa_{lim} will define the limits of a valid assessment, as described in Paragraph A5.2 of R/H/R6 Rev. 2, it may not be sufficient to identify the crack length at which disruptive failure occurs. In order to illustrate this, the analysis will be continued to larger crack extensions by extrapolating the experimental data base in Fig. 5.1 to $\Delta a = 15$mm.

5.3 Stress Analysis

Applied pressure, P = 5.7 MPa
The hoop stress due to the applied pressure is given by

$$\frac{PR}{t} = \frac{5.7 \times 0.380}{0.013}$$

$$= 166.6 \text{ MPa}$$

The applied pressure contributes to plastic collapse, hence
$\sigma^P = 166.6$ MPa

5.4 Evaluation of S_r

The plastic collapse stress, σ_1, of a pipe containing an axial through thickness flaw of length $2a$ is given by Hahn, G.T., Sarrate, M. and Rosenfield, A.R., 1969, Int. Jnl of Fract. Mechs. 5, 187 as

$$\sigma_1 = \bar{\sigma} \left(1 + 1.61 \frac{a^2}{Rt}\right)^{-1/2}$$

From paragraph 5.2 of R/H/R6
$\bar{\sigma} = 1/2 (\sigma_y + \sigma_u) = 1/2 (500 + 700) = 600$ MPa

Hence $\qquad S_r = \dfrac{\sigma^P}{\bar{\sigma}\left(1 + 1.61 \dfrac{a^2}{Rt}\right)^{-1/2}}$

$$= 0.277(1 + 326a^2)^{1/2}$$

Because ductile crack extension may occur before failure the procedures of Appendix 5 of R/H/R6 Rev. 2 will be followed. Hence the crack length to be used is the instantaneous crack length making allowance for any (postulated) crack extension, Δa.

Thus $a = a_0 + \Delta a$.

5.5 Evaluation of K_r

From Appendix 5 of R/H/R6 – Rev. 2

$$K_r = \frac{K_I^p}{K_Q}$$

The maximum value of the applied stress intensity factor for an axial through thickness crack in a thin pipe was obtained from Fig. 5.2, taken from D P Rooke and D J Cartwright, 1976, "Compendium of Stress Intensity Factors", HMSO.

$$K_I^p = \sigma^p \sqrt{\pi a} \left[G_m + G_b \right]$$

where G_m and G_b are obtained from Fig. 5.2.

K_Q is obtained from Fig. 5.1 and is dependent on the amount of crack growth, Δa.

5.6 Assessment

A ductile crack growth locus is constructed by following the method of paragraph A5.1 of Appendix 5 of R/H/R6. Values of S_r and K_r are listed as a function of postulated crack extension, Δa, in Table 5.1 and the locus is shown in Fig. 5.3. Note that the point for the initial 0.2 m long crack, $a_0 = 0.1$ m, lies outside the Assessment Line while the remaining points all fall inside the line. This indicates that for an applied pressure of 5.7 MPa crack growth will occur but the crack will stabilise after only a small amount of crack extension, indicated by the point X on Fig. 5.3. A number of methods of determining the maximum tolerable load or maximum tolerable initial flaw size are given in paragraph A5.2 of R/H/R6. These are described in the subsections below which should be followed in conjunction with the appropriate section of Appendix 5 of R/H/R6.

5.6.1 Determination of Maximum Tolerable Load

(a) Using paragraph A5.2.1(a) of R/H/R6 Rev. 2

At a given crack length, both K_r and S_r are proportional to applied pressure so that the calculations of K_r and S_r can readily be repeated for other pressures to produce loci similar to that shown in Fig. 5.3. In Fig. 5.4 such ductile crack growth loci are plotted at pressures of P=5.7, 7.0 and 8.9 MPa. The locus at P = 8.9 MPa is tangential to the assessment line and so this locus defines the maximum tolerable pressure

P_t. The crack extension at the point of tangency, Δa_t, is $> \Delta a_{lim}$.

The crack growth permitted by the conditions of paragraph A5.4 of R/H/R6 Rev. 2 is demonstrated in paragraph 5.2.1 above to be 0.8mm. To ensure that the assessment is performed within the restrictions imposed by paragraph A5.4 of R/H/R6 Rev. 2 the tolerable load P_{lim} is the load giving rise to the locus which intersects the Assessment Line when $\Delta a = \Delta a_{lim}$. Hence from Fig. 5.4

$$P_{lim} = 6.0 \text{ MPa}$$

which corresponds to a factor on applied pressure $F^P = 6.0/5.7 = 1.05$.

(b) <u>Using paragraph A5.2.1.1(b) of R/H/R6 Rev. 2</u>

The crack growth locus of Fig. 5.3 is reproduced in Fig. 5.5. The load factor F^P is defined by the length ratios OA'/OA, OB'/OB etc. These have been listed in Table 5.2 and plotted as a function of crack extension in Fig. 5.6. The maximum value of F^P is 1.56 at a crack extension $\Delta a = \Delta a_t = 10$mm at each tip. This corresponds to a pressure of 8.9 MPa and a total crack length of 220mm.

The crack growth permitted by the conditions of paragraph A5.4 of R/H/R6 Rev. 2 is demonstrated in paragraph 5.2.1 above to be 0.8mm. Hence, $\Delta a_t > \Delta a_{lim}$. To ensure the assessment is performed within the restrictions imposed by paragraph A5.4 of R/H/R6 Rev. 2, the tolerable load P_{lim} is obtained from the value of F^P corresponding to $\Delta a = \Delta a_{lim}$. In Fig. 5.5 this is defined by the ratio OX'/OX, which is also listed in Table 5.2 and shown in Fig. 5.6. The pressure P_{lim} is therefore given by

$$P_{lim} = OX'/OX \times P$$

$$= 1.05 \times 5.7 \text{ MPa}$$

$$= 6.0 \text{ MPa}$$

(c) <u>Using paragraph A5.2.1.1(c) of R/H/R6 Rev. 2</u>

The load factor, F^P, is evaluated from the equation

$$F^P = (2/\pi S_r)\cos^{-1}\left(\exp -\pi^2 S_r^2/8K_r^2\right)$$

where K_r and S_r are evaluated for the applied pressure P = 5.7 MPa. The results are given in Table 5.3 for a number of values of postulated crack growth Δa for $a_o = 0.1$m. The values are plotted in Fig. 5.6. The maximum value of F^P is 1.56 at a crack length of 220 mm, equivalent to a crack extension $\Delta a = 10$mm at each crack tip. Hence

$$P_t = 1.56 \times 5.7 = 8.9 \text{ MPa}$$

The crack growth permitted by the conditions of paragraph A5.4 of R/H/R6 Rev. 2 is demonstrated in paragraph 5.2.1. above, to be $\Delta a = \Delta a_{lim} = 0.8$mm. Hence $\Delta a_t > \Delta a_{lim}$. To ensure the assessment is performed within the restrictions imposed by paragraph A5.4, of R/H/R6 Rev. 2 the tolerable load P_{lim} is obtained from the value of F^P corresponding to $\Delta a = \Delta a_{lim}$. The value of F^P at Δa_{lim} is listed in Table 5.3 and plotted in Fig. 5.6. The pressure P_{lim} is given as

$$P_{lim} = F_{lim} \times P$$

$$= 1.055 \times 5.7 \text{ MPa}$$

$$= 6.0 \text{ MPa}$$

5.6.2 Determination of Maximum Tolerable Initial Flaw Size

In a similar manner to the results given in Table 5.1, values of K_r and S_r may be calculated for other values of initial flaw size a_o. Following paragraph A5.2.2 of R/H/R6 Rev. 2 the locus of the resultant assessment points may be plotted as a function of flaw size on the assessment diagram. The intersection of the locus for $\Delta a = 0$ with the Assessment Line is the minimum initial flaw size to initiate ductile crack growth at $P = 5.7$ MPa. The locus for $\Delta a = 0$ is plotted on Fig. 5.7 and this intersects the assessment line at $a_o = 72$ mm. Hence the minimum initial flaw size liable to initiate crack growth at $P = 5.7$ MPa is 144 mm.

(a) Using paragraph A5.2.2(a) of R/H/R6 Rev. 2

A series of crack growth loci for different initial flaw sizes a_o has been plotted on Fig. 5.7. The locus which is tangential to the assessment line defines the maximum tolerable initial flaw size, a_t, at the applied pressure of 5.7 MPa. From Fig. 5.7 this is $a_o = a_t = 175$ mm. The crack extension at the point of tangency is $\Delta a = \Delta a_t = 13$ mm at each crack tip.

The crack growth permitted by the conditions of paragraph A5.4 of R/H/R6 Rev. 2 is demonstrated in paragraph 5.2.1 above to be $\Delta a = \Delta a_{lim} = 0.8$ mm. Hence $\Delta a_t > \Delta a_{lim}$. To ensure the assessment is performed within the restrictions of paragraph A5.4 of R/H/R6 Rev. 2, the tolerable initial flaw size, a_{lim}, is that which gives rise to the locus which intersects the assessment line at $\Delta a = \Delta a_{lim}$. From Fig. 5.7 this is $a_o = a_{lim} = 106$mm. thus the maximum initial flaw size which satisfies the conditions of paragraph A5.4 of R/H/R6 Rev. 2 and which remains stable is 212mm.

(b) Using paragraph A5.2.2(b) of R/H/R6 Rev. 2

For $a_o = 100$mm the load factor F^P was determined as a function of Δa in paragraphs 5.6.1(b) and (c) and plotted in Fig. 5.6 as a function of semi-crack length, $a(= a_o + \Delta a)$. In Fig. 5.8 the load factor F^P has been plotted as a function of semi-crack length for a series of initial flaw sizes, a_o. Loci have been constructed on Fig. 5.8 by connecting points on this series of curves corresponding to:

(i) $\Delta a = 0$, to define the load factor at initiation F_i. $F_i = 1$ for $a_o = 72$mm, hence the minimum initial flaw size to initiate crack growth at $P = 5.7$ MPa is 144mm.

(ii) The maximum value of F^P for each initial flaw size considered. This defines the load factor F_t and the maximum tolerable initial flaw size is that for which $F_t = 1$. This corresponds to $a_o = 175$mm or a total crack length of 350mm. The amount of crack growth corresponding to $F_t = 1$ is $\Delta a = \Delta a_t = 13$mm.

(iii) $\Delta a = \Delta a_{lim}$, to define F_{lim}, with $\Delta a_{lim} = 0.8$mm as demonstrated in paragraph 5.2.1 above. As $\Delta a_t > \Delta a_{lim}$, paragraph A5.2.2(b) defines the tolerable initial flaw size, a_{lim}, as that for which $F_{lim} = 1$. From Fig. 5.8 this corresponds to $a_o = a_{lim} = 106$mm. Hence the maximumm initial flaw which both remains stable and satisfies the conditions of paragraph A5.4 of R/H/R6 Rev. 2 is 212mm long for an applied pressure $P = 5.7$ MPa.

5.7 Sensitivity

The sensitivity of the assessment to variations in the input data can be examined in the same way as in previous examples.

5.8 Observations

Fig. 5.8 provides a comprehensive picture of the potential for disruptive failure of the pipe following ligament failure. The full analysis assumes that the pressure is maintained during ligament failure and that crack extension follows the resistance curve of Fig. 5.1.

The criterion adopted for acceptance of the analysis has a marked effect on the size of defect which can be tolerated. Thus the initial defect size must be less than 144mm to prevent initiation under the pressure of 5.7 MPa. If growth may be accepted up to the limits of paragraph A5.4 of R/H/R6 Rev. 2 then the initial defect size may be 212mm. Acceptability of 5mm growth at each crack tip increases the initial allowable defect size to 316mm but disruptive failure would be predicted only for defect lengths above 350mm.

Turning to the 200mm defect of this example, a reserve factor of 1.05 on pressure has been found to coexist with a reserve factor of 1.06 on defect length if the limits imposed by paragraph A5.4 of R/H/R6 Rev. 2 are observed. These factors rise to 1.56 on pressure and 1.75 on flaw length by allowing crack extension well beyond this range.

In considering the significance of these reserve factors the following points should be considered:

(1) The analysis was performed for a defect initially 200 mm long assuming that no ductile tearing had occurred axially prior to the development of the through thickness crack. Fig. 5.8 demonstrates that

cracks shorter than 200 mm will not extend significantly and will have a higher pressure tolerance than the 200 mm crack. It is only for cracks initially longer than 250 to 300 mm that tearing can substantially influence the pressure tolerance of the pipe. Thus the assumption of no significant axial ductile tearing prior to ligament failure is a satisfactory one.

(2) The fracture toughness data was obtained from compact tension specimens of the same thickness as the pipe material. Thus in the important respect of thickness, the specimens can be considered to be a good model of the pipe. Because of this it is reasonable to extend the analysis beyond the limits imposed by paragraph A5.4 of R/H/R6 - Rev. 2 without compromising its inherent conservatism.

(3) The fracture toughness data base was collected over a crack extension of 5 mm (Fig. 5.1) equivalent to a 10 mm increase in the length of the crack in the pipe. Thus reasonable confidence can be expected in the predictions up to this extent of cracking. A locus depicting this limit has been plotted in Fig. 5.8, demonstrating reserve factors of 1.50 on pressure and 1.58 on flaw length at this limit.

TABLE 5.1

a_o (m)	Δa (m)	a (m)	K_Q (MPa$\sqrt{}$m)	$G_m + G_b$	K_1 (MPa$\sqrt{}$m)	S_r	K_r
0.1000	0.0000	0.1000	160	2.3	215	0.572	1.342
0.1000	0.0008	0.1008	250	2.3	216	0.575	0.864
0.1000	0.0050	0.1050	460	2.4	230	0.594	0.500
0.1000	0.0100	0.1100	595	2.45	240	0.616	0.403
0.1000	0.0150	0.1150	680	2.48	248	0.638	0.365

TABLE 5.2

Δa (mm)	Ratio	F^p
0.0	OA'/OA	0.72
5.0	OB'/OB	1.50
10.0	OC'/OC	1.56
15.0	OD'/OD	1.54
0.8	OX'/OX	1.05

TABLE 5.3

Δa	F^p
0.0	0.718
5.0	1.495
10.0	1.565
15.0	1.544
0.8	1.055

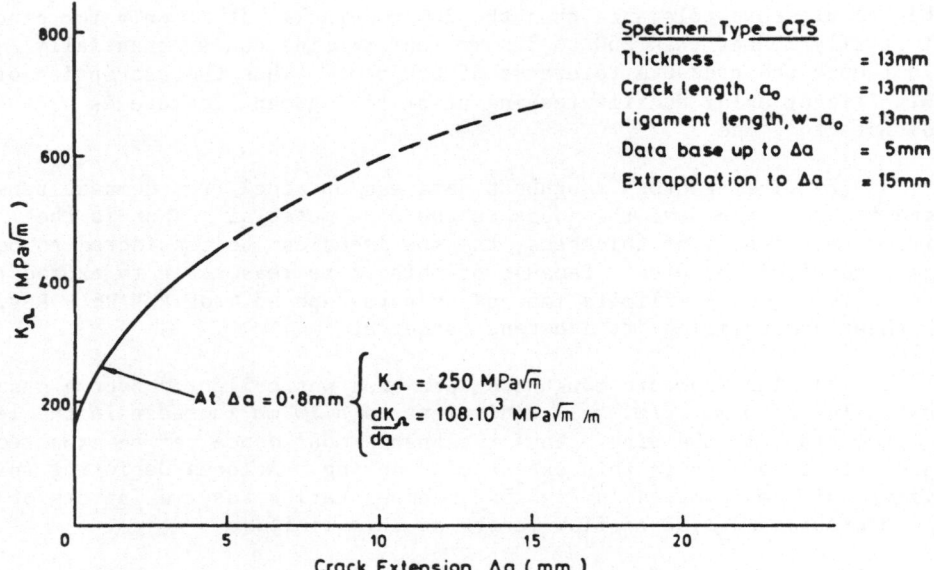

Specimen Type – CTS
Thickness = 13mm
Crack length, a_o = 13mm
Ligament length, $w-a_o$ = 13mm
Data base up to Δa = 5mm
Extrapolation to Δa = 15mm

At $\Delta a = 0.8$mm $\begin{cases} K_{\mathfrak{R}} = 250 \text{ MPa}\sqrt{m} \\ \dfrac{dK_{\mathfrak{R}}}{da} = 108.10^3 \text{ MPa}\sqrt{m} /m \end{cases}$

FIG. 5·1   Resistance Curve Data.

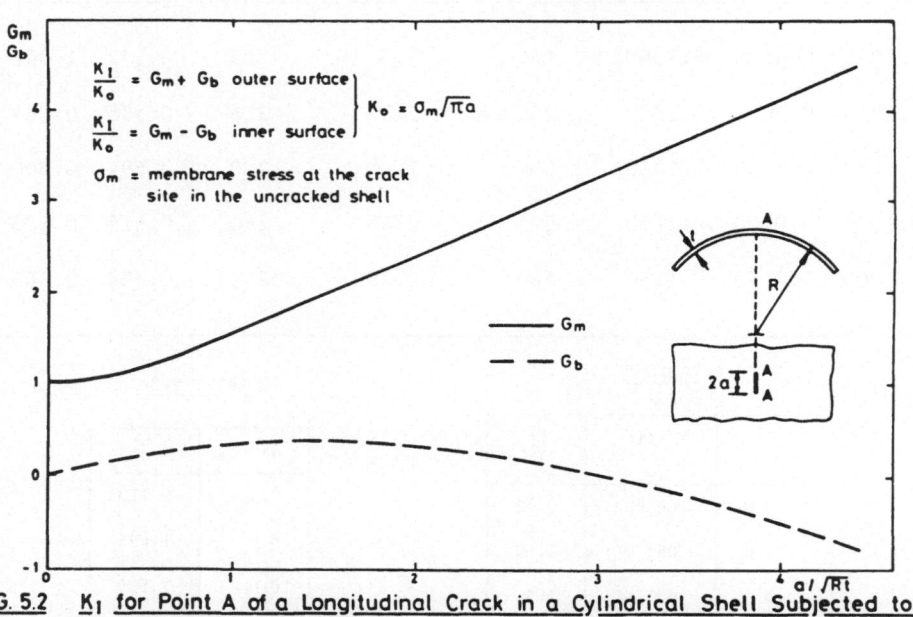

$\dfrac{K_I}{K_o} = G_m + G_b$ outer surface
$\dfrac{K_I}{K_o} = G_m - G_b$ inner surface $\Big\}$ $K_o = \sigma_m\sqrt{\pi a}$

σ_m = membrane stress at the crack site in the uncracked shell

— G_m
- - - G_b

FIG. 5.2 K_I for Point A of a Longitudinal Crack in a Cylindrical Shell Subjected to a Uniform Membrane Stress (from Rooke and Cartwright, 1976,"Compendium of Stress Intensity Factors",H.M.S.O.).

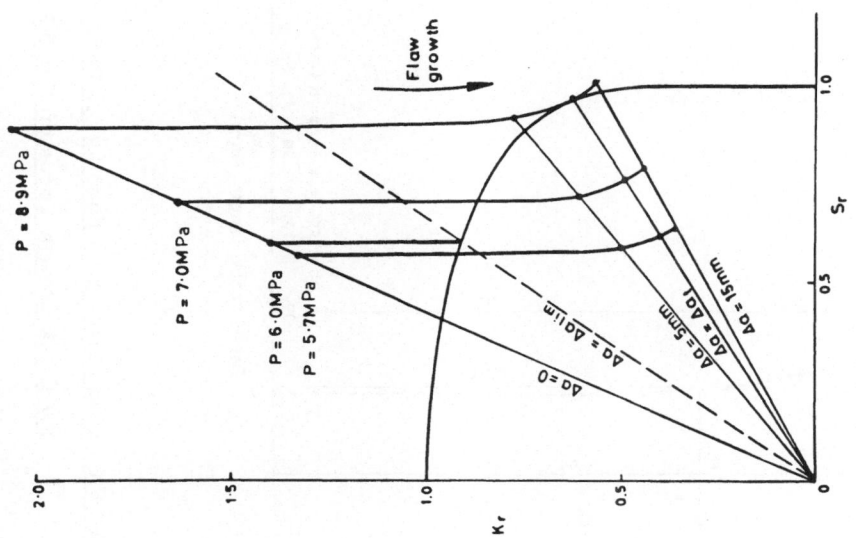

FIG. 5.4   Crack Growth Loci for a Number of Applied Pressures ($a_o = 0.1m$).

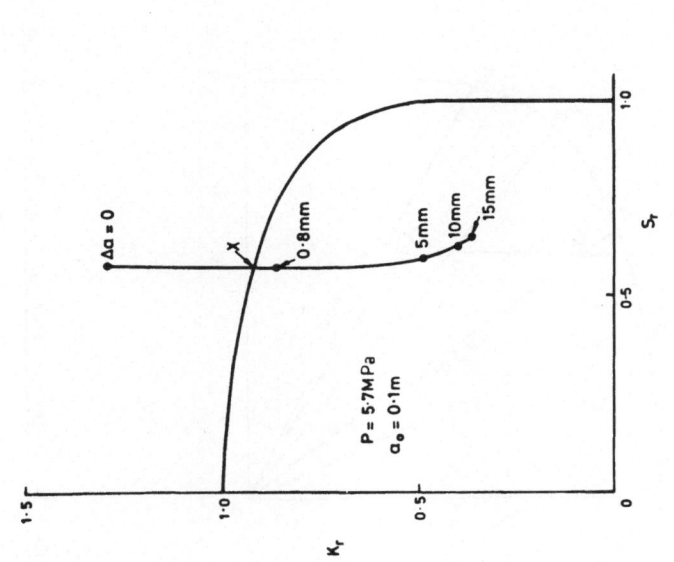

FIG. 5.3   Locus of Assessment Points with Increasing. Postulated Crack Growth.

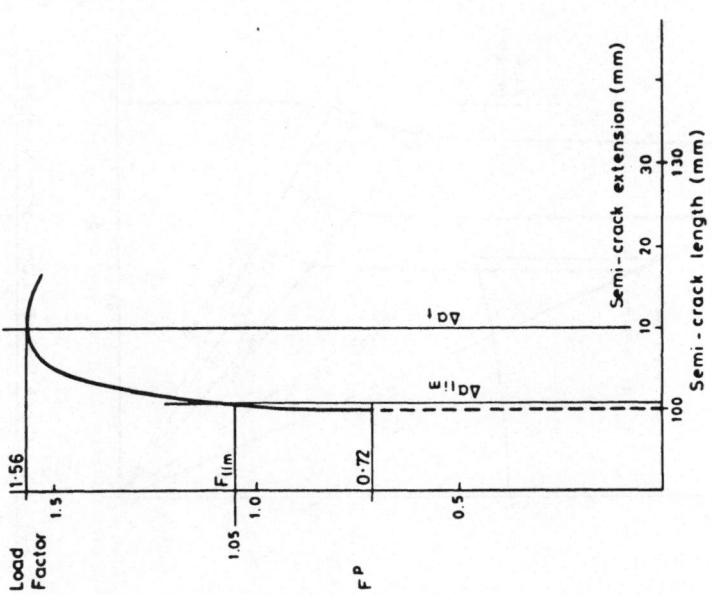

FIG. 5.6   Load Factor as a Function of Position on
the Crack Growth Locus of Fig. 5.5

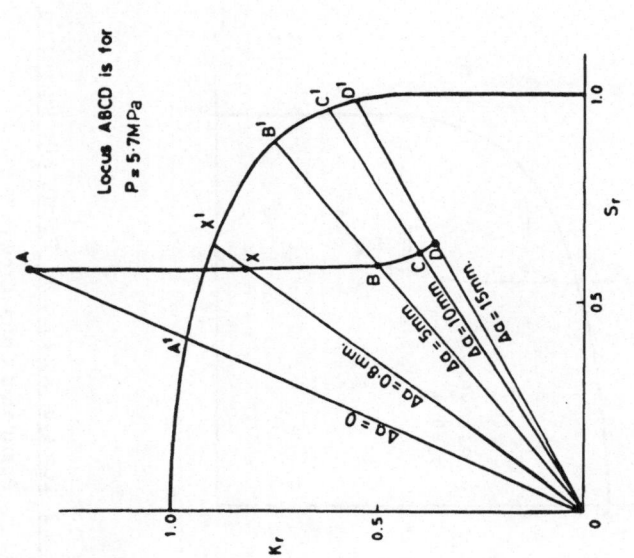

FIG. 5.5   Determination of Load Factors for a Range
of Postulated Crack Growth ($a_0 = 0.1$m).

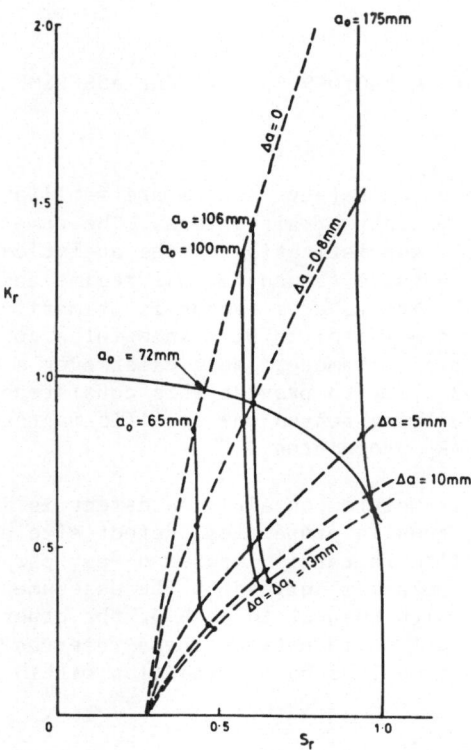

FIG. 5.7 Crack Growth Loci for a Range of Initial Flaw Size (P = 5·7MPa).

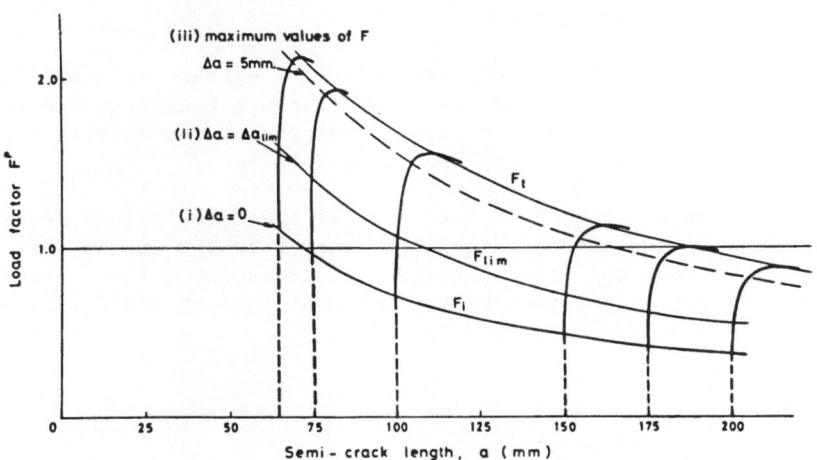

FIG. 5.8 Load Factor as a Function of Position on Crack Growth Loci for a Range of Initial Flaw Size. (Load factor based on P = 5·7MPa).

EXAMPLE 6

A COMPARISON OF RESERVE FACTORS FOR TYPICAL ASSESSMENT CONDITIONS

Introduction

Specific reserve or safety factors are familiar to and welcomed by engineers in a wide variety of situations. The numerical value of such a factor depends on the sophistication of the analytical treatment, the likelihood of going outside the analytical regime and the uncertainty of material properties. The reserve factor is primarily a measure of confidence. For such a factor to be meaningful a consistent relationship between all influential parameters must exist over a wide range of conditions. The inability to provide this consistency in fracture analyses is the underlying reason why specific factors have not been included in the R/H/R6 procedures.

Failure of a structure containing a defect is determined by fracture toughness, tensile properties, defect size and the applied loads. When a specific factor is imposed on any parameter in an assessment a unique point is defined on the assessment locus. This point defines the reserve with respect to each of the other relevant parameters. The relationship between these reserves depends on the component geometry as well as on the position within the assessment diagram.

The present example demonstrates how a factor may be calculated for any specific input and how its sensitivity to changes in other input parameters may be investigated.

The interdependence of reserve factors and the influence of the region of the assessment diagram on the relationships between them are also demonstrated and discussed.

The example is based on a semi-circular surface defect in a flat plate. By assuming different levels of fracture toughness two particular conditions are considered in detail. These cases demonstrate (i) K_r-controlled and (ii) S_r-controlled situations.

The example begins by presenting all the calculations necessary to produce loci of assessment points from zero flaw size up to that sufficient to take the locus outside the assessment curve. Reserve factors are then calculated for various positions on these loci and compared.

Solution

6.1 Flaw Characterisation

The flaw is defined as a semi-circular surface flaw in a flat plate subjected to uniform tension. The plate is considered to be infinitely wide and 108 mm thick. The flaw is assumed to maintain a constant length to depth ratio of 2.

6.2 Material Properties

Lower bound yield stress	=	430 MPa
Lower bound ultimate stress	=	580 MPa
Case (i) Lower bound fracture toughness	=	50 MPa√m
Case (ii) Lower bound fracture toughness	=	200 MPa√m

6.3 Stress Analysis

The applied stress is = 169 MPa

The applied stress contributes to plastic collapse hence
$$\sigma^p = 169 \text{ MPa}$$

6.4 Evaluation of S_r

From Appendix 2 of R/H/R6, equation A2.1.2 with $\sigma_{bc} = 0$

$$S_r = \frac{\sigma_{mc}(c/w) + \sigma_{mc}\left[(c/w)^2 + (1 - c/w)^2\right]^{1/2}}{\bar{\sigma}\,(1 - c/w)^2}$$

where $\quad c/w = \pi a \ell / 4t(\ell + t)\quad$ Appendix 2, paragraph A2.3 of R/H/R6 Rev.2.

$\qquad \sigma_{mc} = 169 \text{ MPa}$

and $\qquad \begin{aligned} \bar{\sigma} &= (\sigma_y + \sigma_u)/2 \\ &= (430 + 580)/2 \\ &= 505 \text{ MPa} \end{aligned}$

Values of S_r for a range of flaw sizes are shown in Table 6.1.

6.5 Evaluation of K_r

$$K_r = K_r^p = K_I^p / K_{Ic}$$

and $\qquad K_{Ic} = 50 \text{ or } 200 \text{ MPa√m}$

Using ASME XI Article A3000, only σ_m applies

$$K_I = K_I^p = \sigma_m M_m (\pi a/Q)^{1/2}$$

for $\qquad a/\ell = 0.5$, $M_m = 1.1$, and $Q = 2.5$

$\qquad K_I^p = 208.4\sqrt{a}$

For case (i) $K_r = \dfrac{208.4\sqrt{a}}{50}$

and for case (ii) $K_r = \dfrac{208.4\sqrt{a}}{200}$

Values of K_r are given in Table 6.1.

6.6 Assessment

Values of S_r, K_r are plotted as functions of postulated defect size in Fig. 6.1.

6.7 Calculation of Reserve Factors

It is stated in Section 9 of R/H/R6 that reserve factors associated with particular input parameters should be assessed. These may be obtained by constructing the locus of the assessment point as a function of that parameter. The procedure is demonstrated in Fig. 6.2.

6.7.1 Reserve factor on K_{Ic} (F^K)

The locus due to K_{Ic} variation is a vertical line (S_r=constant)

$$F^K = \frac{BA}{BP}$$

6.7.2 Reserve factor on flow stress ($F^{\bar{\sigma}}$)

The locus due to $\bar{\sigma}$ variation is a horizontal line (K_r=constant)

$$F^{\bar{\sigma}} = CD/PD$$

6.7.3 Reserve factor on flaw size (F^a)

The locus due to variation of flaw size is generally non-linear.

$$F^a = \frac{\text{Flaw size at E}}{\text{Flaw size at P}}$$

6.7.4 Reserve factor on applied load (F^P)

For primary loads only, the locus due to variation of applied load can be simply constructed as a straight line from the origin through the assessment point.

$$F^P = \frac{OF}{OP}$$

6.8 Sensitivity Analysis

The choice of what to include in sensitivity analyses depends upon the status of input data and the way the results of an assessment will be used. The reserve factor of most interest in the majority of cases is that on applied load.

The following are typical of sensitivity tests which may be performed.

6.8.1 Case (i) K_r controlled (K_{Ic} = 50 MPa$\sqrt{}$m)

Assume the defect of interest is of depth 0.02 m. The assessment point taken from Fig. 6.1 is shown in Fig. 6.3.

$$F^p \text{ at } A = {}^{OB}/_{OA} = 1.59$$

What is the effect of uncertainty in defect sizing?

Assume the defect depth can be measured to ± 25%
i.e. =0.015 to 0.025 m
F^p varies from 1.79 to 1.42 i.e. +13% to −11%

Would it be worth striving for a material of superior toughness?

Assume the lower bound fracture toughness is increased by 100%
i.e. from 50 to 100 MPa$\sqrt{}$m
F^p is increased from 1.59 to 2.47 i.e. by 56%

How sensitive is the assessment to tensile properties?

Assume the flow stress range is ± 10% of 505 MPa
i.e. from 455 to 555 MPa
F^p varies from 1.56 to 1.62 i.e. ± 2% of 1.59

6.8.2 Case (ii) S_r controlled (K_{Ic} = 200 MPa$\sqrt{}$m)

Assume the defect of interest is of depth 0.05 m. The assessment point taken from Fig. 6.1 is shown in Fig. 6.4.

$$F^p \text{ at } A = {}^{OB}/_{OA} = 1.99$$

What is the effect of defect sizing uncertainty?

Assume the defect depth can be measured to ± 10%
i.e. the defect depth is between 0.045 and 0.055 m
F^p varies from 2.10 to 1.85 i.e. +5.5% to −7% of 1.99

Need fracture toughness be so tightly controlled?

Assume the fracture toughness is decreased by 50%
i.e. from 200 to 100 MPa√m
   F^p decreases from 1.99 to 1.69 i.e. by 15%

What is the effect of tensile property variability?
   Assume the flow stress range is ± 25% of 505 MPa
   i.e. 380 to 630 MPa
      F^p varies from 1.51 to 2.47 i.e. ± 24% of 1.99

   6.8.3 The variations of F^p with changes in parameters a, $\bar{\sigma}$ and K_{Ic}
are presented graphically in Figs. 6.5, 6.6 and 6.7. The initial
assessment points from the cases (i) and (ii) are shown on these figures
as A(i) and A(ii).

6.9 The Significance of Specifying Reserve Factors

   Using the two loci in Fig.6.1 a factor of 2 has been imposed in
turn on fracture toughness, flow stress, defect size and applied load.
Tables 6.2 and 6.3 illustrate the reserves which result from such a
process. In each column the boxed 2 represents the chosen factor, the
remaining three numbers in the column being the resulting reserves on the
other parameters. Comparison of Tables 6.2 and 6.3 reveals that the
imposed reserve factors provide the only common ground.

   It must be emphasised that the values in this example are relevant
only to the one component. If similar calculations were performed on
different geometries or even on the same geometry with different material
properties, different relationships would exist between the various
reserve factors.

6.10 Comments

   Interdependence of factors has been demonstrated by use of a simple
component under straightforward loading conditions. This interdependence
is inherent in fracture analysis and is not due to the complexity of any
particular problem. There are quite different relationships between the
factors arising from different conditions. For example, if a factor of 2
on applied load is considered to be desirable, a factor on flaw size of
1.6 is implied in the present case (ii) but 4.7 in case (i).

   Fig. 6.5 demonstrates that the maximum possible reserve on load for
the sample component is 3.0. This can only be achieved by having large
factors on flaw size with consequent NDT sensitivity problems. Hence
unless the full relationship between factors is considered, it is
possible, by specifying a particular reserve, inadvertently to require
unrealistically high reserves on other parameters.

   Figure 6.6 shows that the load factor does not change with fracture
toughness at high levels of toughness. In these relatively ductile

situations therefore a factor on fracture toughness is almost meaningless. A similar insensitivity to one parameter is demonstrated by the case (i) curve of Fig. 6.7 where changes in flow stress have little effect on the load factor at the higher material flow stress levels.

The most important message however is that by proper use of the Failure Assessment Diagram it is possible to decide which parameters have most influence on structural integrity. It is then possible for the user to select a suitable reserve factor on a parameter of interest in the full knowledge of what it really means.

TABLE 6.1

a (m)	ℓ (m)	c/w	S_r	K_I (MPa√m)	(i) K_r for K_{Ic} = 50 MPa√m	(ii) K_r for K_{IC} = 200 MPa√m
0	0	0	0.335	0	0	0
0.005	0.01	0.0031	0.337	14.7	0.294	
0.01	0.02	0.0114	0.342	20.8	0.416	0.104
0.02	0.04	0.039	0.363	29.5	0.590	
0.03	0.06	0.078	0.395	36.1	0.722	0.181
0.04	0.08	0.124	0.440	41.7	0.834	
0.05	0.10	0.175	0.492	46.5	0.930	0.232
0.06	0.12	0.230	0.583	51.0	1.020	
0.07	0.14	0.287	0.695	55.1		0.276
0.085	0.17	0.378	0.957	60.8		0.304

Case (i) K_r controlled (K_{Ic} = 50 MPa$\sqrt{m}$)

In each column the boxed 2 represents the chosen factor, the remaining three numbers in the column being the resulting reserves on the other parameters.

	Value of Reserve Factor			
F^K	[2]	1.2	1.5	2.3
$F^{\bar{\sigma}}$	2.7	[2]	2.5	2.8
F^a	3.2	1.4	[2]	4.7
F^p	1.7	1.2	1.4	[2]

TABLE 6.3

Case (ii) S_r controlled (K_{Ic} = 200 MPa$\sqrt{m}$)

In each column the boxed 2 represents the chosen factor, the remaining three numbers in the column being the resulting reserves on the other parameters.

	Value of Reserve Factor			
F^K	[2]	4.0	4.5	4.0
$F^{\bar{\sigma}}$	1.0	[2]	2.2	2.0
F^a	1.0	1.6	[2]	1.6
F^p	1.0	2.0	2.2	[2]

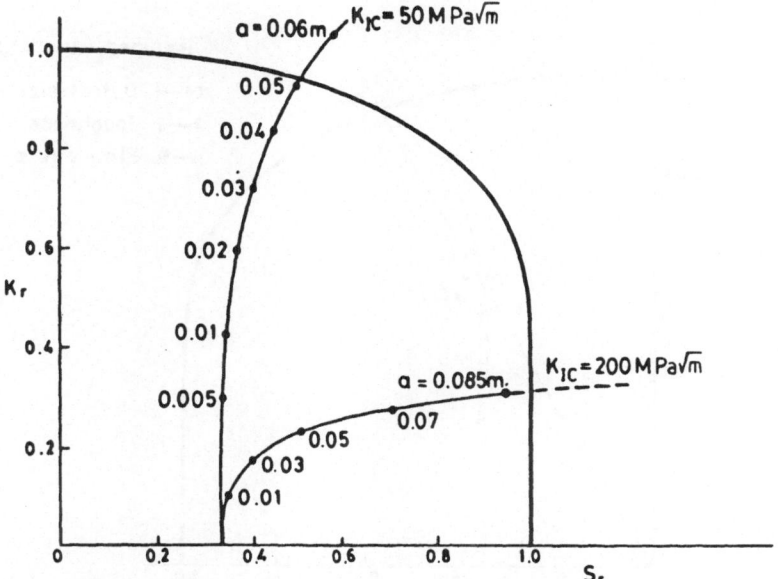

FIG. 6.1   Loci of Assessment Points.

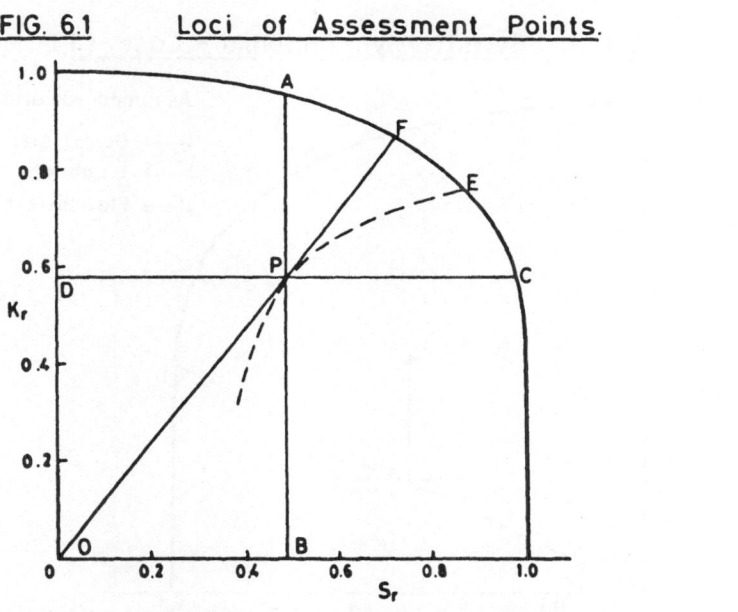

FIG. 6.2   Calculation of Reserve Factors.

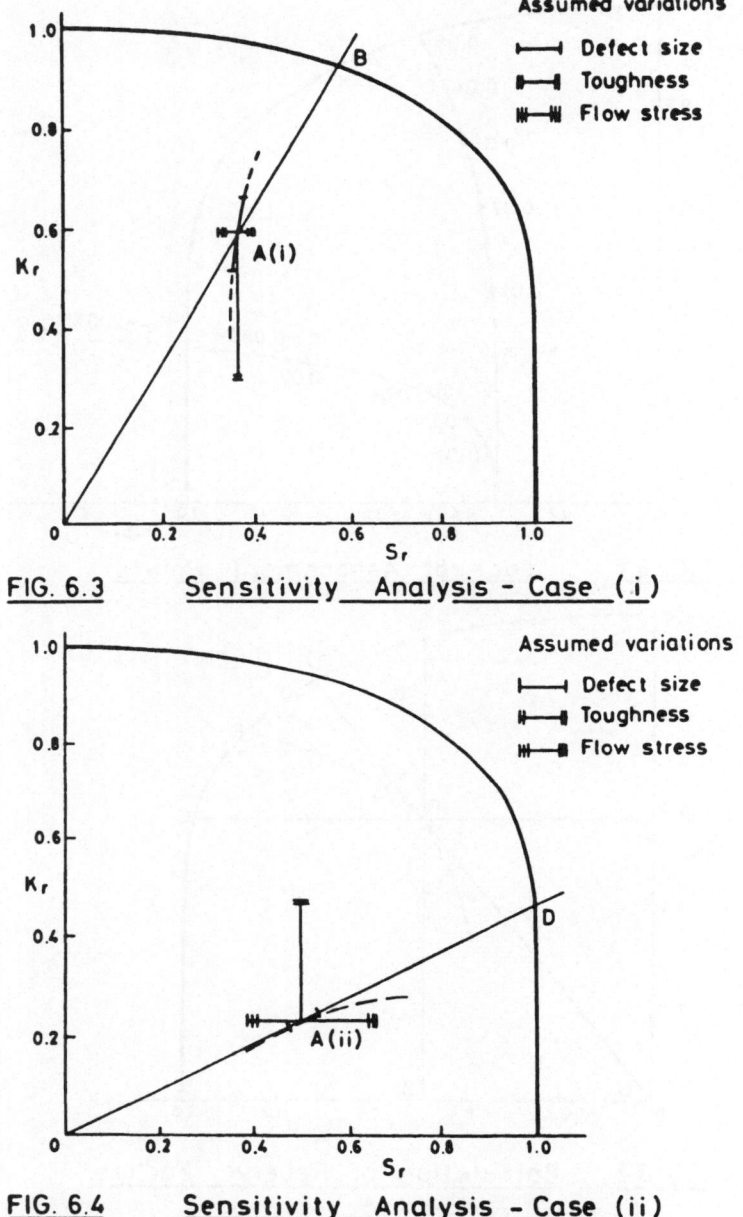

FIG. 6.3   <u>Sensitivity Analysis - Case (i)</u>

FIG. 6.4   <u>Sensitivity Analysis - Case (ii)</u>

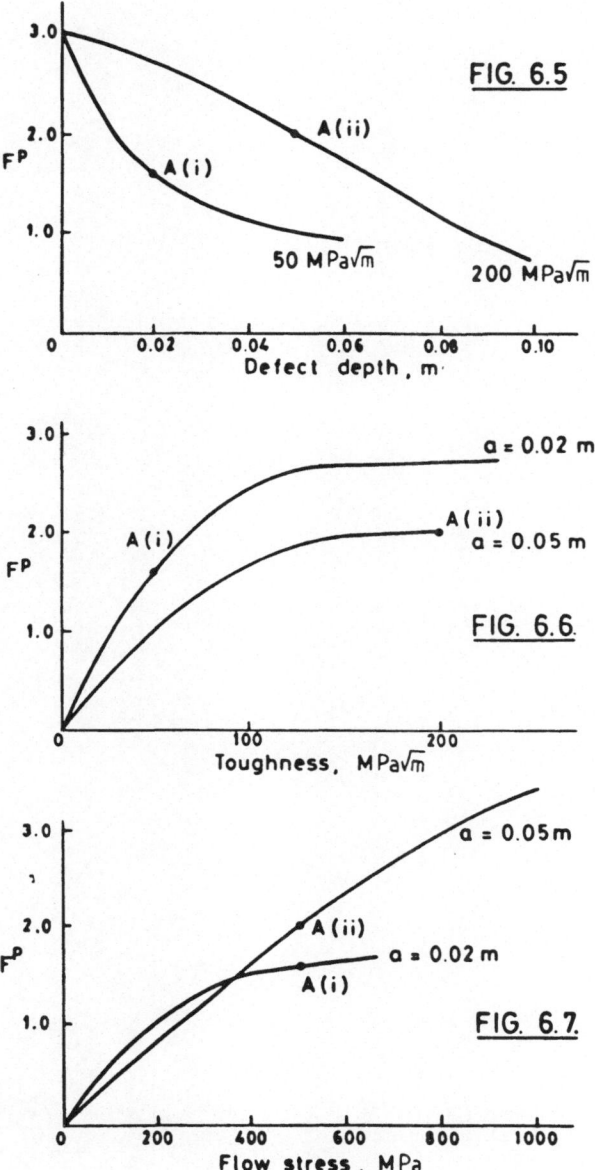

FIG. 6.5

FIG. 6.6.

FIG. 6.7.

WORKSHOP 2 :

THE EPRI* METHOD FOR DUCTILE FRACTURE ANALYSIS

G. Rousselier
Electricité de France
Département Étude des Matériaux
Les Renardières
77250 Moret-sur-Loing, France

ABSTRACT. The EPRI method for ductile fracture analysis, based on the J-integral, is an engineering approach which permits fracture evaluations of flawed structures to be carried out, without finite element calculations, by engineers who are not specialists in fracture mechanics or inelastic analysis. The method is exposed and then applied to fracture mechanics tests on large welded plates. Some difficulties are encountered and a good modelling of the tests cannot be obtained ; nevertheless the EPRI method gives rather conservative results for maximum load and stable crack growth.

1. INTRODUCTION

1.1. Grounds for an engineering approach

When the material behaviour is dominantly elastic, the demonstration of the structural integrity of components is based on linear elastic fracture mechanics concepts. Elastic solutions for a large selection of cracked geometries are available in handbooks [1]. In case a numerical calculation would be necessary, it is now accessible to almost all laboratories and utilities, even for 3D-geometries.

At the other extreme, in some components, for example forged austenitic stainless steel pipes, fracture is associated with intense through-the-thickness striction. In that case a simple net-section collapse load criterion for failure gives quite reasonable results [2].

However, in the more general intermediate conditions, an elastic-plastic fracture mechanics analysis is necessary. A number of key developments have occurred over the past several years in this field [3]. The most frequently used crack tip characterizing parameter is the J-integral, and the tearing modulus derived from J for stability analyses. The application of the developed methodology requires expensive and time-consuming finite element computations. Moreover elastic-plastic numerical analysis is far more critical than elastic analysis and cannot be considered as a routine calculation, accessible to non-expert engineers. That is why an engineering approach was developed by General Electric for EPRI [4]. This approach, based on the J-integral, permits fracture evaluations of flawed structures to be carried out by engineers who are not specialists in fracture mechanics or inelastic analysis.

The engineering approach does not at all lessen the usefulness of sophisticated finite

* Electric Power Research Institute, Palo Alto, California, USA.

L. H. Larsson (ed.), Elastic-Plastic Fracture Mechanics, 373–396.
© 1985 ECSC, EEC, EAEC, Brussels and Luxembourg.

element analyses. As the J-integral concept suffers serious limitations :

— the resistance curve J-Δa is not an intrinsic property of materials [5],
— the J-integral calculation is sometimes problematic (3D, non radial and thermal loads),

it is necessary to perform verifications on reference problems, as far as possible, in connection with experimental results. The use of local criteria, that do not suffer the above limitations, would be particularly appropriate in reference problems. These reference problems would permit the verification of the conservatism of the simple and inexpensive engineering approach.

1.2. The EPRI method

In the engineering approach the various stages of ductile fracture process :

— crack initiation,
— stable crack growth,
— instability,

are quantified through the crack tip characterizing parameter, J-integral. In summary, the approach employs the following principal elements [4] :

1 - A handbook-style compilation of J-integral solutions for fully plastic geometries containing cracks. The compilation also includes solutions for the crack-mouth opening and load-point displacements.

2 - An estimation procedure which enables the construction of elastic-plastic solutions (crack driving force estimates) for cracked geometries through the combination of results from the handbook for fully plastic conditions and existing elastic solutions.

3 - Simple methods for predicting crack initiation, stable crack growth and instability by combining the crack driving force estimates with a resistance curve determined from standard specimen tests.

In section 2 the conditions for J-controlled crack growth and the resistance curve concept are recalled. The theoretical bases for deriving the fully plastic solutions, and the elastic-plastic estimation procedure, are given. The solutions for a compact tension (CT) specimen are given in tabular and graphical form, and an example of «crack driving force diagram» is given for this specimen.

In section 3 a «case study» is presented : very large welded CT specimens and center-cracked plates of a C-Mn structural steel (E36-4) are considered. The analysis of the specimens is conducted according to the EPRI method, and the results concerning the load-displacement curve and the stable crack growth are discussed.

In the EPRI method, several different analysis diagrams can be generated [4]. The «stability assessment diagram» defines regions of loading where stable and unstable behaviour exist, and the «failure assessment diagram» allows one to perform a quick analysis to determine whether or not a fracture problem exists. «Failure assessment diagrams» were also derived by BLOOM and MALIK [6] from the EPRI estimation procedure. These diagrams link the EPRI method with the CEGB procedure, the so-called R-6 method, derived from the original two criteria approach (see Workshop 1). Also note that different estimation formulae of the stress-strain curve can be used with the EPRI method [4]. These developments are not discussed in this workshop.

2. ELASTIC-PLASTIC ESTIMATION PROCEDURE AND CRACK DRIVING FORCE DIAGRAM

2.1. J-integral as a crack tip characterizing parameter

The validity of the J-integral as a crack tip characterizing parameter was discussed by Bakker in his first lecture. This validity is questionable, and it is not even demonstrated that a single parameter can characterize the crack tip region in all situations. Nevertheless, for some cracked geometries and dimensions, a nearly unique relationship between the J-integral and the amount of stable crack growth is found. A necessary condition is that the region of dominance of the HRR field, the dimension of which is a fraction of the uncracked ligament length b, is large enough to include the fracture process zone. As the dimension of the fracture process zone is related to the crack tip opening displacement, proportional to J/σ_y, the condition is :

$$\rho = b/(J/\sigma_y) \gg 1 \tag{1}$$

where σ_y is the yield stress. The same condition implies that the crack growth is limited :

$$\Delta a \ll b \tag{2}$$

A second condition is that during crack growth the radial increments in the strain field due to an increment in the applied load or displacement, proportional to dJ/r in a non hardening material, overwhelm the non radial strain increments associated with the advance of the crack tip, proportional to $(J/r)(da/r)$. The radial coordinate r is a fraction of the ligament b ; so the condition $dJ/r \gg Jda/r^2$ can be restated as :

$$\omega = \frac{b}{J_{Ic}} \frac{dJ}{da} \gg 1 \tag{3}$$

(taking $J = J_{Ic}$ instead of J, as in [4]).

Systematic finite element investigations [7] with the J_2 flow theory of plasticity led to an estimated quantification of the conditions (1) through (3). For members subjected primarily to bending, the conditions are :

$$\rho > 25 \quad \Delta a < 0.06 \, b \quad \omega > 10 \tag{4}$$

For configurations subjected to tensile loads the conditions for J-controlled growth are more stringent :

$$\rho > 200 \quad \omega > 80 \tag{5}$$

For high hardening materials (n < 3) the size requirements may be reduced somewhat.

When J-controlled crack growth is applicable, the condition for continued crack growth is :

$$J_{app} = J_R (a - a_o) \tag{6}$$

The J_R curve is a function of crack growth $\Delta a = a - a_o$ and is obtained experimentally. The applied J, J_{app}, is a function of crack length a and of load P. Consider the particular example of a compact specimen shown in figure 1. The total displacement, δ_{tot}, can be separated into a part due to the crack, δ, and a remaining part related to the compliance of the structure :

$$\delta_{tot} = \delta + C_M P \tag{7}$$

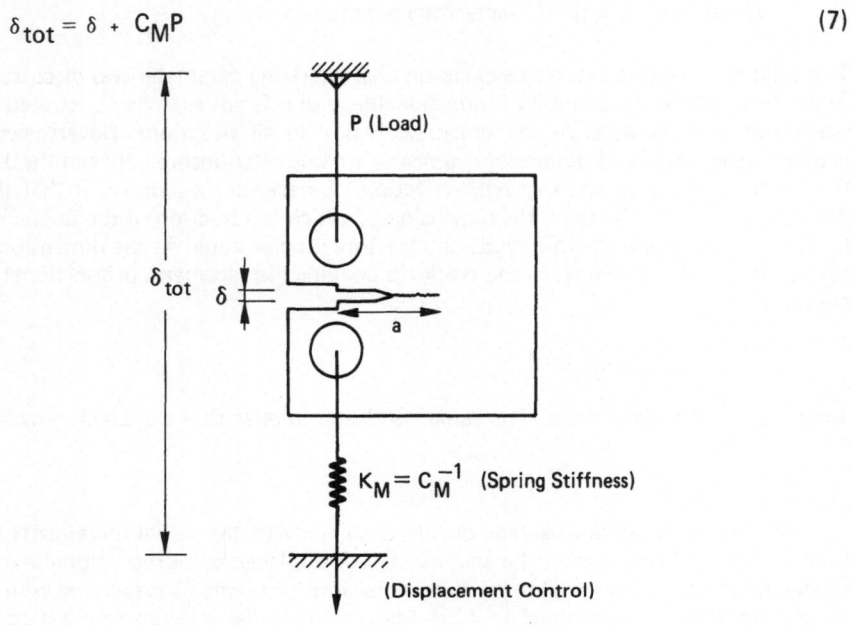

Figure 1 — Particular example of a structure comprising a compact tension specimen in series with a spring.

Crack growth is unstable if the slope of the resistance curve dJ_R/da becomes smaller than the rate of increase in applied J even with a non increasing δ_{tot} :

$$(\partial J/\partial a)_{\delta_{tot}} \geqslant \partial J_R/\partial a \tag{8}$$

The subscript in equation (8) denotes a partial derivative with $d(\delta_{tot}) = 0$. Equation (8) is equivalent to the tearing modulus approach of PARIS et al. [8] :

$$T_{app} = \frac{E}{\sigma_y^2} \left(\frac{\partial J}{\partial a}\right)_{\delta_{tot}} \geqslant T_{mat} = \frac{E}{\sigma_y^2} \frac{\partial J_R}{\partial a} \tag{9}$$

The three «crack driving force diagrams» of figure 2 illustrate that the extent of stable crack growth is strongly dependent on the system compliance. The first diagram corresponds to the usual displacement controlled test on a compact specimen ($C_M = 0$). At a given displacement $\delta = \delta_{tot}$ the applied J decreases with crack length ; the left-hand side of equation (8) is negative and the crack growth is stable. The second diagram illustrates the dead load system ($C_M = \infty$). The applied J (a, P) is a steep function of crack length and the point of instability coincides with the maximum of the load-displacement curve. The intermediate case ($C_M \neq 0$) is shown in the third diagram.

The analysis can be repeated for different initial crack lengths by sliding the J_R curve along the crack length axis, if one assumes that the J_R curve is independent of a_0.

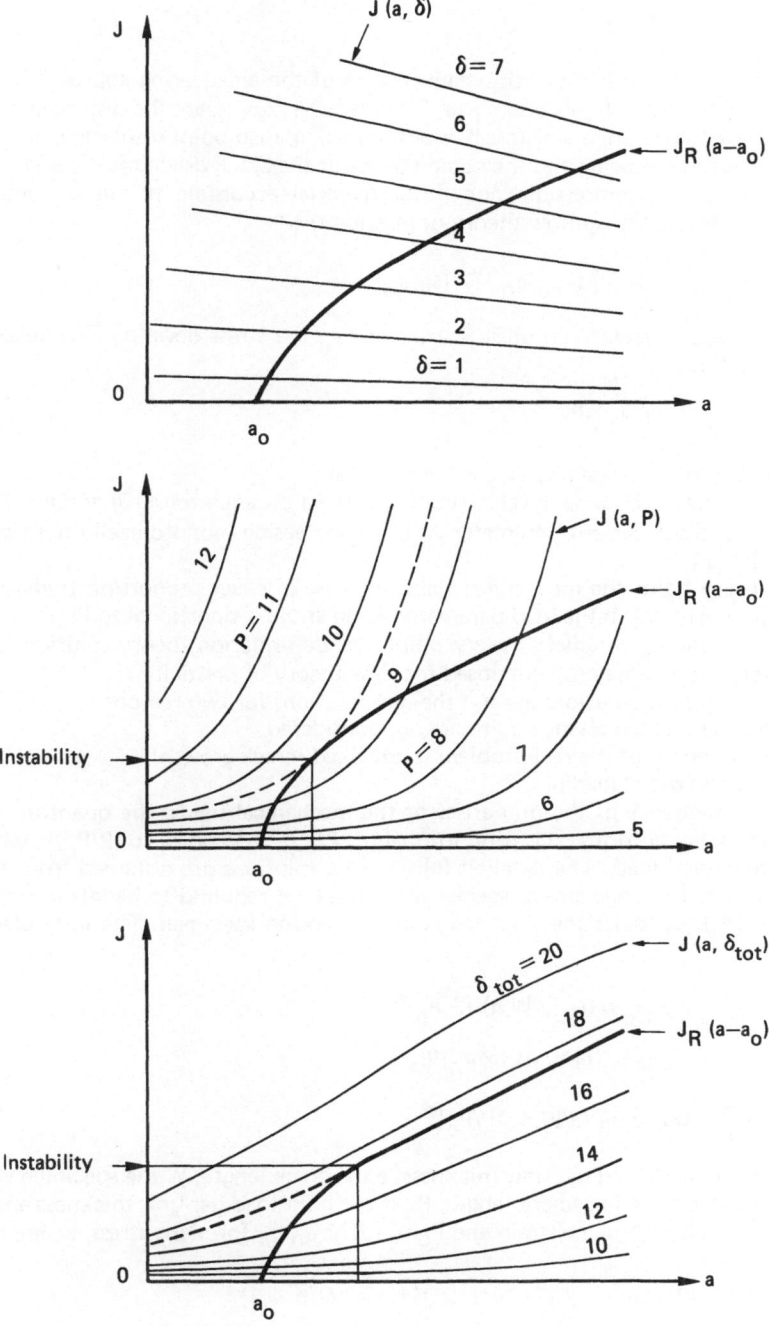

Figure 2 — J-integral crack driving force diagrams for a compact tension specimen in displacement controlled (top), load controlled (middle) and compliant system conditions (bottom).

2.2. Fully plastic solutions

As indicated in the introduction, the main feature of the engineering approach is the tabulation of the functions J (a, P) and Δ_i (a, P), where $\Delta_1, \Delta_2, ...$ are the displacements of important points of the component (crack-mouth opening, load-point displacement, ...). These functions permit the drawing and the complete use of the crack driving force diagram.

Consider an incompressible non linear material according to the J_2 deformation theory (as opposed to the J_2 <u>flow</u> theory of plasticity) :

$$\epsilon_{ij}/\epsilon_o = (3\alpha/2) \, (\sigma_{eq}/\sigma_o)^{n-1} \, (s_{ij}/\sigma_o) \tag{10}$$

where σ_{eq} is the equivalent von MISES stress and s_{ij} the stress deviator. The uniaxial form of equation (10) is :

$$\epsilon/\epsilon_o = \alpha \, (\sigma/\sigma_o)^n \tag{11}$$

where σ_o is a reference stress and ϵ_o a reference strain.

The solution of boundary value problems based on equation (10) and involving only a single load or displacement parameter which is increasing monotonically has two important properties [9] :

— at every point the mechanical fields increase in exact proportion (radial loading), with stress proportional to the load parameter P and strain proportional to P^n,

— as the loading is radial at every point, the deformation theory solution is also the exact solution to the same problem posed for flow theory of plasticity.

The fully plastic solutions are not the real solutions for two reasons :

— the compressible elastic strains are not considered,

— the boundary of the real problem is not fixed (crack growth),

both inducing non radial loading.

As a consequence of the properties of the mechanical fields, the quantities J and Δ_i have simple functional forms : J is proportional to $(P/P_o)^{n+1}$ and Δ_i to $(P/P_o)^n$, where P_o is a limit or reference load. The detailed fully plastic solutions are obtained from finite element calculations (in plane strain, special techniques are required to handle incompressible deformation [4]). Consider the standard compact tension specimen. The fully plastic solutions are [4] :

$$J^P = \alpha \, \sigma_o \epsilon_o \, b \, h_1 \, (a/W, n) \, (P/P_o)^{n+1} \tag{12}$$

$$\delta^P = \alpha \, \epsilon_o a \, h_2 \, (a/W, n) \, (P/P_o)^n \tag{13}$$

$$\Delta_L^P = \alpha \, \epsilon_o a \, h_3 \, (a/W, n) \, (P/P_o)^n \tag{14}$$

where P is the applied load per unit thickness, a the crack length, W the specimen width and $b = W - a$ the uncracked ligament length. P_o is the limit load per unit thickness and is given by $P_o = 1.455 \, \sigma_o \, b\eta$ for plane strain and $P_o = 1.071 \, \sigma_o \, b\eta$ for plane stress, where :

$$\eta = [(2 \, a/b)^2 + 2 \, (2a/b) + 2]^{1/2} - (2a/b + 1) \tag{15}$$

δ is the mouth opening displacement of the crack at the outer edge and Δ_L at the load line. h_1, h_2, h_3 are functions of a/W and n alone ; they are given in table I for plane strain and table II for plane stress.

TABLE I

h_1, h_2 and h_3 for the compact specimen in plane strain

		$n = 1$	$n = 2$	$n = 3$	$n = 5$	$n = 7$	$n = 10$	$n = 13$	$n = 16$	$n = 20$
a/W = 1/4	h1	2.23	2.05	1.78	1.48	1.33	1.26	1.25	1.32	1.57
	h2	17.9	12.5	11.7	10.8	10.5	10.7	11.5	12.6	14.6
	h3	9.85	8.51	8.17	7.77	7.71	7.92	8.52	9.31	10.9
a/W = 3/8	h1	2.15	1.72	1.39	0.970	0.693	0.443	0.276	0.176	0.098
	h2	12.6	8.18	6.52	4.32	2.97	1.79	1.10	0.686	0.370
	h3	7.94	5.76	4.64	3.10	2.14	1.29	0.793	0.494	0.266
a/W = 1/2	h1	1.94	1.51	1.24	0.919	0.685	0.461	0.314	0.216	0.132
	h2	9.33	5.85	4.30	2.75	1.91	1.20	0.788	0.530	0.317
	h3	6.41	4.27	3.16	2.02	1.41	0.888	0.585	0.393	0.236
a/W = 5/8	h1	1.76	1.45	1.24	0.974	0.752	0.602	0.459	0.347	0.248
	h2	7.61	4.57	3.42	2.36	1.81	1.32	0.983	0.749	0.485
	h3	5.52	3.43	2.58	1.79	1.37	1.00	0.746	0.568	0.368
a/W = 3/4	h1	1.71	1.42	1.26	1.033	0.864	0.717	0.575	0.448	0.345
	h2	6.37	3.95	3.18	2.34	1.88	1.44	1.12	0.887	0.665
	h3	4.86	3.05	2.46	1.81	1.45	1.11	0.869	0.686	0.514
a/W → 1	h1	1.57	1.45	1.35	1.18	1.08	0.950	0.850	0.730	0.630
	h2	5.39	3.74	3.09	2.43	2.12	1.80	1.57	1.33	1.14
	h3	4.31	2.99	2.47	1.95	1.79	1.44	1.26	1.07	0.909

TABLE II

h_1, h_2 and h_3 for the compact specimen in plane stress

		$n = 1$	$n = 2$	$n = 3$	$n = 5$	$n = 7$	$n = 10$	$n = 13$	$n = 16$	$n = 20$
a/W = 1/4	h1	1.61	1.46	1.28	1.06	0.903	0.729	0.601	0.511	0.395
	h2	17.6	12.0	10.7	8.74	7.32	5.74	4.63	3.75	2.92
	h3	9.67	8.00	7.21	5.94	5.00	3.95	3.19	2.59	2.023
a/W = 3/8	h1	1.55	1.25	1.05	0.801	0.647	0.484	0.377	0.284	0.220
	h2	12.4	8.20	6.54	4.56	3.45	2.44	1.83	1.36	1.02
	h3	7.80	5.73	4.62	3.25	2.48	1.77	1.33	0.990	0.746
a/W = 1/2	h1	1.40	1.08	0.901	0.686	0.558	0.436	0.356	0.298	0.238
	h2	9.16	5.67	4.21	2.80	2.12	1.57	1.25	1.03	0.814
	h3	6.29	4.15	3.11	2.09	1.59	1.18	0.938	0.774	0.614
a/W = 5/8	h1	1.27	1.03	0.875	0.695	0.593	0.494	0.423	0.370	0.310
	h2	7.47	4.48	3.35	2.37	1.92	1.54	1.29	1.12	0.928
	h3	5.42	3.38	2.54	1.80	1.47	1.18	0.988	0.853	0.710
a/W = 3/4	h1	1.23	0.977	0.833	0.683	0.598	0.506	0.431	0.373	0.314
	h2	6.25	3.78	2.89	2.14	1.78	1.44	1.20	1.03	0.857
	h3	4.77	2.92	2.24	1.66	1.38	1.12	0.936	0.800	0.666
a/W → 1	h1	1.13	1.01	0.775	0.680	0.650	0.620	0.490	0.470	0.420
	h2	5.29	3.54	2.41	1.91	1.73	1.59	1.23	1.17	1.03
	h3	4.23	2.83	1.93	1.52	1.39	1.27	0.985	0.933	0.824

2.3. Elastic-plastic estimation procedure

The elastic-plastic behaviour of a material can be characterized by :

$$\epsilon/\epsilon_0 = \sigma/\sigma_0 + \alpha \, (\sigma/\sigma_0)^n \tag{16}$$

with $\epsilon_0 = \sigma_0/E$. The plastic strain corresponding to σ_0 is $\alpha\epsilon_0$; the total strain is $\epsilon = \epsilon_0 + \alpha\epsilon_0$
An approximate solution can be proposed for the quantities J and Δ_i, for example :

$$J(a, P) = J^e \, (a, P) + J^p \, (a, P) \tag{17}$$

where J^e is the elastic contribution :

$$J^e = K_I^2/E' \tag{18}$$

where $E' = E/(1 - \upsilon^2)$ for plane strain and $E' = E$ for plane stress. J^e is proportional to $(P/P_0)^2$ and J^p to $(P/P_0)^{n+1}$. The hardening exponent is for example $n \cong 10$ in nuclear pressure vessel ferritic steels and $n \cong 5$ in austenitic stainless steels. Thus, for $P < P_0$, the plastic contribution is small while experimental and numerical investigations show that the deviation from linearity is important. That is why a plastic correction is introduced in the elastic contribution, through a modified IRWIN's effective crack length [4] :

$$a_{eff} = a + \emptyset \, r_y \tag{19}$$

$$r_y = \frac{1}{\beta\pi}\left(\frac{K_I}{\sigma_0}\right)^2 \cdot \frac{n-1}{n+1} \tag{20}$$

$$\emptyset = 1/[1 + (P/P_0)^2] \tag{21}$$

For plane strain $\beta = 6$ and for plane stress $\beta = 2$. The resulting estimation procedure is :

$$J \, (a, P) = J^e \, (a_{eff}, P) + J^p \, (a, P) \tag{22}$$

$$\Delta_i \, (a, P) = \Delta_i^e \, (a_{eff}, P) + \Delta_i^p \, (a, P) \tag{23}$$

Consider the example of the standard compact tension specimen. The plastic contributions are given by equations (12), (13) and (14). The elastic contributions are available in handbooks [1] :

$$J^e = (aP^2/E'W^2)f_1 \, (a/W) \tag{24}$$

$$\delta^e = (P/E') \, f_2 \, (a/W) \tag{25}$$

$$\Delta_L^e = (P/E') \, f_3 \, (a/W) \tag{26}$$

$$f_1 \, (a/W) = [29.6 - 185.5 \, (a/W) + 655.7 \, (a/W)^2 - 1017.0 \, (a/W)^3 + 638.9 \, (a/W)^4]^2 \tag{27}$$

The functions f_2 and f_3 are plotted in figure 3.

In equation (20) : $K_I^2 = (aP^2/W^2)f_1(a/W)$, where a is the physical crack length.

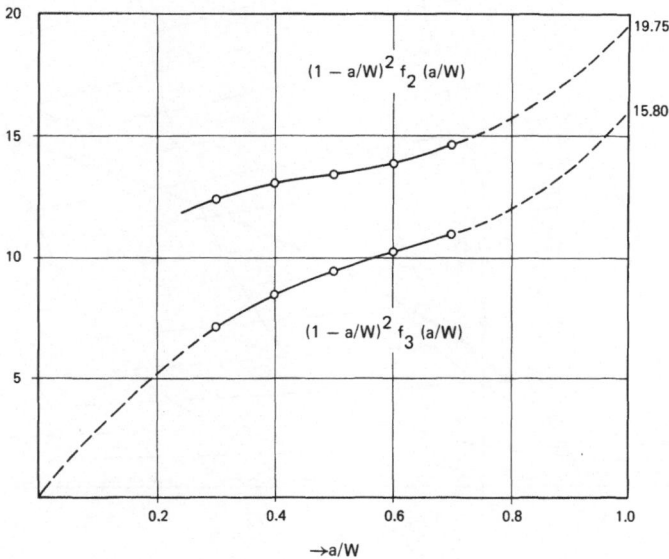

Figure 3 — f_2 and f_3 for the compact tension specimen

For lack of well-established theoretical bases, the agreement of the estimation procedure with elastic-plastic J_2 flow theory and experimental results has to be verified.

2.4. Verification of the estimation procedure

Four compact tension (CT) specimens of A533B steel were tested at 93°C by General Electric [10]. The material constants are E = 207 GPa, v = 0.3, σ_0 = 414 MPa, $\epsilon_0 = \sigma_0/E$ = 0.002, α = 1.12, n = 9.71. As the 100 mm thick standard compact tension specimens are 25% side-grooved, plane strain behaviour is assumed in the analysis, and the comparison of the predicted load per unit thickness with the experimentally measured load is made on the basis of the net thickness B_{net} = 75 mm.

The elastic-plastic estimation formulae of sections 2.2 and 2.3 permit the computation of J (a, P) and Δ_L (a, P). J versus a curves for a fixed P and for a fixed Δ_L are then generated and combined into one diagram, as illustrated in figure 4. An experimentally determined J-resistance curve [10] is superimposed on figure 4 at the initial crack length of 115.4 mm (a_0/W = 0.577) corresponding to one of the four specimens. The analysis can be repeated for the other specimens (with a_0/W = 0.615, 0.659 and 0.801) by sliding the J_R curve along the crack length axis (actually the range of values for crack length, load and displacement in figure 4 is not fitted for the analysis of the other specimens). The J_R curve is a mean curve obtained from a number of tests on compact tension specimens.

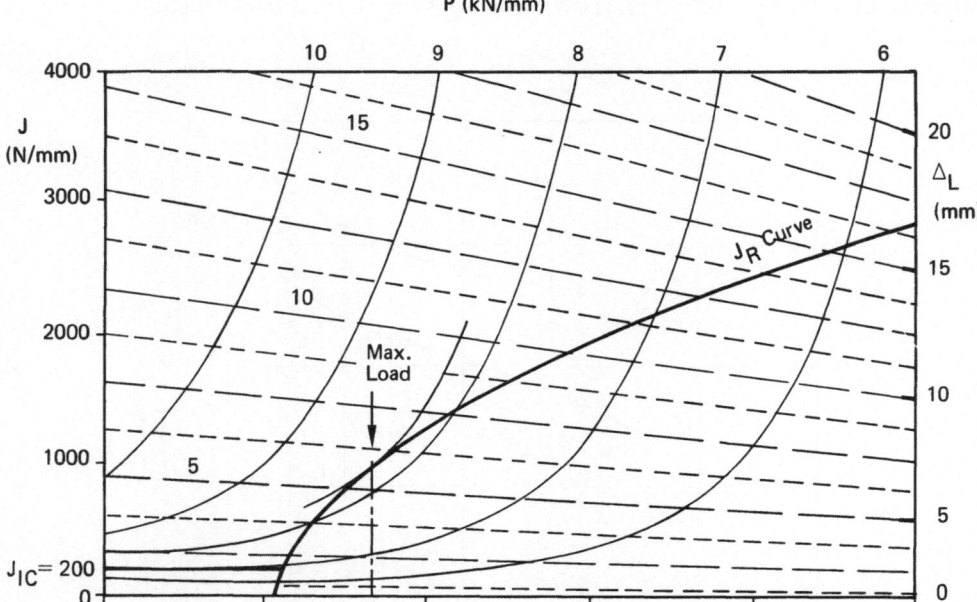

Figure 4 — Crack driving force diagram for a plane strain CT specimen of A 533 B steel
(W=200 mm, a_o = 115.4 mm). The constant load curves are shown by
solid lines and the constant displacement curves by dashed lines.

Equilibrium of crack growth requires that the applied crack driving force in terms
of J equals the material resistance to crack growth, J_R. Thus, by examining successive points
along the J_R curve, and reading the load and displacement on the fixed load (solid line) and
fixed displacement (dashed line) curves intersecting at these points, the complete load-
displacement behaviour is obtained.

The predicted load-displacement curve shown in figure 5 is in good agreement with
the experimental data. The maximum load is underestimated by 6% (note that the use of
the net thickness makes this comparison somewhat arbitrary). The agreement is good too
with a plane strain finite element calculation based on J_2 flow theory of plasticity (independently of thickness this time). It is worth to notice that a good prediction of the load-
displacement curve of a CT specimen requires a good prediction of the crack growth behaviour. It is illustrated in figure 5 : a completely erroneous load-displacement curve is predicted for a stationary crack.

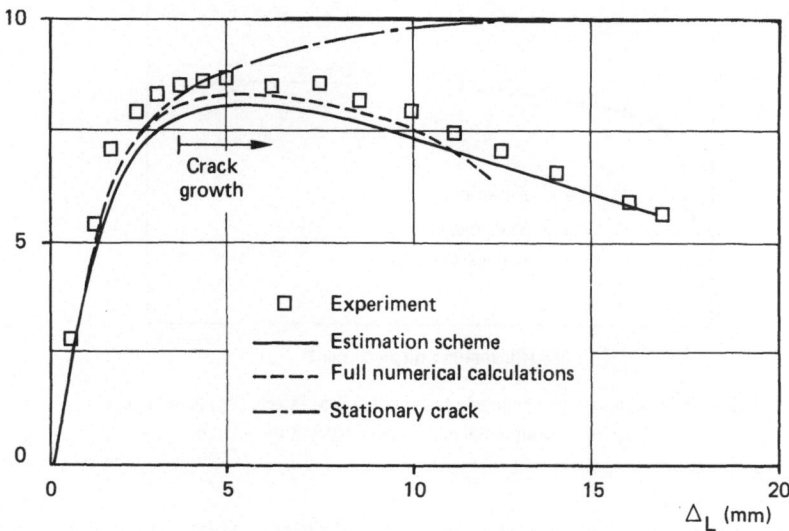

P (kN/mm)

Crack
growth

□ Experiment
—— Estimation scheme
– – – Full numerical calculations
—·— Stationary crack

Δ_L (mm)

Figure 5 — Comparison of load-displacement curves predicted from fig. 4 and experimentally
measured on a 100 mm thick, 25 % side-grooved CT specimen, with P = Load/Bnet
and Bnet = 75 mm. The calculated curve with J_2 flow theory of plasticity and the
predicted curve for a stationary crack are also shown.

Note that the validity conditions (4) given in section 2.1 would limit the crack growth
to 5 mm and the J-integral to 1400 N/mm (the condition $\omega > 10$ is fulfilled along the whole
J_R curve). The corresponding displacement is $\Delta_L \cong 7$ mm. Actually the J-controlled crack
growth is not necessary in the present application since the J_R curve was determined with
similar compact tension specimens. This discussion will be completed in section 2.5.

Other information can also be obtained from figure 4. For a minimum value of J_{IC}
equal to 200 N/mm, the load and displacement at crack growth initiation are $P \cong 7$ kN/mm
and $\Delta_L \cong 2$ mm. The predicted maximum load, P = 8.2 kN/mm, would be the instability
load for an infinitely compliant system (dead load for example).

The other three CT specimens also give a good agreement with the experimental data.
A further verification is given by a radically different crack configuration : the single edge
crack panel loaded in tension [10]. The J_R curve determined with the CT specimens is used
in the crack driving force diagram. The agreement is good again (fig. 6).

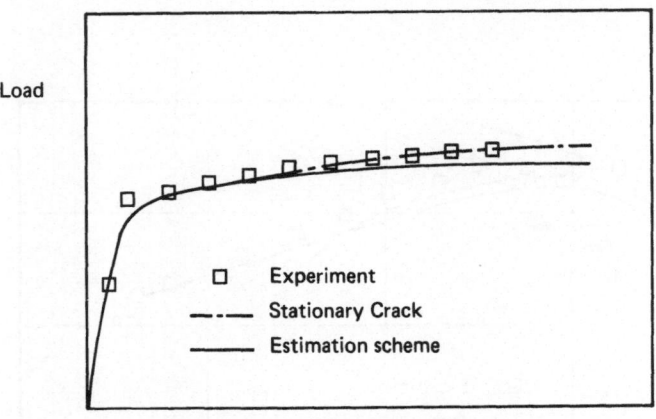

Figure 6 — Comparison of predicted load-displacement curves with test
data for a single edge crack panel loaded in tension.

2.5. Discussion

The examples presented in the preceding section, and other analyses of cracked cylinders in tension or under internal pressure [4], provide a *good (though incomplete) verification of the elastic-plastic estimation procedure as far as the load-displacement behaviour is concerned.*

In some geometries (e.g. bend specimens) a good knowledge of the crack growth behaviour in these geometries is required as a preliminary. In other geometries (e.g. tension cracked panels) the load displacement curve is quasi-independent of the crack growth behaviour (fig. 6).

The ability of the method to predict the crack growth behaviour in all situations has not been verified. The crack growth behaviour used for the CT specimens was determined with, among others, the same CT specimens. The «predicted» crack growth behaviour is inevitably good even well beyond J-controlled crack growth. For the single edge crack panels, the good prediction of the load-displacement curve is not a test for the predicted crack growth behaviour, as the former is quasi-independent of the latter. No stable crack growth measurements are reported or compared with predictions for these specimens [4, 10].

3. CASE STUDY

3.1. Material and specimen geometry

Fracture mechanics tests were conducted at IRSID [11] (Institut de Recherches de la Sidérurgie Française) on electroslag weldments of C-Mn steel plates. The geometry of the welds and of the specimens is given in figure 7. The base metal is a grade E36-4 of C-Mn steel (AFNOR A35-501 specification) ; the chemical composition of the plate is given in table III. Large welds were obtained by a one pass vertical electroslag welding process ; the dimensions of the assemblings are 1000 x 1000 x 19 (mm), with welds perpendicular to the longitudinal orientation of the plates. The chemical composition of the welds is given in table III.

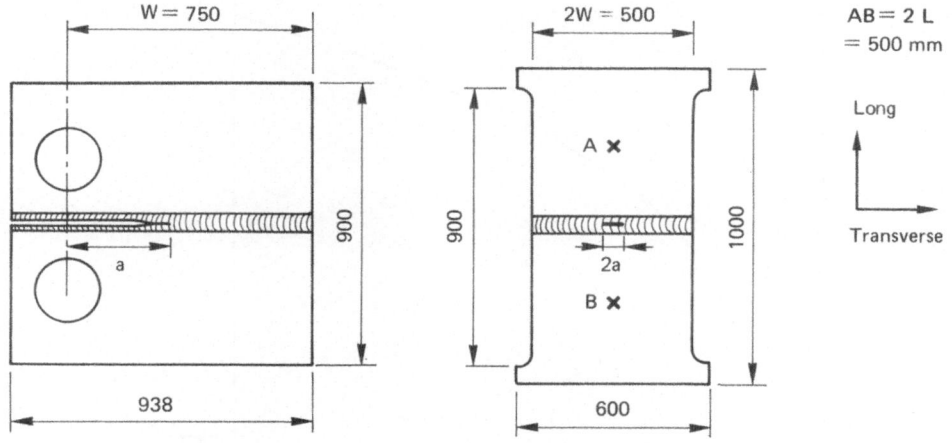

Figure 7 — Geometry of the compact tension specimen and of the center-cracked plate (dimensions in mm, thickness B = 17.75 mm)

TABLE III

Chemical composition of E 36-4 steel plate NB (weight %)

	C	Mn	Si	S	P	Al	Mo	Nb
Base metal	0.218	1.473	0.365	0.0135	0.0217	0.039	–	–
Welds	0.124 −0.148	0.79 −0.86	0.215 −0.241	0.0149 −0.0159	0.0223 −0.0235	0.007 −0.009	0.202 −0.242	–

As the yield stress and the ultimate strength of the base metal in the longitudinal and transverse orientations and of the welds are very similar (table IV), the analysis of the whole specimens can be performed with the stress-strain curve of the base metal in the longitudinal orientation. The stress-strain curve of plate NA is given in figure 8. The cracks are located in the welds (figure 7) ; the Charpy V energy of the welds at 20°C is between 25 and 40 J and the ductile plateau of about 80 J is achieved above 120°C, so a brittle fracture of the specimens is to be expected at 20°C, possibly after some stable crack growth.

TABLE IV

Mechanical properties of E 36-4 steel plates (20°C)

	Orientation	σ_y 0.2% (MPa)	σ_u (MPa)	A (%)	Z (%)
Base metal, plate NA	L	365	540	35	75
Base metal, plate NB	L	392	585	31	75
	T	390	580	26	64
Welds	T	371 − 428	530 − 575	18 − 26	63 − 65

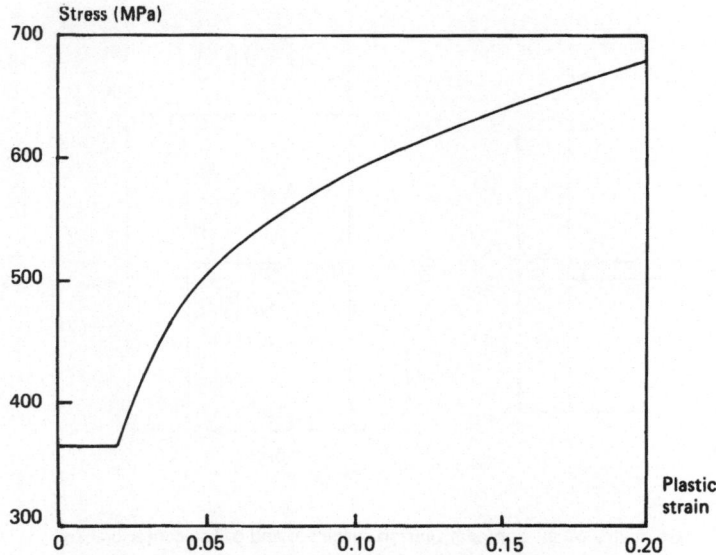

Figure 8 – Stress-strain curve of E 36-4 steel (plate NA, true stress - true strain)

3.2. Fracture mechanics tests

Three tests on large compact tension specimens are considered. The specimens NA8F1 and NA8F2 are taken from E36-4 steel plate NA, and both received a stress relief heat treatment. Specimen NA8F1 is fatigue cracked and specimen NA8F2 has just a mechanical notch. The specimen NB7F1, taken from plate NB, received no heat treatment. The load-load line opening displacement of the three specimens are very similar up to the point of brittle fracture ; it is illustrated in figure 9 for the two NA specimens.

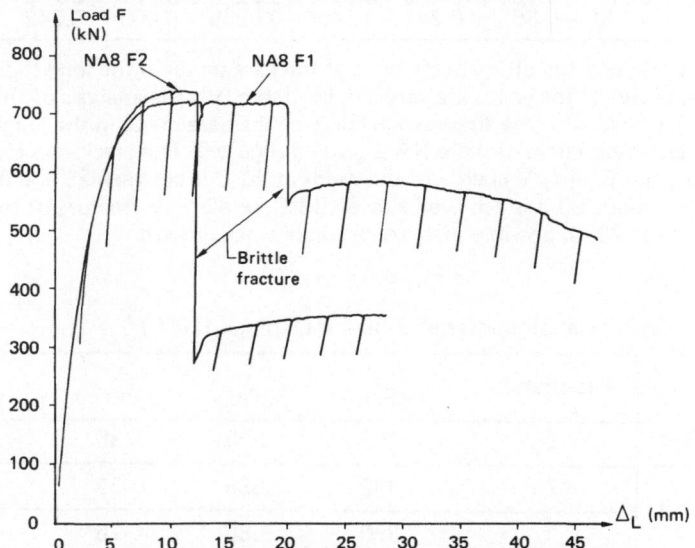

Figure 9 – Experimental load-displacement curves of CT 375 specimens :
W = 750 mm, a_0 = 269 mm (NA8F1) and 275 mm (NA8 F2),
B = 17.75 mm

The stable crack growth was determined with the partial unloading compliance method. Actually a double compliance method was necessary [11] ; the simple compliance method is not accurate enough on account of small frictions with lateral stiffeners. The J-integral was computed from the load-load line displacement curves according to the tentative procedure for J-resistance curve determination [12] (section 9.1.1.). The experimental points are plotted in figure 10 ; the large scatter is not out of the ordinary for common industrial welds. The last unloading measurements are in good agreement with the final stable crack growths measured on the crack surfaces after the brittle fracture that took place in all three specimens.

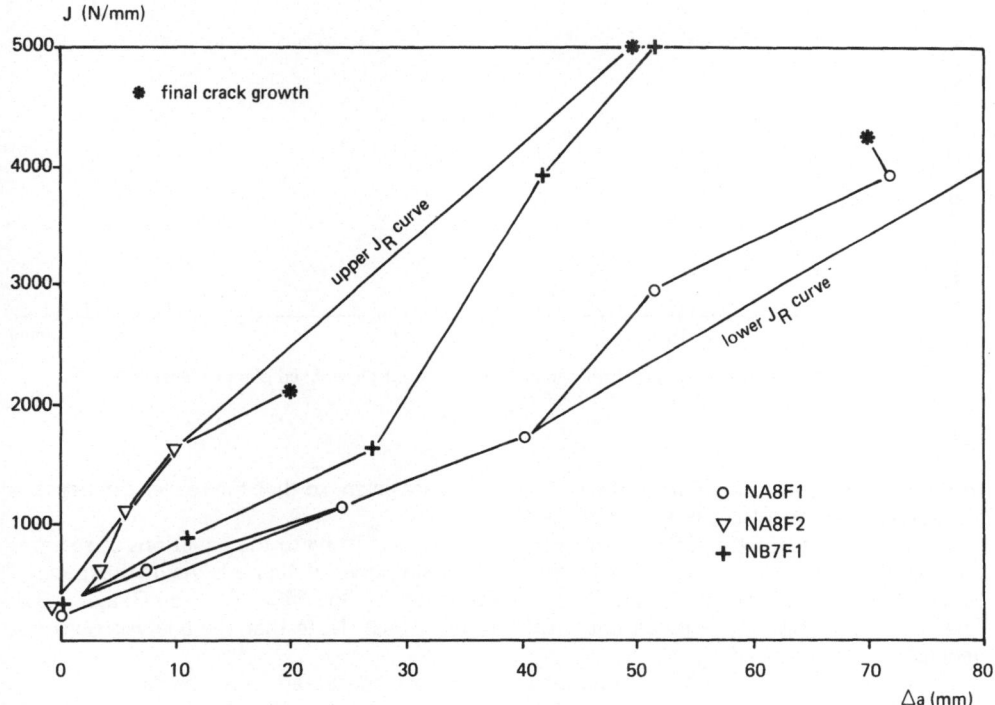

Figure 10 — J-resistance curve of the weld material (partial unloading double compliance method)

A stress relief heat treated plate with a small center crack (CCP : figure 7) was loaded in tension up to brittle fracture (the test was conducted for IRSID in the Institut für Eisenhüttenkunde, Aachen). The elongation was measured on a 500 mm gauge length in the middle of the specimen ; it exceeds the mean elongation by less than 5%. The load-displacement curve is given in figure 11. The final stable crack growths at brittle fracture initiation are 54 mm and 57 mm at the two crack extremities. Note that the stable crack growth, initially flat, is of the slant type as in the CT specimens.

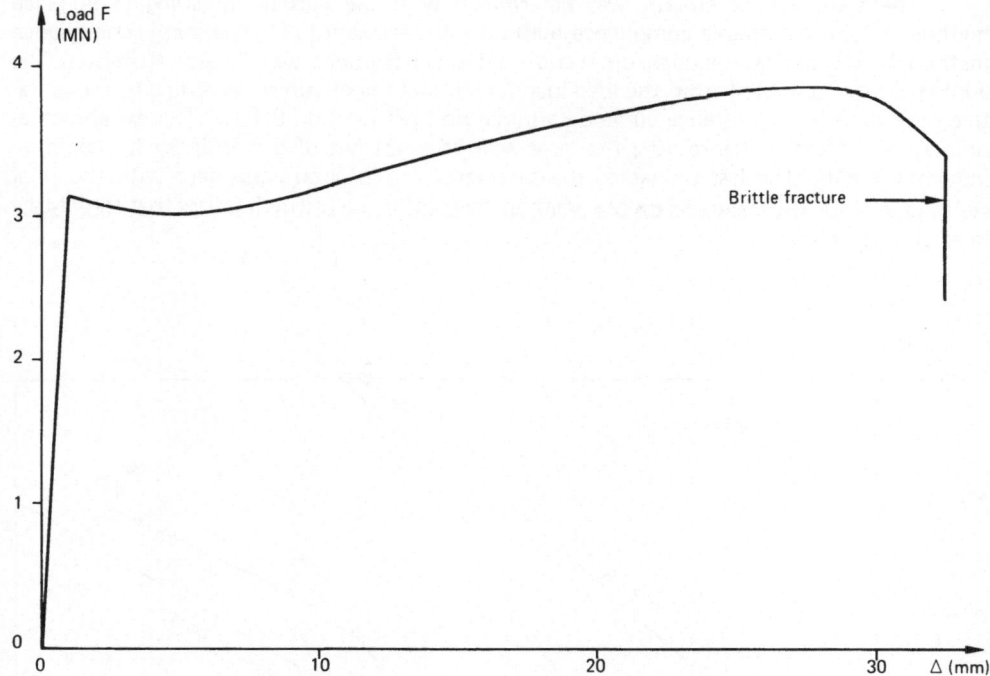

Figure 11 — Experimental load-displacement curve of a center-cracked plate in tension
(W = L = 250 mm, a_o = 8.5 mm, B = 17.75 mm)

3.3. Problem

In this section only the guide-lines for the analysis are given, so that the reader can use it as an exercise. The solution is given in section 3.4.

The analysis of the CT specimens is performed according to the equations of sections 2.2 and 2.3. First a least square fit of the stress-strain curve of figure 8 yields the material constants α and n. The other material constants are σ_o = 365 MPa, E = 206 GPa, v = 0.3. The reference stress σ_o is taken equal to the yield stress. Note that the analysis does not depend on the choice of σ_o.

J-integral and load-line opening displacement are computed from the data of tables I and II (functions h_1 and h_3) and from figure 3 (function f_3). If necessary, interpolation polynomials are to be used to complete the data for the given n and different values of a/W. The analysis can be performed in plane stress only, or in both plane strain and plane stress. A crack driving force diagram similar to that of figure 4 is then constructed.

A predicted load-displacement curve can then be plotted and compared to the experimental curves of figure 9. Three J_R (a − a_o) curves may be considered : stationary crack, upper and lower J_R - curves from figure 10 (take a_o = 275 mm).

3.4. Solution

The least square fit of the stress-strain curve yields α = 5.69, n = 4.77 (see figure 12). Then a quadratic interpolation is conducted to determine the functions h_1 and h_3 for n = 4.77 from the data of tables I and II ; a cubic interpolation is also conducted to determine h_1 and h_3 for a/W = 5/16 and 7/16 ; the results are brought together in table V. The function f_1 is given by equation (27) ; an interpolation formula is used for f_3 ($0.3 < a/W < 0.5$) :

$$f_3 (a/W) = [1.49 + 23.15 (a/W) - 14.5 (a/W)^2] / (1 - a/W)^2 \qquad (28)$$

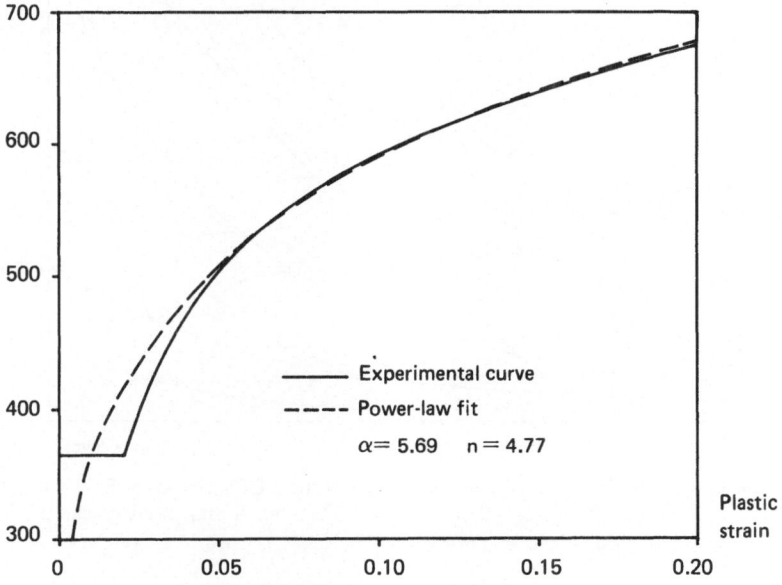

Figure 12 – Stress-strain curve of E 36-4 steel (plate NA).

TABLE V

h_1 and h_3 for the compact specimen and n = 4.77

a (mm)	a/W	Plane stress		Plane strain	
		h_1	h_3	h_1	h_3
187.5	1/4	1.08	6.07	1.51	7.80
234.4	5/16	0.935	4.50	1.16	4.69
281.25	3/8	0.825	3.38	1.01	3.25
328.1	7/16	0.749	2.63	0.962	2.56
375	1/2	0.706	2.18	0.951	2.12
468.75	5/8	0.712	1.86	1.00	1.86

Tables of J and Δ_L versus total load F = BP are then made up for a/W = 5/16, 3/8, 7/16, 1/2 with the help of a personal computer ; from these tables the crack driving force diagrams of figure 13 (plane stress) and 14 (plane strain) can be constructed. Three distinct J_R - curves are considered : stationary crack (A), upper (B) and lower (C) J_R - curves from figure 10.

The predicted load-displacement curves are plotted in figure 15 for initial crack length a_0 = 275 mm. The best agreement with the experimental curve is obtained with the lower J_R - curve in plane strain ; nevertheless, as the specimens are very thin (B/W < 0.024 !), the analysis should be performed in plane stress : a further discussion is presented in section 3.5. Note that the validity conditions (4) correspond to : J < 5000 N/mm, Δa < 20 mm, dJ/da > 10 N/mm/mm ; the second condition only is violated for large crack growth.

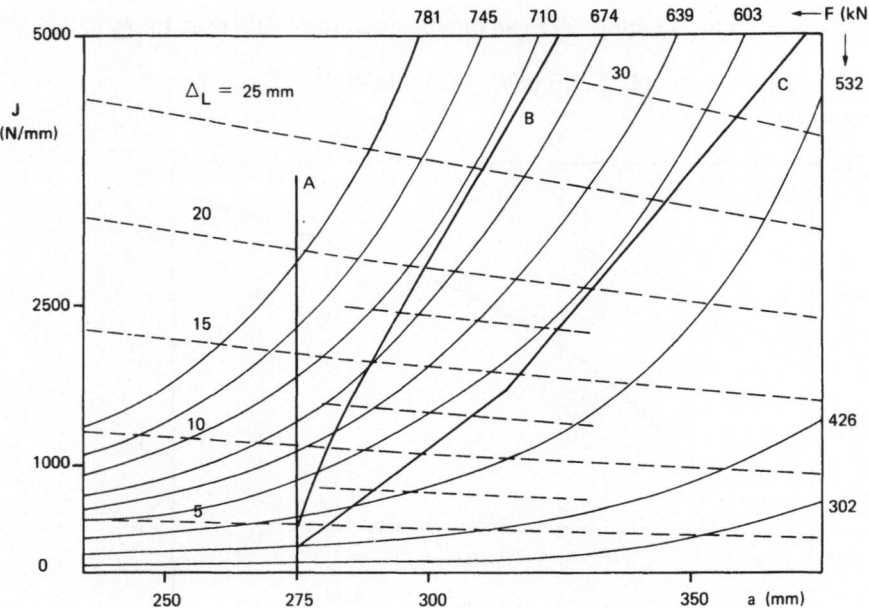

Figure 13 – Crack driving force diagram for a plane stress CT specimen of E 36-4 steel (W = 750 mm, a_0 = 275 mm, B = 17.75 mm) A : stationary crack, B : upper J_R-curve, C : lower J_R-curve.

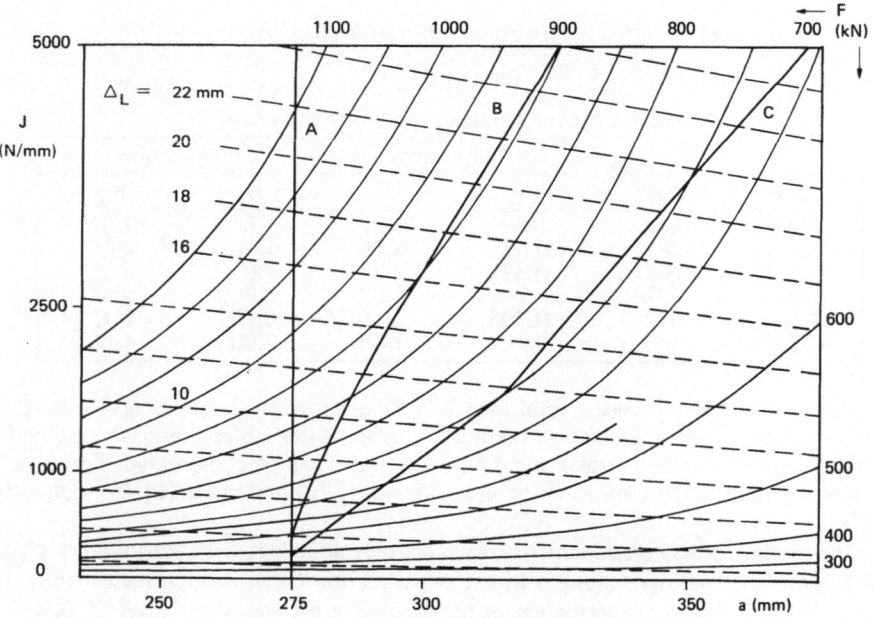

Figure 14 – Crack driving force diagram for a plane strain CT specimen of E 36-4 steel (W = 750 mm, a_0 = 275 mm, B = 17.75 mm) A : stationary crack, B : upper J_R-curve, C : lower J_R curve.

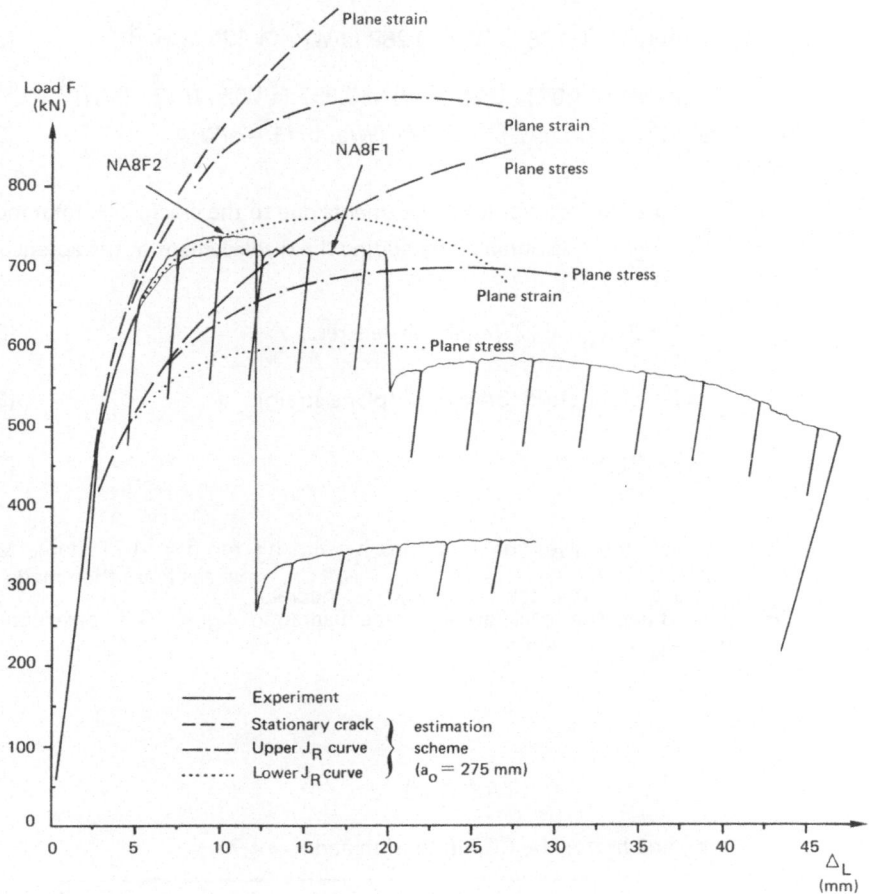

Figure 15 – Comparison of predicted load-displacement curves with experimental curves of CT 375 specimens : W=750 mm, a_O=269 mm (NA8F1) and 275 mm (NA8F2), B = 17.75 mm

For the CCP in tension the following equations are used [4] :

$$J^p = \alpha \, \sigma_O \epsilon_O \; (ab/W) \; h_1 \; (a/W,n) \; (P/P_O)^{n+1} \tag{29}$$

$$\Delta^p_{crack} = \alpha \, \epsilon_O a \, h_3 \; (a/W,n) \; (P/P_O)^n \tag{30}$$

$$P_O = 2b\sigma_O \quad \text{(plane stress)} \tag{31}$$

$$P_0 = 4b\sigma_0/\sqrt{3} \quad \text{(plane strain)} \tag{32}$$

$$J^e = K_I^2/E' = (aP^2/E'W^2)f_1\,(a/W) \tag{33}$$

$$\Delta^e_{\text{crack}} = (P/E')f_3\,(a/W) \tag{34}$$

$$f_1\,(a/W) = (\pi/4)\,[1 + 0.128\,(a/W) - 0.288\,(a/W)^2 + 1.525\,(a/W)^3]^2 \tag{35}$$

$$f_3\,(a/W) = (2a/W)\,[-1.071 + 0.25\,(a/W) - 0.357\,(a/W)^2 + 0.121\,(a/W)^3$$
$$- 0.047\,(a/W)^4 + 0.008\,(a/W)^5 - 1.071\,(W/a)\ln\,(1 - a/W)] \tag{36}$$

Δ^p_{crack} and Δ^e_{crack} are the load-point displacements due to the crack. The total mean displacement on a gauge length 2L is obtained by adding the displacement in the absence of a crack :

$$\Delta^p_{\text{nocrack}}\,(L) = 2\,\alpha\epsilon_0\,L(P/2W\sigma_0)^n \quad \text{(plane stress)} \tag{37}$$

$$\Delta^p_{\text{nocrack}}\,(L) = \sqrt{3}\alpha\epsilon_0 L(P\sqrt{3}/4W\sigma_0)^n \quad \text{(plane strain)} \tag{38}$$

$$\Delta^e_{\text{nocrack}}\,(L) = PL/E'\,W \tag{39}$$

A quadratic interpolation gives the functions h_1 and h_3 for $n = 4.77$ (table VI). Tables of J and $\Delta = \Delta^p_{\text{crack}} + \Delta^p_{\text{nocrack}} + \Delta^e_{\text{crack}} + \Delta^e_{\text{nocrack}}$ versus P are then made up for a/W = 0, 1/8, 1/4, 3/8 and the crack driving force diagram of figure 16 is constructed ; the initial crack length is $2a_0 = 17.1$ mm.

TABLE VI

h_1 and h_3 for the CCP in tension and $n = 4.77$

a (mm)	a/W	Plane stress		Plane strain	
		h_1	h_3	h_1	h_3
31.25	1/8	4.43	1.49	4.33	1.25
62.5	1/4	3.20	1.79	3.29	1.60
93.75	3/8	2.37	1.58	2.53	1.43

The predicted load-displacement curves are plotted in figure 17. As it is expected, reasonable results are obtained in plane stress only ; still there is a large discrepancy between experimental and predicted curves. In figure 17, the points of stable crack growth $\Delta a = 55$ mm are also plotted. Note that the validity conditions (5) correspond to : $J < 500$ N/mm $dJ/da > 80$ N/mm/mm ; the first condition is widely violated.

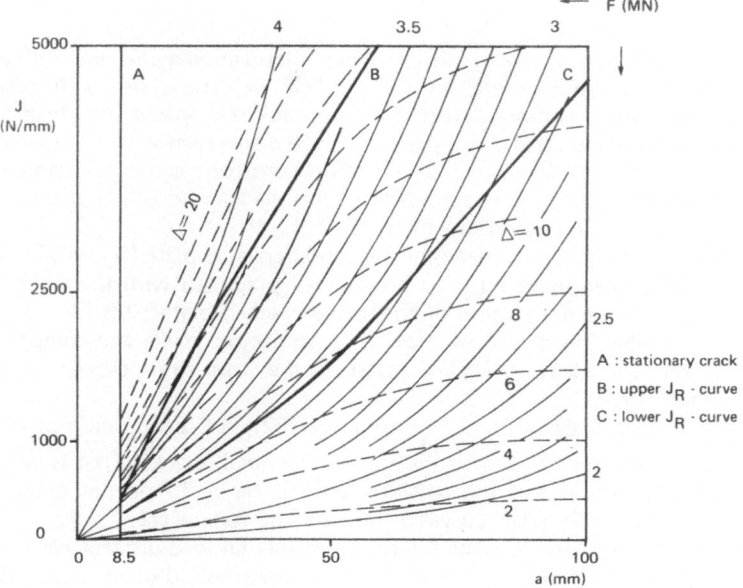

Figure 16 — Crack driving force diagram for a plane stress center-cracked plate
in tension (W = L = 250 mm, a_0 = 8.5 mm, B = 17.75 mm)

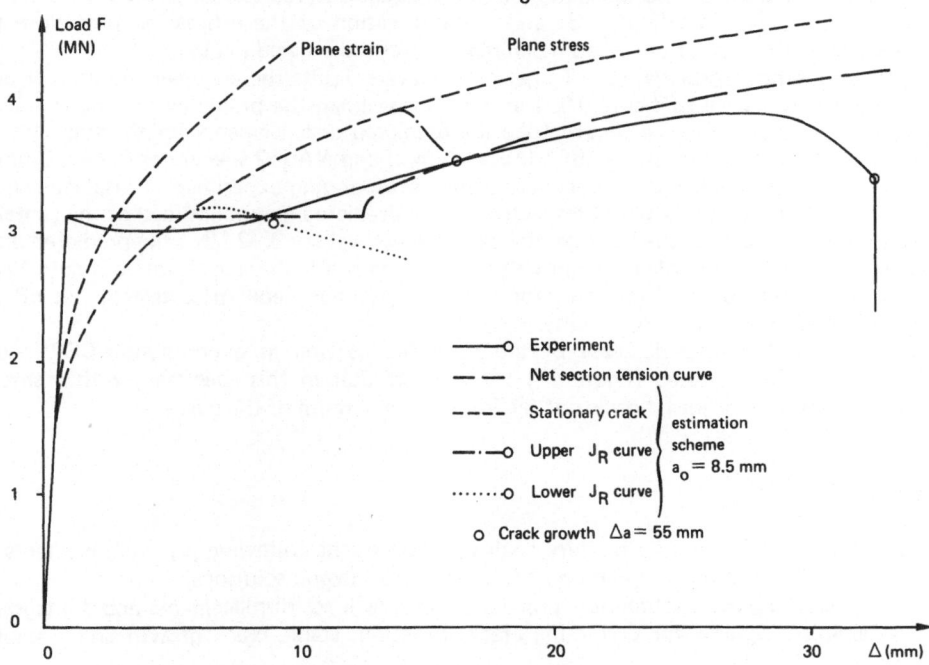

Figure 17 — Comparison of predicted load-displacement curves with experimental curves
of a center-cracked plate in tension (W = L = 250 mm, a_0 = 8.5 mm,
B = 17.75 mm)

3.5. Discussion

From the results of figure 15 and 17 it is obvious that a good agreement with experiments cannot be obtained for both geometries, CT and CCP, in plane strain conditions ; moreover the choice of plane stress is more natural for these very thin specimens. Unfortunately, for the CT specimen in plane stress, the predicted load-displacement curves deviate from the experimental ones at loads down to F = 400 kN (figure 15) ; as this deviation is also observed for the stationary crack, a probable cause of it is the bad modelling of the initial part of the stress-strain curve with a power-law fit (figure 12).

The same comment can be made for the center-cracked plate (figure 17). Actually the experimental load-displacement curve is pretty well predicted with the real tension curve (load-elongation) and the net section of the center-cracked plate : 2 B (W − a) = 2 x 17.75 x (250 − 8.5). Besides the bad power-law fit, two factors make the comparison worse :

— the power-law equation (16) is fitted on the stress-strain curve, not on the load-elongation tension curve,

— the limit load used in $\Delta^p_{nocrack}$, equation (37), is $P/2W\sigma_0$ and not $P/2 (W-a)\sigma_0$, and the crack contribution Δ_{crack} is insufficient for small cracks. That is why the stationary crack predicted curve and the tension curve of figure 17 do not come together at Δ = 30 mm ($\epsilon = \Delta/L$ = 6%) as the curves of figure 12 do for $\epsilon \cong 6\%$.

In conclusion, a better agreement with experimental load-displacement curves would be obtained with a power-law fit of the stress-strain curve limited to about $\epsilon < 5\%$: it would yield a higher value of the hardening exponent n. But the power-law fit should not depend on the plastic strain of the specimen, and of the size of the specimen, if the method aims at a general application. The difficulty of fitting equation (16) to real stress-strain curves is mentioned by AINSWORTH [13] and a modification of the estimation procedure is proposed. The influence of the α and n parameters is also discussed in [14].

Is the method conservative ? To get conservative results the engineer should a priori use the lower J_R curve of figure 10. For the CT specimen the predicted maximum load, about 600 kN, is conservative (figure 15) ; the predicted displacements for the final stable crack growths measured on the NA8F1 ($\Delta a \cong 70$ mm) and NA8F2 specimens ($\Delta a \cong 20$mm) are $\Delta_L \cong 25$ mm and 9 mm respectively ; the corresponding experimental displacements are $\Delta_L = 20.25$ mm and 12.5 mm respectively. The predicted maximum load of the center-cracked plate is 3.25 MN, smaller than the experimental value : 3.92 MN ; the predicted displacement for the final stable crack growth ($\Delta a = 55$ mm) is highly conservative (figure 17).

It is possible to conclude that, for the two specimen geometries tested, the EPRI method gives rather conservative results.

Note that the load decrease of the predicted as well as experimental CCP load-displacement curves is steep (figure 17) ; it suggests that in this specimen, with a small crack, the crack growth initiation is not far from the maximum of the curve.

4. CONCLUSION

The EPRI method for ductile fracture analysis employs the following principal elements :

— a handbook-style compilation of fully plastic J-integral solutions,

— an elastic-plastic estimation procedure for the load, displacements and J-integral,

— simple diagrams for predicting crack initiation, stable crack growth and instability.

The «crack driving force diagram» has been presented for the compact tension specimen. The verification of the method performed by its authors (General Electric [10]), is convincing, though incomplete, as far as the load-displacement behaviour is concerned, but insufficient to demonstrate the ability of the method to predict the crack growth behaviour in all situations.

The method was applied to tests on very large welded compact tension specimens and center-cracked plate of a C-Mn structural steel. The predicted load-displacement curves are in bad agreement with the experimental curves. This is due in part to the bad modelling of the stress-strain curve, that presents an initial plastic plateau, with the power-law curve used in the EPRI method. Nevertheless it is possible to conclude that, for the two geometries tested, the EPRI method gives rather conservative results for maximum load and stable crack growth.

The practical use of the method, in particular by nuclear engineers, needs it to be completed with solutions to *surface and through flaws in cylinders, and thermal stress* ; this work is in progress at General Electric. It should also take into account the *residual stresses*. With these improvements, the method would be a powerful tool in structural analysis, though full numerical calculations might be required under some circumstances.

ACKNOWLEDGEMENTS

The author is grateful to B. MARANDET and P. de ROO of IRSID for the authorization to use their experimental results on large welded plates.

REFERENCES
[1]   TADA, H., PARIS, P.C., IRWIN, G.R., «The stress analysis of cracks handbook», Del Research Corporation, Hellertown (Pennsylvania), 1973.

[2]   KANNINEN, M.F., POPELAR, C.H., BROEK, D., «A critical survey on the application of plastic fracture mechanics to nuclear pressure vessels and piping», *Nuclear Eng. & Design,* 67 (1981) 27-55.

[3]   *Advances in elasto-plastic fracture mechanics*, edited by L.H. LARSSON, Applied Science Publishers, London, 1980.

[4]   KUMAR, V., GERMAN, M.D., SHIH, C.F., «An engineering approach for elastic-plastic fracture analysis», Topical report NP-1931 on EPRI project 1237-1, July, 1981.

[5]   MARANDET, B., PHELIPPEAU, G., de ROO, P., ROUSSELIER, G., «Effect of specimen dimensions on J_{IC} at initiation of crack growth by ductile tearing», 15th ASTM Symposium on Fracture Mechanics, July 7-9, 1982, College Park, Maryland, USA.

[6]   BLOOM, J.M., MALIK, S.N., «Procedure for the assessment of the integrity of nuclear pressure vessels and piping containing defects», Final report NP-2431 on EPRI project 1237-2, June, 1982.

[7]   SHIH, C.F., GERMAN, M.D., «Requirements for a one parameter characterization of crack tip fields by the HRR singularity», *Int. J. of Fracture,* 17 (1981) 27-43.

[8]   PARIS, P.C., TADA, H., ZAHOOR, A., ERNST, H., «The theory of instability of the tearing mode of elastic-plastic crack growth», *ASTM Special Technical Publication 668*, 1979, 5-36.

[9]   ILYUSHIN, A.A., «The theory of small elastic-plastic deformations», *Prikadnaia Matematika i Mekhanika,* 10 (1946) 347.

[10]   SHIH, C.F., GERMAN, M.D., KUMAR, V., «An engineering approach for examining crack growth and stability in flawed structures», *Int. J. Pres. Ves. & Piping,* 9 (1981) 159-196.

[11]   MARANDET, B., de ROO, P., «Éléments d'estimation du risque de rupture d'un joint soudé», IRSID report RE 895, March, 1982.

[12]   ALBRECHT, P., ANDREWS, W.R., GUDAS, J.P., JOYCE, J.A., LOSS, F.J., McCABE, D.E., SCHMIDT, D.W., Van Der SLUYS, W.A., «Tentative test procedure for determining the plane strain J_I-R curve» (ASTM), *J. of testing and evaluation,* 10 (1982) 245-251.

[13] AINSWORTH, R.A., «The assessment of defects in structures of strain hardening material», *Eng. Fracture Mech.,* **19** (1984) 633-642.

[14] STEENKAMP, P., «Applicability of the EPRI estimation scheme for the prediction of load-load line displacement behaviour and for single specimen J—R—curve determination», *Int. J. of Fracture,* **23** (1983) R91—R100.

WORKSHOP 3.

THE COD METHOD FOR ASSESSING THE SIGNIFICANCE OF DEFECTS

C.E.Turner
Mechanical Engineering Department
Imperial College
London, SW7 2BX

ABSTRACT.   The COD method for assessment of the significance of
defects in weldments, BS:PD6493, is described and then applied to a
simple problem of a non-stress relieved butt weld in a two-dimensional
configuration. The sequence of analysis is first outlined for those
who may wish to attempt their own solution. The example is solved
both by following the standard method and by an alternative application
of the principles in which certain simplifications of PD6493 are not
made.   The problem is then extended to deeper cracks and to combined
tension and bending of a three-dimensional defect, both with and without residual stress, in order to illustrate their dominant role in such
problems, and also the judgements that may have to be made outside the
terms of the standard procedures.

INTRODUCTION

The published COD method (1) is entitled "Guidance on some Methods for
the Derivation of Acceptance Levels for Defects in Fusion Welded
Joints", known briefly as PD6493.
     It contains more than just an estimation procedure for avoidance
of brittle fracture. The contents are summarised below but only the
fracture aspects will be discussed here.

General scope of the COD method.

Ref. (1) includes a description of :
Types of defect; planar (cracks, lack of fusion or penetration);
undercut; non-planar (cavities and inclusions).
Modes of failure: brittle, fatigue, yielding, leakage of vessels,
erosion, corrosion, corrosion fatigue, stress corrosion, etc.)
Data: orientation, stress and temperature of defect; material properties
Detailed assessment for brittle fracture. This is outlined below.
Detailed assessment for fatigue; data required, probability of
survival, assessment of known defects, estimation of tolerable sizes
of defects.

L. H. Larsson (ed.), Elastic-Plastic Fracture Mechanics, 397–410.
© 1985 ECSC, EEC, EAEC, Brussels and Luxembourg.

Assessment of other modes of failure; yielding due to overloading of remaining section; leakage in containing equipment; environmental effects, buckling, creep.

The present description of fracture assessment methods thus covers only some one-third or so of the contents of (1).

A sequence of calculation is set out that cannot here be paraphrased. An idealisation of actual defect shape is described, and detailed definition is given of when adjacent defects should be treated as one, or near surface defects treated as surface breaking. These procedures are termed "recategorisation" and involve the primary membrane or bending stresses rather than secondary stress terms, and an average flow stress rather than first yield. For the assessment of known defects, LEFM is recommended if the maximum principal stress is less than the uniaxial yield stress. Linearisation of stress across a cracked section is permitted, provided the result is greater or equal to the uncracked body values over the crack surface itself. Residual stress is treated as a secondary stress and for as-welded components given the value of the yield stress of the material in which the defect lies. For stress relieved structures, a non-zero value should be estimated, if appropriate.

PRINCIPLES OF ESTIMATION USING PD6493

LEFM or COD?

In the COD method use of LEFM is permitted if both the sum of the applied stress components is less than the yield stress, σ_y, for parent and weld metal, and the estimated value of $K < 0.7 K_{Ic}$, implying thereby a factor of safety of about 2 on crack size. In (1) K is expressed as

$$K = \{\sigma_m M_m + \sigma_b M_b)/Q_o\}\sqrt{a} \qquad \text{Eqn.1}$$

where σ_m is membrane stress, σ_b is bending stress, M_m, M_b and Q_o are shape factors given graphically. In outline this is consistent with the ASME Pressure Vessel Code, although the notation differs in that Q_o in Eqn.1 is $\sqrt{Q/\pi}$ from ASME, and the values of M_b available when (1) was derived, were limited.

The COD method.

If the stresses exceed yield or valid K_{Ic} data are not available, then the COD analysis is used, the procedure depending on whether defect size is known or unknown. In both cases an effective crack length $\bar{a}$ is used that conceptually is defined by

$$Y^2 a = \pi \bar{a} \qquad \text{Eqn.2}$$

where

$$K = Y\sigma\sqrt{a} \qquad \text{Eqn.2a}$$

In reality, the terms a and $\bar{a}$ are related by diagrams such as Figs.la, b, in which the implied value of Y is not always the currently recognised best value. In particular, for a part.-through semi-elliptical crack in bending, the tensile value of Y is used, and that is often quite conservative. A correction has been suggested but neither it nor direct use of Eqn.2 has been incorporated into (1) as yet. If the real crack size is known (length, depth, distance below the surface, combination with adjacent defects and so on) then $\bar{a}$ is found from charts, such as Fig.la, b. If $\bar{a} < \bar{a}_m$, the defect is acceptable, where $\bar{a}_m$ is the tolerable defect defined, if LEFM can be used, as

$$\bar{a}_m = C(K_{Ic}/\sigma_y)^2 \qquad\qquad \text{Eqn.3a}$$

or, if COD is to be used, as

$$\bar{a}_m = C(\delta_{crit}/\varepsilon_y) \qquad\qquad \text{Eqn.3b}$$

where ε_y is the yield strain, σ_y/E

For curved shells a correction is applied to $\bar{a}_m$ from tabulated values based on the well-known Folias bulging factor (3).

Relating COD to stress or strain. In principle, the basic equations for assessing the severity of the applied crack opening in steel are functions of the strain, ε, that would exist at the position of the defect is the defect were not there;

$$\Phi = \delta/2\pi\varepsilon_y\bar{a} = (\varepsilon/\varepsilon_y)^2 \quad \text{for } (\varepsilon/\varepsilon_y) < 0.5 \qquad \text{Eqn.4a}$$

$$\Phi = \delta/2\pi\varepsilon_y\bar{a} = (\varepsilon/\varepsilon_y) - 0.25 \text{ for } (\varepsilon/\varepsilon_y) > 0.5 \qquad 4b$$

In (1) these relations are expressed in terms of

$$C = 1/2\pi\Phi \text{ or } \delta/\varepsilon_y\bar{a} \text{ or } \delta E/\sigma_y\bar{a} \qquad\qquad \text{Eqn.5}$$

If the sum of the applied principal stresses, including thermal but excluding residual, does not exceed $2\sigma_y$, then the normalised strain, $\varepsilon/\varepsilon_y$, is replaced by stress, σ/σ_y, and Eqns.4a, b, are inverted to give

$$C = 1/2\pi (\sigma/\sigma_y)^2 \qquad\qquad \sigma/\sigma_y < 0.5 \qquad \text{Eqn.6a}$$

$$C = 1/[2\pi\{(\varepsilon/\varepsilon_y) - 0.25\}] \quad \varepsilon/\varepsilon_y \equiv \sigma/\sigma_y > 0.5 \qquad b$$

Ref.1 limits Eqn.6b (and by implication Eqn.4b) to ferritic steel and for other materials uses

$$C = 1/2\pi (\varepsilon/\varepsilon_y)^2 \qquad\qquad \varepsilon/\varepsilon_y \equiv \sigma/\sigma_y > 0.5 \qquad c$$

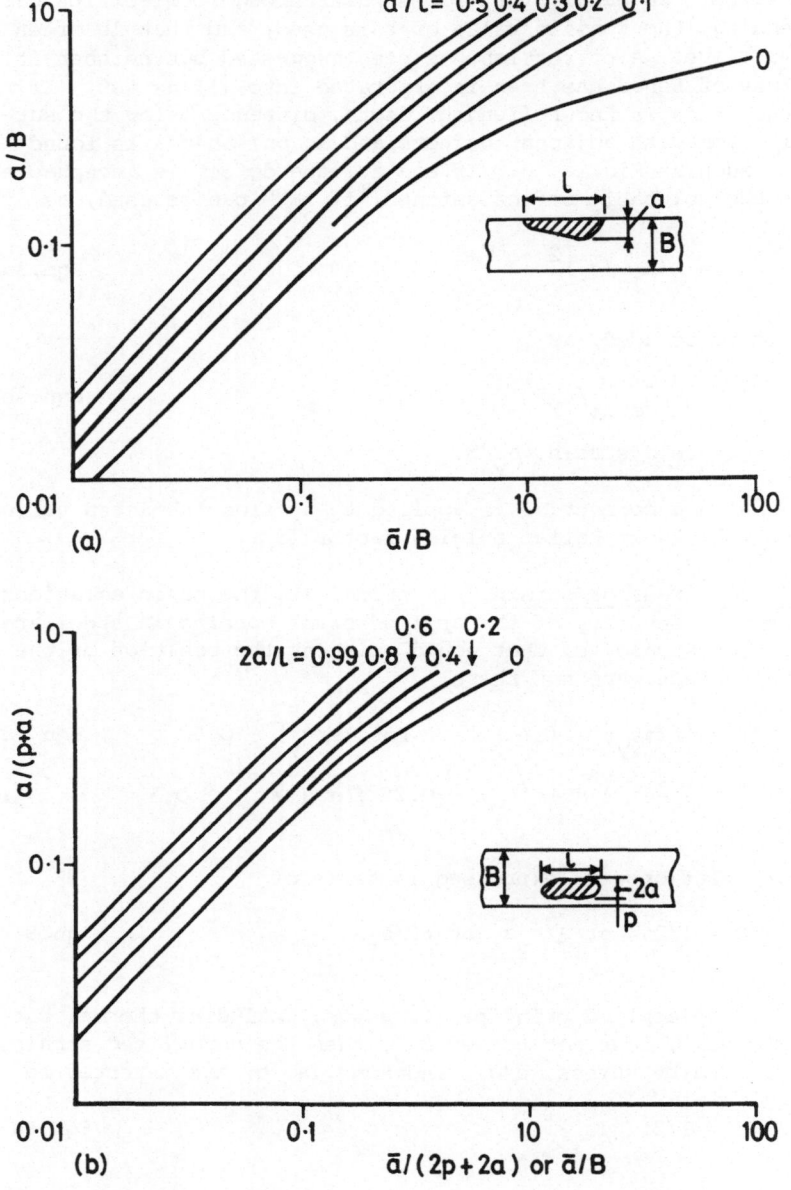

Fig. 1. Equivalent crack sizes for PD6493: the relationship
between actual surface defect dimensions and the parameter a;
    a) for surface defect; b) for embedded defects.
(These diagrams are based on Ref.1. Note the length of defect, 1,
is 2c in the present text).

An equivalent expression in terms of ϕ is just Eqn.4a used without the limitation to $\varepsilon/\varepsilon_y$ < 0.5. These equations are presented (1) on a logarithmic graph, as shown Fig.2, where σ is the sum of P_m, the primary membrane stress, P_b, the primary bending stress, Q, a secondary stress, and F, a peak stress at a stress concentration. If the stresses (excluding residual) exceed $2\sigma_y$, a full elastic-plastic analysis should be made to find the strain ε. F is written by use of the additive peak term, F = peak stress = σ_{nom} (SCF-1), where SCF is the conventional elastic stress concentration factor. For fillet welds, the concentration at the toe is taken to reduce linearly from 3 to unity at a depth of .15B, below which depth stress concentration at the toe is ignored.

If toughness is known . To estimate tolerable sizes of planar defect, a similar analysis is used in reverse, with some caution on assumed levels of stress and strength such as the uncertainties of weld offset, ovality of vessels, differences between test pieces and actual weldments, etc. etc. If the sum of stresses does not exceed $2\sigma_y$, then $\bar{a}_m$ is determined from Eqn.6b and 3b. If the sum is less than σ_y, then LEFM can be used, Eqn.6a, 3a. Where the sum of the stresses (excluding residual) exceeds $2\sigma_y$, plastic strains should be computed for the uncracked body. With $\bar{a}_m$ known, this sets the upper limit to $\bar{a}$ which is then converted to values of true size for through, surface, or embedded defects, using such as Fig.1. Extensive definitions of the relationship between $\bar{a}$ and actual dimensions for the various cases are tabulated. For example, for an embedded or surface defect, recategorised because of high stress in the remaining ligament, $\bar{a}$ = c+a; $[(\ell/2)$ + t in (1)] where 2c x 2a is the size of the enveloping rectangle around one (or several adjacent) defects, $2\bar{a}$ [t in (1)] being in the thickness direction. For through defects, $\bar{a}$ = c $[\ell/2$ in (1)]. For defects at a hole, radius R, with SCF = 3, if crack length $\bar{a}$ < 0.15R, then $\bar{a}$ = (2R+a)/2, or for two diametrically opposite cracks, $\bar{a}$ = R+a.

In converting actual crack configurations to the equivalent crack, $\bar{a}$, Fig.1 is used for all types of loading, whereas tensile values of shape factors are implied in its derivation, as already noted. For through cracks, use of the wide plate relationship , $\bar{a}$ = c, $[\ell/2$ in (1)] K = $\sigma\sqrt{\pi a}$, is optional for LEFM cases but specified for COD cases. This will also lead to differences with other methods that could readily be avoided at the expense of a small further effort by use of the well-known appropriate shape factors. Of course, such through cracks are more often found in test pieces than in actual welded structures.

Recategorisation. Checks for yielding are specified in both Sections 2 and 4 of (1), the former for the cracked section, the latter as a general reminder even when no crack is present. If a crack or group of defects encompassed by the rectangle 2a x 2c (2a in the thickness direction) is such that after allowance for a through defect of length 2c, the remaining cross section would yield (or bulge for a vessel with 2a > $0.1\sqrt{RB}$), then a procedure is entered that may lead to recategorisation of a part through defect as a through defect on the basis of yield of that remaining ligament, using $\bar{\sigma}$ = $(\sigma_y+\sigma_u)/2$. The initial

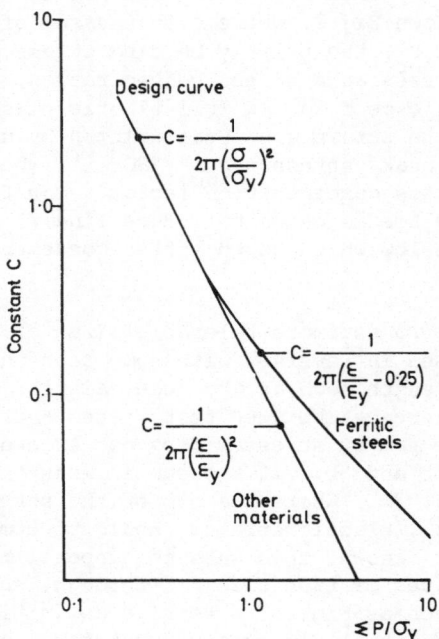

Fig.2. Values of constant C for different loading conditions
defined by the sum of all the stress components, P.
(Note scales are logarithmic).

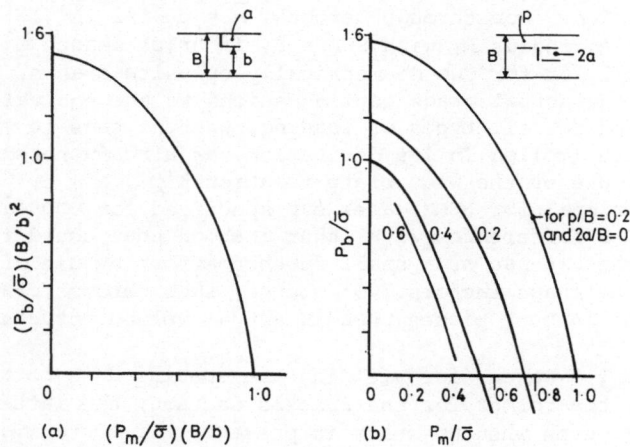

Fig.3. Recategorisation procedure for ligament yield under combined tension & bending; stress states that fall outside
the relevant diagram are recategorised.
   a) surface cracks;   b) embedded cracks.
(Note in (1) separate diagrams are given for different
values of p/B = 0, 0.2 and 0.4).

check is for collapse or bulging of the final structure containing a
(notional) through crack of size 2a and the criterion for recategorisation is the yielding of the ligament (or two, for the buried crack)
adjacent to the defect under consideration. No explicit analysis is
made of the toughness required in the yielding ligament. To assist
recategorisation, diagrams are given, shown schematically, Fig.3, to
allow for the combined effects of bending and tension, crack size to
section thickness and, for embedded defects, of eccentricity of defect.

Material properties

The normalising terms σ_y, E and ε_y (=σ_y/E) are the conventional uniaxial values of yield stress, modulus and yield strain.

Non-ferritic materials.   For non-ferritic materials, Eqn.6c is used
instead of Eqn.6b. By implication such materials are deemed to have
a much higher or much lower degree of work-hardening than the structural C-Mn steels which are the basis of Eqn.6b. That basis is, of
course, experimental, but the argument is that for full hardening the
LEFM relationship, parabolic with strain, would be continued, whilst
for non-hardening (be it a yield plateau or a high strength material)
the linear slopes predicted by (4) occur, whereas the basis of the
COD diagram is an appreciably lesser slope found as relevant by experiment on shallow notch (e.g. a/W < 0.1) test pieces of mildly
hardening material in which yield of the whole body (i.e. the gross
section) occurs. The predictions (4) apply mainly to deep notch
configurations where only yield of the ligament occurs. Note the
change to Eqn.6c is by material, rather than by an explicit degree of
hardening.

Evaluation of toughness. The toughness test"should be made with the
orientation of the crack in the test piece similar to the defect in
the structure and should take account of thickness, welding, temperature and rate of loading in service." The methods are standard K_{Ic}
procedures or the COD procedure which stipulates that the standard
test piece should be of thickness equal to the material under consideration. Alternative assessments of δ_{crit} are possible in (5)
according to the type of test record, δ_i at crack initiation, if
known; δ_c, δ_u or δ_m, as outlined in the Appendix. Where slow stable
crack growth occurs, the selected toughness may, by agreement, be at
onset of maximum load, δ_m, with no direct knowledge of where initiation takes place.

EXAMPLE PROBLEMS

A Welded Tension Member.

The basic problem is taken from Smith et al (6) for a weld detail
on the deck of an offshore platform under repeated tension. The
material is a structural steel, σ_y = 340 MPa, E = 205 GPa, 40mm

thick. The tensile strength is here taken as $1.4\sigma_y$, The fatigue aspects of the problem are not pursued here, so that a crack at the toe of the weld (see sketch) is considered when grown to a length of 5.75mm as already estimated (6) for the fatigue problem. The applied tensile stress is 117 MPa. The weld is not stress relieved; the mechanical properties of the weld are not stated. The objective is to find the minimum required toughness of the plate or weld metal. The treatment in (6) is two-dimensional (2D) implying the crack exists all across the (unstated) width of the member.

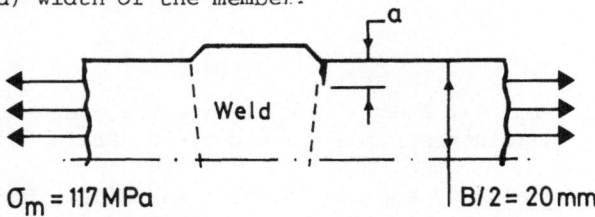

Fig.4. A toe-crack in a butt-weld.

For the benefit of any who wish to work out their own solution, the sequence used for the solution given here is first set out, in terms of the methods and data of PD6493 (1). Following the actual solution, some alternatives are shown, based on the same principles but not using some of the steps in (1).

Outline of solution.

   a) All the stress systems acting, mechanical, thermal (if any)
       and residual must be identified.
   b) Their values are then examined to see whether an LEFM or
       COD procedure is the more appropriate and if the latter,
       whether the sum of the stresses (excluding residual) is
       less than $2\sigma_y$, so that in Eqn.6 normalised strain is in-
       terpreted as normalised stress.
   c) The appropriate equation, Eqn.6a, b, or c, is selected to
       evaluate C, or Fig.2 (Fig. 14 of Ref.1) can be used directly.
   d) The effective defect size must be evaluated in terms of the
       actual defect, using the appropriate diagram, here Fig.1a
       (Fig.12 of Ref.1) with due regard to questions of recate-
       gorisation (Fig.3) or bulging of a pressure vessel where
       such steps are necessary. This defect, if just tolerable,
       is denoted $\bar{a}_m$.
   e) The tolerable defect size $\bar{a}_m$ and C are related by Eqn.3b
       (or Eqn.3a if an LEFM solution were being used) to give the
       critical toughness δ_{crit} (or K_{Ic} for LEFM).
   f) The critical COD value, δ_{crit}, must be interpreted with
       reference to the standard test procedure (5), see Appendix 1.

Conventional solution (using PD6493). The applied stresses are a

primary membrane stress of 117 MPa and perhaps a small peak stress from the weld toe concentration. There is no bending applied to the un-cracked body. A residual stress of yield stress magnitude σ_r=340MPa is assumed for an un-stress relieved weld in the absence of more specific data. The sum of the stresses (including residual) is clearly more than σ_y so that the analysis is conducted in terms of COD; the sum (excluding residual) is less than $2\sigma_y$, so that the COD analysis is conducted in terms of stress rather than strain.

In (1) a procedure is given for cracks at the toe of a fillet weld, as already outlined above in the Principles of Estimation. No remarks are made for the case discussed (6) of a butt weld. If, nevertheless, the same values were used, the stress concentration allowance would operate for cracks up to .15B in depth, i.e. 6mm. Thus, for the given crack, the SCF, $k = 1 + \{2(.15B-a)/.15B\} = 1.083$, so that the peak stress would only be $(0.083)(117) \approx 10$ MPa.

The mechanical stress (here membrane plus peak) and residual stress values are added and normalised to σ_y;

$$\varepsilon/\varepsilon_y \equiv \sigma/\sigma_y = (117 + 10 + 340)/340 = 1.374 \qquad (7)$$

Since this is greater than 0.5, the second COD equation, Eqn.6b, is used(for structural steel), or the upper curve of Fig.2 can be used directly.

$$\therefore \quad C = 1/2\pi\left[(\varepsilon/\varepsilon_y) - 0.25\right] = 1/2\pi(1.124) = 0.142 \qquad (8)$$

The actual and effective crack sizes are related using Fig.1a, with $a/2c\left[(t/\ell \text{ in } (1)\right] \rightarrow 0$ for a two-dimensional treatment. Entering at a/B $\left[(t/e \text{ in } (1)\right] = 5.751 / 40 = 0.144$, then $\bar{a}/B\left[(\bar{a}/e \text{ in } (1)\right] = 0.24$ whence

$$\bar{a} = 9.6mm \qquad (9)$$

Note this method makes no explicit use of the LEFM shape factor, Y, since it is implied by the curves of Fig.1. No distinction can there-fore be made between the nature of membrane, bending, peak or resi-dual stress, all of which are treated as if the various components of σ are uniform over the crack depth with the particular factor implied in Fig.1.

The stress level as implied by C and the effective crack size $\bar{a}$ are then related by Eqn.3b. If $\bar{a}$ is interpreted as the maximum crack size $\bar{a}_m$, then the value found for COD, δ, will of course be the value δ_{crit}

$$\delta_{crit} = 9.6(340)/0.142(205 \times 10^3) = 0.113mm \qquad (10)$$

This is the minimum toughness required for the defect to be tolerable. The toughness should be measured according to (5) on a full thickness piece but the value may refer to a post-initiation point in the test, as outlined Appendix 1, according to the type of test record obtained.

Alternative solution. This problem is now re-worked using the same principles but using Eqn.2 instead of Fig.1a. A solution for the LEFM shape factor Y is required. One simple model would be to treat the component as an SEN tension piece, $a/W = 0.144$, or use Eqn.1 with $M_m = 1.24$; $Q_o = 0.564$ (where values are given (1) for use in an LEFM solution). This gives Y = 2.21 but neglects the stress concentration effect other than by including the peak stress in the summation of the σ terms. The value of Y implied by Fig.1a is Y = 2.3. An approximate weight function solution to include the stress concentration (not detailed here) gives Y = 2.44. For this latter case the peak stress is not included in the summation of σ since the effect of the concentration is included in the Y factor. Thus,

$$\sigma/\sigma_y = 1.344; \quad C = 0.145 \text{ (or } \phi = 1.094); \quad \delta_{crit} = 0.120 \text{mm} \quad (11)$$

In (6) a finite element solution gave Y=2.5, so δ_{crit} = 0.13mm. The difference betwen use of Fig.1 or Eqn.2 is greater for cracks subjected to bending and the difference between the treatment of stress concentration cases would be more appreciable if the peak stress were greater. Note the possibility exists when using Eqn.2 to use a different shape factor for the mechanical and residual patterns. This point is examined in the following paragraph.

An Extension of the Problem. The problem is now extended beyond that of Smith et al (6) to illustrate certain other features. Suppose the crack has grown to a depth of 14mm, i.e. a/B = 0.35. The effect of the stress concentration is not then relevant. The mechanical stress remains σ = 117MPa = 0.344σ_y, but an engineering judgement is required on two related aspects: is the member deforming with "ends parallel" or in "pin loaded" style and what residual stress is appropriate? The mechanical loading is considered first.

End Load Conditions. If "pin loading" is assumed an estimate of limit state must allow for the eccentricity of loading with respect to the ligament, so that collapse is governed as much by bending as by tension. An estimate (here based on Fig.A.2.1 of R-6) suggests limit state at σ/σ_{fl} = 0.38, i.e. a margin of some 35% so that collapse due to overload is not an immediate risk, up to σ/σ_y = 0.46. Using Fig.1a with a/2c (t/ℓ in (1)] $\rightarrow$ o and a/B (t/ℓ in (1)]$_y$ = 0.35, gives $\bar{a}$/B [$\bar{a}$/e in (1)] = 1.25. The corresponding value of $\bar{a}$ = 50mm is greater than B, but as far as the writer is aware, this is acceptable within the method. Using Eqn.4a for Φ or 6a (or Fig.2) for C, gives Φ = $(0.344)^2$ = 0.118, or C = 1.35, whence from Eqn.3b δ_{crit} = $\Phi(2\pi\epsilon_y \bar{a})$ = $\epsilon_y \bar{a}$/C = 0.0625mm. The value of $\bar{a}$ and the remainder of the calculation agrees with that found using Eqn.2, instead of Fig.1a, with Y in Eqn. 2 $\simeq$ 3.3 appropriate to the "pin loaded" SEN tension, which solution allows for the significant degree of bending implied.

Assumptions for Residual Stresses. If the member were held sufficiently rigid by its surrounding structure that the "ends parallel" mode is judged more relevant, σ_{net} = 0.334σ_y/0.65 = 0.53σ_y, so that net section yield is not approached. A lower value of Y is appropriate, perhaps about 2.5 (there seem few reliable data for this case, Fig.55 of (7)), leading to $\bar{a}$ = 28mm and δ_{crit} 0.035mm. Using (1), however, Fig.1a is still appropriate, there being no distinction between the two cases of "pin loaded" or "ends parallel".

For the previous discussion residual stress was neglected. In this modified problem with a deeper crack, the acceptance of the same tensile shape factor for the residual stress, as for the mechanical stress, is open to question. Unless a reaction load is induced, the residual stress must be in equilibrium, and so cannot be purely tensile across the whole section, although it could be up to a depth of say, B/4. If the residual stress is included as 340MPa tension, at least over the depth of the crack and (1) is used, then $\varepsilon/\varepsilon_y$ = 1.344 and Φ = 1.094, as for the first case, except for loss of the small concentration effect, but now $\bar{a}$ = 50mm instead of 9.6mm. Thus, for the deeper crack, δ_c = 0.552mm. In principle, it is possible to separate any tensile and bending components, using Eqn.1, so that, if thought more appropriate, the residual stress could be treated as reducing linearly across the section or following any other pattern thought to be realistic. Use of Fig.1a prevents this, so that the Alternative Solution method involving Eqn.2 has to be employed. For a/B = 0.35, using Figs. 9 & 11 of (1), M_m = 1.96, M_b = 1.18, Q = 0.564, so that if the residual stress is treated as tensile, $K = \{[(340+117) (1.96) + O (1.18)]/0.564\}\sqrt{a}$ = (1587 + O)$\sqrt{a}$ = 1587$\sqrt{a}$. If residual stress is taken as bending, $K = \{[(117(1.96)+340 (1.18)]/0.564\}\sqrt{a}$ = (406 + 712)$\sqrt{a}$ = 1118$\sqrt{a}$. If this is written K = $Y\sigma_{total}\sqrt{a}$ with σ_{total} = $\sigma_m + \sigma_r$ = 457MPa , then in the first case Y (with σ_r tensile) = 3.47, and in the second, Y (with σ_r bending) = 2.45, whence $\bar{a}$ = 54mm and $\bar{a}$ = 26.7mm respectively from Eqn.2 and δ_{crit} = 0.61mm or 0.301mm respectively from Eqn.3b. The former value, δ_{crit} = 0.61mm, corresponds to the value δ_{crit} = 0.552mm found using (1) for this modified problem, the difference arising from the use of Fig.1 or Eqns. 1 & 2, but the final value δ_{crit} = 0.301mm relating to use of a different distribution of mechanical and residual stress across the section has no counterpart in the COD method of (1), although it is acceptable in (1) if the problem is within the scope of LEFM.

A Three-Dimensional Crack. In yet another version of the problem, the surface length of the crack, 2c, might be limited, so that a semi-elliptical part through crack of proportions 2c x a (ℓ x t in (1)) would be treated. As an example, suppose 2c = 5a (or 2c could be taken as a fixed value, such as 20mm, to be found by NDE). For a deep crack the question of recategorisation as a 'through' crack arises. In the problem as originally specified there is no bending stress on the uncracked section, and whether or not bending stress is induced in the cracked section depends upon the nature of the loading, as just discussed. The present crack with its tensile load alone would not need recategorisation according to Fig.3a(Fig.6 of (1)) because $(\sigma_m/\sigma_y)(B/b)$= 0.53 (where the membrance stress $\sigma_m \equiv$ 117MPa) and σ_b = 0. This is obviously inside the permitted region, Fig.3. Secondary bending action, if any, is not included in the recategorisation process because it does not contribute to collapse. For the sake of illustration, a bending stress of 180MPa is superposed as an applied moment additional to the tensile load. In the recategorisation process hardening is allowed for. Using $\sigma_u = 1.4\sigma_y$, then $\sigma_{fl} = (\sigma_y+\sigma_u)/2 = 1.2\sigma_y$. The abscissa is reduced to .53/1.2 = .46 and the ordinate is $[(180)/1.2(340)] (40/26)^2$ =

1.04. The point (.46, 1.04) falls just within the boundary of Fig.3, so there is no need to recategorise. If the crack size or stress level had been larger so that recategorisation were necessary, a through crack of length 2c would be reconsidered.

For the three-dimensional (3D) cases the equivalent crack size is read from Fig.1a for a/B $[t/e]$ = 0.35 and $a/2c$ $[t/\ell]$ = 0.2 to give $\bar{a}/B$ $[\bar{a}/e]$ = 0.37, i.e. $\bar{a}$ = 14.8mm. As already noted, the bending or tensile nature of the applied stress does not enter into the relation between a and $\bar{a}$ in (1), except for an LEFM usage. The combined stress is now $\sigma_m + \sigma_b$ = 297 MPa (if residual is not involved) or $\sigma_m + \sigma_b + \sigma_r$ if it remains relevant, i.e. 637MPa, so that Eqn.4b (or 6b) is entered at σ/σ_y = 0.873 if no residual stress, to give Φ = 0.623 (or C = 0.25), or σ/σ_y = 1.873 if residual stress is retained, to give Φ = 1.623 (or C = 0.98). Using Eqn.3b, δ_c = 0.096 mm without residual stress or δ_c = 0.249mm, with tensile residual stress. If, as in the second part of the example, there were neither bending nor residual stress, then, as there found, σ/σ_y = 0.344, Φ = 0.118 (or C = 1.35) but $\bar{a}$ is now only 14.8mm for the 3D case, instead of 50mm, as it was for the 2D, so that δ_c = 0.019 mm. If the tension and bending stresses were treated (for no residual stress) using Eqn.2, with appropriate LEFM shape factors from (7), instead of using Fig.1a, then $\bar{a}$ = 9.7mm, whence δ_{crit} = 0.0635mm.

TABLE 1:   Summary of Related Results

Crack size mm	Shape †	Stresses: MPa Primary ∅		Residual *	Result δ_{crit} mm	Method
		σ_m	σ_b	σ_r		
a) 5.75	2D	117 pl	0	340 t	0.113	Ref.1
		pl			0.120	Eqn.2
b) 14	2D	117 pl		0	0.063	Ref.1
		ep			0.035	Eqn.2
c) 14	2D	117 pl	0	340 t	0.552	Ref.1
		pl		340 b	0.301	Eqn.2
d) 14 x 70	3D	117	0		0.019	Ref.1
e) 14 x 70	3D	117	180	0	0.096	Ref.1
					0.063	Eqn.2
f) 14 x 70	3D	117	180	340 t	0.249	Ref.1

* Assumed pattern t, tensile; b, bending
∅ Assumed end conditions, pl = pin-loaded as in all Ref.1 solutions;
ep = ends parallel.

† 2D = line crack across the plate

3D = semi-elliptical part-through crack, aspect ratio 2c/a = 5/1

DISCUSSION

In the Summary Table, the differences between d), e) and f), and a), b) and c) reflect the 3D as opposed to 2D nature of the problem, d) and b) being directly comparable. The difference between b) and a) or between e) and f) reflect the dominant role of residual stress over mechanical stresses of (roughly) half yield magnitude. The two results under b) reflect the engineering judgement of use of pin-loaded or ends parallel for the distribution of stress in the structure as a crack spreads. The two results under (c) reflect the engineering judgement for the distribution of residual stress (whilst also containing the differences as in a)). The two results under e) reflect the conservatism in (1) in not using shape factors now known to be appropriate for semi-elliptical cracks under bending.

Reference has been made several times to the use of the same shape factor for different distributions of applied or residual stress. Indeed, it is the use of the tensile Y value previously noted as conservative that here gives a safe answer for the 2D problem, if indeed, the member is pin-loaded, since it is the conventional two-dimensional SEN value of Y that accommodates the bending effect, at least in the LEFM range. This problem is only acute for $a/2c \to 0$ (or $2c \to W$) and a/B large, where structural bending is induced without a moment being applied to the uncracked body. for multiple cracks, the rules in PD6493 for deriving an equivalent crack, which might, for example, embrace two adjacent cracks, must be recalled. This sort of rule must be included in any "design procedure" as distinct from "estimated procedure", but has not been examined here.

The acceptance of a post-initiation value of toughness under suitable conditions in (1) without a detailed R-curve analysis, has already been noted. It is not easy to give simple rules when such a value is acceptable for structural steels in or near the transition region but it must be recalled that (say) a maximum load value of toughness is associated with the often mentioned conservatism of the COD procedure, as set out in (1). In short (1) is a "package deal" justified by experience, which alterations, however logical, undermine unless separately verified.

REFERENCES

1) 'Guidance on some methods for the derivation of acceptance levels for defects in fusion welded joints.' PD6493, 1980, BSI (London).

2) Kamath, M.S. 'The crack tip opening (CTOD) design curve: some prospects for incorporating stress gradient effects.' Fitness for Purpose Validation of Welded Construction. Welding Institute, Cambridge, Paper 23.

3) Folias, E.S. 'The effect of initial curvature on cracked sheets,' Eng. Fract. Mech. 2, 1970, pp. 151-169.

4) Bucci, R.J. et al. 'J-integral estimation procedures.' Fracture Toughness, ASTM STP514, 1972, pp.40-69.

5) 'Methods for crack opening displacement (COD) testing'. BS5762, 1979, BSI (London).

6) Smith, I. et al. 'Assessment of fracture toughness requirements for the deck structure of the Hutton Tension Leg Platform, using finite elements and the CTOD design curve.' Offshore Tech. Conf., Texas, 1982.

7) Rooke, D.D. & Cartwright, D.C. 'Compendium of stress intensity factors'. HMSO (London) 1979.

8) Towers, O.L. & Garwood, S.J.. 1980. 'The geometry-dependence and significance of maximum load toughness values'. ECF-3, Ed. Radon, J.C., Pergamon, pp.57-68.

9) Towers, O.L. & Garwood, S.J. (1980). 'Maximum load toughness', Int. Jour.Fract. 16, pp.R85-R91.

APPENDIX 1

In the COD test method (5), the critical value of toughness, δ_{crit}, may refer to initiation of cracking or tearing or to a later stage of the test according to the nature of the test record, illustrated schematically, Fig.5. A resistance curve expressed in terms of COD ν Δa can be obtained if desired, for the purpose of defining the COD at initiation, δ_i, or a value of δ at some specific amount of crack growth. However, the use of a post-initiation value of toughness is "by agreement" between user and supplier, not usually related to R-curve data. It must be recalled that (5) normally requires test pieces of full thickness. Some justification for use of post-initiation data without R-curve or instability analysis is offered Refs. 8 and 9.

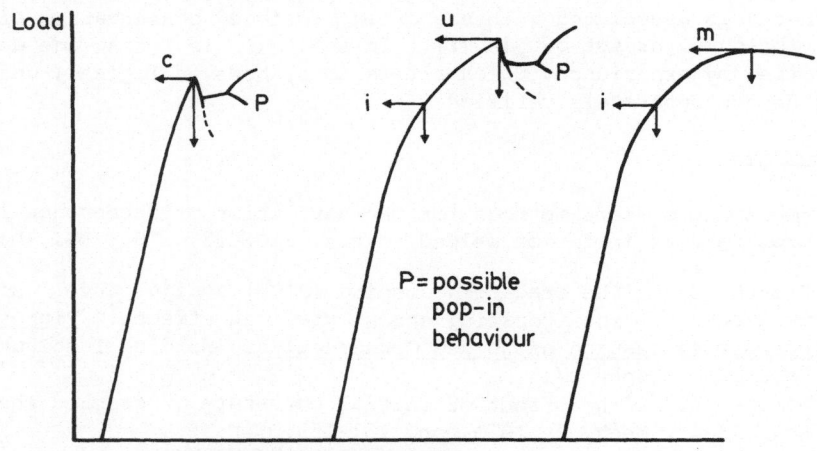

Fig.5. Values of δ_{crit} may be based on δ_c, δ_u, δ_m, according to the test record, or on δ_i if initiation is determined.

WORKSHOP 4
A J-BASED FRACTURE SAFE ESTIMATION PROCEDURE, EnJ., WITH
APPLICATIONS INCLUDING ESTIMATION OF THE MAXIMUM LOAD FOR
DUCTILE TEARING.

C.E. Turner
Mechanical Engineering Department
Imperial College of Science and Technology
London SW7 2BX

ABSTRACT.   A simple method, called EnJ, to denote the engineering use
of J for the assessment of defects, is described and then applied to
example problems. The method is broadly comparable to the well known
COD and $R-6$ methods but, contrary to the recent $EPRI$ method, retains
one diagram for all estimates of J. Three cases taken from the
previous Workshops are solved for purposes of comparison, one a
pressure vessel in the near $LEFM$ regime already trated by $R-6$, one a
butt weld with high residual stresses already treated by COD and
finally an estimate of maximum tearing load in a compact tension test
piece already treated by the $EPRI$ method. The results provide an
alternative treatment that in each case seems as satisfactory as the
original solutions.

INTRODUCTION

Several attempts have been made to define a quick, simple, yet
realistic route for assessing the significance of defects where $LEFM$
seems inadequate because of either the ductility of the material or the
presence of stresses near yield level. The most used methods, at
least in the U.K., have been the COD method (1), and the so-called $R-6$
method (2); more recently a J-based method, called EnJ, has been presented, (3), (4). A much more detailed method, derived under the
auspices of the Electric Power Research Institute, has also been published (5) with a simplified version (6) that is in essence a modified
form of $R-6$. The $EPRI$ method is particularised towards nuclear grade
pressure vessels and the simplified version (6), is restricted to
certain grades of steel over 102mm thick. The general background of
$EPFM$ relevant to all these studies is given in books such as (7) (8)
(9) and others. These methods are by no means as dissimilar as they
may at one time have seemed, and to emphasise the broad similarities
certain leading features of EnJ are first related to all the other
methods.
     The EnJ method uses J, as do (5) (6) and (2) whereas (1) uses COD;
it uses a 'simple design curve' estimate as do (1) (2) and (6) whereas

L. H. Larsson (ed.), Elastic-Plastic Fracture Mechanics, 411–426.
© *1985 ECSC, EEC, EAEC, Brussels and Luxembourg.*

(5) **has a** different curve for each configuration and material; it does not use plastic limit load as an inherent part of the procedure, as do (2) and (6), but treats collapse as a separate study, as does (1); it includes residual stress on the fracture ordinate as do (2) and (6) rather than on the applied stress abscissa as in (1). The ordinate of the diagram, Fig. 1, is the normalised term J/G_y where G_y is the value of G at $\sigma = \sigma_y$, i.e.

$$J/G_y = JE/Y^2\sigma_y^2\, a \qquad\qquad \text{Eqn. 1}$$

where σ_y is the yield stress, E is the modulus, a is crack size and Y is the LEFM shape factor defined by $K = Y\sigma\sqrt{a}$. The ordinate of the R-6 (2) and EPRI (6) methods is $\sqrt{G/J}$, which can be obtained by dividing the EnJ ordinate by $(\sigma/\sigma_y)^2$, inverting and square rooting. The ordinate of the COD method (Workshop 3, Eqn. 4) is $\phi = \delta/2\pi\varepsilon_y a$. Using the effective crack concept (Workshop 3, Eqn 2), $\pi\bar{a} = Y^2 a$, and also writing $J = m\sigma_y\delta$ where m is a constant, $1 \leqslant m \leqslant 3$, then the ordinates of PD 6493 (1) and EnJ are related by

$$J/G_y = 2m\phi \qquad\qquad \text{Eqn. 1a}$$

The abscissa of the EnJ diagram is notionally normalised strain, following the COD usage, whereas R-6 and EPRI methods use a normalised load Q/Q_c, where Q_c is a 'collapse load' local to the remaining ligament. Both COD and EnJ accept that in the near LEFM regime, the uncracked body stress, σ/σ_y, which is often the only known design term, shall be used for an estimate of the abscissa. Whilst recognising that an analysis of the forces in a cracked body is not normally made, an attempt is made in EnJ to estimate an effective applied strain, once the ligament has yielded.

In all the methods it is intended that J (or δ) is estimated and then restricted to a critical value, the fracture toughness, J_{crit} (δ_{crit}) although the practical meaning of the term will be discussed later.

THE EnJ METHOD

Estimation of J

The relation between J and applied loading is expressed through an effective strain ratio $\varepsilon_f/\varepsilon_y$. It is defined more fully in the next Section but for the moment can be taken as just the applied stress ratio, σ/σ_y.

For $\sigma/\sigma_y \leqslant 1.2$; $\qquad J/G_y = (\varepsilon_f/\varepsilon_y)^2 \, (1 + 0.5(\varepsilon/\varepsilon_y)^2)$ $\qquad$ Eqn.2a

For $\sigma/\sigma_y > 1.2$; $\qquad J/G_y = 2.5((\varepsilon_f/\varepsilon_y) - 0.2)$ $\qquad\qquad$ Eqn.2b

If $\sigma/\sigma_y > b/W$ or $Y/\sqrt{\pi} > W/b$ then plastic collapse must be
$\qquad\qquad\qquad\qquad\qquad\qquad$ checked explicitly $\qquad$ Eqn. 2c

Note: it is to be understood that the ligament ratio b/W in two dimensions also implies b/B in three dimensions, where W is width and B is thickness.

When Eqn. 2c calls for assessment of collapse, the user must decide whether the cracked body is liable to collapse soon after net ligament yield. The first warning in Eqn. 2c is a general precaution, the second relates to possible collapse in a mode other than that implied by the nominal stress (e.g. a pin-loaded or compact tension piece collapsing in bending). If collapse is possible then closer study would have to be made. If collapse will not occur then the $cbs\varepsilon$ term described below can be used, usually with Eqn. 2b, to accommodate the localisation of strain that is a feature of ligament yielding with loading controlled by adjacent elastic material, instead of the re-characterisation procedures of (1) and (2).

The effective stress and strain, $cbs\sigma$ and $cbs\varepsilon$. The 'cracked body structural stress', $cbs\sigma$, is the physically dominant stress acting on the cracked region due to the forces in the cracked body, expressed in terms of the elastic stress in the un-cracked body, so that $K = \bar{Y}(cbs\sigma)\sqrt{a}$. Usually $cbs\sigma = \sigma$ and $\bar{Y} = Y$ but for some configurations that is not so. Such cases are often indicated by $Y/\sqrt{\pi} > W/b$. Usually the force in the cracked member is taken as the force in the uncracked body but there are some cases where that assumption is not reasonable and further structural analysis is required to identify a meaningful value. An example is a statically indeterminate system where the tensile to bending ratio of the forces acting change from the un-cracked to cracked configuration. When the ligament approaches yield the plastic strain, $\varepsilon_{p\ell}$ is partly concentrated into the notch by a slip-line type action, with an elastic component still transmitted by the ligament. This augmented strain is termed the 'cracked body structural strain', $cbs\varepsilon$, and can be evaluated approximately for two-dimensional cases by slip-line analyses. The general form is $cbs\varepsilon$ = (nominal $\varepsilon/\varepsilon_y$ + local $\varepsilon_{p\ell}/\varepsilon_y$) = $\oint(\sigma/\sigma_y, b/W)$*. An empirical correction is made for the effects of three dimensional support and for high work hardening. Suggested values of the $cbs\varepsilon$ are given in Table 1.

If there are no elastic load paths in parallel then the problem must be examined carefully to see whether it is controlled by external deformation, by hardening or whether plastic collapse is imminent. This use of $cbs\varepsilon$ obviates the need for the re-characterisation procedures of (1) and (2). The $cbs\varepsilon$ augmentation factor is applied to mechanical stress only and not to residual or thermal stress unless a reaction force is induced. It becomes important above about 0.8 of net section yield but to ensure a smooth transition from $cbs\sigma$ to $cbs\varepsilon$ as σ/σ_y increases, the larger term can be adopted for the value of effective strain, $\varepsilon_\oint$, at which Eqns. 2 are entered.

*Footnote. It will be understood that a term such as b/W in two dimen-sions (2D) implies b/B in three dimensions (3D),since the governing crack dimension, a, and remaining ligament, b, are then in the thick-ness sense.

<u>TABLE 1</u>   Suggested values of the $cbse$ augmentation factor where significant plasticity occurs but collapse is not imminent.

General formula: $cbse/\varepsilon_y = h.g. \left[(cbs\sigma/\sigma_y) \ (1 + \alpha) - (\beta_t + \beta_b) \right]$

$\alpha = \bar{W}/\bar{b}$; $\beta_t = 0.8\sigma_t/\sigma$; $\beta_b = (1.2\ \bar{b}/\bar{W})\ \sigma b/\sigma$.

$\sigma = \sigma_t \pm \sigma_b$ where t is tension, b is bending ($\pm$ according to sense).

For tension only, $\sigma = \sigma_t$, $\sigma_b = 0$; for bending only $\sigma = \sigma_b$, $\sigma_t = 0$; neglect σ if $-ive$.

For edge cracks, a = crack length, $\bar{b} = b$ = ligament $W - a$; $\bar{W} = W$ =width.

For buried cracks, $2a$ = crack length, $\bar{b}$ = tensile ligament; $\bar{W} = b + a$ (if both ligaments are tensile use the larger ratio of W/b)

g is a geometric factor; for $2D$, $g = 1$; for $3D$, $g = \left[2 + (B/W) \right]/\left(2 + (B/\ell) \right)$
        B is thickness and ℓ is surface crack length in the W direction.
h is a hardening factor; for low hardening $(n \to 0)$, $h = 1$; for high hardening $(n \to 1)$, $h = 1/(n + 1)$.

<u>Secondary stresses</u>. For stress concentrations, residual stresses or thermal stresses, allowance is made on the ordinate J/G_y, in a manner comparable to that used in R-6 rather than on the abscissa of strain, as used in the COD design curve.
        For residual stress, σ_r, or thermal stress, σ_{th}, J is estimated from Eqn. 2a, b, separately for these cases or for mechanical stress, σ_m, to give separate terms, J_r, J_{th} and J_m. The terms are combined to give the total J by adding J^γ values.

$$J = (J_m^{\ \gamma} + J_r^{\ \gamma} + J_{th}^{\ \gamma})^{1/\gamma}$$ Eqn.3

where $0.5 < \gamma < 1$ according to $\gamma \simeq 0.5 + (\sigma_m + \sigma_r + \sigma_{th})/2\sigma_y$ (but $\leqslant 1$), i.e γ is estimated between $1/2$ for elastic and 1 for plastic conditions. The $cbse$ augmentation concept may be necessary if restraint can induce reaction stresses of general yield level, but it is not necessary for self-equilibrating compatibility stresses. Thermal effects may clearly lead to strains in excess of yield. Since collapse does not occur, Eqn. 2b is appropriate. For the thermal cases, biaxiality appears to be a relevant factor, since for example, local surface cooling of a thick body would produce an equi-biaxial thermal strain field on the surface, whereas general cooling of a bar restrained in one direction would not. On the evidence of the biaxial mechanical equal displacement case (Ref. 9, p.83) it is suggested that

$$\varepsilon_\theta/\varepsilon_y = \alpha\Delta T/\varepsilon_y$$ (uniaxial restraint)   Eqn.4

$$\varepsilon_\theta/\varepsilon_y = \alpha\Delta T/\varepsilon_y(1 - \nu)$$ (biaxial restraint)   Eqn.4

with $\nu = 0.5$ if yield is exceeded.
        For biaxial mechanical loading the important point is the correct estimation of the ligament limit load but assuming passage beyond that

is feasible, for example, because of surrounding elastic material, then for displacement control effects similar to those just discussed are relevant, and it is suggested

$$\varepsilon_{\delta}/\varepsilon_y = \varepsilon/(1 - \nu Z)\varepsilon_y \qquad \text{Eqn. 5}$$

where ε_t (transverse) $= Z\varepsilon$ (axial). A value of $\nu = 0.5$ is implied where plasticity occurs which is, of course, the case under discussion. This term reduces to the above thermal case if $Z = + 1$.

For stress concentrations, Eqn. 2a is used with the appropriate $le\delta m$ value of Y whilst $cb\delta e < 1$. If $cb\delta e > 1$ an estimate must be made of the uncracked body strain at the point of concentration to give the effective strain $\varepsilon_{\delta}/\varepsilon_y$.

Use of the Neuber relationship, $k_\sigma k_\varepsilon = kt^2$, was originally suggested (10) for this purpose and adopted in the early formulation of EnJ. Further examination $(9; 2^{N}E\delta)$ suggested that the Neuber estimate would only serve for contained yield where $le\delta m$ corrected for plasticity as Eqn. 2a is simpler and better. As yield becomes uncontained, the Neuber estimate of strain is no longer adequate for the moderate hardening case studied.

Maximum tearing load and ductile instability.

If a low energy mode of fracture, such as cleavage, does not intervene, then ductile instability will always be preceded by attainment of maximum load. The maximum load may be due to plastic collapse of the ligament (i.e. the conventional limit load) or stable ductile tearing. To evaluate the maximum tearing load, it is supposed that a unique $J-R$ curve exists for a given material or a curve that is directly relevant to the configuration and orientation in question can be found or a lower bound curve assumed. Values of J_R, $(a + \Delta a)$ and the shape factor Y up-dated to $(a + \Delta a)$ are then used to assess J/G_y and the value of σ/σ_y derived, usually from Eqn. 2a. It will be found that if a maximum load behaviour is relevant, then the values of J/G_y and σ/σ_y at first increase with J_R but then decrease as the effect of changes of Y and a dominate. This procedure is identical in concept to the tangency construction of the $R-6$ method, although no diagram is actually drawn. If J/G_y exceeds 2.5, the result will give $\sigma/\sigma_y > 1.2$, i.e. beyond Eqn.2a, thus warning that collapse must be examined and, for a two-dimensional problem with minimal hardening, may already have intervened. For those cases already noted where $Y/\sqrt{\pi} > (W$ or $B/b)$, then $cb\delta\sigma$ rather than σ must be used for the abscissa. For those cases where the $cb\delta e$ would normally be used (and likewise for those cases in $R-6$ where yielding of the ligament would lead to re-characterisation) distinction must be made between a local maximum in the ligament, which is predicted by the present procedure, and the global maximum load for the whole structure. The two will be the same only if the local 'cracked body region' does indeed dominate the behaviour of the whole structure. If R-curve usage is restricted to J-controlled growth, then appropriate checks must be made that the relevant crack growth is not exceeded during the estimation of maximum load. It is a separate issue whether or not the often quoted restrictions on J-controlled growth (such as $\omega \gg 40$ and $\Delta a/b < 0.06$) are in fact necessary; the present writer inclines to the view that they are not as essential as usually stated, if the R-curve is determined appropriately.

Although EnJ and other methods offer estimates of J, it may well be that these estimates are not adequate for prediction of dJ/da in connection with ductile instability. That might clearly be true, for example, if the $cbse$ procedure were being used to estimate J. The EnJ procedure does not itself go beyond the estimation of maximum load just described. The writer has elsewhere proposed an energy rate method for predicting ductile instability (9) (11) but that is independent of the EnJ method, and is not further described here.

Selection of critical toughness values

The estimation procedures cannot be compared sensibly without discussion of the selection of an estimate of toughness. In the following, δ or J_i, implies a direct measurement of the start of growth of a crack or tear, in a high constraint test such as deep notch bending or compact tension. J_{1c} is taken as a formal evaluation of J_i by the standard test method (12) unless further qualified. δ_{crit} is used loosely to embrace several possibilities (13) from δ_i to δ_m at first onset of maximum load in the test piece, and J_{crit} implies a similar broad definition from J_i to J_m. Lower bound values of toughness or R-curve with respect to scatter and orientation are implied throughout, without being rigorously defined. The use of parent metal, weld metal or HAZ data, must depend upon the metallurgical effects of welding particular materials and whether these zones provide a continuous crack path. It is supposed that time-dependent effects are not important; it may be necessary to obtain a long term R-curve for any of the methods where sustained load occurs close to ligament yield (14). The further use of conventional factors of safety based on engineering judgement of the reliability of the estimated load and on the consequences of failure is, of course, possible, and may well differ according to the perception of conservatism or otherwise in the fracture procedure chosen.

Some doubt may be expressed on the role of secondary stress systems in contributing to possible ductile failure. It seems clear that secondary systems (with no follow up characteristic) can contribute to initiation and to complete failure by brittle cleavage. In so far as plastic flow causes a permanent strain but releases the secondary stress system, the mechanics is somewhat similar to plastic collapse where secondary terms cannot contribute to failure. However, in a multiply-connected body, such as a buried crack separated from the surface by a ligament that may be small in relation to the extent of a secondary stress system, it seems that there could be sufficient elastic recovery to cause the ligament to "snap through" if (conceptually) the ratio of ligament ductility to elastic strain were less than the ratio of extent of ligament slip field to the extent of secondary stress system. These are an extension of the arguments used to deduce the $cbse$ factors in EnJ and suggest a tentative rule that for 'snap through' of a fully yielded ligament in a multiply-connected body or other local or 'secondary' tearing, the secondary stress systems <u>should</u> be included, but for overall (or 'primary') tearing instability under sustained loading, then secondary systems can be neglected.

The possibility of several 'secondary tears' combining to reduce the cross-section left to resist the primary system seems implausible, but not conceptually impossible. The presumably beneficial effect of loss of constraint on the material R-curve for a small ligament is not quantified.

All these issues are resolved if a rigorous interpretation of initiation toughness is used. Thereafter, an instability analysis should, strictly be made. In fact most users would intuitively accept that J_{Ic} (12) is an adequate interpretation of J_i for ductile materials without an explicit analysis of instability. It is less clear whether a toughness, J_{crit} corresponding to a specified amount of growth or even a maximum load value, can be accepted as is feasible in (13) and argued in (15) unless, as in (1), there is a degree of conservatism deliberately used in the method of assessing applied severity. The advantage of using a maximum load toughness is, of course, simplicity of testing. Following the above analysis for maximum load in a structure, it seems "any toughness J_R, on the R-curve" can be used with a "simple design curve" method without conducting an explicit analysis for instability, and without acceptance of counter-vailing conservatism in the estimation method, provided the crack length shape factors are updated in the estimation procedure, a factor which is clearly more important when the ligament is small, and also that the resulting load is less than the maximum (tearing or plastic limit) load of the structure. A rigorous simple statement for this requirement has not yet been made but for most cases $b_{structure} > b_{testpiece}$ is sufficient.

The EnJ proposals. Space permits only the briefest statement of how the above arguments are used in EnJ. If cleavage is possible and impulsive loading is a design feature or the consequence of failure is catastrophic, then thick section dynamic or running crack test data are necessary. If some lesser degree of dynamic loading is plausible, or consequence of failure is incapacitating rather than catastrophic, a static design philosphy usually seems adequate and an initiation value $J_{crit} = J_{Ic}$ may be selected, whereas if dynamic events are implausible and failure inconvenient rather than incapacitating, a static toughness $J_{crit} = J_{max}$ may be selected.

If cleavage is not a risk, initiation of ductile tearing under very occasional severe load is not seen as catastrophic, provided ductile instability does not ensue, and subsequent loading takes account of the possible consequences of the exceedance of initiation.

Avoidance of ductile instability may rely on either the use of

$$i) \quad J < J_{Ic} \text{ from (12)} \qquad\qquad \text{Eqn. 7a}$$

or

$$ii) \quad J < J_{max} \qquad\qquad \text{Eqn. 7b}$$

from a full section test in which the corresponding crack advance Δa_m is also measured , and for which $b_{structure} > b_{test\ piece}$. The former

guards against fracture by avoiding initiation. The latter tolerates
initiation but limits $\partial J/\partial a|_{Q\text{(applied)}} < dJ/da_{\text{(material)}}$ at onset of
maximum load in the test by ensuring maximum load is not exceeded,
provided growth up to Δa_m is incorporated in the analysis.

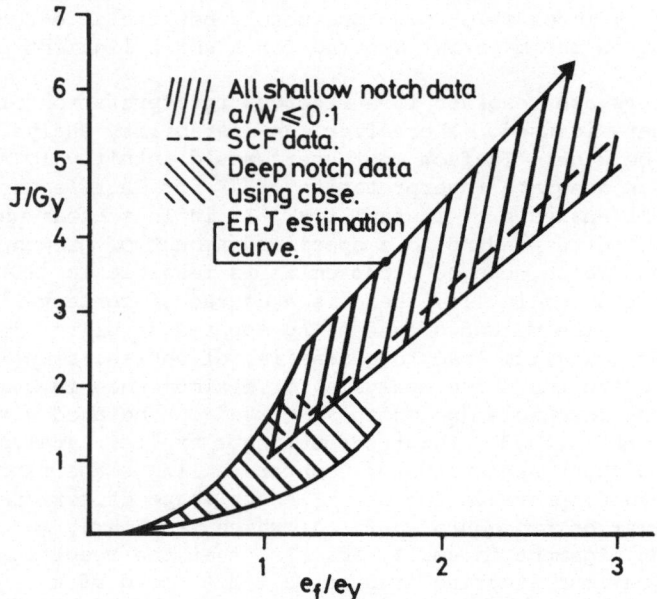

Fig. 1: The EnJ curve with summary of supporting data

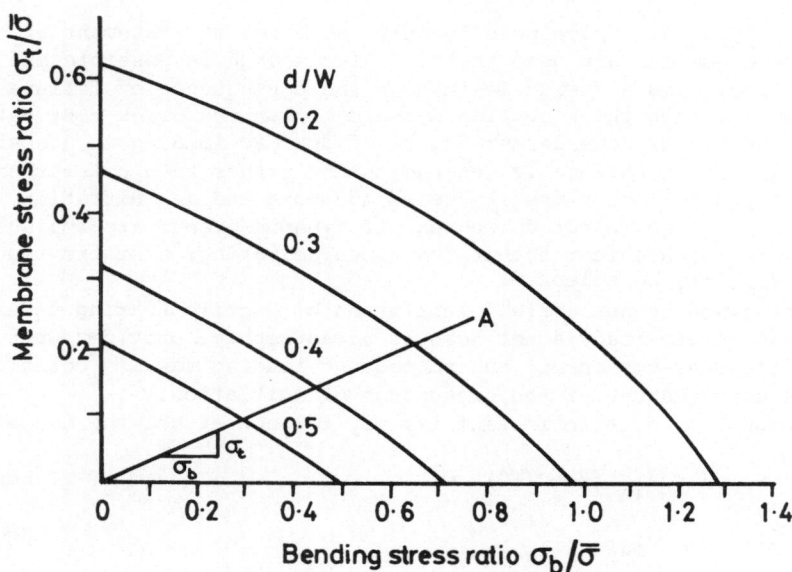

Fig. 2: Limit state for combined bending and tension (after R-6).
Note d/W is the ratio of effective crack size to thickness.

Summary   The data on which EnJ is based is summarised in Figure. 1, including use of the $cbse$ concept in $2D$. Plastic collapse must be considered separately. **Fig. 2** is an excerpt from (2), suitable for combined bending and tension, which is used in one of the following examples.

EXAMPLE PROBLEMS

To illustrate the use of EnJ, examples from within the previous work-shops are selected. From Workshop 1, on R-6, problems 2 and 3 are selected; from Workshop 2, on the EPRI method, the data for the CT pieces are examined with respect to maximum load prediction; from Workshop 3, on COD, cases (e) and (f) are worked to show some of the features that lead to conservatism in the COD method.

From Workshop 1: An axial crack in a pressure vessel.

The object is to assess whether a particular cracked vessel, described in Workshop 1, Problems 2 and 3, is safe. The only difference in the R-6 treatment of these two cases is the inclusion or not of the residual stress in the estimate of S_h. Limit load is not required within the EnJ estimation method, and a cursory check shows plastic collapse is not in question. The LEFM shape factor Y is taken from (16) but all the other data are used from Workshop 1 without further comment.

The EnJ Solution. J_{mech} and J_{res} are found separately using Eqn. 2a. For mechanical stress,

$$\sigma_\theta/\sigma_y = 169/430 = 0.393$$

$$\therefore J_{mech}/G_y = (0.393)^2 \{1 + 0.5 (0.393)^2\} = 0.166$$

from (16), $Y = 0.935\sqrt{\pi} = 1.66$

$$\therefore J_{mech} = 0.166 (1.66 \times 430)^2 (0.007)/205 \times 10^3$$
$$= 0.00289 \ MN/m$$

For residual stress,

$$\sigma_{res}/\sigma_y = 43/430 = 0.1$$

$$\therefore J_{res}/G_y = (0.1)^2 \{1 + 0.5 (0.1)^2\} = 0.010$$

The residual stress is small, so the assumed pattern of stress cannot be of great importance and the same shape factor Y is used.

$$\therefore J_{res} = 0.01 (1.66 \times 430)^2 (0.007)/205 \times 10^3$$
$$= 0.000 \ 17 MN/m$$

The two terms are added according to Eqn. 3.

i.e

$$J_{total} = \{J_{mech}^{\gamma} + J_{res}^{\gamma}\}^{1/\gamma}$$

where $\gamma = 0.5 + (169 + 43)/860 = 0.747$

$\therefore \quad J_{total} \approx 0.00336$ or $K_J = \sqrt{EJ} = 26.0$ MPa$\sqrt{m}$

In Workshop 1, the R-6 method gave $K = 30.3$ MPa$\sqrt{m}$ (Problem 2) or 30.1 MPa$\sqrt{m}$ (Problem 3).

If K values, corrected for plastic zone size as above, are directly added (i.e. $\gamma = 1/2$); $K = 30.2$ MPa$\sqrt{m}$.

If LEFM were used with no plastic zone correction,

$$K = K_{mech} + K_{res} = 23.4 + 6.0 = 29.4 \text{MPa}\sqrt{m}.$$

Discussion. The close similarity of the uncorrected LEFM solution and either the R-6, or that inferred with a (plane stress) plastic zone correction (about 3% differences) reflects the fact that $\sigma_{mech} + \sigma_{res} \simeq 0.49\sigma_y$, suggesting that EPFM is not in fact necessary for this problem. The fact that K_J from EnJ is slightly smaller than from LEFM alone reflects that EnJ essentially adds displacements for discontinuity residual or thermal terms, which terms are relieved rather than enforced by the plasticity effect. The EnJ solution is not, of course, less than the value for K_{mech} alone. Any differences in the methods are little more than between the values found according to the different sources of Y in Workshop 1 and here.

All methods, including LEFm, show $K \ll K_{Ic}$ $(K_{Ic} = 100$ MPa$\sqrt{m})$ so that the vessel is safe.

From Workshop 3: A Welded Joint

The object is to apply EnJ to cases (e) and (f) of Workshop 3 in which combined tension and bending exist for a semi-elliptical surface crack, firstly without and then with yield level residual stresses, to find the required toughness. The values used for the LEFM shape factor, Y, are the same as those used for the part of the COD solution when Eqn.2 was used in Workshop 3, taken from (16).

The EnJ Solution. For a part through crack subjected to tension and bending first find Y using:-

$$Y\sigma = Y_t \sigma_t + Y_b \sigma_b; \text{ where } \sigma = 117 + 180 = 297 \text{ MPa};$$

$$\sigma_t = 117, Y_t = 1.685; \sigma_b = 180, Y_b = 1.155, \text{ whence } Y = 1.36$$

Check on the cbse term for combined bending and tension, with the 3D and hardening correction factors. For this problem $cbs\sigma \equiv \sigma/\sigma_y = 297/340 \simeq 0.874$. From Table 1,

$$cbse/\varepsilon_y = \{1/(n+1)\}\{[2 + (B/W)]/[2 + (B/\ell)]\}\{(cbs\sigma/\sigma_y)(1 + \alpha) - (\beta_t + \beta_b)\}$$

For the mechanical terms only; assuming $n = 0.1$ (not known); $B = 40mm$, $\ell = 70mm$, $W \to$ very large, $a = 14\,mm$,

$$cbse/\sigma_y = \{0.908\}\{2/2.57\}\{0.874(1 + 1.54) - [0.8 (0.394 + 1.2 (0.65) (0.606)]\} = 1.01$$

This is more than σ/σ_y, so the member is sufficiently near ligament yield to warrant use of $cbse/\varepsilon_y$ for $\varepsilon_\delta/\varepsilon_y$ although $cbse/\varepsilon_y < 1.20$ so Eqn. 2a is still used. Thus, for mechanical stresses only,

$$J(205 \times 10^3)/(1.36 \times 340)^2(0.014) = (1.01)^2(1 \times 0.5(1.01)^2)$$

$$\therefore J_{mech} = 0.0225\,MN/m$$

(Note if the $cbse$ augmentation factor were not used, entering Eqn. 2a at $\sigma/\sigma_y = 0.874$ gives $J_{mech} = 0.0156$ MN/m). If residual stress $\sigma_\hbar = 340MPa$ is also acting; $\sigma_\hbar/\sigma_y = 1$

$$J_{res}/G_y = 1 (1 + 0.5) = 1.5$$

Assuming the stress is tensile over the cracked region, $Y_t = 1.685$

$$\therefore J_{res} = 1.5(1.685 \times 340)^2(0.014)/205 \times 10^3 = 0.0224\,MN/m$$

Since $\sigma_{mech} + \sigma_{res} > \sigma_y$, the values of J are added

$$\therefore J_{total} = J_{mech} + J_{res} = 0.0225 + 0.0224 = 0.0449\,MN/m$$

Discussion. All the results relate to an acceptable crack by the COD method, Workshop 3, and a 'best estimate' for the J method. For comparison, J and COD results are expressed as $\alpha = m\sigma_y\delta/J$ where m is the constant relating J to COD, $J = m\sigma_y\delta$, here taken as 2. α is the apparent 'factor of safety' in the COD method, judged against the EnJ estimate of J.

      For case (e) i) $\alpha = 2.44$ (based on the PD6493 solution in Workshop 3)

                  ii) $\alpha = 1.60$ (based on use of Eqn. 2 in Workshop 3)

      For case (f) iii) $\alpha = 3.42$

If m is indeed 2, then α is well over 2 for results i) and iii) which make direct use of PD6493, and about 1.5 for result ii) which uses Eqn. 2. This lower 'factor of safety' is because ii) uses a more realistic value for Y_b for the bending stress than does PD6493 and that stress component is the largest term in problem (e).

    The value of J required according to EnJ should relate to a full thickness test if cleavage is a possibility. If initiation is to be avoided, then clearly J_i (or J_{Ic} as a conventional measure of initiation) must exceed J as estimated. If the conditions used for the calculations refer to a 'worst ever' loading for which some amount of ductile tearing is acceptable, then a post-initiation value of J can be used, either in conjunction with a maximum load analysis (as in the following example) or as a 'spot' value J_R at a known amount of crack growth (Δa) which value must then be added to the value of crack length (here 14mm) used in the structural analysis, whilst also checking that

$b_{structure} \geqslant b_{test\ piece}$

From Workshop 2: A Compact Tension Piece

The object is to show how EnJ is used to predict the maximum load of a structure after some stable tearing. The tearing of the stress re-lieved weld metal, Workshop 2, for the compact tension piece of Fig. 7, at left, is studied, using the mechanical properties, Table IV. No attempt is made to determine either the complete load-deflection relationship or the final instability point, since neither are part of the conventional 'simple design curve' methods, COD, R-6 or EnJ. Note that if the R-curve data relates to the thickness in question then the degree of plane stress or plane strain is not directly relevant to the prediction of maximum tearing load. However, if the R-curve data, Workshop 2, Fig. 10, truly reflects material scatter, then there is no way to predict any tearing behaviour closely. The curves are also unusual in showing an increasing slope which, if genuine, might reflect an increasing degree of shear-lip due to crack tunnelling, which would be surprising just prior to the cleavage failures shown. The rising trend of the un-heat treated piece, NB7F1, might be due to residual stresses in the weldment. For the present purposes only, the stress-relieved data are considered and one curve is drawn, Fig. 3 . here. This assumption allows a useful exercise to be constructed although it may undermine direct comparison with the test result.

Does plastic collapse occur ? If the conventional LEFM shape factor is used for the initial crack length, then $a/W = 275/750 = 0.366$; $Y = 11.0$ and $Y/\sqrt{\pi} > W/b$, so that Eqn. 2c warns that collapse must be checked explicitly. A particular solution for the compact tension piece could be used but an estimate is made here for combined tension and bending, based on R-6. For this two-dimensional problem the effective crack size, d/W, Fig. 3, is just a/W. The nominal tensile stress, for load Q, is $\sigma_t = Q/BW$; the bending stress is $\sigma_b = 6M/BW^2$, where even for the un-cracked body, $M_O = QW/2$, i.e. $\sigma_b = 3\sigma_t$. For a crack length, a, the moment increases to $M_a = Q\ [a + (b/2)]$ so σ_b becomes $3A\sigma_t$ where $A = M_a/M_O = (a + (b/2))/(W/2)$. Fig. 3 is then entered at the appropriate value of σ_b/σ_t which defines the slope of a line such as OA, and σ_t is read off as $\sigma_t/\bar{\sigma} = \alpha$ for the corresponding size d/W. The limit load Q_l is $\alpha\ BW\ \bar{\sigma}$. Values for all these terms are given in Table 2. with $\bar{\sigma} = 470MPa$.

TABLE 2. Estimate of plastic collapse load (from R-6).

$a$mm	Δamm	a/W	A	$3A$	α	Q_l kN
275	0	0.366	1.36	4.11	0.14	920
315	40	0.420	1.42	4.26	0.115	755
345	70	0.460	1.46	4.38	0.09	530

For $B = 17.75mm$ the pieces are near LEFM valid plane strain at initiation, $J = 0.25MN/m$ but well towards plane stress if a value of J is
used after appreciable crack growth. As far as the writer knows the
basis of Fig. 2 is plane stress with some empirical allowance towards
plane strain. It is inferred that the behaviour shown, Workshop 2,
Fig. 15, is likely to be due to tearing, although as the crack length
increases limit state of the ligament may well be reached so that a
closer estimate than made here is really desirable. A limit load prediction based on initial dimensions would certainly not be conservative
so that crack growth must be considered.

The EnJ estimate of maximum tearing load.

The ratio of ultimate to
yield strength is about 1.5 or rather less for the weld metal. This
degree of hardening is quite typical of the structural steels for
which EnJ is intended. A mean yield stress value of $\sigma_y = 380$ MPa
is used. If there were a high degree of hardening a larger value could
be used for an effective yield stress. The scheme for calculating
maximum load is simply to use the EnJ equations, updating the geometric
terms (crack length and shape factor) against J values from the R-curve,
in steps, with stress as the unknown term to be found. A maximum is
found for the load when the change in geometric terms exceed the
increase in the J value. In order to conduct the calculation a value
of the shape factor, Y, is required and the peculiarity of the compact
tension piece must be noted. As already seen the ratio of tension to
bending stress must be allowed for in any plasticity analysis and as
noted in connection with the meaning of the terms in the EnJ equations,
the stress must be the physically meaningful term, already seen to be
$\sigma = \sigma_b + \sigma_t = (3A + 1)\sigma_t$. Thus the effective shape factor, $\bar{Y}$, is
$\bar{Y}(3A + 1) = Y_t + 3AY_b$ where Y_t is the LEFM shape factor for tension
with ends parallel and Y_b is the factor for bending. Values are taken
from (16) (in which they are given in the form $Y/\sqrt{\pi}$). The data used
are given in Table 3 in which Y_{ct} is the conventional value for the
compact-tension piece. It is seen that the present 'structural'
analysis agrees to within about 6% with the value of Y_{ct}, $(a/w \leqslant 0.5)$ i.e.
$\bar{Y} \approx Y_{ct}/(3A + 1) = Y_{ct} \sigma_t/\sigma$, and it is re-iterated that all that is
required for EnJ is a reasonable estimate of the stress systems acting
expressed in terms of the uncracked body stress as is conventional for
most LEFM solutions, with the notable exception of one or two cases
including compact tension.

Thus, using Eqn. 2a and taking $J = 0.25$ MN/m at initiation,
(Fig. 3 here), $a = 275mm$, $a/W = 0.366$, $\bar{Y} = 2.17$

Thus $\quad 0.25/G_y = (\sigma/\sigma_y)^2(1 + 0.5(\sigma/\sigma_y)^2)$ $\quad \sigma_y = 380$ MPa, $E = 205$GPa

so $\quad G_y = (2.17 \times 380)^2(275 \times 10^{-3})/205 \times 10^3 = 0.91$ MN/m

$\quad \therefore \sigma/\sigma_y = 0.495$; $\quad \sigma_t/\sigma_y = 0.097$; and $Q = 490$ kN.

This procedure is continued but as the stress level increases the $cb\delta e$
term must be checked and used when it exceeds σ/σ_y. Because of the
steep rise of the R-curve this occurs by 10mm growth. A sample
calculation for 40mm of growth is given: $W/b = 1.72$; $b/W = 0.58$;

$\sigma_b/\sigma = 4.26/5.26$; $\sigma_t/\sigma = 1/5.26$. From Table 1 the $cbse$ is

$$cbse/\varepsilon_y = \{1/(n + 1)\}\{(\sigma/\sigma_y)(1 + 1.72) - [(0.8/5.26) + 1.2(0.58)$$
$$(4.26/5.26)]\}$$
$$= \{1/(n + 1)\}\{2.72(\sigma/\sigma_y) - 0.716\}$$

From Fig. 3 for $\Delta a = 40mm$, $J_R = 3.05\ MN/m$.
For $a = 315mm$, $a/W = 0.366$, $\bar{Y} = 2.32$

$$\therefore\ G_y = (2.32 \times 380)^2(315 \times 10^{-3})/205 \times 10^3 = 1.195MN/m$$

From Eqn. 2b, with $J = J_R$

$$J/G_y = 2.5\{(cbse/\varepsilon_y) - 0.2\} = 3.05/1.195$$

$$cbse/\varepsilon_y = \{3.05/(1.195 \times 2.5)\} + 0.2 = 1.22$$

If hardening is neglected, $n = 0$, $\sigma/\sigma_y = 0.714$, $\sigma_t/\sigma_y = 0.136$, $Q = 685kN$.

If hardening is taken as $n = 0.1$, $\sigma/\sigma_y = 0.755$, $\sigma_t/\sigma_y = 0.143$, $Q = 725kN$.

If hardening is taken as $n = 0.2$, $\sigma/\sigma_y = 0.800$, $\sigma_t/\sigma_y = 0.152$, $Q = 766kN$.

For other crack lengths the results for $n = 0.1$ and the previous limit loads are:

Crack growth, Δa	0	10	30	40	50	70	mm
J value from R-curve	0.25	1.65	2.60	3.05	3.35	3.90	MN/m
Max. load, tearing	490	650	715	725	720	690	kN
Max. load, limit state,	920			755		530	kN

Discussion. The good agreement with the experimental maxima (Workshop 2. Fig.15) of 725 and 738 kN may be fortuitous in view of the assumptions made about the R-curve. On the results as presented, tearing starts and continues up to about 40mm growth where a maximum load is reached, the exact value depending on the hardening and of course on the R-curve. Thereafter, limit state is also reached, so that, if the estimates were indeed precise, tearing would cease since limit load is the lower. Note that brittle fracture by cleavage is not discussed since a separate value of cleavage toughness is not quoted. If the value from test NA8F2, at the asterisk, is used, $\Delta a = 20mm$, $J = 2MN/m$ (Fig.3 here as in Workshop 2, Fig. 10) then it is correctly predicted since the R curve used passes through that point. The second result cannot of course be predicted since there is a two-fold difference in J value, presumably due to material variability. However realistic the numbers may or may not be, the example serves to demonstrate the method of using EnJ to predict maximum load. It may be noted that, because $cbse$ reflects the different response to tension and bending and because the ratio of σ_t to σ_b alters in this problem, maximum load does not coincide with maximum $cbse$ and some simpler analysis predicting the latter would be unconservative, particularly in respect of amount of growth.

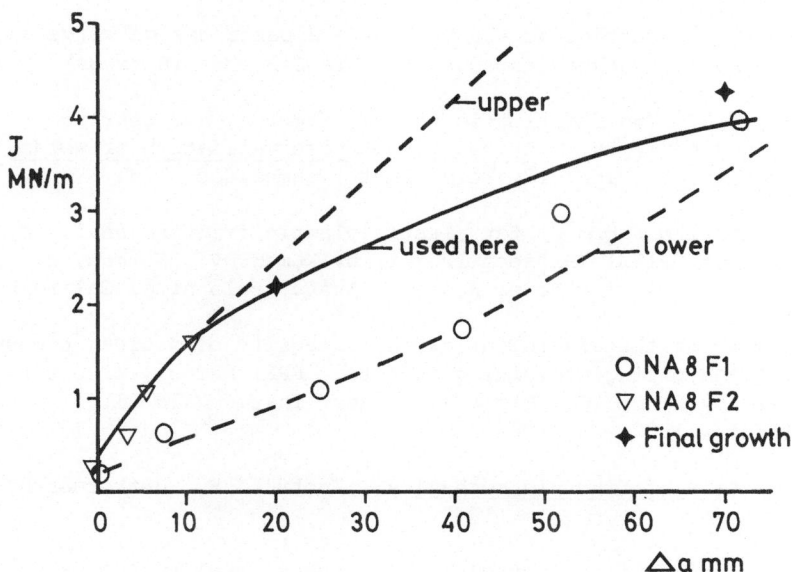

Fig. 3: R-curve used here, based on data from Workshop 2.

TABLE 3   Effective shape factor, $\bar{Y}$, for compact tension

a/w	Y_t^*	Y_6	3A	$\bar{Y}$	$(3A+1)\bar{Y}$	Y_{ct}
0.35	2.32	2.09	4.05	2.13	10.78	10.90
0.40	2.39	2.22	4.20	2.25	11.71	11.54
0.45	2.48	2.41	4.35	2.42	12.96	12.12
0.50	2.58	2.65	4.50	2.63	14.50	13.58

* from (16) p.85. For deeper notches, better agreement
  is found with Y_t based on (16) p.89.

REFERENCES

1) Guidance on some methods for the derivation of acceptance loads in
    defects in fusion welded joints. PD6493, British Standards
    Institution, 1980.

2) Assessment of the integrity of structures containing defects, CEGB
    Report R/H/R6 by Harrison, R.P., Loosemore, K & Milne, I., 1976,
    and supplements, 1979, 1981.

3) Turner, C.E. Further development of a J-based design curve and its relationship to other procedures. ASTM STP 803 (in press).

4) Turner, C.E. The J-estimation curve, R-curve, and tearing resistance concepts. Fitness for purpose validation of welded constructions. Paper 17, Weld. Inst., Cambridge, 1981

5) An engineering approach for elastic-plastic fracture analysis. EPRI Report NP1931 (Project 1237-1) by Kumar, V., German, M.D., & Shih, C.F. Electric Power Research Inst., Palo Alto, California.

6) Procedure for the assessment of the integrity of nuclear pressure vessels and piping containing defects. EPRI NP2431, (Project 1237-2) by Bloom, J., Elec. Power Res. Inst., Palo Alto, California, 1982.

7) Fundamentals of fracture mechanics. Knott, J.F., Butterworth, 1973.

8) Fracture and fatigue control in structures. Rolfe, S.T. & Barson, J.M. Prentice-Hall, 1977.

9) Post-yield fracture mechanics. Ed. Latzko, D.G.H. Applied Science Publ., 1979. (and 2nd Edition, to be published 1984).

10) Begley, J.A. et al. An estimation model for the application of the J-integral, ASTM STP560, pp.155-168, 1974.

11) Turner, C.E. Description of stable and unstable crack growth in the elastic-plastic regime in terms of J_R resistance curves. Fracture Mechanics, ASTM STP677, pp.614-678, 1979.

12) ASTM E813-81. A standard test for J_{Ic}: a measure of fracture toughness. ASTM, Philadelphia, 1981.

13) BS5762. Methods for crack opening displacement (COD) testing. British standards Inst., London, 1979.

14) Tsuru, S. & Garwood, S.J. Some aspects of time-dependent ductile fracture of line pipe steels, ICM-3, Cambridge, 3, pp.519-528,1979.

15) Towers, O.L. & Garwood, S.J. The geometry-dependence and significance of maximum load toughness values, ECF-3, ED: Radon, J.C., pp.57-68, Pergamon, 1980.

16) Rooke, D.P and Cartwright D.J. Compendium of stress intensity factors. HMSO, London, 1976.

DYNAMIC FRACTURE MECHANICS – RAPID CRACK GROWTH IN LINEAR AND NON-
LINEAR MATERIALS

Fred Nilsson
Swedish Nuclear Power Inspectorate
Box 27106
S-102 52 Stockholm

Björn Brickstad
The Swedish Plant Inspectorate
Box 49306
S-100 28 Stockholm

ABSTRACT. With dynamic fracture mechanics is meant analysis of cracked
structures, where inertia effects cannot be ignored. It may include
stationary cracks subjected to impact loading and the conditions are
sought for such cracks to start growing. It also includes propagating
cracks and the analysis is then focused on the conditions for the continuous growth of the cracks and the possibility for a crack arrest to
occur.

    First some general tools are introduced in describing moving
cracks in linear elastic materials and also some basic theoretical and
numerical solutions to crack problems are reviewed. These solutions
demonstrate how inertia effects often will considerably raise the risk
of fracture or contribute to further propagate a growing crack.
    A discussion of small scale yielding (SSY) for fast propagating
cracks is held and based on SSY, the requirements are stated for the
possibility to determine the crack-tip motion and especially the point
of crack arrest.
    General aspects of the physics of a growing crack in a non-linear
material are discussed. Two types of description are reviewed in more
detail, i.e. the cohesive zone model and the singular crack-tip model.
Some features of different fracture criteria used under large scale
yielding are discussed, of which an energy criterion seems to possess
certain advantages. Then the merits of some specific, commonly used
constitutive material models are discussed, where we distinguish
between elastic/plastic, hardening/non-hardening and rate-independent/
rate-dependent materials.
    Rate-effects are expected to be important, since rapid crack
growth is associated with very high strain rates, of the order 10^4s^{-1},
near the crack tip. It appears that in some cases the inclusion of
rate-effects resolves some of the basic difficulties of the rate-independent models and some experimental results are presented to support
this statement.

L. H. Larsson (ed.), Elastic-Plastic Fracture Mechanics, 427–478.
© 1985 ECSC, EEC, EAEC, Brussels and Luxembourg.

1. LINEAR FRACTURE DYNAMICS

1.1 Introduction

Dynamic behaviour of cracked structures has during the past years been
the object of intense studies. This is partly due to an understanding
that in many cases dynamic conditions may cause failure for a specimen
or a structural component at a lower load level than for a corresponding static case. Here failure is defined as the sudden creation of
additional free surfaces in the body.

In the following we are going to discuss situations where stress
analysis involving the full equations of motion is required, i.e.
inertia effects cannot be ignored. Two main frames of question are considered. Firstly, a body containing one or more stationary cracks is
subjected to an impact loading. Under what conditions does such a crack
begin to propagate? Secondly, a fast propagating crack is considered.
The analysis is now focused on the conditions for the continuous growth
of the crack and the possibility for a crack arrest to occur. The inherent time-dependence of the process results in mathematical models
which are more complex than quasi-static treatments. Despite this complexity, a number of analytical solutions to the above stated problems
are available in the literature, of which a few will be reviewed in
this lecture. In most cases, however, these analytical results are
restricted to geometries of infinite extension in at least one direction. For finite geometries one is often directed to numerical techniques and some methods for determination of the dynamic stress-intensity factor for stationary and moving cracks by means of finite elements are described in Appendix 2. Only peripheral mention will be
given of experimental methods in this field, these are the object of
the lecture by Kalthoff in this seminar.

Most of the results outlined in the first section of this lecture
refer to two-dimensional bodies under plane stress or plane strain,
behaving linear elastically except for a small region in the vicinity
of the crack tip. Furthermore, it is assumed that the geometrical configuration and the loading conditions are such that symmetry prevails
with respect to the crack plane, i.e. only mode I crack growth is considered.

The second section is devoted to growth of singular crack tips
under mode I or mode III conditions in non-linear materials. The non-
linearities may include plastic as well as rate effects in the deformation process.

In Appendix 1 a brief summary of elastodynamic concepts are
given.

1.2 The state at a moving crack tip

We consider a two-dimensional body containing a sharp crack along the
x-axis (Fig. 1). The loading conditions could be arbitrary static or
dynamic but allowing the crack to extend in mode I only. The position
of the moving crack tip is given by a time dependent crack length a(t).
Surrounding the crack tip, and assumed to be fixed with respect to the
local polar coordinate system (r, θ) with the origin at the crack tip,
is a small loop denoted by c which may have arbitrary shape.

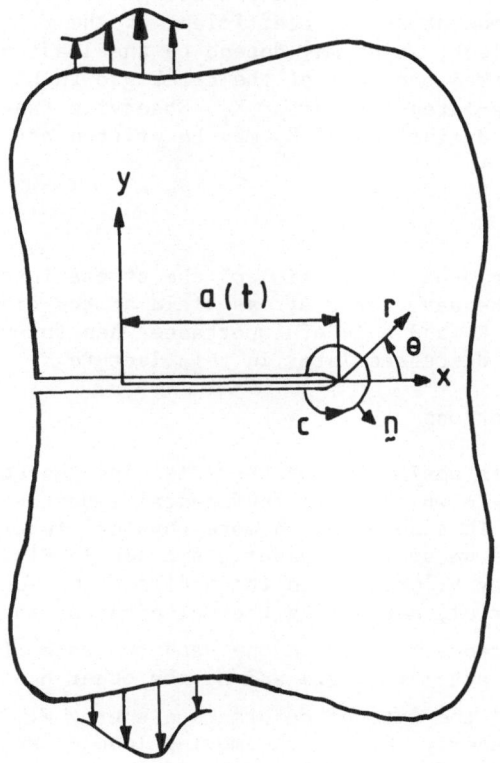

Figure 1. A plane body containing a moving crack at a fixed instant of time.

The crack may extend with an arbitrary crack velocity denoted by $\dot{a}$, which is assumed less than any characteristic wave speed of the material. It is now possible to show that the stresses in the vicinity of the crack tip can be expressed as (Nilsson $\lfloor 1 \rfloor$, Freund and Clifton $\lfloor 2 \rfloor$)

$$\sigma_{ij} = \lim_{r \to 0} \frac{K_I(t)}{\sqrt{2\pi r}} \, f_{ij}(\dot{a}/c_1 \, , \, \dot{a}/c_2 \, , \, \theta) \tag{1}$$

Here c_1 and c_2 are the velocities for longitudinal and shear waves, respectively, f_{ij} is a universal function, independent of the magnitude of the applied load and of the geometrical configuration of the body. Thus the dependence of the stress field on spatial coordinates in the crack-tip region has the same form for all crack problems, and the interaction with the applied load for the particular geometry is fully characterized by a time-dependent scalar coefficient of the $r^{-1/2}$ term in the stress. This coefficient, which may depend on the loading conditions, the shape of the cracked body and of the crack-growth history, is called the dynamic stress-intensity factor K_I. Observing that $f_{yy}=1$ at the crack plane $\theta = 0$, a definition of K_I may be written as

$$K_I = \lim_{r \to 0} \sqrt{2\pi r} \cdot \sigma_y(r,t) \tag{2}$$

which is consistent with the usual definition of the stress-intensity factor. This result, that the asymptotic stress field at the crack tip essentially depends only on K_I and $\dot{a}$ is of importance when formulating a crack-growth criterion as discussed later in this lecture.

1.3 Energy-balance considerations

Besides the dynamic stress-intensity factor there is also the concept of dynamic energy-release rate which is of fundamental importance in dynamic fracture mechanics. It also offers a more physical interpretation of K_I. For this purpose we again consider the crack in Fig. 1, propagating with an arbitrary velocity $\dot{a}$ in the x-direction. Generalizing to a non-elastic material, neglecting thermal effects, the interest is focused on the difference between on one hand the rate of work $\dot{P}$ of tractions working on the outer boundary and on the other hand the rate of kinetic energy $\dot{T}$ and the rate of deformation energy $\dot{W}$. The net change is then the flux of energy through the moving loop c, surrounding the traction-free crack surface, and is consumed in advancing the crack forward. This energy flux is denoted by $\dot{\gamma}$ and may be formulated as

$$\dot{\gamma} = \dot{P} - \dot{T} - \dot{W} \tag{3}$$

Here the deformation energy is defined as $W = \int_0^\varepsilon \sigma_{ij} \, d\varepsilon_{ij}$

which reduces to the elastic strain-energy density for a linear elastic material. In that case one may write

$$\dot{\gamma} = \dot{a} \, G \tag{4}$$

where G is the dynamic energy-release rate which expresses the energy flow to the crack tip for a growing crack in a linear elastic material. All real non-elastic deformations are then assumed to be associated with crack growth only and confined to a region around the crack tip with negligible extension. This is in accordance with the concept of small-scale yielding in which the singular stresses at the crack tip, the strength of which is measured by the stress-intensity factor, are relieved by plastic flow in this region. It should be observed that the energy-rate balance itself, eq. (3), does not offer any possibility to predict crack growth. It is merely a relation which must be satisfied for any crack trajectory.

If the quantities in (3) are expressed in terms of their field quantities, i.e. stresses σ_{ij} and rate of displacements $\dot{u}_i$, the rate of energy flow to the crack tip through the vanishingly small loop c may also be formulated as an integral expression

$$\dot{\gamma} = \lim_{c \to 0} \int_c \left\{ \left(W + \frac{1}{2} \rho \dot{u}_i \dot{u}_i \right) \dot{a} \; n_x + \sigma_{ij} n_j \dot{u}_i \right\} dc \tag{5}$$

where ρ in the inertia term is the mass density. This result has been given similar forms by Kostrov and Nikitin $|3|$ and Strifors $|4|$ and for the elastodynamic case by Atkinson and Eshelby $|5|$ and Freund $|6|$. In general the resulting energy flux is dependent on the particular choice of integration loop, but for the special case of steady-state crack growth in an elastic material the result is path-independent as noted by Sih $|7|$. Furthermore, eq. (5) reduces to the well-known form of the J-integral if the inertia term is omitted in (5) for a path-independent material.

Perhaps a few comments are appropriate regarding possible singularities of the field quantities entering in $\dot{\gamma}$ in order to obtain a meaningful measure of the energy flow to the crack tip. If c is taken as a circular loop with $dc = rd\theta$, then it is easily seen that the integrand in (5) must be of the order r^{-1} for a bounded but non-zero contribution to the integral. This condition is met for a linear elastic material since then both the stress and strain near the crack tip are proportional to $K_I/(2\pi r)^{1/2}$. Thus the energy terms are singular as $K_I^2/2\pi r$, which integrated over the arc length $2\pi r$ yields a non-zero energy rate as $r \to 0$. Judging from asymptotic solutions of dynamic crack growth in materials with conventional plastic properties, they suggest singularities that do not have the required strength, cf. Achenbach and Dunayevsky $|8|$, Achenbach et al $|9|$. In that case the energy flow to the crack tip cannot be resolved in the form as stated above. These arguments are of importance when choosing a suitable dynamic crack-growth criterion based on an appropriate characterizing fracture parameter for various material models.

By use of the asymptotic stress in eq. (1) and the corresponding result for the displacement rate, it is possible to evaluate $\dot{\gamma} = \dot{a} \, G$ in terms of the dynamic stress-intensity factor for an elastic material, see e.g. Freund $|10|$. For plane strain conditions this relation is

$$G = \frac{1-\nu^2}{E} K_I^2(t) \cdot f_1(\dot{a}/c_1 \, , \, \dot{a}/c_2) \tag{6}$$

where

$$f_1(\dot{a}/c_1 \, , \, \dot{a}/c_2) = \frac{\beta_1 (\beta_2^2 - 1)}{(1-\nu) \left\lfloor (1+\beta_2^2)^2 - 4\beta_1\beta_2 \right\rfloor} \tag{7}$$

$$\beta_i = \left(1 - (\dot{a}/c_i)^2 \right)^{1/2} \qquad i = 1,2 \tag{8}$$

Here E and ν are the elastic constants and the function f_1 is shown in Fig. 2 for $\nu = 0.3$.

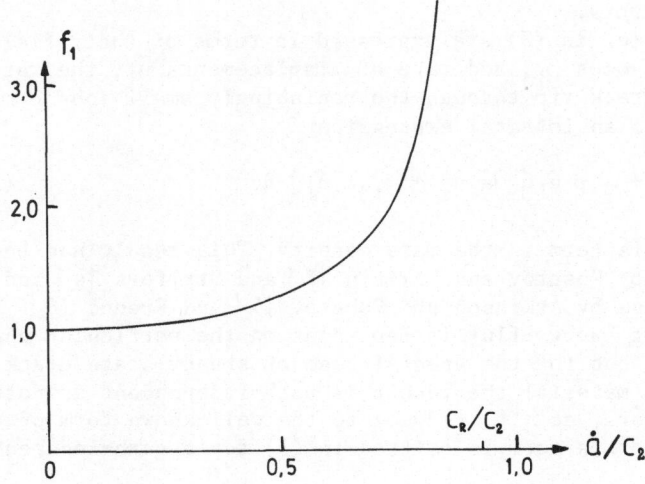

Figure 2. The velocity-dependent function f_1 appearing in the relation between G and K_I.

The function f_1 is unity at zero crack velocity and thus eq. (6) reflects the well-known static relation between G and K_I. In this case $\dot{\gamma}$ = 0 according to eq. (5) since the crack is stationary. The limit in eq. (6) does exist, however, and G should be interpreted as a generalized crack-extension force for a virtual crack-growth increment. f_1 tends to infinity as $\dot{a} \to c_R$ corresponding to a root of the denominator of (7), where c_R is the Rayleigh wave speed of the material. This does not necessarily mean that G tends to infinity since in general $K_I \to 0$ for $\dot{a} = c_R$. The relation (6) is valid for arbitrary crack growth, loading conditions and geometries and demonstrates the equivalence between the two concepts of G and K_I. Thus, for some situations of crack propagation for which G is obtained by a simple energy-balance consideration, K_I may be easily determined.

1.4 Some results on stationary cracks under transient conditions

In this section we consider a stationary crack in a body subjected to dynamic loading, by which is meant that the loading rate is sufficiently high for the effect of stress waves on the local crack-tip field to be important. This generally requires the solution of the elastodynamic equations of motion with careful considerations to the influence from the crack. Analytical solutions to this problem do generally imply infinity in some direction of the outer geometry and/or the crack.

Finite cracks are in this sense more attactive since then the corresponding static solution always exists and may generally be recovered by letting the time to infinity. Thus the influence from dynamic effects on the stress-intensity factor is easy to survey.

In the first example a finite crack of length 2a in an unbounded body is subjected to a normal impact with stress of magnitude σ_0 applied as a step-load. This problem has been studied by Chen and Sih [11] and the result is shown in Fig. 3 in form of normalized stress-intensity factor as function of dimensionless time.

First a sharp rise of $K(t)$ is noted, reaching a maximum of approximately 1.23 times the static stress-intensity factor $\sigma_0\sqrt{\pi a}$ at $c_2 t/a \approx 3$. Then the curve decays in amplitude and will oscillate around the static value.

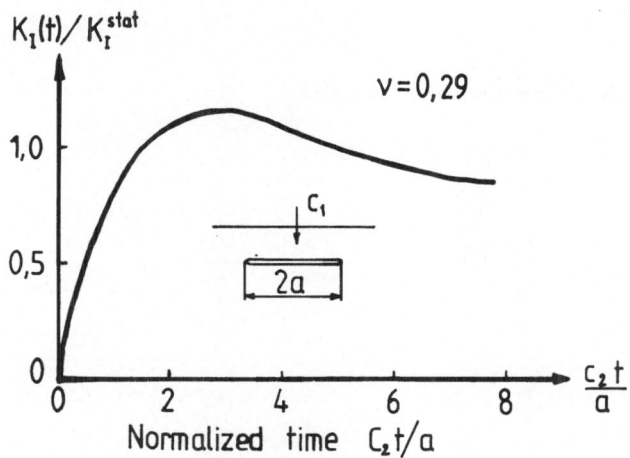

Figure 3. Dynamic stress-intensity factor for an incoming plane wave hitting a finite crack. Ref. [11].

Next consider an infinite strip of finite height 2h containing a finite crack of length 2a (Fig. 4). At t = 0 the crack faces are loaded with a step in stress σ_0, which is equivalent with stressfree crack faces and instead the horizontal boundaries subjected to a sudden impact load σ_0.

This problem is also analyzed in ref. [11]. The dynamic stress-intensity factor $K_I(t)$ is plotted in Fig. 5 for two different ratios of

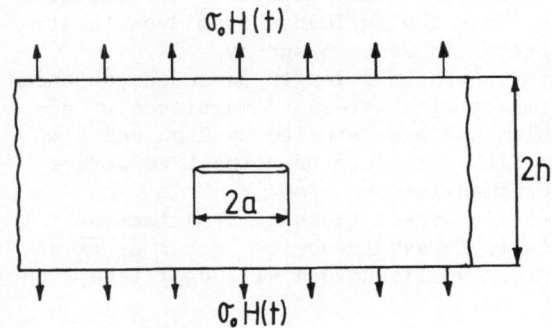

Figure 4. A finite crack in an infinite strip.

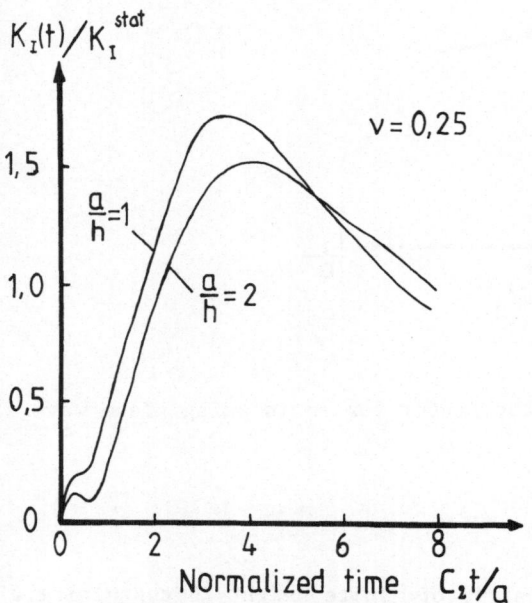

Figure 5. Normalized stress-intensity factor for the crack problem in Fig. 4. Ref. |11|.

a/h. A normalization is done with respect to the corresponding static values which in this case depend on the geometrical configuration, i.e. the quotient a/h.

The general behaviour of $K_I(t)$ is similar to the previous case. An intensification of $K_I(t)$ is seen as compared with the static solution, the peak values are predominantly caused by the interaction of waves emanating from the crack tips. The local maxima observed for small times are generated by wave reflections from the strip boundaries.

For the third example we consider a centrally cracked rectangular strip subjected to suddenly applied and maintained tension σ at each end (Fig. 6). The finiteness in outer geometry and crack length makes it difficult to solve this problem analytically. Therefore, various numerical methods have been used (Chen ⌊12⌋, Aberson et al ⌊13⌋, Mall ⌊14⌋, Brickstad ⌊15⌋) to obtain the plane strain response with respect to the stress-intensity factor as function of time (Fig. 6).

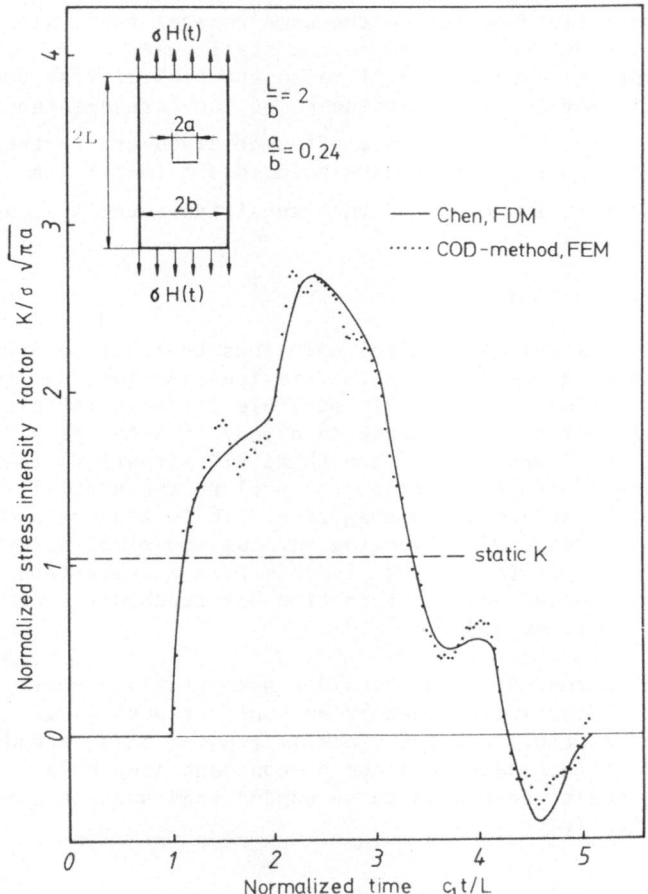

Figure 6. Normalized stress-intensity factor as function of dimensionless time for a centrally cracked strip. From ref. ⌊15⌋.

These numerical techniques utilize the relation between the stress-intensity factor and the stresses or displacements in the vicinity of the crack tip. The dynamic K is seen to exceed the static solution (as indicated by the broken line) by a factor reaching a maximum of 2.6 occurring when a shear wave hits the crack tip after being reflected from the nearest boundary. One may also observe the very high values of $\dot{K}$, the time derivative of K, which sometimes is used as measure of the rate of deformation.

In summary, these examples demonstrate that transient loading on stationary cracks causes a higher stress-intensity factor, during some time interval, than the corresponding static solution. The intensification $K(t)/K_{stat}$ is generally higher for finite geometries than for problems involving infinity in some direction. The dynamic stress-intensity factor is a function of the applied load, crack length and the time for the transient loading situation. However, the dynamic singular stress field at the crack tip does retain the same general form, with respect to angular variation, as compared to the static case.

The high values of the dynamic K will raise the risk of fracture for materials which fail for a critical value K_c of the stress-intensity factor. The high values of $\dot{K}$ which are often observed may further increase this risk since investigations have pointed out that K_c in general is a decreasing function of $\dot{K}$ for rate-sensitive materials, see for instance Klepaczko ⌊16⌋.

1.5 Crack-propagation solutions

We are now going to discuss some analytical solutions to fracture problems involving a propagating crack. As in the previous section, due to the mathematical difficulties, this is only possible for semi-infinite crack growth or extension of a finite crack in a body of infinite size in at least one direction. Despite this "non-physical" situation, these analytical solutions are of great importance to explore the special phenomena associated with rapid crack propagation. It is also valuable to have an analytical reference when checking various numerical solutions for growing cracks in finite bodies. In this case a comparison is possible within a time-interval until information has reached the crack tip(s) of the finiteness of the body.

1.5.1 __Steady-state solutions.__ First we consider some problems where the field quantities are constant as seen by an observer moving with the same pace as the crack tip, i.e. the crack is growing under steady-state conditions. Yoffe ⌊17⌋ studied a crack of constant length 2a moving with a constant crack speed $\dot{a}$ in an unbounded body subjected to uniaxial tension σ_o, Fig. 7.

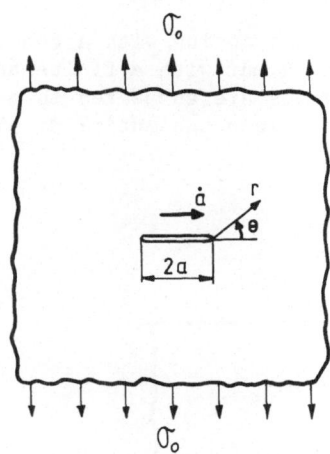

Figure 7. Geometry for Yoffe's problem

The dynamic stress-intensity factor for this problem turns out to be constant, equal to the static value $\sigma_o \sqrt{\pi a}$. By aid of relation (6), this implies that the energy-release rate G will be unbounded as the crack velocity $\rightarrow c_R$. Obviously, this result is not physically acceptable and the inadequacy of the solution is tied to the steady-state assumption. However, the correct velocity dependence of the singular crack-tip stress field (eq. (1)) was retained in ref. ⌊17⌋. Yoffe found that the maximum for the polar stress $\sigma_{\theta\theta}$ as $r \rightarrow 0$ is moved out from $\theta = 0$ to angles $\theta = 60^o - 90^o$ as the crack speed exceeds about $0.7\ c_R$. This result is not tied to the particular problem and has been suggested as a mechanism for crack branching.

Craggs ⌊18⌋ considered a semi-infinite crack moving steadily in an unbounded medium and subjected to time-independent tractions on the crack surfaces moving with the crack tip, see Fig. 8.

The same result as in ref. ⌊17⌋ of the stress distribution in the vicinity of the crack tip was retained. Also a stress-intensity factor independent of the crack velocity was obtained, against which the same objections may be raised as in the previous case.

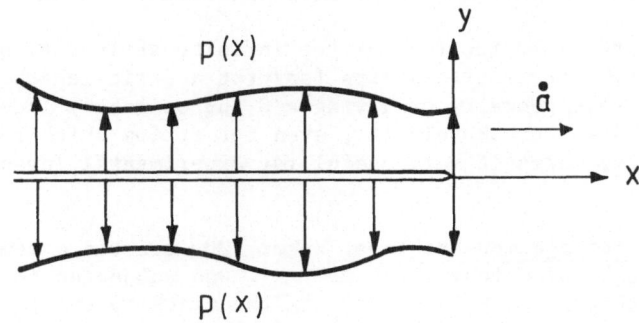

Figure 8. Craggs' problem.

Next consider a semi-infinite crack moving with a constant velocity $\dot{a}$ in a strip of infinite length but with a finite height 2h (Nilsson ⌊19⌋). The horizontal boundaries are subjected to a prescribed uniform normal displacement $v = \pm v_o$ and zero tangential displacements $u = 0$, see Fig. 9.

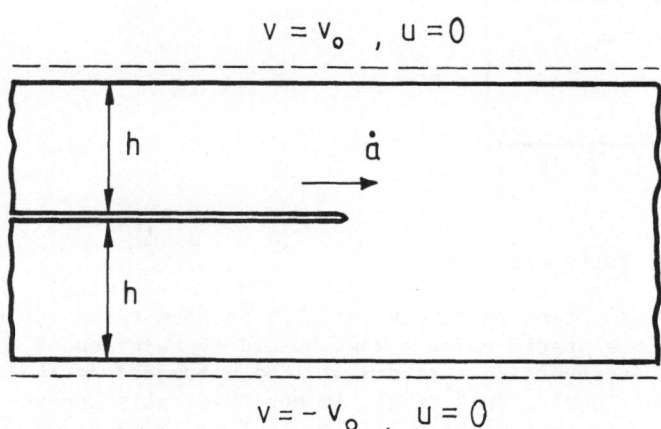

Figure 9. The strip problem considered in ref. ⌊19⌋.

The resulting energy-release rate for the plane-strain case is

$$G = G_{stat} = \frac{v_o^2 E}{h} \cdot \frac{1-\nu}{(1+\nu)(1-2\nu)} \tag{9}$$

that is G is constant, independent of the crack velocity. With the relation (6), a velocity-dependent stress-intensity factor is obtained with the more physically acceptable behaviour of K approaching zero when $\dot{a} \to c_R$. This difference in the results as compared with the steady-state solutions by Yoffe and Craggs may be attributed to the finite height of the strip and the displacement controlled boundary conditions.

It appears that the solution (9) for the infinite strip serves as a good approximation also for finite strips (say for a strip length >4h). The property of this geometry of giving a G approximately independent of crack length and crack velocity, even for strips which can be realized in practice, makes it very useful for experimental investigations.

1.5.2 <u>Transient crack-propagation problems</u>. Baker ⌊20⌋ treated a semi-infinite crack in an unbounded body which at $t = 0$ was subjected to prescribed tractions at the crack faces causing the crack to start growing with a constant velocity $\dot{a}$. This is equivalent with tensile stress waves with speed c_1 hitting the crack at $t = 0$, which then starts to grow, see Fig. 10.

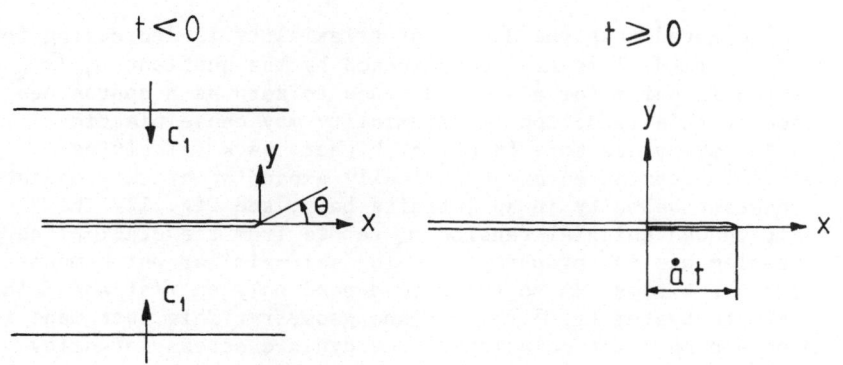

Figure 10. The problem studied in ref. $\lfloor 20 \rfloor$.

The variation of the dynamic stress-intensity factor may be written as

$$K_I(t,\dot{a}) = K_I(t,\dot{a}=0) \cdot f_2(\dot{a}) \tag{10}$$

where $f_2(\dot{a})$ is shown in Fig. 11.

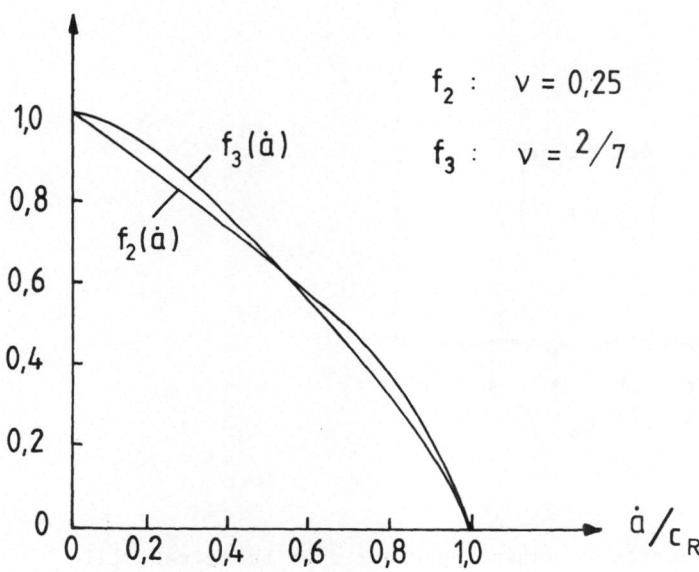

Figure 11. The velocity factors $f_2(\dot{a})$ and $f_3(\dot{a})$ appearing in eqs. (10) and (11).

Baker also noted that the degree of triaxiality is decreasing for increasing crack speed. This may be indicated by the quotient σ_{yy}/σ_{xx} for $\theta = 0$, which is unity for $\dot{a} = 0$ and tends to zero as $\dot{a}$ approaches c_R. The effect of this reduction in triaxiality may cause plastic deformations to take place more freely at higher crack velocities.

Broberg [21] considered a symmetrically expanding crack propagating with a constant velocity in an infinite body, see Fig. 12. The loading condition was uniaxial tension σ_o remote from the crack plane.

This problem has the property of being self-similar which means that the field quantities can be shown to depend only on $r/\dot{a}t$ and θ in a polar coordinate system (r, θ) for a plane geometry. This fact made it possible to obtain an exact solution of the dynamic stress-intensity factor $K_I(t)$ as

$$K_I(t) = \sigma_o \sqrt{\pi \dot{a}t} \cdot f_3(\dot{a}) \tag{11}$$

where the velocity factor $f_3(\dot{a})$ is shown in Fig. 11.

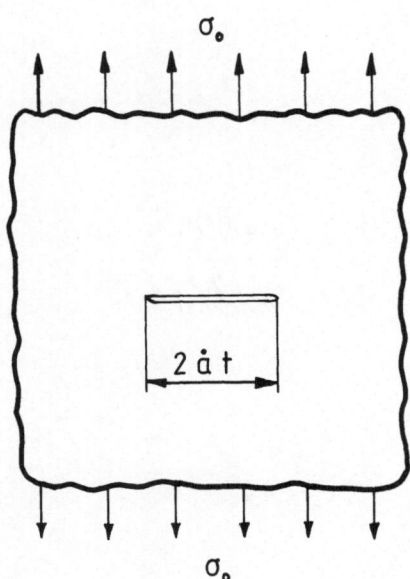

Figure 12. A symmetrically expanding crack studied in ref. [21]

Finally, we discuss a useful result, derived by Freund [22]. He considered a semi-infinite crack in an infinite plate which at $t = 0$ begins to propagate with an arbitrary crack motion $a(t)$ with $a(0) = 0$, see Fig. 13. It is assumed that arbitrary time-independent loads are prescribed.

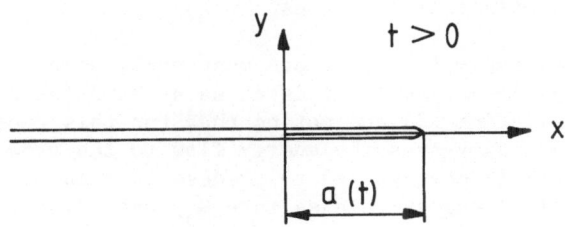

Figure 13. A semi-infinite crack moving with an arbitrary rate of extension.

The dynamic stress-intensity factor for this case is

$$K_I = f_4(\dot{a}) \int_0^{a(t)} \sqrt{\frac{2}{\pi}} \cdot \frac{p(x)dx}{\sqrt{a(t) - x}} \tag{12}$$

where $p(x)$ is the tensile stress along the plane $y = 0$ acting at $t=0$. Thus, K_I depends essentially only on the instantaneous crack-tip position and the current crack velocity, and not on the crack-growth history. The velocity factor $f_4(\dot{a})$ is identical with $f_2(\dot{a})$ appearing in eq. (10). The integral in (12) is equal to the static stress-intensity factor for the crack length considered at a certain time instant. As pointed out by Smith ⌊23⌋ this is tied to the particular loading conditions and it is not true for prescribed time-independent displacements.

By aid of eq. (6) we could equally well formulate the result in terms of the energy release rate G according to

$$G = G_{stat}(a) \cdot f_5(\dot{a}) \tag{13}$$

where for all practical purposes the velocity-dependent function could be approximated by

$$f_5(\dot{a}) \approx 1 - \dot{a}/c_R \tag{14}$$

In ref. ⌊24⌋ the above result was generalized to dynamic loads with a resulting stress-intensity factor of the same form as eq. (10).

Perhaps a few comments are appropriate on the maximum crack speed $\dot{a}_{max}$ which could be reached in practical situations. The theoretical results on moving cracks in infinite bodies denounce the Rayleigh wave speed c_R as an upper limit. This limit, however, is not reached in practice and crack-growth ratios $\dot{a}/c_R > 0.6$ are seldom observed in experiments. The reason for this is not fully understood but it has been suggested that crack branching may set an upper limit to the speed of fracture. Also other effects, which may be connected to crack branching, are probably influencing $\dot{a}_{max}$. For instance, increased surface roughening and heat dissipation observed for high crack velocities may all contribute in making the process of fracturing less efficient.

1.6 Criterion for crack propagation and arrest

In order to determine the crack-tip motion one must apply some fracture criterion. For linear elastic materials it is close at hand to start from the energy-balance relation (3) and noting that for this case the energy flux $\dot{\gamma} = \dot{a}G$, where G expresses the energy flow to the growing crack tip. The crack growth is now assumed to proceed in a manner so that G is equal to a critical energy-release rate G_{pc} according to

$$G(t) = G_{pc}(\dot{a}) \qquad \dot{a} > 0$$
$$G(t) < G_{pc}(\dot{a} = 0) \qquad \dot{a} = 0 \tag{15}$$

The last relation corresponds to a condition for a crack arrest to occur. The material property G_{pc} is assumed only to be dependent on the instantaneous crack velocity $\dot{a}$. By aid of (6) it is completely equivalent to formulate a fracture condition in terms of K_I as

$$K_I(t) = K_{pc}(\dot{a}) \qquad \dot{a} > 0$$
$$K_I(t) = K_{pc}(\dot{a} = 0) \qquad \dot{a} = 0 \tag{16}$$

where K_{pc} is denoted the crack-propagation toughness of the material.

Apart from thickness and temperature, G_{pc} and K_{pc} may also depend on geometry, crack length a, and higher time derivatives of a. This matter is worth some discussion. For a fracture criterion of the type (16) being relevant, i.e. involving essentially only K_I and $\dot{a}$, it is assumed that a sufficiently large K-dominated zone is developed surrounding the inelastic zone which is present in the vicinity of the crack tip for all real materials. In this K-dominated zone the stresses may be adequately described by eq. (1). Events within the small inelastic region are then controlled by the K-field and may be expected to be a functional of K(t) and $\dot{a}$(t) only, at least if these quantities do not change much during a time interval when a fixed particle in front of the moving crack tip is deforming non-elastically. The conditions for this have been derived by Dahlberg and Nilsson ⌊25⌋ and can be expressed as

$$\frac{2R}{\dot{a}K} \left| \frac{\partial K}{\partial t} \right| \ll 1$$

$$\frac{2R}{\dot{a}^2} \left| \frac{\partial^2 a}{\partial t^2} \right| \ll 1 \tag{17}$$

where R is the radius of the inelastic zone. If these conditions are satisfied, small-scale yielding (SSY) can be said to prevail and then it is expected that the criteria (15) or (16) may adequately describe the crack-propagation process. All inelastic effects are then lumped into the material parameter K_{pc} or G_{pc}. In practice it is difficult to decide if SSY is valid for a particular case. An operating definition may then be that a state of SSY prevails when the crack growth can be satisfactorily described by (15) or (16). The validity of this description can only be checked by careful experiments.

The result of some crack-propagation experiments on edge cracked sheets of a high strength steel is shown in Fig. 14, illustrating the velocity dependence on K_{pc} (Brickstad and Nilsson [26]).

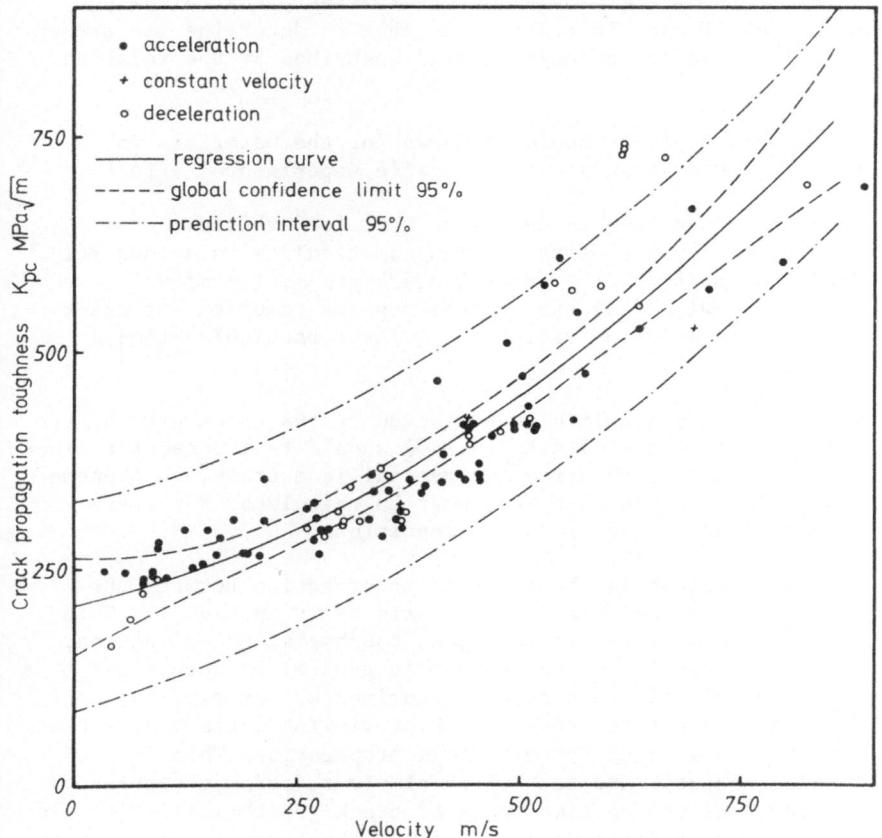

Figure 14. Crack-propagation toughness versus crack velocity, ref. [26].

K_{pc} was found to be a strictly increasing function of $\dot{a}$, and no significant effect of crack-tip acceleration was seen. For higher crack velocities corresponding to higher load levels an increased scatter may be observed indicating the presence of non-linear effects. This can be manifested as a geometry dependence. A comparative study reported by Dahlberg et al $\lfloor 27 \rfloor$ did also result in deviations from a unique relationship between K_{pc} and $\dot{a}$ for different specimen geometries in this high-velocity region in Fig. 14.

Several different effects may contribute to the increase of K_{pc} with $\dot{a}$. These include for instance increased roughening of the crack surface, tendency to crack branching and a decrease of triaxiality in the stress state with increasing crack velocity. Also the strain-rate sensitivity of certain materials may have an increasing effect on K_{pc}.

If we now turn to the problem of crack arrest, the following requirements should be met in order to be able to determine the crack-tip motion if the fracture process is well described by the relation (16):

i) The K_{pc}-$\dot{a}$ relation should be known for the material. In general, this demands a considerable experimental effort.

ii) It must be possible to determine $K(t)$ for arbitrary crack trajectories. This often implies numerical calculations for finite geometries. The special crack-tip motion $a(t)$ consistent with (16) then represents the solution. At crack arrest this solution yields $\dot{a} = 0$ for a particular time instant t_1.

iii) To check for a possible reinitiation of the crack growth, the arrest condition $K(t) < K_{pc}(\dot{a} = 0)$ should be checked for times $t > t_1$. Since crack arrest in general is a transient phenomenon, this requires that $K(t)$ must be calculated for a stationary crack under transient conditions.

Of special interest is the point of intersection between the regression curve $K_{pc}(\dot{a})$ and the vertical axis $\dot{a} = 0$ in Fig. 14. This value, $K_{pc}(0)$, represents the crack-arrest toughness. By definition, $K_{pc}(0)$ is a material property and it must in general be determined by a fully dynamic analysis of crack-arrest experiments, see e.g. Kalthoff et al $\lfloor 28 \rfloor$. This is a natural consequence of viewing a crack arrest as the termination of a general dynamic crack propagation. This is contrary to a quasistatic approach which simply regards the arrest event as the reversal on the time scale of crack-growth initiation. In that case it would be sufficient to establish the arrest capability of a material by determining the arrest value of the static stress-intensity factor, confer the ASME Code procedure $\lfloor 29 \rfloor$ for pressure vessels. It has been argued that a comparison with the solution for a semi-infinite crack propagating in an unbounded body, eq. (12), provides some justification to this quasistatic approach since then the dynamic stress-intensity factor at $\dot{a} = 0$ equals the static K. This is, however, a highly idealized situation since the presence of free surfaces and wave reflections will soon invalidate this solution.

In Fig. 15, the result of a numerical elastodynamic analysis of a crack-arrest experiment is shown in form of the dynamic K as function of time before and after the arrest event (Brickstad |15 |). It is evident from this figure that dynamic effects are influencing the post-arrest behaviour a considerable time after the arrest has occurred. It was found in ref. |15 | that the arrest toughness K_a, calculated as the dynamic stress-intensity factor at the time of arrest, proved to be a useful quantity to assess the capability of the material to arrest a running crack. The quasistatic K_a turned out to be dependent on the

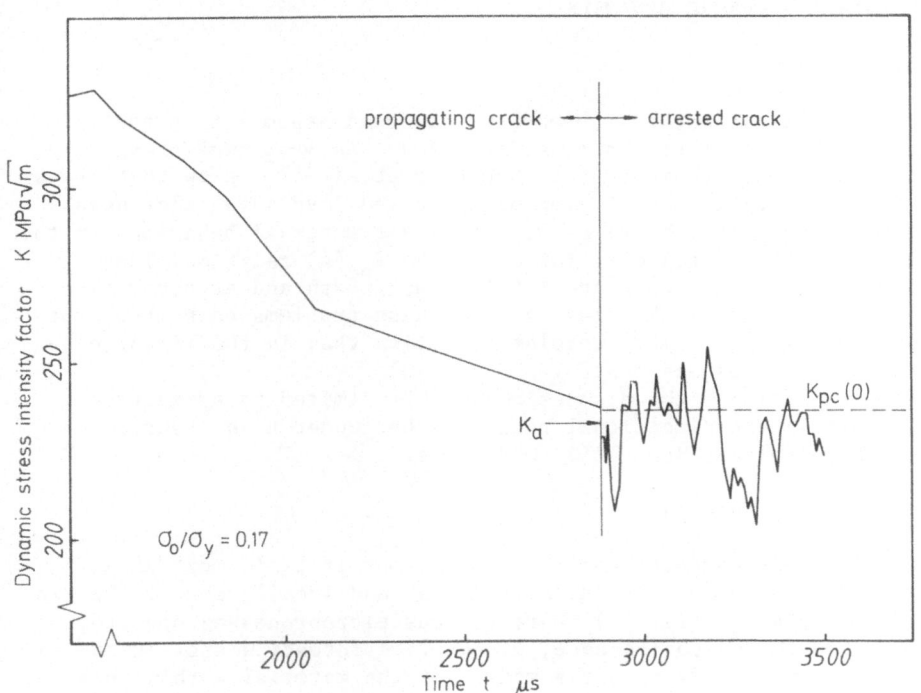

Figure 15. The dynamic stress-intensity factor as function of time before and after crack arrest. Ref. |15 |.

crack-growth history and it is thus of less importance as a general property related to the arrest event.

In practice to achieve a crack arrest it is perceived that either $K_I(t)$ should be kept as low as possible, and/or a material with a high resistance to rapid crack growth, i.e. a high K_{pc}-level, should be chosen. For the latter alternative, it should be pointed out that it generally requires knowledge of the complete K_{pc}-$\dot{a}$ relation of the material, or at least the arrest toughness. That the static fracture toughness K_{Ic}, governing crack-growth initiation, is high for a certain material does not necessarily imply that this material is better with respect to arrest capability than a material with a somewhat lower K_{Ic}-value. The first alternative, to prevent high K_I-levels, is a matter of structural design. It may involve the use of auxiliary structural elements such as stiffeners, reinforcing rings etc. to produce a decrease in stress-intensity factor. A propagating crack may then be brought to a standstill in these "arrestors".

2. NON-LINEAR FRACTURE DYNAMICS

2.1 Introduction

In section 1 linear material behaviour has been assumed to prevail in almost the whole part of the considered body. As mentioned previously experimental investigations, e.g. Dahlberg et al [27], show that deviations from the linear description occur as the load level increases. Furthermore, even if SSY prevails, non-linear material behaviour in the near vicinity of the tip will influence the $K_{pc}(\dot{a})$ relation. These facts motivate studies of crack initiation, growth and arrest under non-linear conditions. The treatment of these problems constitute for obvious reasons a much more complex situation than in the linear case and explicit results are scarce.

As in section 1 the discussion will be limited to symmetrical crack growth in two-dimensional bodies either under plane (strain or stress) or antiplane (Mode III) conditions.

2.2 Some general aspects

In materials that exhibit non-linear behaviour it is in most cases unlikely that the crack tip is well-defined and ideally sharp. The tip is instead a region (Fig. 16) where various microprocesses operate, e.g. void growth and coalescence, micro-crack formation etc. It may in some cases be necessary to use a model for the material within this process zone (cf. Broberg [40]) that differs from the model of the "normal" material. In [40] such a modelling is discussed in qualitative terms and some conclusions regarding the energy consumption in the tip vicinity are drawn.

Depending on the characteristic sizes d_1 and d_2 we can distinguish between three main classes of crack-tip region models. This region is here understood to be where different governing equations are assumed.

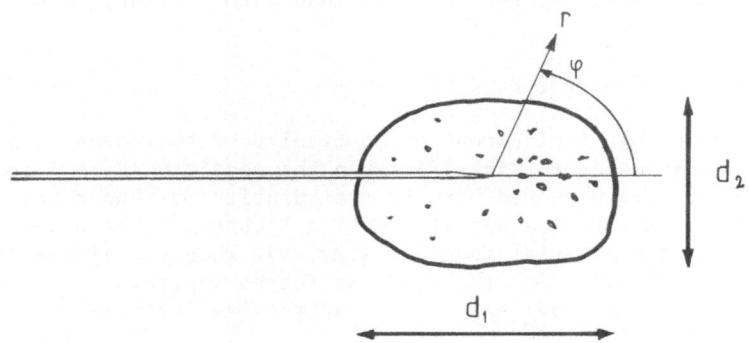

Figure 16. The process region around a crack tip.

Type I Both d_1 and d_2 are assumed to be finite. So far, no model of this kind seems to have been considered in quantitative terms together with dynamic effects.

Type II $d_2 \to 0$ and d_1 finite. These are the so-called cohesive zone models which have been used extensively in cases when the remaining material is assumed to be linearly elastic. Few results are available when the surrounding material behaves non-linearly. However, it is believed that the cohesive zone models are a promising way of description and they will therefore be considered in more detail in the subsequent section.

Type III $d_1 = d_2 \to 0$. In this case the crack tip region shrinks to zero and we call this the singular crack-tip description. The bulk of hitherto performed work is concerned with this model.
     In some cases combinations in a certain sense of the features of type II and type III models are used. This is for instance the case when the boundary problem is set as being of the third type, but a finite length is retained when specifying the crack-growth criterion. It should also be pointed out that in numerical analyses finite dimensions of the crack-tip region may more or less unintentionally be specified by the particular nature of the discretization process.

2.3 Cohesive zone models

The concept of a cohesive zone at the crack tip was first proposed by Prandtl |41|. Strifors |42-43| has discussed basic aspects of cohesive zones within the framework of a thermo-mechanical theory. The only efforts to solve boundary-value problems that combine dynamic effects and cohesive zones in a non-linear material seem to have been performed by Nilsson and Oldenburg |44|. In their analysis the material is assumed to behave according to its normal constitutive relation as long as the criterion (18), here written for pure mode I conditions, is satisfied.

$$F_c(\sigma_{ij}, \kappa) = \sigma_y - g(0, \kappa) < 0 \tag{18}$$

g is a function of the displacement discontinuity of the opened crack and a hardening parameter κ (Fig. 17). When the condition $F_c = 0$ is met, decohesion occurs measured here by the quantity Δv. Under the decohesive process there is a normal traction T_c between the crack surfaces given by the material function $g(\Delta v, \kappa)$. When the displacement Δv exceeds a certain value Δv^o the cohesive forces vanish.

The cohesive zone model has several attractive features.

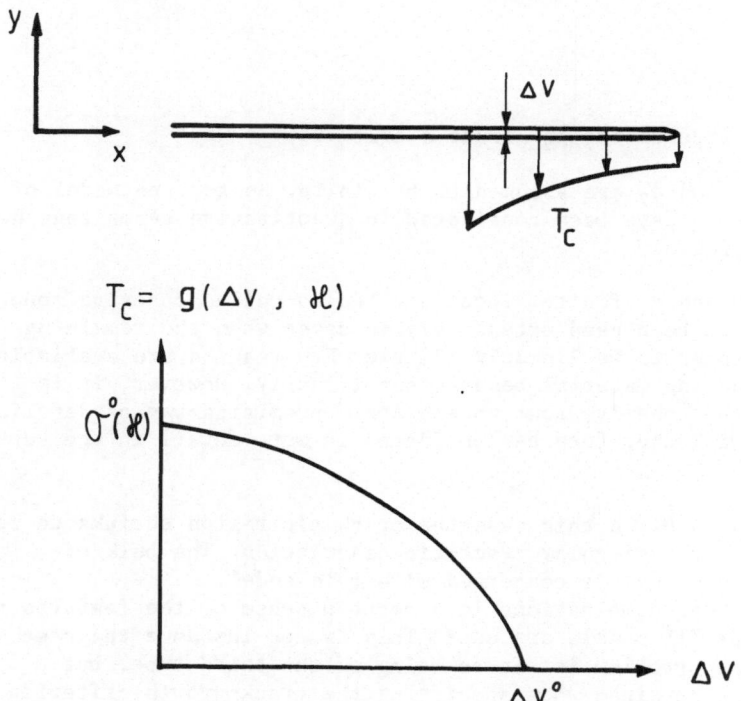

Figure 17. The cohesive zone model.

It is conceptually simple and self-consistent. No additional fracture
criteria are needed once the criterion (18) is specified. Thus it is
free from the difficulties that adhere to models with a singular crack
tip (cf. Rice |45|). It is also applicable to the problem of crack
initiation in an uncracked body. The cohesive zone can also be used to
model fracture process zones where void growth and similar processes
operate provided that the dimension of the zone perpendicular to the
crack line is small.

There are, however, drawbacks, the first one being the computational difficulties. Numerical methods must be used in almost all
problems. Since the size of the cohesive zone in general is small compared to other dimensions of the problem, a fine mesh must be used to
resolve the stress distribution accurately. Nilsson and Oldenburg |44|
try to overcome this difficulty by converting the displacement discontinuity Δv into continuum strains by an energy balance argument. It is
not clear to the present authors to what extent the properties of the
original problem are retained and if the equivalent continuum problem
converges to the real solution when the element mesh is made infinitely
dense. Nilsson and Oldenburg apply their model to a case of impact on a
concrete rod where the material is assumed to be elastic-viscoplastic-
plastic. They claim good agreement with experimental results for the
axial strain measured at a few places in the rod.

The other main difficulty is how to choose the function g since
it is not directly measurable. One possibility is to determine the
appropriate form of this function by physical modelling of the microprocesses.

In the present authors' opinion the cohesive zone model is a
promising alternative and as computers and computer programs develop
the computational difficulties can certainly be overcome.

Beside the above cited papers there seems to be little work combining dynamic effects, cohesive zones and non-linear material behaviour. In the remaining part of this lecture we will turn to the third
type of crack modelling, i.e. the singular crack tip.

2.4 Crack-growth criteria for singular crack tips

Strifors has extensively investigated the basic relations for singular
crack tips within the theory of large deformations |42| and within the
theory of linearized strain measures |43|. Here the discussion will be
restricted as before to small deformation theory.

An important result for the kinematics of fields in the vicinity
of a singular crack tip is given by eqs. (19) and (20) [Strifors
|43|).

$$\dot{u}_i \to -\frac{\partial u_i}{\partial x} \dot{a} \qquad \text{as } r \to 0 \tag{19}$$

$$\ddot{u}_i \to \frac{\partial^2 u_i}{\partial x^2} \dot{a}^2 \qquad \text{as } r \to 0 \tag{20}$$

These equations hold provided that (21) is satisfied.

$$u_i \sim r^{1-\alpha} \qquad \text{as } r \to 0 , \qquad 0 < \alpha < 1 \tag{21}$$

For many cases of interest the relation (21) is satisfied. Since (19) and (20) contain only time derivatives in form of the crack-tip velocity $\dot{a}$, the singular crack tip can be said to be in a state of <u>asymptotically steady motion</u>. This fact is shown by eq. (1) for the linearly elastic case.

Another quantity of interest is the energy flow to the tip γ. The relation (5) given previously is also valid for non-linear problems provided that thermal effects are neglected. For the cohesive zone model no further assumptions than those made when specifying $g(\Delta v, \kappa)$ are needed to solve the problem of determining the crack growth. This is not so for the singular crack-tip model and a crack-growth criterion has to be formulated separately. This subject is complex (cf. Rice [45]) and only a few aspects will be discussed here.

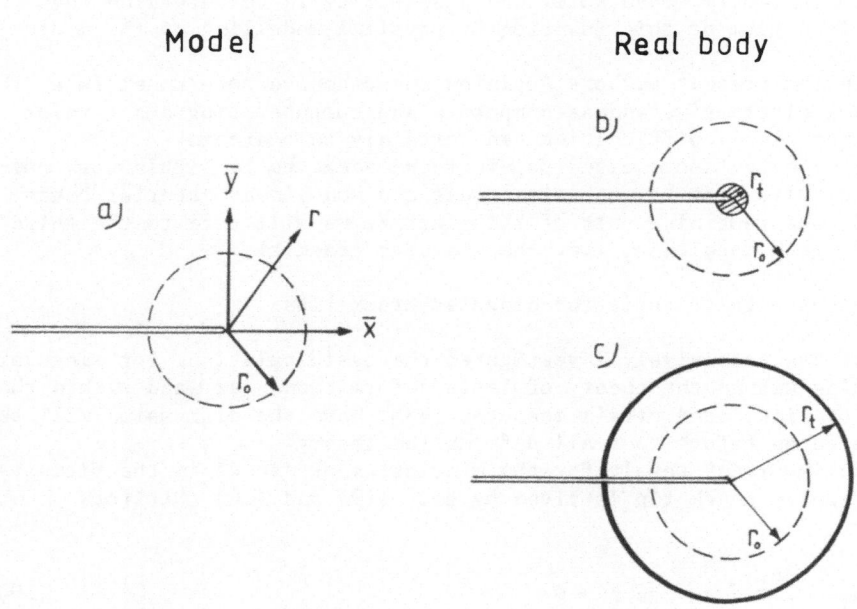

Figure 18. Illustration of possibility of a one-parameter description.

Consider a situation (Fig. 18) where the material behaviour is modelled by some specific constitutive relation. Suppose now that for this model the stress and strain fields in the near-tip vicinity are completely determined by a scalar parameter Q, the tip velocity $\dot{a}$ and the constitutive parameters of the material model. Q will of course depend on the entire boundary value problem. Suppose further that within a region $r < r_o$ the state is determined by this near-tip field to some specified accuracy. Such a characterization is, as discussed above, possible for linearly elastic materials and is also possible for several non-linear material models.

A growth law which is consistent with such a model is to assume a unique relation between Q and $\dot{a}$, viz.

$$g_c(Q, \dot{a}) = 0 \qquad\qquad (22)$$

In order to verify if a criterion of this form is valid or not for a particular class of materials in a certain range of situations, experiments have to be performed. As mentioned earlier we cannot obtain more guidance from the particular model used. Only a higher order model can provide additional information. We can, however, make some conjectures on the question under which circumstances it is likely that an equation of type (22) is valid.

Suppose that in the real body for which the modelling has been made the situation is in a particular case as sketched in Fig. 18b. Within the region $r < r_t$ the governing equations differ from those assumed in the model. This may occur because of different material behaviour, large deformation effects or thermal effects. If r_t is much smaller than r_o the criterion can be expected to be valid. For several reasons we cannot say this definitely. It may be so that the disturbances created by the different behaviour within r_t propagate outwards and have effects at larger distances. This happens if the governing equations within r_o become hyperbolic. This question can in general only be answered if a higher-order model incorporating the different behaviour within r_t is used. Without actually modelling this region only qualitative conclusions of the type shown in eq. (17) can be obtained. This problem is analogous to that of a quantitative formulation of Saint Venant's principle for non-linear bodies. Another reason why eq. (22) may not be valid is that effects of the different history during the time a particle is swept by the region $r < r_t$ may be of importance. However, the question if eq. (22) is valid under the circumstances sketched in Fig. 18b can, as mentioned above, only be answered by experiments or partly by more elaborate modelling. We cannot expect to find a better criterion than those covered by (22) within the frame of the chosen model. It may of course also be difficult to ensure whether the situation is as sketched in (22). This problem can often be qualitatively treated by dimensional analysis.

The other case that may occur is when r_t is of the same order or larger than r_o. In this case there is no particular reason to expect that a criterion of the type (22) is valid. This case may then be viewed in the same manner as when no one-parameter description exists. Either the model must be changed or some other criterion must be used.

In section 1 fracture criteria for the linearly elastic case were discussed. It was pointed out that an energy criterion of type (15) is completely equivalent to a stress-intensity factor criterion (16). If a one-parameter description of the type defined above exists the same is true also for non-linear material behaviour. This is immediately obvious from the energy expression (5). Thus a criterion of the type (22) is completely equivalent to an energy criterion provided of course that the singularity permits a non-zero energy flow.

For the other case, i.e. $r_t \gtrsim r_o$ (Fig. 18c), various fracture criteria are possible. The following way has been suggested by several authors. The crack is assumed to start to grow if a critical value of some strain measure is reached at a certain structural distance δ_f ahead of the tip. During growth this critical level of the strain measure is maintained at the distance δ_f ahead of the moving tip. The strain measure may be the ε_y (McClintock |46| or the effective plastic strain ε_e^p (Rice et al |47|). Other criteria of this type are possible.

Aboudi and Achenbach |48| for instance suggested the use of the dissipated plastic work instead of a strain measure. Experimental verification of criteria of this type is, however, non-existent and for this reason it is not possible to express any preference.

It can be noted that in this type of fracture criterion a finiteness as in models of type II is introduced.

2.5 Solutions to boundary-value problems

The number of available solutions for non-linear behaviour is much less than in the linear elastic case. For obvious reasons these solutions show a greater variety and less common features than linear solutions. In the following sections a number of solutions for different constitutive laws will be briefly reviewed. Appendix 3 contains a presentation of the constitutive relations mentioned.

The solutions fall into two classes. In the first class only the asymptotic field in the vicinity of the tip is sought. Such solutions are of interest for setting up proper fracture criteria. In the second class we have complete solutions to boundary-value problems. These have exclusively been obtained by numerical methods, in most cases the finite element method (FEM).

2.5.1 Rate-independent elastic perfectly plastic material. A number of authors have studied the near-tip field for a mode III crack growing in an elastic perfectly plastic material. To the knowledge of the authors no such solution has been obtained for the case of transient loading of a stationary crack. This may prove difficult judging from the anomalies that occur for velocities tending to zero in the solutions for moving cracks.

The solutions are for the case of a steadily moving crack in a material obeying the Prandtl-Reuss flow rule (see Appendix 3) and contributions have been given by Slepyan ⌊49⌋, Achenbach et al ⌊50⌋, Achenbach and Dunayevsky ⌊8⌋, Dunayevski and Achenbach ⌊51⌋, Freund and Douglas ⌊52⌋. Let $\bar{x}, \bar{y}, \bar{z}$ denote a coordinate system fixed to and moving with the tip (Fig. 19).

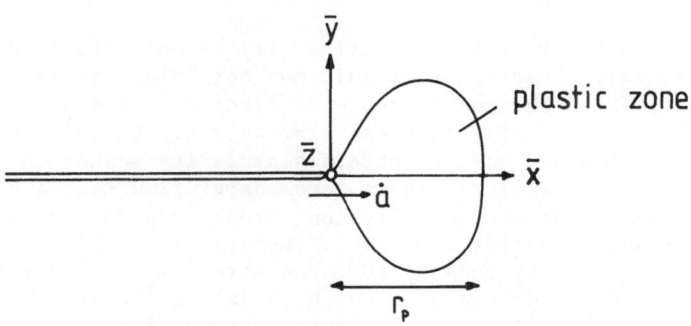

Figure 19. Notation for the problem of steady growth.

It is found that the stresses $\sigma_{\overline{zy}}$ and $\sigma_{\overline{zx}}$ are bounded and of the order of the yield stress in pure shear τ_0. The strain $\varepsilon_{\overline{zx}}$ is also bounded and of the order τ_0/μ where μ is the elastic shear modulus. The only singular strain is $\varepsilon_{\overline{zy}}$ for which Freund and Douglas ⌊52⌋ derived an exact implicit solution valid for $\bar{y} = 0$ within the plastically deforming region. Their solution contains a parameter r_p (the extent of the plastic zone Fig. 19) which has to be determined by a complete solution.

$$\varepsilon_{\overline{zy}} = \frac{(1-m^2)}{4\ m^2}\ \ln\ \left(\frac{1-m^2\tan^2\omega}{1-m^2}\right) + 1\ ,\ \text{for}\ \overline{y} = 0 \tag{23}$$

$$m = \dot{a}/c_2 \tag{24}$$

$$\overline{x}/r_p = \frac{I(m\ \tan\ \omega)}{I(m)} \tag{25}$$

$$I(z) = \int_0^{\frac{1-z}{1+z}} \frac{s^{\frac{1-m}{2m}}}{1+s}\ ds \tag{26}$$

In the limit $\overline{x} \rightarrow 0$ this solution agrees with the asymptotic form (27) found by the other authors cited above

$$\varepsilon_{\overline{zy}} \rightarrow -\frac{\tau_o}{2\mu}\ \frac{1-m}{m}\ \ln\ \left(\frac{r}{r_p}\right)\quad \text{as}\ \overline{x} \rightarrow 0 \tag{27}$$

It can first be noted that the criterion (21) is not satisfied and that thus the locally steady-state result may not hold. We also note that this solution has the one-parameter characterization property, r_p being the scalar variable in these expressions. Furthermore, since the stresses are bounded and the strain-singularity weaker than r^{-1}, insertion into eq. (5) will result in zero energy flow to the tip, thus precluding the use of an energy criterion. Freund and Douglas also point out that the range of validity of the asymptotic form (27) is extremely small, typically less than $r_p/100$. The strain $\varepsilon_{\overline{zy}}$ becomes unbounded as $\dot{a} \rightarrow 0$ and does not converge to the quasi-static result obtained by Chitaley and Mc Clintock [53]. This fact was discussed by Dunayevsky and Achenbach [51]. These arguments taken together suggest that a fracture criterion ought not to be based on the asymptotic solution.

Freund and Douglas [52] favoured the criterion that a critical plastic strain $\varepsilon_{\overline{zy}}^p = \varepsilon_f$ should be maintained at a distance δ_f ahead of the tip. This necessitated a determination of the elastic-plastic boundary. They considered the SSY-problem, i.e. the boundary conditions at large distances from the tip are given by the elastic field with the intensity K_{III}. The solution was accomplished by a finite element technique especially derived for the particular problem. In Fig. 20 their result for the shape of the plastic zone is shown. The approximate expression (28) was developed for r_p from the numerical results.

$$r_p = (0.295 - 0.5 \, m^2) \, (K_{III}/\tau_o)^2 \tag{28}$$

δ_f can be eliminated in favour of K_{III}^C which is the critical stress-intensity factor for initiation of growth of a <u>stationary</u> crack assuming the same fracture criterion. The resulting expressions become as follows, [52].

$$\left(\frac{K_{III}}{K_{III}^C}\right)^2 = \frac{I(m) \, I(q)}{\pi(2\varepsilon_f+1)(0.295-0.5 \, m^2)} \tag{29}$$

$$q^2 = 1 - (1-m^2) \, \exp\lfloor-4 \, \varepsilon_f m^2/(1-m^2) \rfloor \tag{30}$$

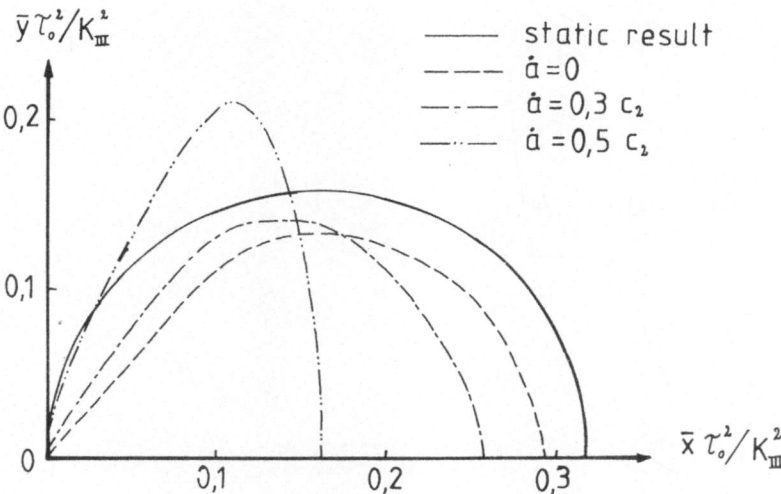

Figure. 20. The shape of the plastic zone. After Freund and Douglas [52].

The functions are monotonously increasing with $\dot{a}$ and show the same general behaviour as has been observed in experiments (Fig. 21).

Calculations of the near-tip field for the corresponding problem in <u>plane strain</u> were performed by Achenbach and Dunayevsky [8]. They assumed the Tresca yield-criterion. Since the plane strain case is much more difficult than the antiplane shear case they had to employ a perturbation approach valid for small values of m. The stresses and the strains exhibit qualitatively the same features as in the mode III case. Only one strain component $\varepsilon_{\overline{xy}}$ has a singularity for finite values of m. In the limit m = $\dot{a}/c_2 \to 0$ several of the strain components become unbounded and the fields do not converge to the quasi-static results by Rice and Tracey [54]. Achenbach and Dunayevsky [8] conjectured that this may be the consequence of a vanishing convergence radius for the dynamic solution as m → 0.

Several features of dynamic crack growth in elastic perfectly plastic materials remain to be investigated. Some of the peculiarities mentioned here suggest, however, that the ideally plastic model is perhaps not suitable to use for description of dynamic crack-growth problems. As will be seen later the inclusion of strain-rate effects to some extent resolves these difficulties.

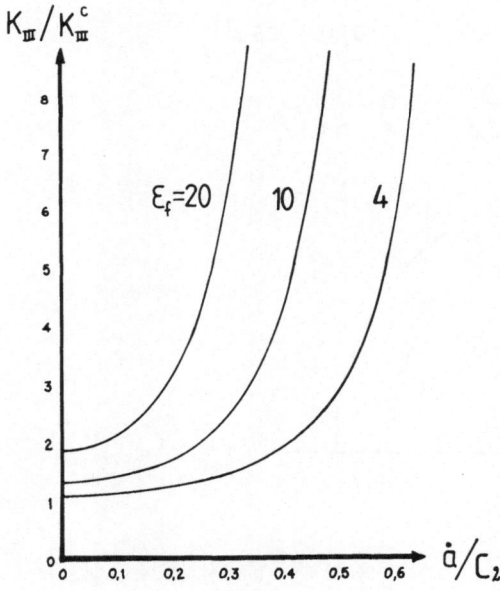

Figure 21. Critical stress-intensity factor as function of velocity for a perfectly plastic material. After Freund and Douglas [52]

2.5.2 Rate-independent elastic-plastic hardening material.

For stationary cracks Achenbach [55] has discussed the singular solution assuming a deformation theory with power-law hardening. This solution can thus only be expected to be valid during the initial loading part as a deformation theory does not model unloading properly. Achenbach finds that the singularity is of the well known HRR-type [56,57].

$$\varepsilon_{ij} \rightarrow k_\varepsilon(t) E_{ij}(\theta) r^{-1/(N+1)} = k_\varepsilon \, \hat{\varepsilon}_{ij} \qquad \text{as } r \rightarrow 0 \tag{31}$$

$$\sigma_{ij} \rightarrow k_\sigma(t) \Sigma_{ij}(\theta) r^{-N/(N+1)} = k_\sigma \, \tilde{\sigma}_{ij} \qquad \text{as } r \rightarrow 0 \tag{32}$$

k_ε and k_σ are scalars to be determined from a complete analysis, the functions E_{ij} and Σ_{ij} can be found in [56,57]. N is the hardening exponent. k_ε and k_σ are related by (33).

$$k_\sigma(t) = k_\varepsilon(t)^N \tag{33}$$

By aid of a path-independent integral Achenbach [55] derived the following relations for the SSY case.

$$\bar{k}_\sigma(p) \, \bar{k}_\varepsilon(p) \, J \, (\hat{\varepsilon}_{ij}, \, \tilde{\sigma}_{ij}) = \frac{\bar{K}(p)^2 (1-\nu^2)}{E} \tag{34}$$

$\bar{k}_\sigma$, $\bar{k}_\varepsilon$ and $\bar{K}$ are here Laplace-transforms with p as transformparameter. J is the conventional J-integral calculated with the fields $\hat{\varepsilon}_{ij}$ and $\tilde{\sigma}_{ij}$. By combining (33) and (34) the functions $k_\sigma(t)$ and $k_\varepsilon(t)$ can be solved for a given time history of the stress-intensity factor K(t).

Numerical studies with finite element methods have been performed. One such example is the paper by Ahmad et al [58]. In [58] no particular modelling of the tip was made, standard isoparametric elements were used. The crack-tip opening displacement and the J_1-integral (see eq. (A2.4)) were evaluated. It was found that the elastic-plastic analysis differs from the purely elastic when analysing the impact of a 3-point bend specimen. No general conclusions, however, can be drawn from this example.

Since the equations for <u>growing</u> cracks in hardening materials are under some circumstances elliptic, separable solutions are possible in contrast to the perfectly plastic case. Achenbach and Kanninen ⌊59⌋ studied the asymptotic field of steady-state mode III crack growth in a <u>linearly</u> strain-hardening material obeying the von Mises yield criterion and Prandtl-Reuss flow rule. Achenbach et al ⌊9⌋ investigated the corresponding problem for mode I plain strain and plane stress conditions. In both mode I and mode III it was found that the particle velocities and the stresses can be written in the following forms

$$\dot{u}_i \rightarrow Q \, \dot{a} \, \dot{U}_i^p \, (\theta) \, r^s \qquad \text{as } r \rightarrow 0 \qquad\qquad (35)$$

$$\sigma_{ij} \rightarrow Q \dot{a} \, E \, \Sigma_{ij}^p \, (\theta) \, r^s \qquad \text{as } r \rightarrow 0 \qquad\qquad (36)$$

Q is a scalar and the functions $\dot{U}_i^p$ and Σ_{ij}^p depend on θ, the mode of growth, the velocity $\dot{a}$ and the constitutive parameters of the material. These functions as well as the exponent s were calculated numerically in the cited papers. The results for s are summarized in Fig. 22. E_t and μ_t are the tangent moduli of the strain-hardening curve in tension and shear, respectively. It is noted that for $E_t/E < 1$ we have $0 < -s < 0.5$. Thus the condition (21) for asymptotically steady growth is satisfied. The order of the singularity gives <u>zero</u> energy-flow to the tip also for this case. The sensitivity of the results to the moduli ratio casts some doubts on the usefulness of these since it is in practice difficult to define a unique value of the tangent modulus at large strains. The applicability of eqs. (35)-(36) is also limited to the following velocity range.

$$\dot{a} < \sqrt{\mu_t/\rho} \qquad \text{mode III} \qquad\qquad (37)$$

$$\dot{a} < c_R(E_t) \qquad \text{mode I} \qquad\qquad (38)$$

$c_R(E_t)$ is the velocity of surface waves corresponding to a fictitious elastic modulus E_t. Above these limits the governing equations become hyperbolic and no separable solution exists.

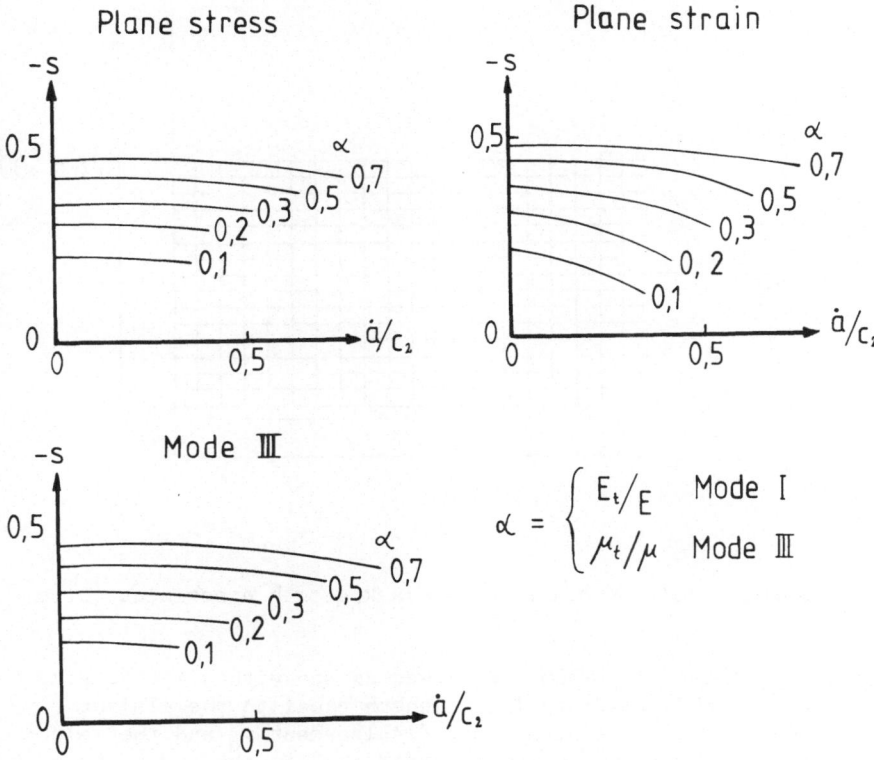

Figure 22. Singularity exponent for linear strain-hardening.

Contrary to the ideally plastic case, these solutions converge to the appropriate quasi-static results. Also a region of elastic un-loading exists for angles $\theta > \theta_p$ where θ_p depends on velocity and constitutive parameters. In the ideally plastic case the loading region surrounds the tip entirely.

Few complete solutions have been published where a strain-harden-ing model is assumed, Dahlberg [60] analysed an experiment with the geometry shown in Fig. 23 by finite element calculations. The crack growth between the nodes along the crack path was simulated by the gradual relaxation method suggested by Rydholm et al [36] (see Appendix 2 for details). According to the previous discussion no energy flow would be expected in an exact solution to the problem so that a non-zero energy flow is rather a consequence on the discrete modelling.

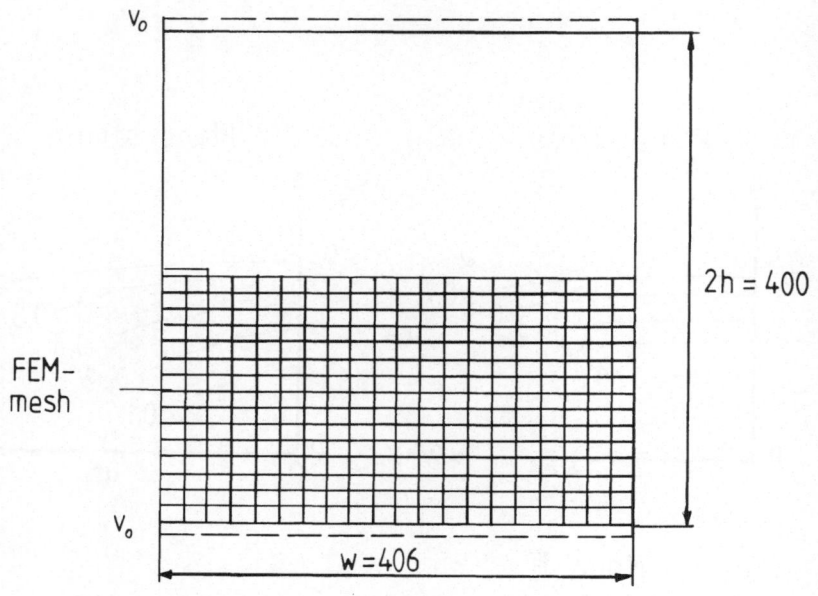

Figure 23. Geometry and FEM-model for a crack-growth experiment (from Dahlberg [60])

This kind of model should perhaps be viewed as one with a finite size crack-tip region with a characteristic length equal to the element division. The values of the prescribed displacement v_o and the velocity variation was taken from an actual experiment in the series performed by Dahlberg et al [27]. The velocity range in this particular experiment was 300-600 m/s. Since E_t/E for this material is about 0.05, the pseudo-Raleigh velocity is about 650 m/s and thus the actual velocities are close to limit (38). Dahlberg [60] found that the energy flow calculated with the element model was about half of the corresponding values from a purely elastic analysis. He also showed graphs of the plastic zone and the unloading zone. The conclusion was that for this experiment, plastic deformation plays an important role which was to be expected from the high nominal stresses of about 0.95 of the initial yield stress. Although no more specific conclusions can be drawn from this analysis it illustrates an important principle in verifying dynamic crack-growth criteria in that conclusions can only be reached by analysing real experiments with real growth and loading histories. That this has to be done may seem self-evident, but one finds a lot of examples in the literature where it is attempted to draw conclusions about fracture criteria on basis of arbitrarily guessed crack-growth histories.

In quasi-static fracture mechanics it is often assumed that the material is hardening according to a power law. It was pointed out by Achenbach and Kanninen [59] that in this case no singular tip solution can be found in the dynamic growth case. For sufficiently large strains the limit velocity based on the tangent modulus is always lower than any finite crack-tip velocity so that the governing equations become hyperbolic.

2.5.3 <u>Rate-dependent elastic-viscoplastic material</u>. An important step to the understanding of crack growth in non-linear materials is the inclusion of rate effects. This resolves some of the difficulties in the earlier discussed rate-independent models. Furthermore, it can be expected that rate effects are of importance because of the exceedingly high strain rates that exist in the vicinity of a running crack tip.

In the investigations discussed here the authors have chosen to use as material model the special case of Perzyna's [61] visco-plastic material that is described in Appendix 3. The singularity at a quasi-statically growing tip in such a material has been studied by Hui and Riedel [62] and McCartney [63]. They found that for n < 3 (see Appendix 3) the near-tip field is of the <u>elastic</u> type (eq. (1)). For n $\geq$ 3 the HRR-singularity develops. The result for n < 3 is also valid when inertia effects are included (Brickstad [64]). If n $\geq$ 3 no separable singular solution exists for the same reasons as in the rate-independent case. This means that for n < 3 it is possible to use the same criteria as in the elastic case (eqs. (15) – (16)).

Douglas [65] performed numerical calculations for the mode III SSY-case in a similar way as in [52]. He did not utilize the observation of elastic asymptotic predominance for n < 3 but instead applied the critical strain criterion. He derived curves for (K_{III}/K_{III}^{C}) in the same way as in [52] for different values of the material parameters.

Brickstad [64] analysed the experiments performed by Dahlberg et al [27] using the nodal relaxation technique described in Appendix 2. The material parameters were determined from uniaxial tests. A number of interesting observations were made in [64]. The elastic-viscoplastic analysis (n = 2.0) was compared to the elastic-plastic analysis by Dahlberg [60]. For the same element division as used in [60] the difference between these two analyses was fairly small. Typically, the energy flow to the tip was about one-half of the elastic value. A convergence study was then performed in which the element size was successively diminished. This did not appreciably affect the results of the elastic calculation. The calculated energy flow γ, calculated from the viscoplastic analysis, did on the other hand decrease monotonously with element size, seemingly convergent to a value of about one-quarter of the elastic value. Brickstad also performed a convergence study for simulated rate-independent behaviour. In this case the convergence rate was much slower and it could not be judged if a finite limit value existed. According to the previous discussion this should not be expected. In Fig. 24 the results from a number of experiments on three different geometries (cf. Fig. 23) are summarized and compared to the elastic analysis. It is noted that the viscoplastic analysis gives a much better correlation of the results for the energy flow to the crack tip. However, more research is needed before one is allowed to draw more general conclusions.

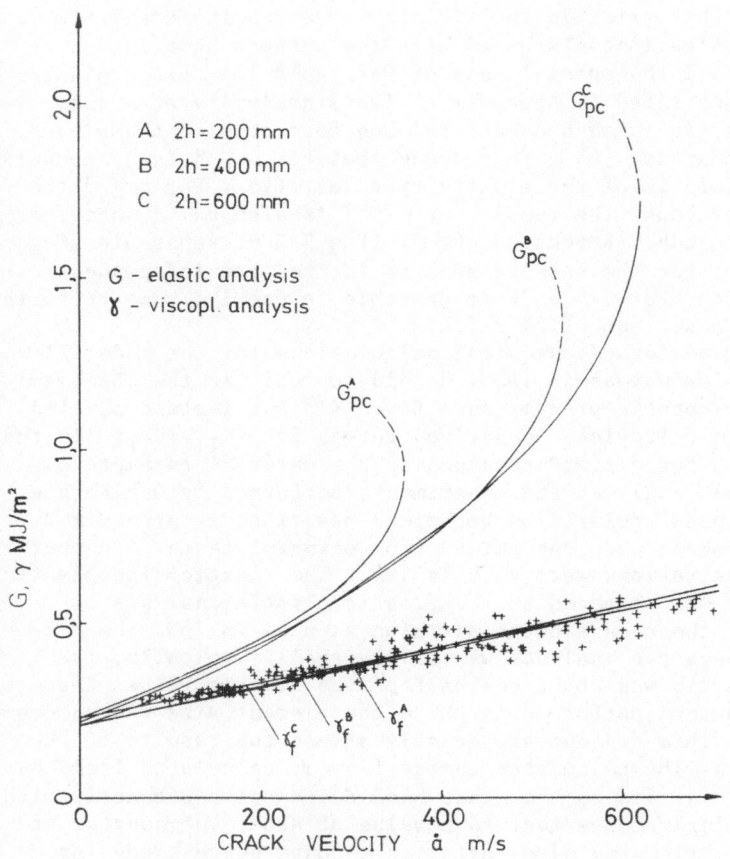

Figure 24. Critical energy flow to the crack tip versus tip velocity å for different specimen geometries indicated by the superscripts A, B and C (from Brickstad ⌊64⌋)

3. REFERENCES

⌊ 1 ⌋ Nilsson, F., "A note on the stress-singularity at a non-uniformly moving crack tip", J. Elasticity 4, 73-75, 1974.

⌊ 2 ⌋ Freund, L.B. and Clifton, R.J., "On the uniqueness of plane elastodynamic solutions for running cracks", J. Elasticity 4, 293-299, 1974.

⌊ 3 ⌋ Kostrov, B.V. and Nikitin, L.V., "Some general problems of mechanics of brittle fracture", Archives of Mechanics 6, No. 22, 749, 1970.

⌊ 4 ⌋ Strifors, H.C., "On constitutive properties at singular crack borders", Proc. ICF4, Vol. 3, p. 63, Waterloo, 1977.

⌊ 5 ⌋ Atkinson, C. and Eshelby, J.D., "The flow of energy into the tip of a moving crack", Int. J. Fracture Mechanics 4 3, 1968.

⌊ 6 ⌋ Freund, L.B., "Energy flux into the tip of an extending crack in an elastic solid", J. Elasticity 2, 341, 1972.

⌊ 7 ⌋ Sih, G.C., "Dynamic aspects of crack propagation", in 'Inelastic Behaviour of Materials' ed. Kanninen, M.F. et al, McGraw-Hill, New York, pp. 607-639, 1970.

⌊ 8 ⌋ Achenbach, J.D. and Dunayevsky, V., "Fields near a rapidly propagating crack tip in an elastic perfectly-plastic material", J. Mech. Phys. Solids 29, 283, 1981.

⌊ 9 ⌋ Achenbach, J.D., Kanninen, M.F. and Popelar, C.H., "Crack-tip fields for fast fracture of an elastic-plastic material", J. Mech. Phys. Solids 29, 211, 1981.

⌊10⌋ Freund, L.B., "The analysis of elasto-dynamic crack tip stress fields", Mechanics Today, Vol. III, ed. S. Nemat-Nasser, Pergamon, p. 55, 1976.

⌊11⌋ Chen, E.P. and Sih, G.C., "Transient response of cracks to impact loads", in 'Mechanics of Fracture' Vol. 4, ed. Sih, G.C., Noordhoff, Leyden, 1977.

⌊12⌋ Chen, Y.M., Eng. Fracture Mech. 7, 653-660, 1975.

⌊13⌋ Aberson, J.A., Anderson, J.M. and King, W.W., "Dynamic analysis of cracked structures using singularity finite elements", in 'Mechanics of Fracture' Vol. 4, ed. Sih, G.C., Noordhoff, Leyden, 1977.

⌊14⌋ Mall, S., "Finite element analysis of stationary cracks in time dependent stress fields", Numerical Methods in Fracture Mechanics, Swansea, 1980, 539-551.

|15| Brickstad, B., "A FEM analysis of crack arrest experiments", Int. J. Fracture 21, 177, 1983.

|16| Klepaczko, J., "Application of the split Hopkinson pressure bar to fracture dynamics", Inst. Phys. Conf. Ser. No. 47, 1979.

|17| Yoffe, E.H., "The moving Griffith crack", Philosophical Magazine 42, 739, 1951.

|18| Craggs, J.W., "On the propagation of a crack in an elastic-brittle solid", J. Mech. Phys. Solids 8, 66, 1960.

|19| Nilsson, F., "Dynamic stress intensity factors for finite strip problems", Int. J. Fracture Mechanics 8, 403, 1972.

|20| Baker, B.R., "Dynamic stresses created by a moving crack", J. Applied Mechanics 29, 449, 1962.

|21| Broberg, K.B., "The propagation of a brittle crack", Arkiv för Fysik 18, 159-192, 1960.

|22| Freund, L.B., "Crack propagation in an elastic solid subjected to general loading II. Non-uniform rate of extension", J. Mech. Phys. Solids 20, 141-152, 1972.

|23| Smith, E., "The crack-tip stress intensification associated with crack propagation and arrest", Proc. ICF5, Vol. 5, p. 2215, Cannes, 1981.

|24| Freund, L.B., "Crack propagation in an elastic solid subjected to general loading III. Stress-wave loading", J. Mech. Phys. Solids 21, 47-61, 1973.

|25| Dahlberg, L. and Nilsson, F., "Some aspects of testing crack propagation toughness", Proc. Int. Conf. on Dynamic Fracture Toughness, The Welding Institute, Cambridge, p. 281, 1976.

|26| Brickstad, B. and Nilsson, F., Int. J. Fracture 16, No. 1, 71-84, 1980.

|27| Dahlberg, L., Nilsson, F. and Brickstad, B., "Crack arrest methodology and applications", ASTM-STP 711, 89-108, 1980.

|28| Kalthoff, J.F., Beinert, J., Winkler, S, and Klemm, W., "Crack arrest methodology and applications", ASTM-STP 711, 109-127, 1980.

|29| ASME, ASME Boiler and Pressure Vessel Code (1975). American Society of Mechanical Engineers, Section XI, Article A-5300

|30| Walsh, P.F., "Stress intensity factors by calibrated finite element method", Technical Note, Journal of the Engineering Mechanics Division, ASCE, Vol. 98, No. EM6, 1611-1614, 1972.

|31| Bazant, Z.P., Glazik Jr, J.L. and Achenbach, J.D., "Finite element analysis of wave diffraction by a crack", Mech. Div. ASCE 102-EM3, 479-496, 1976.

|32| Kishimoto, K., Aoki, S. and Sakata, M., "Dynamic stress intensity factors using J-integral and finite element method", Eng. Fracture Mech. 13, 387, 1980.

|33| Thau, S.A. and Lu, T.H., "Transient stress intensity factors for a finite crack in an elastic solid caused by a dilatational wave", Int. J. Solids Struct. 7, 731-750, 1971.

|34| Kelley, J.W. and Sun, C.T., "A singular finite element for computing time dependent stress intensity factors", Eng. Fracture Mech. 12, 13, 1979.

|35| Aoki, S., Kishimoto, K., Kondo, H. and Sakata, M., "Elastodynamic analysis of crack by finite element method using singular element", Int. J. Fracture 14, 59-68, 1978.

|36| Rydholm, G., Fredriksson, B. and Nilsson, F., "Numerical investigations of rapid crack propagation", Numerical Methods in Fracture Mechanics, Swansea, pp. 660-672, 1978.

|37| Malluck, J.F. and King, W.W., "Fast fracture simulated by a finite element analysis which accounts for crack-tip energy dissipation", Numerical Methods in Fracture Mechanics, Swansea, pp. 648-659, 1978.

|38| Atluri, S.N., Nishioka, T. and Nakagaki, M., "Numerical modeling of dynamic and non-linear crack propagation in finite bodies, by moving singular elements", in Nonlinear and Dynamic Fracture Mechanics, ASME-AMD 35, 37-67, 1979.

|39| Kishimoto, K., Aoki, S. and Sakata, M., "Computer simulation of fast crack propagation in brittle material", Int. J. Fracture 16, 3, 1980.

|40| Broberg, K.B., "On the behaviour of the process region at a fast running crack tip", IUTAM symposium on 'High Velocity Deformation of Solids', Tokyo, Aug. 1977.

|41| Prandtl, L., "Ein Gedankenmodell für den Zerreiss-Vorgang spröder Körper", Z. Angew. Math. Mech., Vol. 13, p. 129, 1933.

[42] Strifors, H., "Thermomechanical theory of fracture, kinematics and physical principles", Report 27, Dept. of Strength of Materials, Royal Institute of Technology, Stockholm, 1980.

[43] Strifors, H., "Thermomechanical theory of fracture based upon the linear strain tensor", Report 28, Dept. of Strength of Materials, Royal Institute of Technology, Stockholm, 1980.

[44] Nilsson, L. and Oldenburg, M., "Non-linear wave propagation in plastic fracture materials. A constitutive modelling and finite element analysis", IUTAM Symposium on Nonlinear Deformation Waves, Tallin, Aug. 22-28, 1982.

[45] Rice, J.R., "The mechanics of quasi-static crack growth", Proc. 8th U.S. National Congress of Applied Mechanics, U.C.L.A., p. 191, 1979.

[46] McClintock, F.A. and Irwin, G.R., "Plasticity aspects of fracture mechanics", In 'Fracture Toughness Testing and its Applications', ASTM-STP 381, p. 84, 1965.

[47] Rice, J.R., Drugan, W.J. and Sham, T.L., "Elastic-plastic analysis of growing cracks", ASTM-STP 700, p. 189, 1980.

[48] Aboudi, J. and Achenbach, J.D., "Arrest of fast mode-I fracture in an elastic-viscoplastic transition zone", Eng. Fracture Mech., Vol. 18, pp. 109-119, 1983.

[49] Slepyan, L.I., "Crack dynamics in an elastic plastic body", Izv. Akad. Nauk SSSR MTT II, p. 126, 1976.

[50] Achenbach, J.D., Burgers, P. and Dunayevsky, V., "Near-tip plastic deformations in dynamic fracture mechanics", ASME AMD, Vol. 35, p. 105, 1979.

[51] Dunayevsky, V. and Achenbach, J.D., "Boundary layer phenomenon in the plastic zone near a rapidly propagating crack tip", Int. J. Solids Structures, Vol. 18, p. 1, 1982.

[52] Freund, L.B. and Douglas, A.S., "The influence of inertia on elastic-plastic antiplane-shear crack growth, J. Mech. Phys. Solids, Vol. 30, p. 59, 1982.

[53] Chitaley, A.D. and McClintock, F.A., "Elastic-plastic mechanics of steady crack growth under anti-plane shear", J. Mech. Phys. Solids, Vol. 19, p. 1, 1971.

[54] Rice, J.R. and Tracey, D.M., "Computational fracture mechanics", in 'Numerical and Computer Methods in Structural Mechanics', p. 585, 1973.

|55| Achenbach, J.D., "Dynamic crack-tip fields according to deformation theory", J. Appl. Mech., Vol. 46, p. 707, 1979.

|56| Hutchinson, J.W., "Singular behaviour at the end of a tensile crack in a hardening material", J. Mech. Phys. Solids, Vol. 16, p. 13, 1968.

|57| Rice, J.R. and Rosengren, G.F., "Plane strain deformation near a crack tip in a power-law hardening material", J. Mech. Phys. Solids, Vol. 16, p. 1, 1968.

|58| Ahmad, J. et al., "Elastic-plastic finite element analysis of dynamic fracture", Eng. Fract. Mech., Vol. 17, p. 235, 1983.

|59| Achenbach, J.D. and Kanninen, M.F., "Crack tip plasticity in dynamic fracture mechanics", in 'Fracture Mechanics', Proc. of Naval Structural Mechanics Conference, eds. N. Perrone et al., p. 649, 1978.

|60| Dahlberg. L., "Plastic effects in dynamic crack propagation", in 'Advances in Fracture Research', Proc. ICF5, p. 2195, Cannes, 1980.

|61| Perzyna, P., "The constitutive equations for rate sensitive plastic materials", Quarterly of Appl. Math., Vol. 20, p. 321, 1963.

|62| Hui, C.Y. and Riedel, M., "The asymptotic stress and strain field near the tip of a growing crack under creep conditions", Int. J. Fracture, Vol. 17, p. 409, 1981.

|63| McCartney, L.N., "On the energy balance approach to fracture in creeping materials", Int. J. Fracture, Vol. 19, p. 99, 1982.

|64| Brickstad, B., "A viscoplastic analysis of rapid crack propagation experiments in steel", J. Mech. Phys. Solids, Vol. 31, No. 4, 1983.

|65| Douglas, A.S., "Dynamic fracture toughness of ductile materials in antiplane strain", Div. of Engineering, Brown University, CME77-15564/2, 1981.

APPENDIX 1

A1.1 Elastic waves in plane bodies

Consider a two-dimensional isotropic elastic body for which the following equations can be formulated in the dynamic plane strain case:

Equilibrium $\qquad \sigma_{ij,j} = \rho \ddot{u}_i$ $\qquad\qquad\qquad\qquad\qquad$ (A1.1)

Compatibility $\qquad \varepsilon_{ij} = \frac{1}{2}\left(u_{i,j} + u_{j,i}\right)$ $\qquad\qquad\qquad$ (A1.2)

Hooke's law $\qquad \sigma_{ij} = 2\mu\left(\varepsilon_{ij} + \frac{\nu}{1-2\nu}\,\varepsilon_{kk}\delta_{ij}\right)$ $\qquad\qquad$ (A1.3)

where σ_{ij} and ε_{ij} are the stress- and strain-tensor, respectively, u_i is the displacement vector, ρ is the mass density, μ is the shearing modulus and ν is Poisson's constant.

If these equations are combined to eliminate σ_{ij} and ε_{ij} one arrives at Navier's equation

$$\mu\,u_{i,jj} + \frac{\mu}{1-2\nu}\,u_{j,ji} = \rho\ddot{u}_i \qquad\qquad\qquad\qquad (A1.4)$$

Assuming the body being located in the x-y-plane, the two displacement components u (in the x-direction) and v (in the y-direction) are functions of x and y only. Now introduce two displacement potentials ϕ and ψ according to

$$u = \frac{\partial\phi}{\partial x} + \frac{\partial\psi}{\partial y}$$
$$v = \frac{\partial\phi}{\partial y} - \frac{\partial\psi}{\partial x} \qquad\qquad\qquad\qquad\qquad\qquad (A1.5)$$

By insertion of (A1.5) into (A1.4), it is easy to see that Navier's equation will be satisfied if the functions ϕ and ψ are solutions to the wave equations

$$\nabla^2\phi = \frac{1}{c_1^2}\,\frac{\partial^2\phi}{\partial t^2} \qquad\qquad\qquad\qquad\qquad (A1.6)$$

$$\nabla^2\psi = \frac{1}{c_2^2}\,\frac{\partial^2\psi}{\partial t^2} \qquad\qquad\qquad\qquad\qquad (A1.7)$$

where ∇^2 is the two-dimensional Laplacian operator. They indicate that two types of disturbances with velocities c_1 and c_2 may be propagated through an elastic solid, defined for the plane strain case as

$$c_1^2 = \frac{E}{\rho} \cdot \frac{1-\nu}{(1+\nu)(1-2\nu)} \tag{A1.8}$$

$$c_2^2 = \frac{\mu}{\rho} \tag{A1.9}$$

c_1 is known as the speed of longitudinal or dilatational waves, characterized by displacements parallel to the direction of propagation. c_2 is the shear wave speed, also known as the velocity of equivoluminal or rotational waves. The displacements in these waves are perpendicular to the direction of propagation.

In an elastic body, it is also possible to have another type of wave, propagating along free surfaces. They are characterized by the amplitude of the displacements diminishing exponentially with increasing distance from the boundary. The simplest surface waves are called Rayleigh waves with a speed c_R. c_R is obtained from the real root of the following equation

$$\left(2 - c_R^2/c_2^2\right)^2 = 4\left\lfloor \left(1 - c_R^2/c_1^2\right)\left(1 - c_R^2/c_2^2\right)\right\rfloor^{1/2} \tag{A1.10}$$

An approximative solution to (A1.10) is given by

$$c_R \approx \frac{0.862 + 1.14\nu}{1+\nu} \cdot c_2 \tag{A1.11}$$

which is somewhat below c_2. Between the three wave speeds the following relation holds

$$c_R < c_2 < c_1 \tag{A1.12}$$

Rayleigh waves can for instance appear in fracture problems where free surfaces are created by moving cracks.

APPENDIX 2

A2.1 Determination of the dynamic stress-intensity factor by FEM.

For the solution of crack problems one often has to rely on numerical methods. This is especially so for dynamic problems where the presence

of wave propagation effects may introduce difficulties not encountered
in the corresponding static case. In this Appendix some methods will be
described to calculate the dynamic K by aid of the finite element
method (FEM).

The spatial descretization of the elastic continuum results in
the following system of dynamic equilibrium equations

$$\lfloor M \rfloor \lfloor \ddot{u} \rfloor + \lfloor C \rfloor \lfloor \dot{u} \rfloor + \lfloor K \rfloor \lfloor u \rfloor = \lfloor F \rfloor \qquad (A2.1)$$

where $\lfloor u \rfloor$ is vector of nodal displacements and a dot corresponds to
successive time derivatives. $\lfloor M \rfloor$, $\lfloor C \rfloor$ and $\lfloor K \rfloor$ are the global mass matrix, damping matrix and stiffness matrix, respectively. $\lfloor F \rfloor$ is the
vector of applied forces. Together with appropriate boundary conditions
the solution of this system gives an approximation of the time-dependent field quantities (displacements and stresses) describing the propagation of elastic waves in the body. In this context it should be
pointed out that the explicit time integration method based on a central difference approximation, combined with a lumped mass formulation,
often is the best technique for the temporal discretization. This
choice implies few numerical operations and a minimum of computer stor-
age. The conditional stability with respect to the time-step, imposed
by this explicit method, is usually not a severe restriction. It appears that in analysis of wave propagation a relatively small time-step
is necessary anyhow, in order to avoid large wave-front distorsions.

A2.1.1 Stationary cracks under transient loading conditions

A2.1.1.1 COD-method. Consider a stationary Mode I crack of length a in
a plane elastic body, subjected to an arbitrary time-dependent stress-
field (Fig. A2.1).

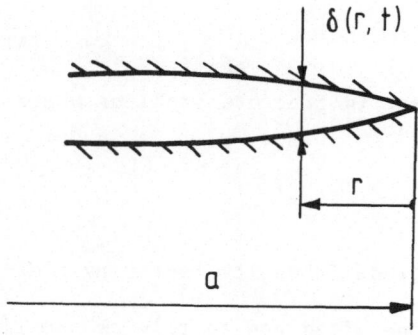

Fig. A2.1. Crack tip in an elastic body.

Let $\delta(r,t)$ be the time-dependent crack-opening displacement COD at a distance r behind the crack tip. In the vicinity of the crack tip the following relation is valid between the dynamic K and δ:

$$K_I(a,t) = \delta(r,t) \cdot \frac{\mu}{\kappa+1} \sqrt{\frac{2\pi}{r}} \qquad (A2.2)$$

where μ is the shearing modulus and κ is defined by

$$\kappa = \begin{array}{ll} 3-4\nu & \text{plane strain} \\ (3-\nu)/(1+\nu) & \text{plane stress} \end{array} \qquad (A2.3)$$

Thus, taking r to be a distance d to a nodal point behind the crack tip, K_I is proportional to the quotient $\delta/\sqrt{d}$ evaluated from a FEM-run. However, errors are introduced because of the approximative nature of the FEM-results and the relation (A2.2) only being accurate for small values of r. Thus, for an accurate result of K_I a fairly strong refinement of the finite element grid is necessary in the vicinity of the crack tip. This disadvantage can be overcome by using a form of calibrated nodal displacement technique, first introduced by Walsh |30| for static problems. A correction factor is introduced based on the difference between the static K, evaluated from the static variant of (A2.2) in a FEM-grid and the K taken from standard solutions. This correction factor is then assumed to be valid also in a dynamic case for the particular FEM-grid. This validity has been verified by e.g. Bazant et al |31| and Mall |14|. Also in Fig. 6 the FEM-solution of the dynamic K has been obtained with this technique.

The main advantage with this method is its simplicity. It does not require any special crack-tip element or strong refinement near the crack tip. Of course, the corresponding static solution must be known exactly but this is the case for many crack geometries and loading conditions.

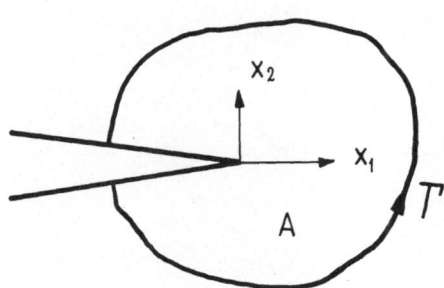

Fig. A2.2. A path Γ surrounding the crack tip.

A2.1.1.2 <u>Modified J-integral method</u>. Consider a fixed crack in an elastic material as shown in Fig. A2.2. The rate of energy release per unit crack advance in the x_1-direction is given by taking account of inertia effects as (Mall [14], Kishimoto et al [32])

$$J_1 = \int_\Gamma \left[W dx_2 - \sigma_{ij} n_j \frac{\partial u_i}{\partial x_1} ds \right] + \int_A \left(\rho \ddot{u}_i - F_i \right) \frac{\partial u_i}{\partial x_1} dA \qquad (A2.4)$$

where W is the strain-energy density, F_i is the body force, Γ is an arbitrary curve surrounding the traction-free crack surfaces and A is the area enclosed by this path. The first integral in (A2.4) is the conventional J-integral and the second integral accounts for kinetic energy contributions. The resulting J-integral is path-independent and $K_I(t)$ may be determined from

$$J_1 = \frac{\kappa+1}{8\mu} \cdot K_I^2(t) \qquad (A2.5)$$

The field quantities appearing in (A2.4) are obtained from a dynamic FEM-run for a suitable integration path surrounding the crack tip.
The validity of the J_1-concept can be seen in Fig. A2.3 taken

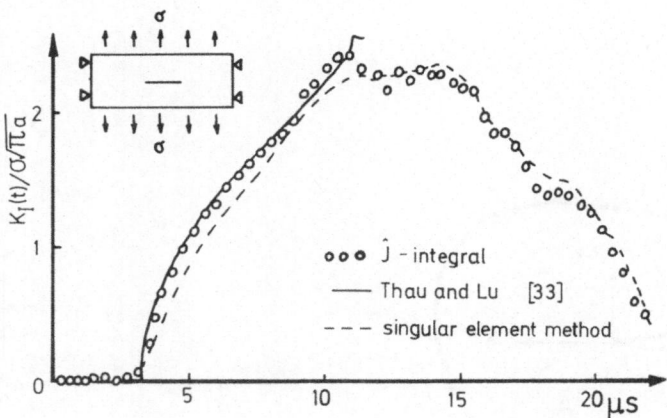

Fig. A2.3. The time variation of $K_I(t)$. Ref. [32].

from ref. [32] in which a rectangular strip with a central crack was subjected to sudden tension. The normalized dynamic stress-intensity factor as function of time obtained by aid of the J -integral is represented by open circles in the figure. A comparison is made with an analytical solution by Thau and Lu [33] which is valid before the time (marked by an arrow in Fig. A2.3) for reflected stress waves to arrive at the crack tip.

The J_1-integral method is easily extended to mixed mode cases and does not require a fine element grid near the crack tip.

The above methods employ regular finite elements to obtain the stresses and displacements in the body. Some authors, Aberson et al [13], Kelley and Sun [34], Aoki et al [35], have developed special singular crack-tip elements for determination of the dynamic K. For these elements the displacement-shape function is taken directly from the analytical crack-tip expansions.

A2.1.2. Propagating cracks. In this section, a method of simulating rapid crack growth and determination of the dynamic K will be described using regular finite elements. It was introduced by Rydholm et al [36] and successfully used in a series of experimental investigations, refs. [15], [26].

Consider a Mode I crack moving in a plane elastic body of unit thickness modelled by finite elements. At a particular time instant the crack tip is located at a distance δ within the finite element distance d, see Fig. A2.4.

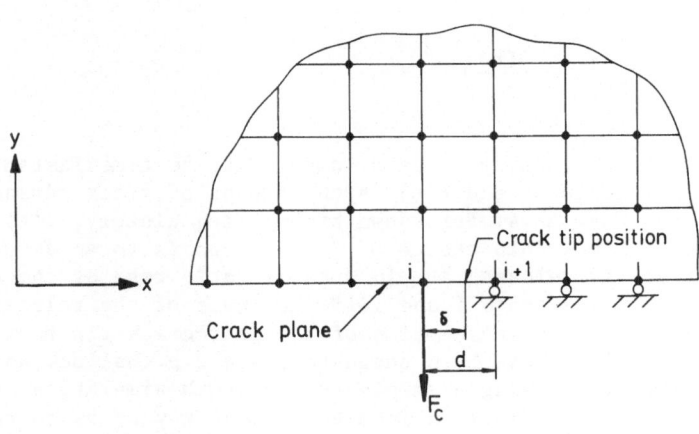

Figure A2.4. Crack-growth model in the FEM-mesh.

The successive crack growth along the element is simulated by introducing a closing force F_c behind the crack tip which is gradually relaxed to zero as the crack is propagating from node i to node i+1. Consider now the energy balance as the crack tip moves between the equivalent times t_i and t_{i+1} during which the nodal relaxation occurs.

$$\int_{t_i}^{t_{i+1}} F_c(t) \; \dot{u}_{iy} dt = \int_{t_i}^{t_{i+1}} \left[\dot{P}(t) - \dot{T}(t) - \dot{U}(t) \right] dt \qquad (A2.6)$$

u_{iy} is the vertical displacement of node i, P is the work of tractions working on the outer boundary, T is the kinetic energy and U is the strain energy. From a comparison of the elastic form of eq. (3) and the definition of the energy-release rate G, it can then be concluded that the net work done by the relaxation force in advancing the crack an element distance d is

$$W_r = 2 \int_{t_i}^{t_{i+1}} F_c(t) \; \dot{u}_{iy} \; dt = \int_{o}^{d} G(\delta) \; d\delta \qquad (A2.7)$$

The factor 2 arises beause there are two crack surfaces and here we consider only one half of the body assuming symmetry with respect to the crack plane. If the element size is not too large G is approximately constant over the length increment d and one may write $G \approx W_r/d$. The dynamic K is then obtained from eq. (6).

The question now remains how the relaxation force F_c shall be chosen. In ref. |37| the reasons for the following choice were stated

$$F_c = F_o \left(1 - \frac{\delta(t)}{d} \right)^{1/2} \qquad (A2.8)$$

where F_o is the closing force of node i at the time instant when the crack tip reaches it and $\delta(t)$ is the amount of crack advance at the time t, see Fig. A2.4. For known propagation history, $\delta(t)$ is a known function of time. This choice of F_c corresponds to an assumption of constant energy-release rate in a quasistatic case as the crack propagates between the nodes i and i+1. The shape of the relaxation function is not linked to any assumption of the near-crack-tip behaviour and other forms of F_c have been suggested, see e.g. Malluck and King |37|.

Apart from giving a smooth crack-growth simulation, this relaxation technique provides an inherent simple way of calculating G. It is also advantageous for numerical reasons. No special crack-tip elements are needed and usually coarse finite element grids are sufficient for linear elastic materials.

The validity of this technique was checked against analytical solutions in ref. $\lfloor 36 \rfloor$. One of the test runs consisted of a quadratic plate of side 2w subjected to uniaxial tension and containing a sym-metrically expanding central crack with a costant velocity $\dot{a}$. Denote t = 0 for the start of the crack growth (zero initial length). During a time interval $t < 2w/(\dot{a}+c_1)$, before reflected stress waves from the boundaries have returned to the crack tip, this problem is equivalent with the solution by Broberg $\lfloor 21 \rfloor$ who considered an infinite body (see Fig. 12). His solution for the energy-release rate G may be expressed as

$$G = G^{\infty}_{stat} \cdot g(\dot{a}) \tag{A2.9}$$

i.e. the static G for an infinite plate times a velocity-dependent function g for which $g(0) = 1$ and $g(c_R) = 0$. The result of the comparison is shown in Fig. A2.5 for $\dot{a}/c_1 = 0.318$.

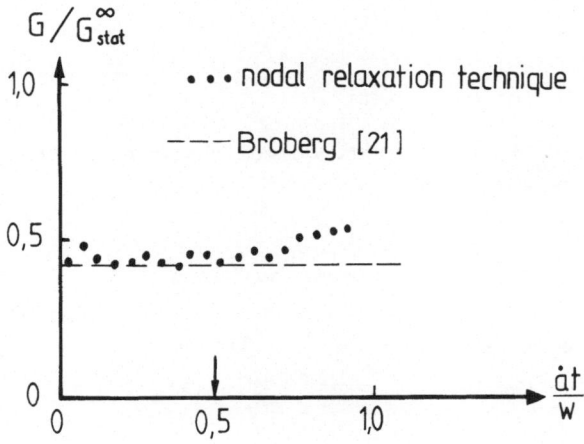

Figure A2.5. Non-dimensional energy-release rate for expanding central crack. Ref. $\lfloor 36 \rfloor$.

The arrow in the figure corresponds to the time interval within which the solution (A2.9) is expected to be valid. The relaxation technique is here shown to exhibit good agreement with Broberg's result. Another example of this method is shown in Fig. 15, where it has been used to determine the dynamic K for the propagation phase.

There exists also alternative methods for the numerical evaluation on of K for moving cracks. Several authors, Atluri et al |38|, Kishimoto et al |39|, have used singular finite elements near the crack tip, moving with the same speed as the crack tip. A powerful method for a direct experimental measurement of K for running cracks is offered by the shadow optical methods of caustics, see e.g. Kalthoff et al |28| and Kalthoff's lecture in this seminar.

APPENDIX 3

A3.1. Summary of constitutive relations

A3.1.1 <u>Elastic relations</u>. The expression for the elastic strains is the inverse of eq. (A1.3). We have

$$\varepsilon^e_{ij} = \frac{1}{2\mu} \left(\sigma_{ij} - \frac{\nu}{1+\nu} \sigma_{kk} \delta_{ij} \right) \tag{A3.1}$$

A3.1.2. <u>Elastic-perfectly plastic materials.</u> In a perfectly plastic material yielding occurs when the yield criterion A3.2 is met.

$$f(\sigma_{ij}) - \sigma_o = 0 \tag{A3.2}$$

f is the yield function and σ_o is a constant. Common choices for the yield function are the following.

$$f = \left(\frac{3}{2} s_{ij} s_{ij} \right)^{1/2} \qquad \text{v. Mises}$$

$$f = \max \left(\sigma_1 - \sigma_2, \ \sigma_2 - \sigma_3, \ \sigma_1 - \sigma_3 \right) \qquad \text{Tresca} \tag{A3.3}$$

Here

$$s_{ij} = \sigma_{ij} - \frac{1}{3} \delta_{ij} \sigma_{kk} \tag{A3.4}$$

σ_1, σ_2, σ_3 denote the principal stresses.

The plastic strain rates follow (A3.5).

$$\dot{\varepsilon}^{p}_{ij} = \lambda \frac{\partial f}{\partial \sigma_{ij}} \qquad (A3.5)$$

Here λ is an unspecified positive parameter. Eq. (A3.5) implies that the strain rates are perpendicular to the yield surface defined by (A3.2) in the stress space but indeterminate in size. The total strain rates are then obtained.

$$\dot{\varepsilon}_{ij} = \dot{\varepsilon}^{e}_{ij} + \dot{\varepsilon}^{p}_{ij} \qquad (A3.6)$$

During unloading the plastic strain rates are zero.

A3.1.3. **Elastic-plastic hardening materials.** For materials that harden isotropically the yield criterion can be written as eq. (A3.7).

$$f(\sigma_{ij}) - \sigma_f(\varepsilon^{p}_{e}) = 0 \qquad (A3.7)$$

Here σ_f is a function of the effective plastic strain ε^{p}_{e}.
The stress-strain relations become (A3.8).

$$\dot{\varepsilon}^{p}_{ij} = \begin{cases} \dfrac{3}{2} \dfrac{\frac{\partial f}{\partial \sigma_{ij}}}{\left(\frac{\partial f}{\partial \sigma_{mn}} \frac{\partial f}{\partial \sigma_{mn}}\right)} \dfrac{\dot{f}}{\sigma'_f(\varepsilon^{p}_{e})} & ; \ f = \sigma_f \ \text{and} \ \dot{f} > 0 \\[4mm] 0 & ; \ f < \sigma_f \ \text{or} \ f = \sigma_f \ \text{and} \ \dot{f} \leq 0 \end{cases} \qquad (A3.8)$$

The derivative σ'_f is the tangent modulus.
   For **linearly** strain-hardening material σ'_f is constant equal to E_t.
   For power-law strain hardening we have eq. (A3.9).

$$\sigma'_f = k \ \sigma_o \left(\varepsilon^{p}_{e}\right)^{k-1} \qquad (A3.9)$$

A3.1.4. **Elastic visco-plastic materials**. Perzyna |61| has suggested the following equations for describing rate-dependent plasticity. The yield condition is the same as for hardening materials, eq. (A3.7). The viscosity effects are introduced by the following flow rule.

$$\dot{\varepsilon}_{ij}^{vp} = \begin{cases} \beta \ \phi \ (f) \ \dfrac{\partial f}{\partial \sigma_{ij}} & f > \sigma_f \\ 0 & f \leq 0 \end{cases} \qquad (A3.10)$$

Here β is a fluidity parameter and the flow rate function is to be determined from experiments. A choice favoured in many investigations is eq. (A3.11).

$$\phi(f) = \left(\frac{f-\sigma_f}{\sigma_f}\right)^n \qquad (A3.11)$$

ON THE MECHANICS OF DYNAMIC FRACTURE PHENOMENA

J.F. Kalthoff
Fraunhofer-Institut für Werkstoffmechanik
Wöhlerstraße 11
7800 Freiburg
W-Germany

ABSTRACT. Influences of time effects on test procedures for measuring dynamic material strength values are analyzed. In particular, procedures for measuring the crack arrest toughness and the initiation fracture toughness at high loading rates are considered. It is shown, that time effects can have a strong influence on the mechanical behavior of the cracks. At arrest the stress condition at the crack tip is still dynamic and not static, although the crack velocity is zero at the moment of arrest. Cracks under drop weight loading, in general, cannot be adequately described on the basis of quasistatic analyses. Static criteria also fail to describe the instability behavior of cracks under short pulse loading. Dynamic material properties based on static evaluation procedures, therefore, can be erroneous. A reliable determination of material strength data, therefore, requires the consideration of dynamic influences on the measuring procedure. Techniques which take the dynamic effects into account in an appropriate form have been outlined: i.e. the reduced dynamic effects crack arrest specimen, the dynamic concept of impact response curves and the short pulse fracture mechanics concept.

1. INTRODUCTION

The concept of fracture mechanics provides an appropriate tool for describing fracture processes in structures or components under various loading conditions. The main emphasis of fracture mechanics analyses is to make predictions on the initiation conditions of cracks which might exist in a component. For constructions with very high safety requirements, however, one is also interested in the failure behavior of the structure after a crack may have initiated since this behavior determines the degree of possible subsequent damage: The damage is limited if the crack is arrested sufficiently fast. The size of the leak which is formed after the crack might have propagated through the wall determines the forces which result from the escaping fluid. A complete separation of a nuclear pressure vessel must be prevented by all means in order to allow to carry through an emergency cool down of the core. In addition, the assessment of the initiation behavior of cracks under the assumption

479

of dynamic loading conditions is often considered to lead to conservative safety predictions for structures which are actually only subjected to quasi-static loading conditions. Furthermore, in other applications, there is a direct interest in the failure behavior or survival behavior of structures under high rate loading conditions.

The description of these dynamic processes in materials which are subjected to temperatures and loads according to service conditions require in general the application of an elasto-plastic fracture dynamics approach. The extension of the usual and well developed concept of quasi-static EPFM into the dynamic regime is necessarily very complicated. Much research has been and is currently carried out in the field of fracture dynamics. The theoretical and numerical aspects are treated in the preceding lecture by Brickstad and Nilsson. Most of the work, in particular efforts for developing measuring procedures for determining dynamic material strength values, however, has been concentrated considering linear elastic material behavior. The reason is twofold. First, the situation becomes less complicated and secondly, the influences of dynamic effects on the mechanical behavior are very large in comparison to equivalent considerations of elasto plastic material behavior. Only preliminary steps are undertaken towards the final goal of combining the results of elasto-plastic and dynamic fracture research.

The principal results on the processes which control various dynamic fracture phenomena and the resulting conclusions for measuring procedures of dynamic fracture properties are discussed in this lecture on the basis of the dynamic fracture research performed over the last years at the Fraunhofer-Institut für Werkstoffmechanik (IWM), Freiburg (see also [1]).

In particular the mechanical behavior of propagating and subsequently arresting cracks, of cracks under impact loading by a pendulum or a drop weight, and of cracks under short pulse loading is discussed. An illustrative view of the problems considered is given in Figure 1. The dynamic influences are demonstrated by comparing the actual dynamic reactions at the crack tip with the equivalent static behavior. Most of the dynamic data reported have been determined by means of the shadow

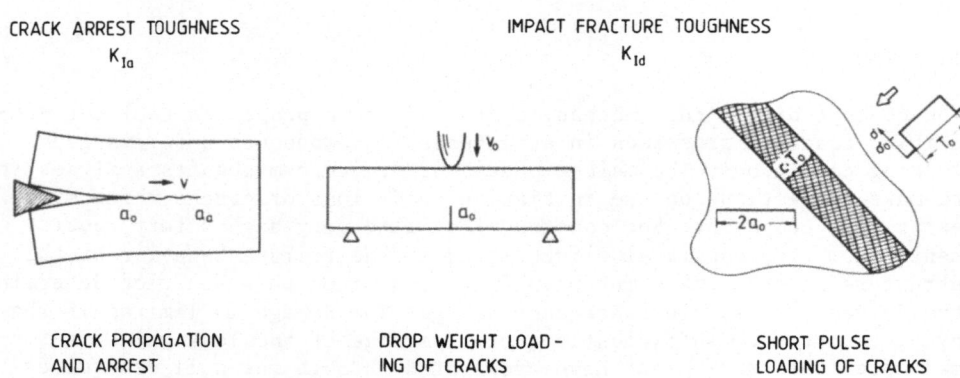

Figure 1. Dynamic fracture problems

optical method of caustics. This technique is briefly described first, the fracture data are presented in the following sections.

2. THE SHADOW OPTICAL METHOD OF CAUSTICS

The method of caustics is an optical tool for measuring stress intensifications. The technique has been introduced by Manogg in 1964 [2,3]. Later on, Theocaris [4] further developed the method. The author and his coworkers extended and applied Manoggs technique for investigating dynamic fracture phenomena [5-7].

The physical principle of the method is illustrated in Figure 2. A precracked specimen under load is illuminated by a parallel light beam. A cross-section through the specimen at the crack tip is shown in Figure 2b for a transparent specimen, and in Figure 2c for a non-

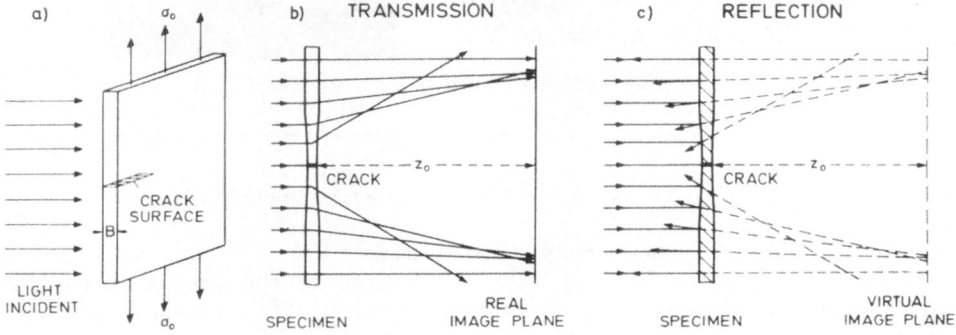

Figure 2.   Physical principle of the shadow optical method of caustics

transparent steel specimen. Due to the stress concentration the physical conditions at the crack tip are changed. For transparent specimens both the thickness of the specimen and the refractive index of the material are reduced. Thus, the area surrounding the crack tip acts as a divergent lens and the light rays are deflected outwards. As a consequence, on a screen (image plane) at a distance z_0 behind the specimen a shadow area is observed which is surrounded by a region of light concentration, the caustic (see Figure 3). Figure 2c shows the situation for a non-transparent steel specimen with a mirrored front surface. Due to the surface deformations light rays near the crack tip are reflected towards the center line. An extension of the reflected light rays onto a virtual image plane at the distance z_0 behind the specimen results in a light configuration which is similar to the one obtained in transmission. Consequently a similar caustic is obtained. The mode I shadow pattern was calculated by Manogg [2] from the linear elastic stress strain field around the crack tip. Figure 3 compares theoretical results with experimentally observed caustics which were photographed in transmission and in reflection with different materials. The single caustic curve obtained for isotropic materials splits up into a double

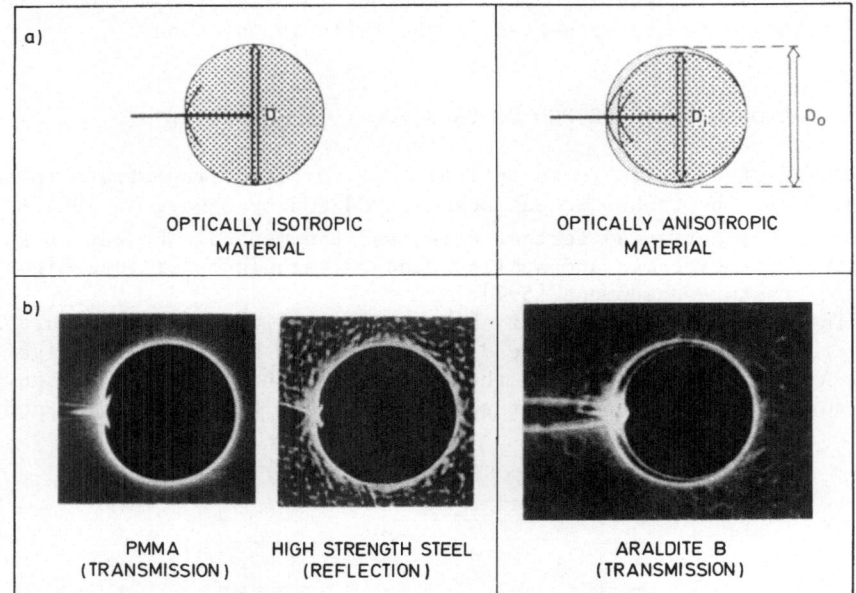

Figure 3. Mode I caustics, a) calculated, b) measured

caustic for optically anisotropic materials.

The size of the shadow pattern is related to the stress intensification at the crack tip. The quantitative correlation between the diameter D of the caustic and the stress intensity factor K_I is given by the relation

$$K_I = \frac{2 \sqrt{2\pi} \cdot F(v)}{3 \cdot f_{o,i}^{5/2} \cdot c \cdot d_{eff} \cdot z_o} D^{5/2} \tag{1}$$

where K_I = mode I stress intensity factor,
         D = diameter of the caustic,
         $f_{o,i}$ = numerical factor for outer/inner caustic,
         c = photoelastic constant,
         d_{eff} = effective thickness of the plate,
                 = d for transparent specimens,
                 = d/2 for reflecting, non-transparent specimens,
         d = physical thickness of the plate,
         z_o = distance between specimen and image plane,
         $F(v)$ = correction factor for non-zero crack velocities,
                 $F(v) \lesssim 1$,
         v = crack velocity.

Numerical values of the constants which appear in the K-evaluation formula are given in Table 1 for different materials. For further details of the shadow optical technique see [5-7].

TABLE 1 - Constants for Caustic Evaluation

| Material | Elastic Constants | | Shadow Optical Constants | | | | | | Effective Thickness |
| | Young's Modulus MN/m² | Poisson's Ratio | for Plane Stress | | | for Plane Strain | | | |
			c m²/N	f_o	f_i	c m²/N	f_o	f_i	d_{eff}
TRANSMISSION: ($z_0 < 0$)									
Optically Anisotropic:									
Araldite B	3660[*]	0.392[*]	-0.970×10^{-10}	3.31	3.05	-0.580×10^{-10}	3.41	2.99	d
CR - 39	2580	0.443	-1.200×10^{-10}	3.25	3.10	-0.560×10^{-10}	3.33	3.04	d
Plate Glass	73900	0.231	-0.027×10^{-10}	3.43	2.98	-0.017×10^{-10}	3.62	2.97	d
Homalite 100	4820[*]	0.310[*]	-0.920×10^{-10}	3.23	3.11	-0.767×10^{-10}	3.24	3.10	d
Optically Isotropic:									
PMMA	3240	0.350	-1.080×10^{-10}		3.17	-0.750×10^{-10}		3.17	d
REFLECTION: ($z_0 > 0$)									
All materials	E	ν	$2\nu/E$		3.17	-		-	d/2

Crack tip caustics are of a simple form and can easily be evaluated, the technique, therefore, is very well suited for investigating complex phenomena, as e.g. in fracture dynamics. A Cranz-Schardin 24 spark high speed camera is utilized in these investigations for photographing shadow patterns under dynamic loading conditions.

3. ARRESTING CRACKS

The usual procedure for measuring the crack arrest toughness K_{Ia} of a material is the following: In a wedge-loaded specimen a fast propagating crack is initiated from a blunted initial notch at an initiation stress intensity factor $K_{Iq} > K_{Ic}$. Figure 4 shows a rectangular double cantilever beam specimen under longitudinal wedge-loading. The crack opening displacement remains constant during crack propagation due to the stiffness of the loading system. Thus the crack propagates into a decreasing stress intensity factor field. It arrests at the length a_a if conditions for crack propagation are not fulfilled anymore. This threshold value represents the crack arrest toughness K_{Ia}.

In order to investigate the influences of dynamic effects on the mechanical behavior of cracks at arrest, the stress condition at the tip of a propagating and subsequently arresting crack was analyzed [8-11]. The results of experiments [8,9] with specimens made from a transparent model material, the epoxy resin Araldite B, are reported. The actual dynamic stress intensity factors, K_I^{dyn}, were measured by means of the shadow optical method of caustics in transmission. These values are compared to the equivalent static stress intensity factors, K_I^{stat}, calculated from the measured crack opening displacement 2δ utilizing

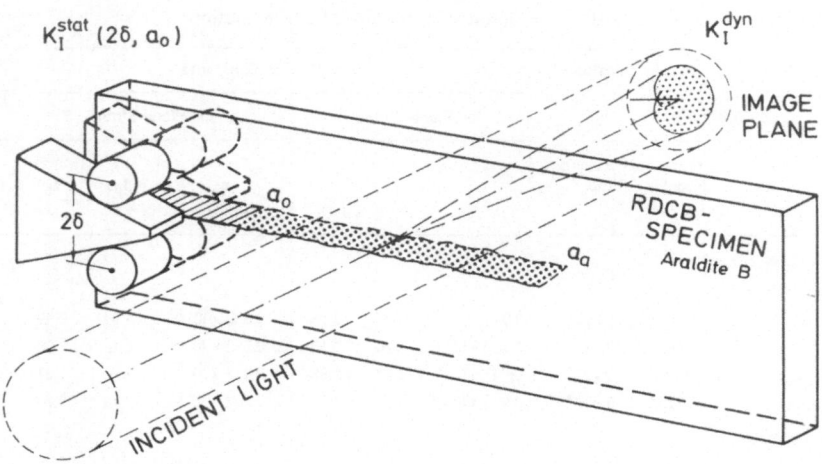

Figure 4. Experimental set-up for a crack arrest experiment and
shadow optical arrangement in transmission (schematically)

conventional stress intensity factor formulas.
    A series of six shadow optical photographs is shown in Figure 5.
Quantitative data for cracks initiated at different K_{Iq}-values are presented in Figure 6. The dynamic stress intensity factors, K_I^{dyn} (experi-

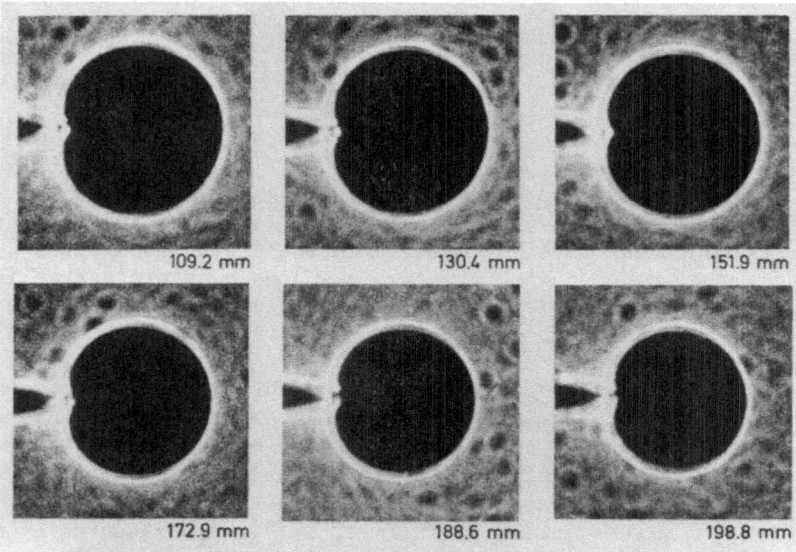

Figure 5. Caustics for a propagating and subsequently arresting crack
(photographed in transmission with an Araldite B specimen)

mental points), are shown as a function of the momentary crack length together with the corresponding static stress intensity factor curves, $K_I^{stat}(a)$. In addition, the measured crack velocities are given in the lower part of the diagram. The following characteristics of the crack arrest process can be deduced from these results: At the beginning of the crack propagation phase the dynamic stress intensity factor K_I^{dyn} is smaller than the corresponding static value K_I^{stat}. At the end of the propagation phase, in particular at the moment of arrest, the dynamic stress intensity factor K_I^{dyn} is larger than the corresponding static value K_I^{stat}. Only after arrest does the dynamic stress intensity factor K_I^{dyn} approach the static stress intensity factor at arrest, K_{Ia}^{stat}. Differences between the dynamic and the static stress intensity factor curves become smaller for cracks initiated at lower K_{Iq} values, i.e. for cracks propagating at lower velocities. The dynamic effects obviously decrease with decreasing velocity, as one might expect.

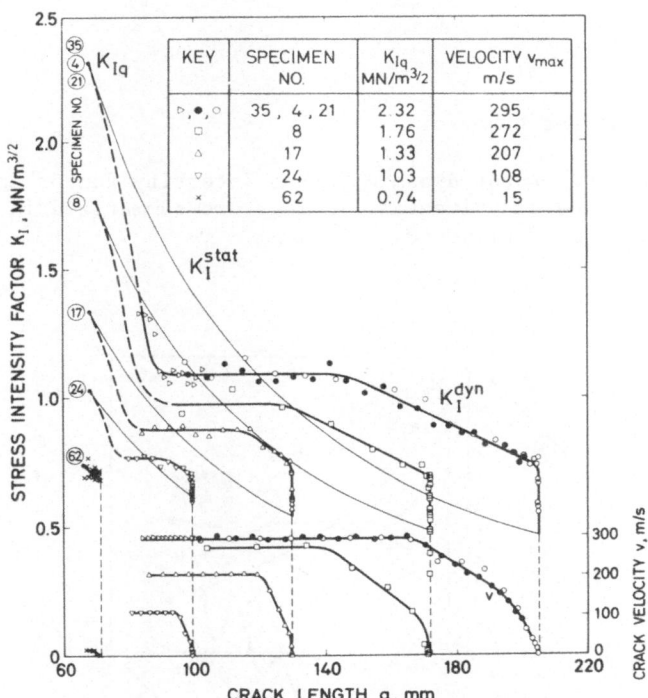

Figure 6. Stress intensity factors and crack velocity for propagating and subsequently arresting cracks

The behavior of the dynamic stress intensity factor in the post-arrest phase is shown in Figure 7. The dynamic stress intensity factor, K_I^{dyn}, is plotted as a function of time. K_I^{dyn} oscillates around the value of the static stress intensity factor at arrest, K_{Ia}^{stat}. Only a

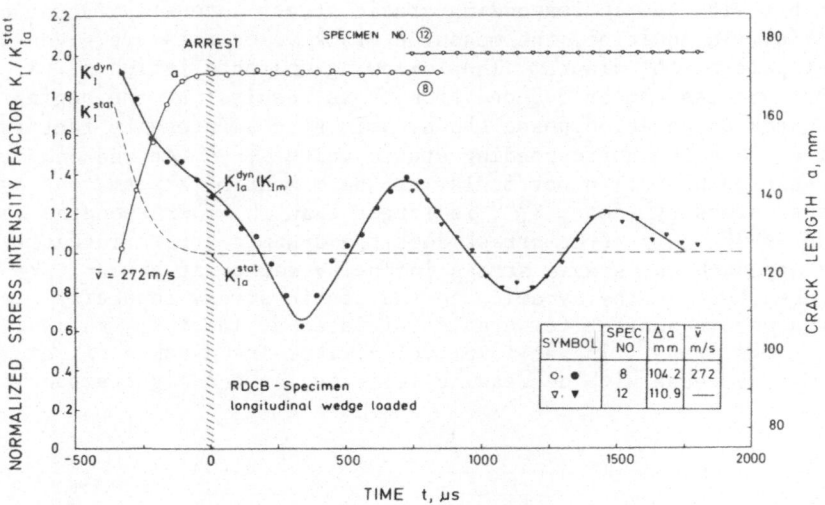

Figure 7.   Post-arrest behavior

large time after arrest the dynamic stress intensity factor approaches
the static value. The experimental findings are summarized in the schematic representation of Figure 8.

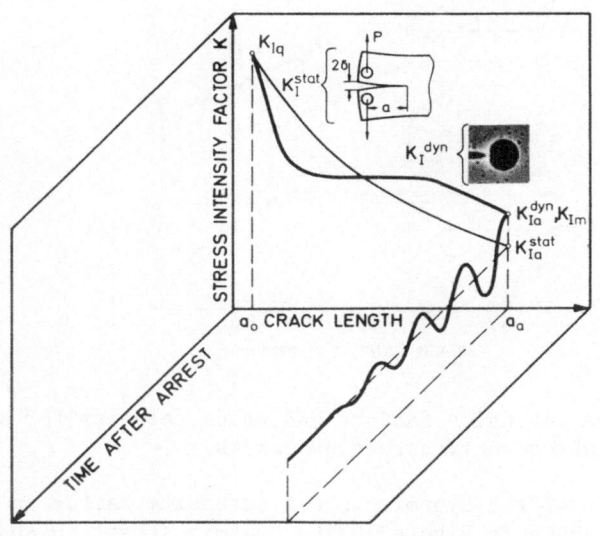

Figure 8.   Crack arrest behavior (schematically)

Wave propagation phenomena explain the observed behavior: elastic waves are produced by the propagating crack, thus kinetic energy is radiated into the specimen and $K_I^{dyn} < K_I^{stat}$. After reflection at the finite boundaries of the specimen the waves then interact with the crack again and contribute to the stress intensity factor, consequently $K_I^{dyn} > K_I^{stat}$. An illustrative view of these processes is given in Figure 9. A fast propagating crack (1000 m/s) in a high strength steel specimen was photographed in a shadow optical reflection arrangement [9]. In addition to the shadow spot at the crack tip the photograph shows the generation of waves at the tip of the propagating crack and the subsequent reflection at the boundaries of the specimen.

Consequently, the stress condition at arrest is not static but dynamic, although the crack velocity is zero at the moment of arrest. Crack arrest toughness values which are determined on the basis of static evaluation procedures [12,13] in principle, therefore, cannot represent a true material property. Statically determined crack arrest toughness values K_{Ia}^{stat} are smaller than the true dynamically determined crack arrest toughness values, K_{Ia}^{dyn} (see Figure 8). The determination of the true crack arrest toughness has to take the dynamic effects into account and requires a fully dynamic analysis, for example the Battelle concept of recovered kinetic energy [14,15]. The differences between statically and dynamically determined crack arrest toughness values depend on the shape and size of the specimen [9] and can be quite remarkable, in particular for larger crack jump distances. Figure 10 compares the results of an ASTM cooperative test program obtained with compact crack arrest test specimens (see data in [16]).

Safety predictions on the basis of K_{Ia}^{stat} data therefore can be erroneous. However, since the dynamic effects in large scale structures are in general smaller than in the relatively small laboratory test specimens which are used for K_{Ia} determination, static crack arrest analyses will yield conservative crack arrest predictions [17]. On the basis of this understanding the static crack arrest concept can be applied by the practical engineer under certain circumstances.

In order to minimize the errors in the static crack arrest concept a special crack arrest specimen with reduced dynamic effects has been developed at IWM. This RDE-(reduced-dynamic-effects)-specimen is shown in Figures 11 and 12. The edges and boundary of the specimen are shaped to reduce wave reflections and to defocus reflected waves. Damping material and additional weights are attached to the "wings" of the specimen to absorb kinetic energy and to increase the period of the eigen-oscillation of the specimen, thus the recovery of kinetic energy is reduced. Details are described by Beinert and Kalthoff [18]. Dynamic effects in the RDE-specimen are three to four times smaller than in the most commonly used compact crack arrest specimen. Thus, with specially designed specimens, as for example the RDE specimen, crack arrest toughness data can be determined by static evaluation procedures with an accuracy which is sufficient for engineering purposes and which avoids over-conservative design.

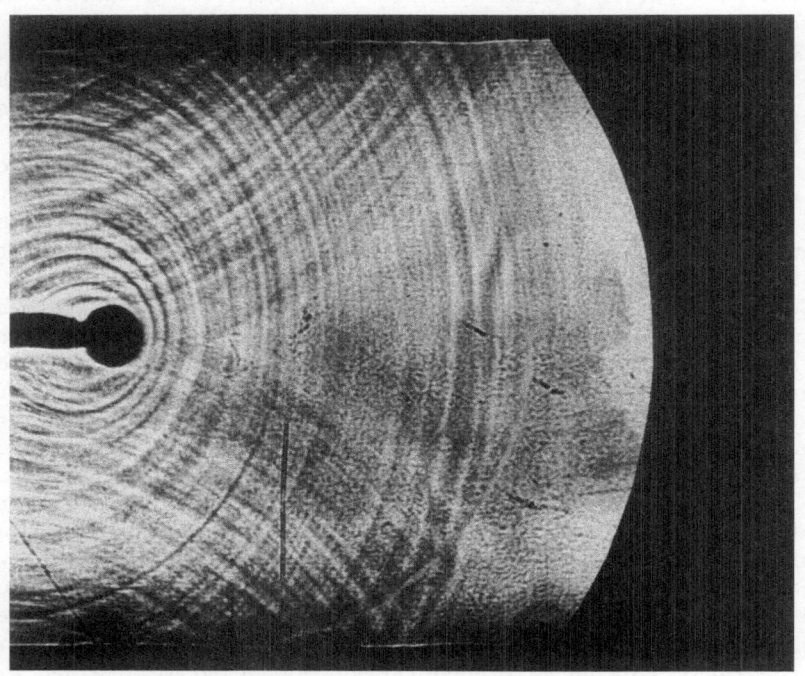

Figure 9.   Shadow optical photograph of a fast propagating
             crack in steel

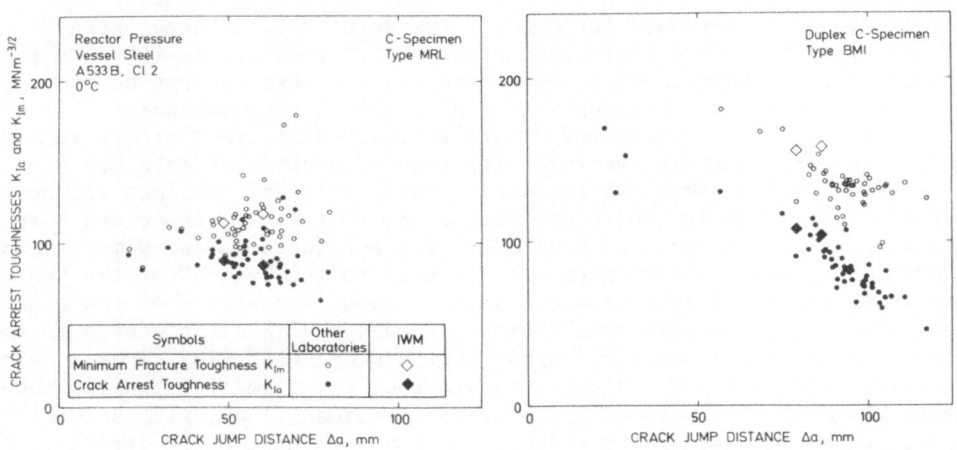

Figure 10. Crack arrest toughness data from ASTM Cooperative Test Program, $K_{Ia} \cong K_{Ia}^{stat}$ - determined according to a static analysis [12,13], $K_{Im} \cong K_{Ia}^{dyn}$ - determined according to a fully dynamic analysis [14,15]

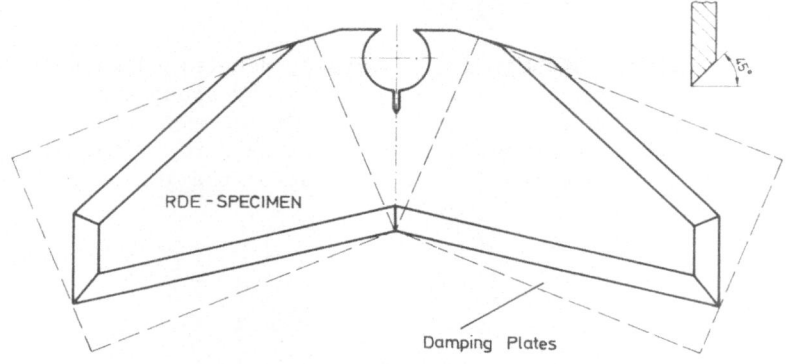

Figure 11.   RDE-crack arrest test specimen

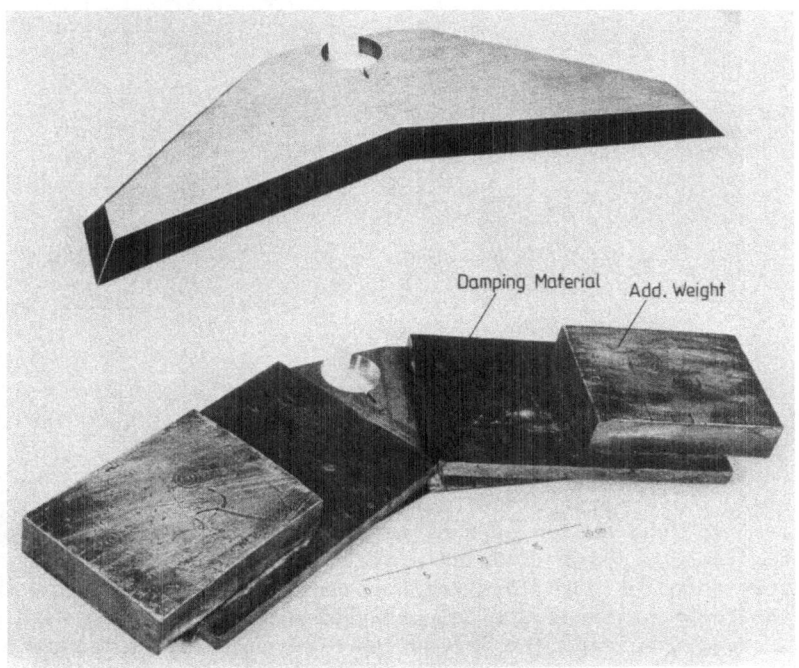

Figure 12.   Photographs of RDE-crack arrest test specimen

4. CRACKS UNDER DROP WEIGHT LOADING

The impact fracture toughness K_{Id} is usually determined with precracked bend specimens in instrumented impact tests. The specimens are loaded by a drop weight or by a pendulum type impact tester. Strain gages at the tup of the striking hammer measure the load during impact. From the critical load at the moment of crack initiation the impact fracture toughness is derived.

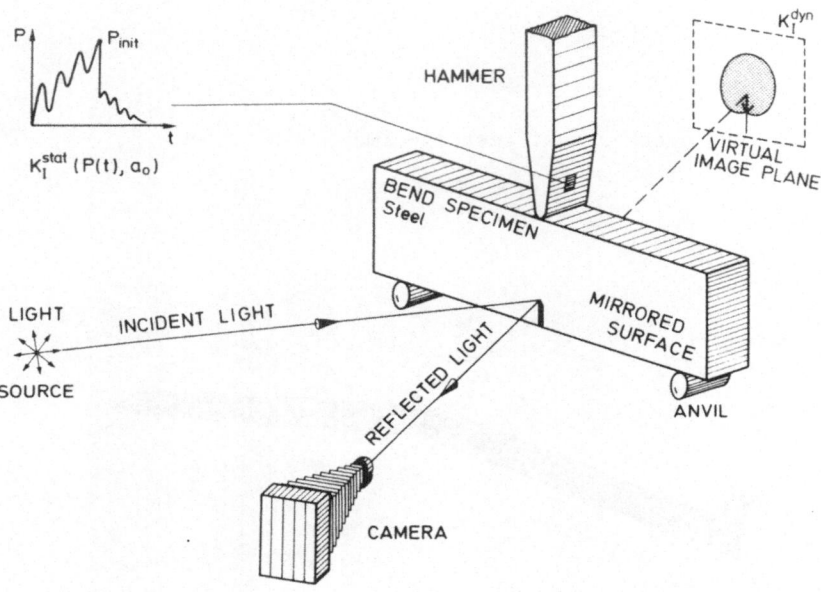

Figure 13. Experimental set-up for a drop weight experiment and shadow optical arrangement in reflection (schematically)

The mechanical behavior of cracks under impact loading was investigated by measuring the dynamic stress intensity factors directly at the crack tip by means of the shadow optical method of caustics [19-23]. Specimens made from the model material Araldite B and a high strength steel were investigated. Figure 13 gives a schematic view of the experimental set-up showing the shadow optical arrangement in reflection with a high strength steel specimen. The influences of dynamic effects were evaluated by comparing the dynamic stress intensity factors, K_I^{dyn}, with equivalent static stress intensity factors, K_I^{stat}. The static values were determined from the measured hammer load P_H utilizing conventional static stress intensity factor formulas from ASTM E 399.

A series of twelve shadow optical photographs of the central part of an impacted steel specimens is shown in Figure 14. Quantitative data for the two materials considered are given in Figure 15. The specimens

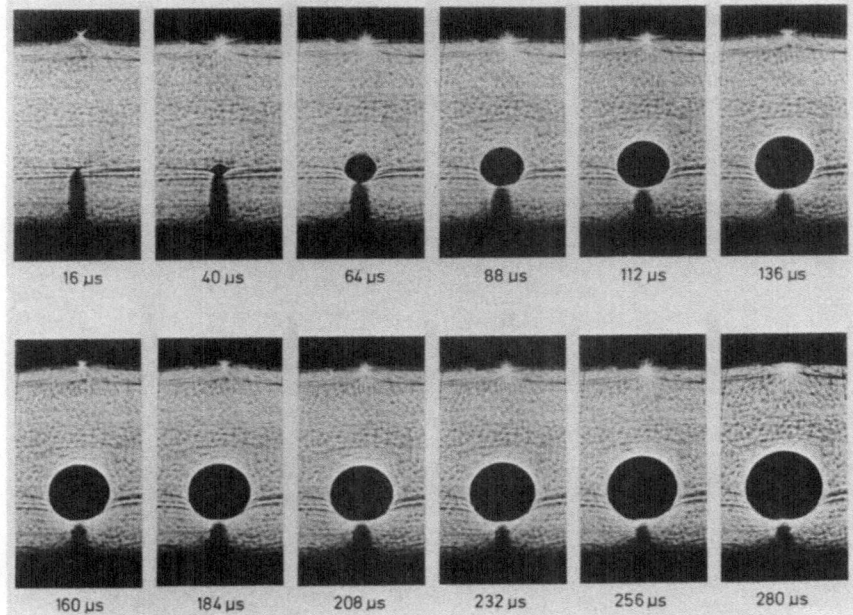

Figure 14.   Shadow optical photographs of a precracked bend specimen
during drop weight loading (photographed in reflection
with a high strength steel specimen)

were impacted at a velocity of 0.5 m/s by a hammer with a mass of
30 kg or 90 kg respectively. The dynamic stress intensity factors K_I^{dyn}
(experimental points) and the corresponding static stress intensity
factors K_I^{stat} (thin curves) are plotted as functions of time. The times
are given in absolute units and also in relative units by normalization
with the period τ of the eigenoscillation of the impacted specimen.

The K_I^{stat}-values show a strongly oscillating behavior, whereas the
actual dynamic stress intensity factors K_I^{dyn} show a more steadily increasing tendency. In the small time range, $t < \tau$, these differences are
very pronounced. The differences become smaller with increasing time,
but even for times larger than 3τ the influences of dynamic effects
obviously have not vanished and there are still marked differences between K_I^{stat} and K_I^{dyn}. Due to a different contact stiffness the effects
are larger with Araldite B specimens than with high strength steel
specimens.

The behavior of specimens under impact loading was further investigated by also measuring the specimen reaction at the anvils [22]. Figure 16 compares the load measured at the striking hammer (a), the
stress intensity factor measured at the crack tip (b), the load measured at the anvils (c), and the position of the specimen ends with
regard to the anvils (d). The data were obtained with Araldite B specimens impacted at 1 m/s. The τ-value of the specimens utilized for

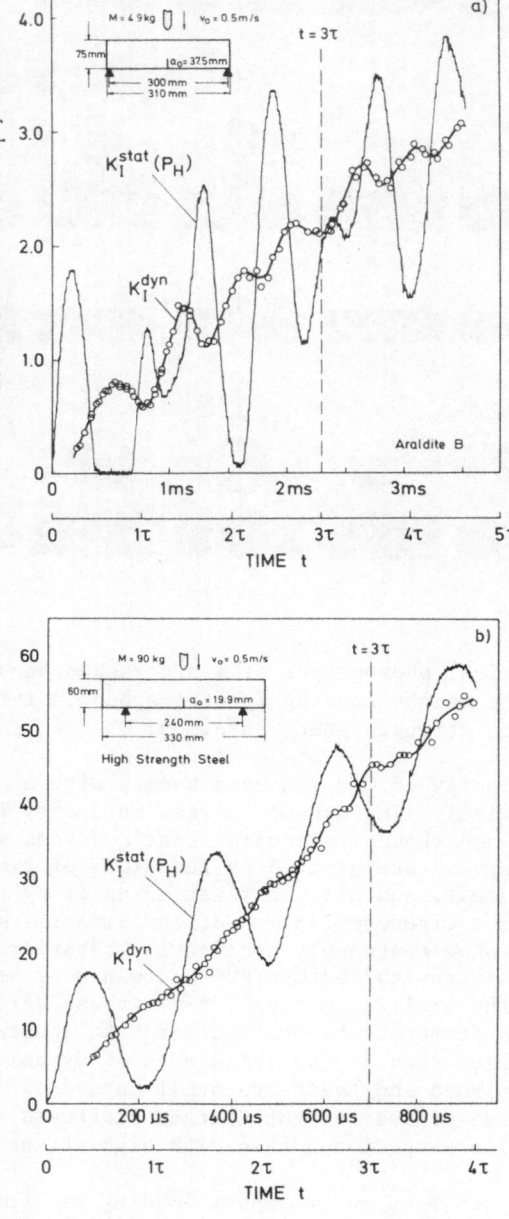

Figure 15.   Stress intensity factors for cracks under drop weight
             loading a) Araldite B specimen, b) high strength steel
             specimen

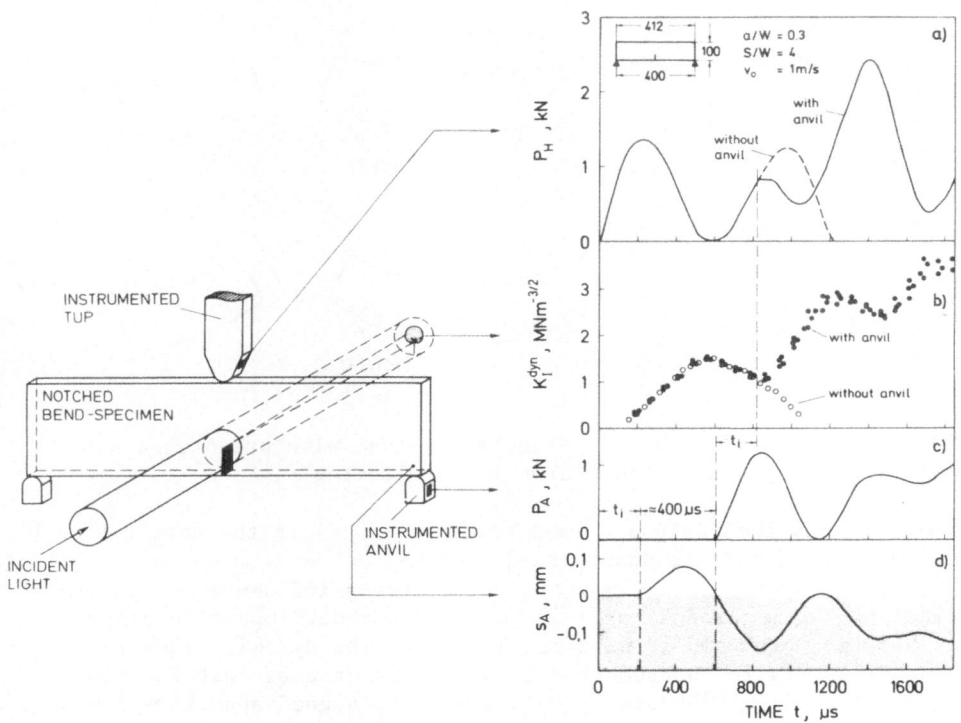

Figure 16.   Mechanical response of a prenotched bend specimen
              to drop weight loading

these investigations is about 700 μs. A comparison of the four signals
indicates that non-zero loads at the anvils were registered only after
a rather long time of about 600 μs. This time is about three times
larger than the time it would take the slowest wave, i.e. a transverse
wave, to travel from the point of impact to the anvils. This unexpected behavior is explained by diagram (d) in Figure 16. A loss of contact is observed between the specimen ends and the anvils. The loss
of contact starts at about 200 μs. This time is in agreement with the
above consideration of wave propagation times. For about 400 μs the
specimen is completely free and only after this time, i.e. at a time
of about 600 μs total, the specimen ends come into contact with the
anvils. In accordance with this observation, load values are then recorded at the anvils. With different test conditions this loss of contact can occur later for a second time and loss of contact can also
take place between the hammer and the specimen. These processes are
illustrated in the schematic representation of Figure 17. Since in the
reported experiments the anvils obviously were of no influence during
the early phase of the impact process, additional experiments have been
performed with unsupported specimens. The results are represented by
the dashed curve and open data points in Figure 16. In accordance with

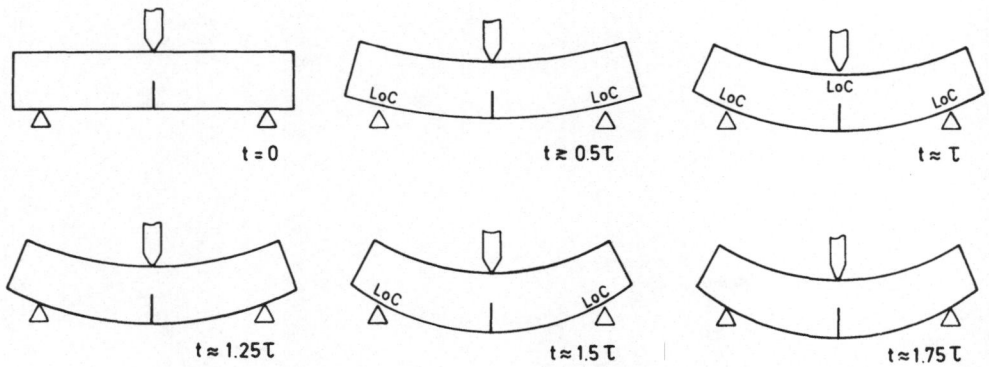

Figure 17. Loss of contact effects observed with prenotched bend
specimens under drop weight loading

speculation, the early specimen reaction $(t < \tau)$ is the same for both,
the supported and the unsupported specimen.

The measured data demonstrate the strong influence of dynamic effects on the mechanical behavior of cracks under impact loading. A
method [24] proposed to ASTM for measuring the dynamic fracture toughness value K_{Id} in instrumented impact tests assumes that for times
larger than 3τ K_I^{stat}-values would represent a good approximation of the
actual dynamic stress intensity factor K_I^{dyn}. However, data from the
presented experiments indicate that a static analysis is not adequate
to describe the loading condition in the specimen under the proposed
conditions except at much later times during the event.

Very large times to fracture, however, cannot always be achieved
[20]: in principle, the time to fracture is increased if the test temperature is increased and if the impact velocity is reduced. The τ-
value of a specimen, on the other hand, is reduced when the specimen
dimensions are reduced. Often both goals cannot simultaneously be reached due to size requirements which in general demand large specimen
dimensions for high test temperatures. The conditions of course become
very unfavorable for testing of brittle materials at high impact velocities [20]. Consequently the determination of reliable impact fracture
toughness data with freedom of choice of test conditions requires a
fully dynamic evaluation procedure.

At IWM, therefore, the dynamic concept of impact response curves
has been developed [25]. The principle of the measuring procedure is
illustrated in Figure 18. For fixed test conditions (i.e. specimen geometry, hammer mass, impact velocity, etc.) the dynamic stress intensity
factor versus time relationship is determined in a pre-experiment by
means of the shadow optical method of caustics with an edge notched
high strength steel specimen (see also Figure 13). The obtained curve,
$K_I^{dyn}(t)$, describes the response of the specimen to the impact process.
This curve, called impact response curve, is controlled by the elastic
properties of the system only. It therefore applies for all steels provided the conditions for small scale yielding are fulfilled. In the

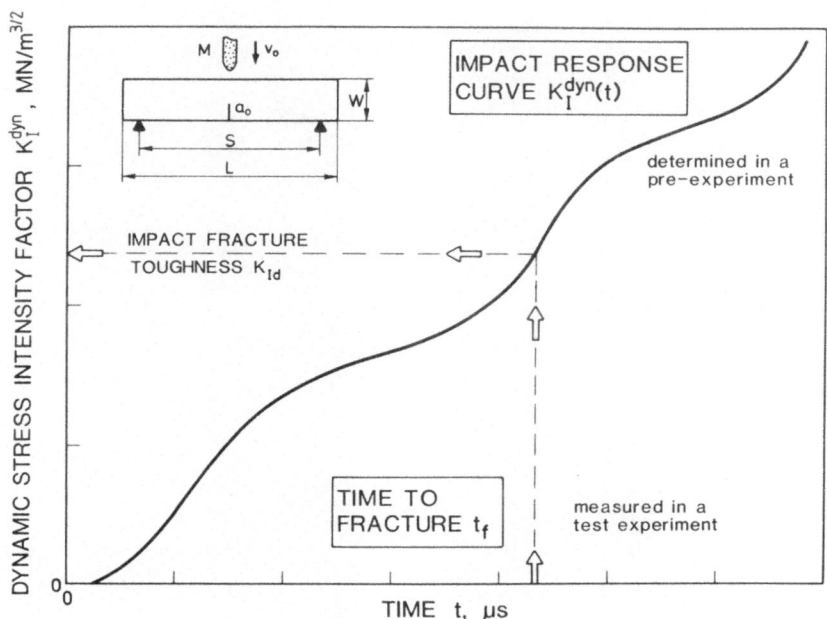

Figure 18. The concept of impact response curves for measuring the impact fracture toughness K_{Id}

real test-experiment a precracked specimen of the steel to be investigated is then tested under the same impact conditions. In this experiment only the time to fracture is measured (e.g. by an uncalibrated strain gage near the crack tip). This time together with the preestablished impact response curve determines the impact fracture toughness value K_{Id}.

The applicability of the measuring procedure has been checked under different test conditions. Figure 19 gives impact response curves for specimens of different sizes and different ratios of specimen length to support span.

The specimens were impacted by a hammer of 90 kg or 260 kg at a velocity of 5 m/s. Specimens which were supported at the very ends of the specimen show a more oscillatory $K_I^{dyn}(t)$-curve. Specimens with a length to support span ratio $L/S = 55/40$ (as with Charpy specimens) are characterized by a more steadily increasing curve (see also [22]). Impact fracture toughness values K_{Id} determined from these response curves are shown in Figure 20 for the steels 30 CrNiMo 8 and 15 MnNi 63. As far as they were available other fracture toughness data for initiation under static loading, K_{Ic}, and for crack arrest, K_{Id} and K_{Im}, (see section 3), are also shown. A comparison of these data indicates that the measured impact fracture toughness values are reasonable [25]. This is not trivial since the specimens failed at relatively small times to fracture, $t_f < \tau$, i.e. at times where the conventional procedure [24] would not be applicable.

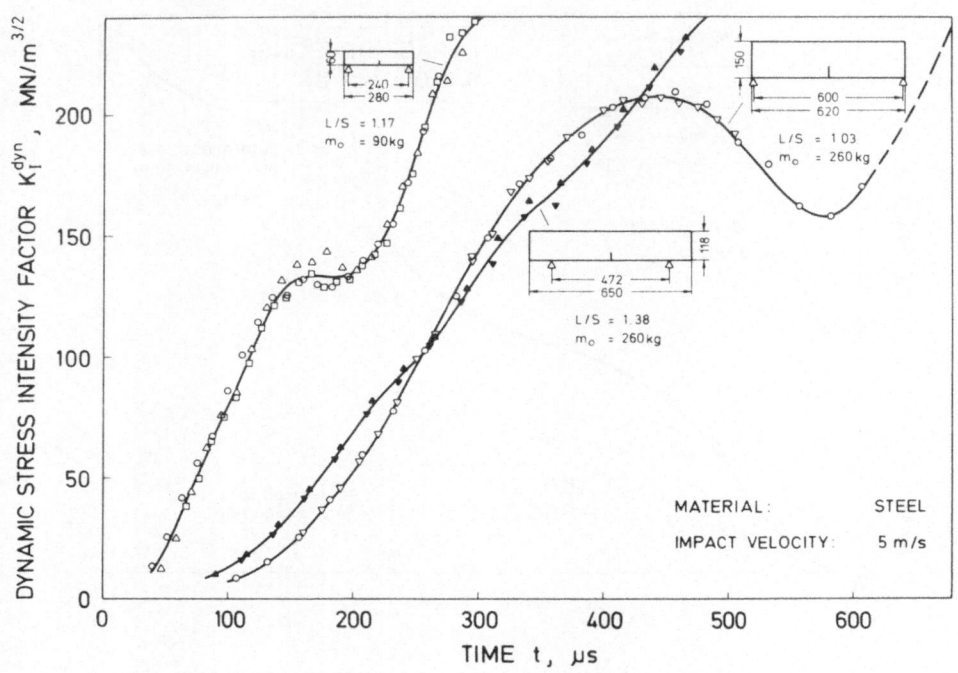

Figure 19. Impact response curves

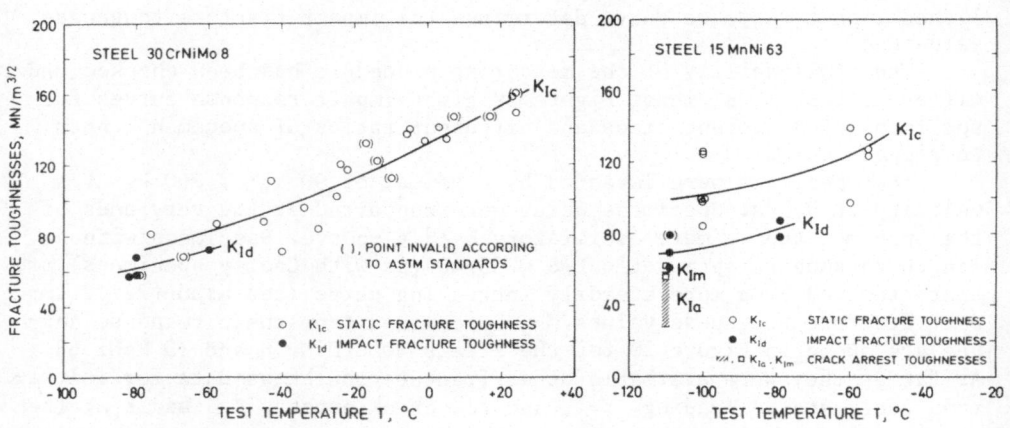

Figure 20. Impact fracture toughness data

Because of the very short times to fracture obtained in these experiments the anvils could not have influenced the specimen reaction. Since at higher loading rates the specimens would fail even earlier, it is natural that a high rate impact test (>5 m/s) for measuring K_{Id}-values can be carried out with unsupported specimens. This loading arrangement is called "one-point-bend" loading.

Typical results of "one-point-bend" experiments [23] are shown in Figures 21 and 22. Steel specimens 270 mm long and 50 mm wide were utilized in these experiments. Impact loading was achieved by a projectile of 1.9 kg mass which was accelerated by a gas gun to a velocity of 17 m/s. The resulting impact response curve is shown in Figure 21. The stress intensity factor capacity of the loading arrangement is about 260 MPa$\sqrt{m}$. Fracture toughness measurements have been performed under the same impact conditions with specimens made from a mild steel. Specimens of 20 mm thickness with sharp initial notches were tested at different temperatures. Complete failure of the specimens was observed in all experiments. Figure 22 shows the fracture surfaces of the broken specimens, the times to fracture t_f, and the resulting dynamic fracture toughness values K_{Id}. These data demonstrate the feasibility of the one-point-bend technique for measuring impact fracture toughness data at high loading rates.

Similar experiments have been performed at a lower impact velocity of 5 m/s with a pendulum type impact tester [23]. The test device was slightly modified to load unsupported specimens. At lower temperatures the specimens broke completely in spite of the only modest impact velocity.

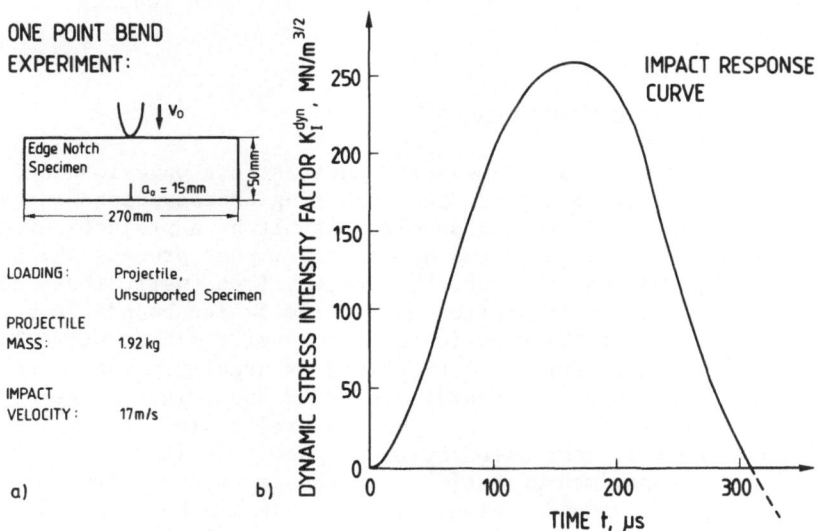

Figure 21. One-point-bend technique, a) test conditions
b) impact response curve

It is evident that a quasistatic evaluation procedure would not be able to describe these experiments. The fully dynamic method of impact response curves, however, can successfully be applied even under these highly dynamic test conditions.

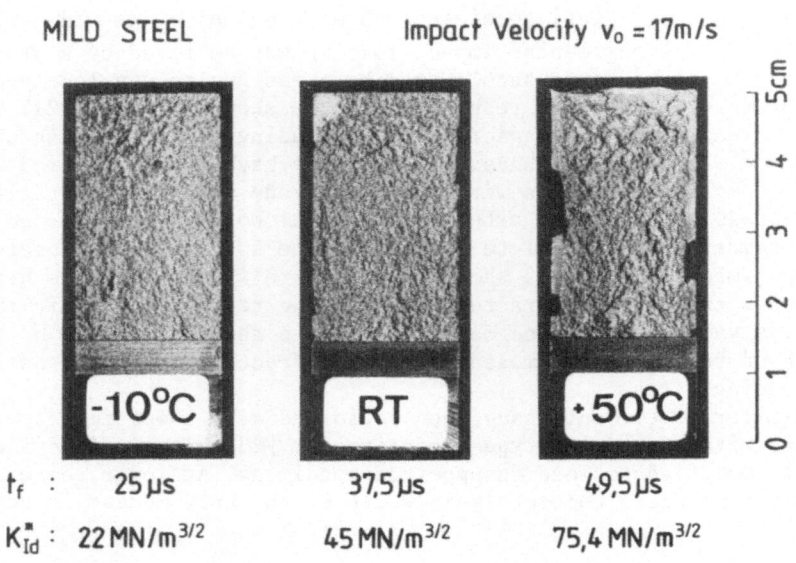

Figure 22.   Fracture behavior under one-point-bend loading

5. SHORT PULSE LOADING OF CRACKS

High rate material data are measured under stress wave loading, e.g. achieved by projectile impact. The resulting stress pulses in general are of short duration. When a specimen is hit by a projectile, compressive stress waves are generated during the impact process which propagate into the specimen and into the projectile as well. These compressive stress pulses, after reflection at the finite boundaries, load the specimen in tension. The duration of the tensile stress pulse T_O is determined by the geometric conditions and is usually given in the form $1/c$, where c is a wave propagation velocity and 1 is a characteristic dimension, for example the length of the projectile. The amplitude of the stress pulse is controlled by the impact velocity. Thus, by varying the projectile-specimen-dimensions and the impact velocity controlled stress pulses of known duration and amplitude can be produced (see Figure 23). For experimental details see for example Shockey et al. [26, 27].

    Applying projectile impact techniques, Homma, Kalthoff, Shockey et al. [27-30] studied the fracture behavior of cracks which were loaded by

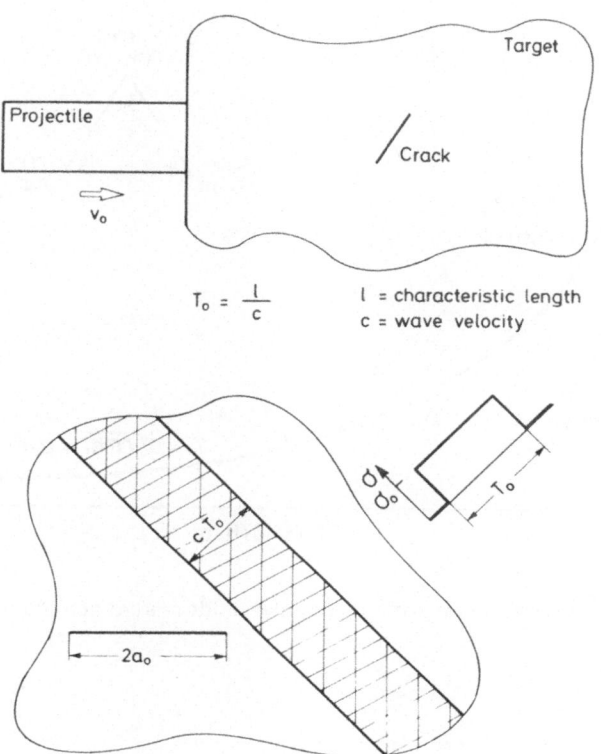

$$T_o = \frac{l}{c}$$

l = characteristic length
c = wave velocity

Figure 23.   Short pulse loading of cracks

stress pulses of durations which are comparable or even smaller than
the time it takes waves to travel the distance given by the crack
length. According to the static fracture mechanics concept, critical
stresses for instability should continuously decrease with increasing
crack length, as shown schematically in Figure 24. Instability stresses
for cracks of different lengths which were subjected to stress pulses
of different durations are shown in Figures 25 and 26 [29,30]. The presented results were obtained with specimens made from an epoxy resin
and 4340 steel. The data indicate the following behavior: For short
cracks the instability stresses decrease with increasing crack length,
similar as in static fracture mechanics. For cracks above a certain
length, however, higher stresses are needed to bring a crack to instability than under static loading conditions. Furthermore, the instability stresses seem not to decrease with increasing crack length but to
stay at the same constant level. The shorter the pulse duration the
larger the instability stress and the smaller the critical crack length
from which on a constant behavior is observed (see Figure 26).

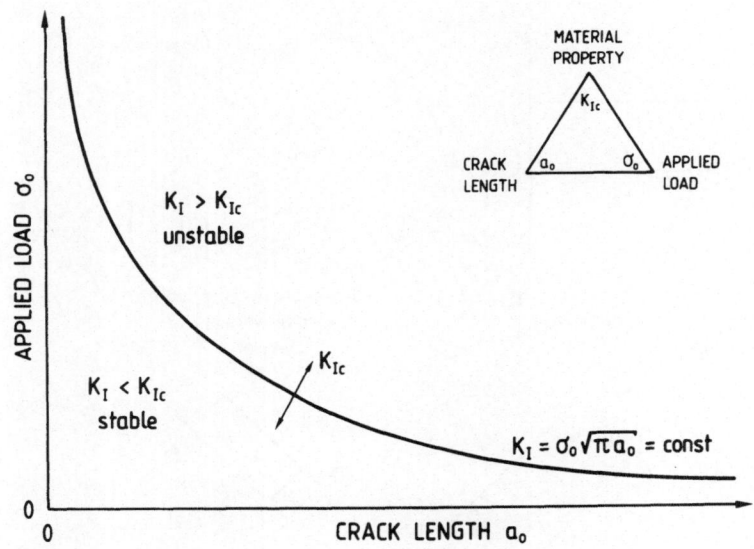

Figure 24. Instability behavior of cracks under static loading
conditions

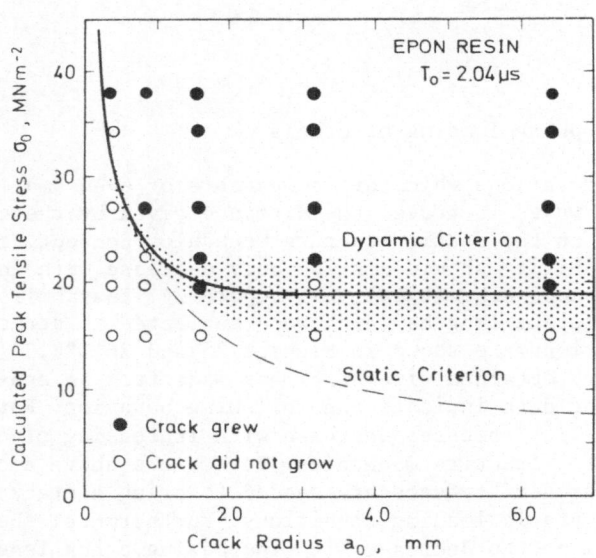

Figure 25. Instability stresses for cracks under short pulse loading

The observed findings demonstrate that a static instability criterion fails to describe the mechanical behavior of cracks under short lived pulses. The formal determination of fracture toughness values on the basis of static evaluation procedures would yield data which would show an increasing tendency with increasing crack length and decreasing pulse duration. Thus, the fracture criterion has to be modified to account for short duration effects.

Obviously a stress intensity factor following the simple $\sigma_0 \sqrt{\pi a_0}$-relationship does not apply anymore for short pulse fracture experiments if the crack length exceeds a certain limit. The actual behavior is indeed controlled by a rather complex stress intensity factor history [28]. The principal behavior for a crack under step function load is shown in Figure 27a [31]. The dynamic stress intensity factor K_I^{dyn} is plotted as a function of the dimensionless time $(c_1/a_0)t$, where c_1 is the longitudinal wave velocity. The stress intensification first increases with time according to a square root of time relationship, overshoots the equivalent static stress intensity factor by a considerable amount, and then reaches a constant static value after damped oscillations. This behavior is plausible: For very early times, $t < a_0/c_1$, only a certain part of the crack can contribute to the stress intensification at the crack tip due to finite "information velocities". Consequently, not the real crack length a_0 but an effective crack length a_{eff}, with $a_{eff} = ct < a_0$, determines the stress intensity factor $K_I^{dyn} = \sigma_0 \sqrt{\pi a_{eff}} = \sigma_0 \sqrt{\pi ct}$, where c is the "information velocity", i.e. an average wave propagation velocity. After the waves have traveled several times along the crack, $t > 10 a_0/c_1$, a static situation is reached and the real crack length a_0 determines the stress intensification, $K_I^{dyn} = \sigma_0 \sqrt{\pi a_0}$.

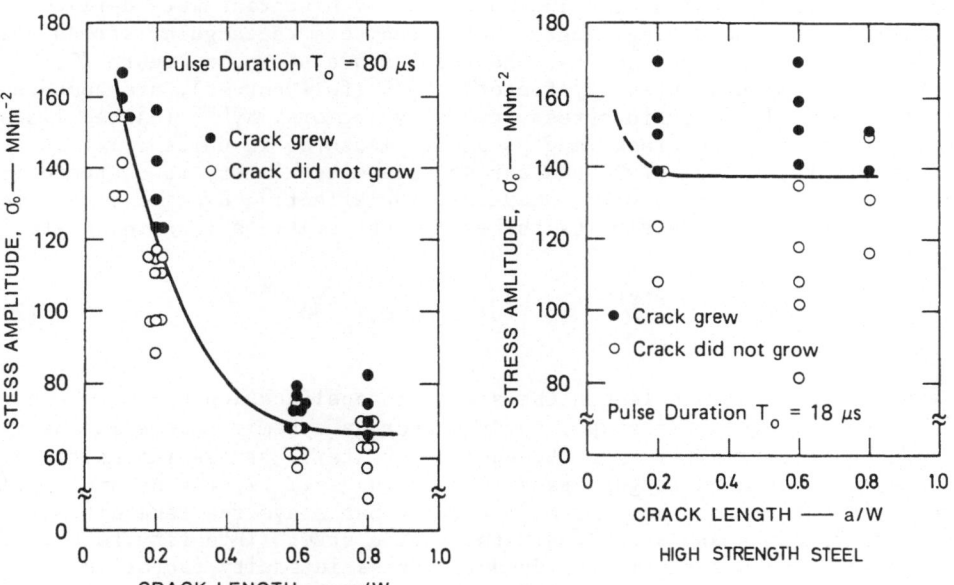

Figure 26. Instability stresses for cracks under short pulse loading

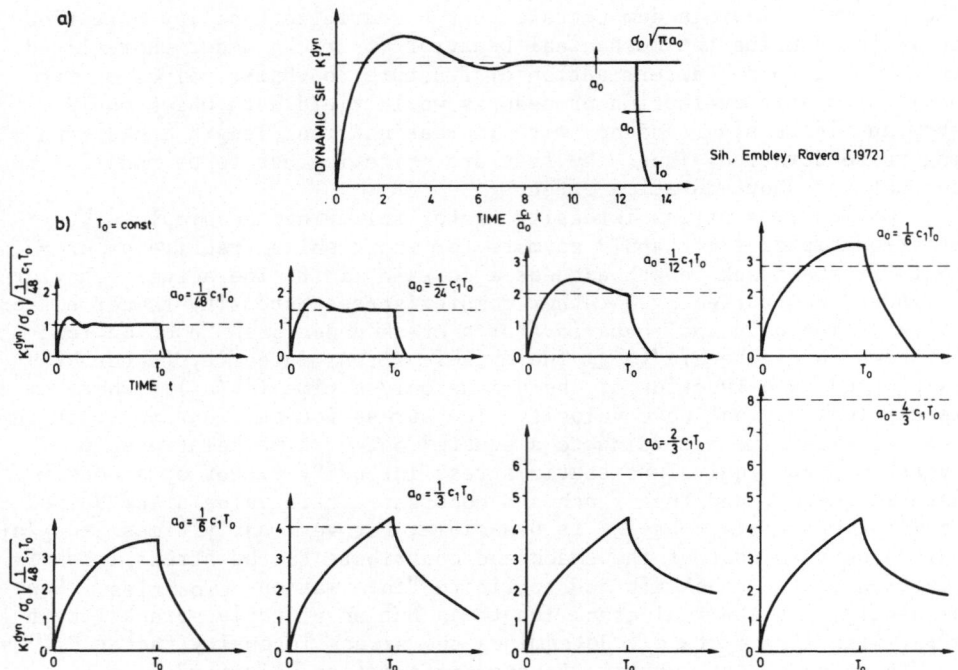

Figure 27. Stress intensity histories for cracks under short
pulse loading

Based on these results, the stress intensity histories were derived
for cracks of increasing lengths subjected to a rectangular stress pulse of finite duration T_0 [28]. The results are shown in Figure 27b.
The dynamic stress intensity factors, K_I^{dyn} (full curves), are compared
to the equivalent static stress intensity factors, K_I^{stat} (dashed lines).
For convenience the crack lengths a_0 are measured in units c_1T_0. For
short cracks, $a_0 \lesssim c_1T_0/30$, the stress intensity history is characterized by an almost rectangular shape and the effective dynamic stress intensity factor is the same as the equivalent static stress intensity
factor value,

$$K_{I,eff}^{dyn} = K_I^{stat} \qquad \text{for } a_0 \lesssim c_1T_0/30. \qquad (2)$$

With increasing crack length the stress intensification becomes larger
and more triangular in shape, but the average dynamic stress intensity
factor becomes smaller than the equivalent static stress intensity factor value. For even larger crack lengths, $a_0 > c_1T_0/2$, the dynamic stress
intensification does not increase anymore but stays the same although
the crack length and accordingly the static stress intensity factor
increases. In this regime the dynamic stress intensity factor is controlled by the effective crack length $a_{eff} < a_0$, and

$$K_{I,eff}^{dyn} = \text{const} < K_I^{stat} \qquad \text{for } a_o \geq c_1 T_o/2. \qquad (3)$$

Based on these stress intensity considerations and the assumption that the crack has to experience a supercritical stress intensity factor for at least a certain minimum time in order to become unstable, the instability behavior was predicted [28]. Results for cracks of different length a_o, loaded by stress pulses of different amplitude σ_o and duration T_o are shown in a three-dimensional $(\sigma_o\text{-}a_o\text{-}T_o)$-diagram in Figure 28. The short pulse fracture behavior is represented in the rear right section of the diagram, the front left regime (long pulse durations, short crack lengths) shows the usual static behavior. For constant pulse durations and cracks above a certain length, $a_o \gtrsim c_1 T_o/30$, higher critical stresses are predicted than in the equivalent static case. Furthermore, for crack length $a_o \gtrsim c_1 T_o/2$ the predicted instability stresses do not depend on crack length anymore. These theoretical predictions are in good agreement with the experimental observations.

The reported findings demonstrate the validity of the developed short pulse fracture criterion. The criterion shows the limitations of static fracture mechanics approaches and represents an appropriate basis for determining high rate fracture toughness data from short pulse fracture experiments.

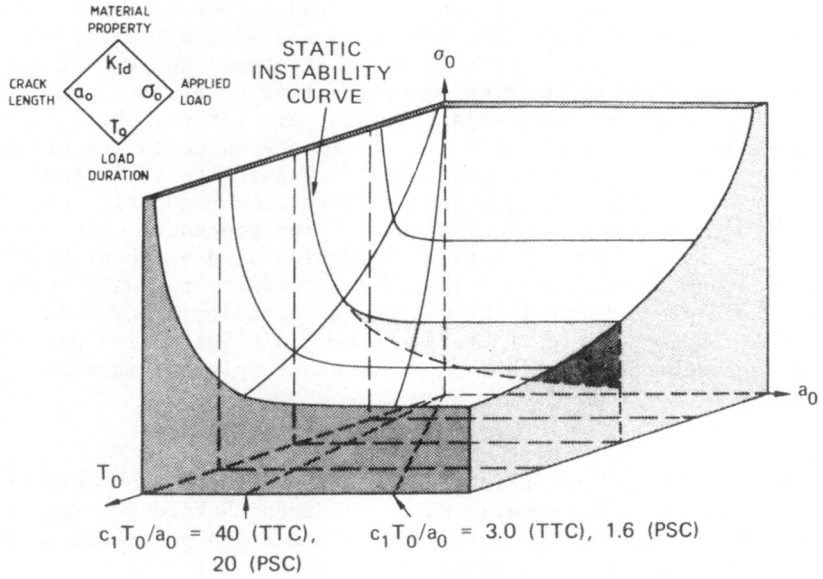

Figure 28. Surface of dynamic instability stresses for cracks under short pulse loading

6. SUMMARY AND CONCLUSION

Influences of time effects on test procedures for measuring dynamic material strength values have been analyzed. In particular, procedures for measuring the crack arrest toughness and the initiation fracture toughness at high loading rates have been considered. It is shown, that time effects can have a strong influence on the mechanical behavior of the cracks. For cracks at arrest the stress condition at the crack tip is still dynamic and not static, although the crack velocity is zero at the moment of arrest. Cracks under drop weight loading, in general, cannot be adequately described on the basis of quasistatic analyses. Static criteria also fail to describe the instability behavior of cracks under short pulse loading. Dynamic material properties based on static evaluation procedures, therefore, can be erroneous. A reliable determination of material strength data, therefore, requires the consideration of dynamic influences on the measuring procedure. Techniques which take the dynamic effects into account in an appropriate form have been outlined: i.e. the reduced dynamic effects crack arrest specimen, the dynamic concept of impact response curves and the short pulse fracture mechanics concept.

Fully dynamic procedures for measuring dynamic material properties in general are more sophisticated and more complicated than for measuring static material properties. This is a necessary consequence of the fact that a dynamic experiment is more than a fast static experiment. From the presented results it can be concluded, however, that dynamic effects in the elasto-plastic fracture regime will be smaller than under the conditions of linear-elastic brittle fracture: Due to the increased toughness levels the velocities of propagating cracks are smaller, and the times to fracture under impact loading are larger. Thus, inertia effects will be of considerably less influence on measuring procedures. Furthermore, the determination of elasto-plastic fracture properties very often is based on energy considerations. Consequently, oscillations in load records cancel when integration procedures are applied and the data are less sensitive to disturbing time effects. Further theoretical and experimental research is necessary to fully understand the processes which control the mechanical behavior of dynamic processes in the elasto-plastic fracture regime and to establish procedures for the measurement of reliable dynamic elasto-plastic material properties.

REFERENCES
[1] KALTHOFF, J.F., "Time Effects and their Influences on Test Procedures for Measuring Dynamic Material Strength Values", Int. Conf. Application of Fracture Mechanics to Materials and Structures, Freiburg, West-Germany, June, 1983

[2] MANOGG, P., "Anwendung der Schattenoptik zur Untersuchung des Zerreißvorgangs von Platten", Dissertation, Universität Freiburg, Germany, 1964

[3] MANOGG, P., "Schattenoptische Messung der spezifischen Bruch-
energie während des Bruchvorgangs bei Plexiglas", Proc. Int.
Conf. Physics of Non-Crystalline Solids, Delft, The Netherlands,
pp. 481-490, 1964.

[4] THEOCARIS, P.S., "Local Yielding Around a Crack Tip in Plexiglas",
J.Appl.Mech., Vol. 37, pp. 409-415, 1970.

[5] BEINERT, J., and KALTHOFF, J.F., "Experimental Determination of
Dynamic Stress Intensity Factors by Shadow Patterns" in Mechanics
of Fracture, Vol. 7, Ed. G.C. Sih, Martinus Nijhoff Publishers,
The Hague, Boston, London, pp. 281-330, 1981.

[6] KALTHOFF, J.F., "Stress Intensity Factor Determination by Caustics",
Int. Conf. Experimental Mechanics, Society for Experimental Stress
Analysis and Japan Society of Mechanical Engineers, Honolulu-Maui,
Hawaii, U.S.A., May 23-28, 1982.

[7] KALTHOFF, J.F., "On Some Current Problems in Experimental Fracture
Dynamics", Workshop on Dynamic Fracture, Ed. W.G. Knauss, California Institute of Technology, Pasadena, Calif., U.S.A., February
17-18, 1983.

[8] KALTHOFF, J.F., BEINERT, J., and WINKLER, S., "Measurements of
Dynamic Stress Intensity Factors for Fast Running and Arresting
Cracks in Double-Cantilever-Beam-Specimens", ASTM STP 627 - Fast
Fracture and Crack Arrest, American Society for Testing and Mate-
rials, Philadelphia, U.S.A., pp. 161-176, 1977.

[9] KALTHOFF, J.F., BEINERT, J., WINKLER, S., and KLEMM, W., "Experimental Analysis of Dynamic Effects in Different Crack Arrest Test
Specimens", ASTM STP 711 - Crack Arrest Methodology and Applications, American Society for Testing and Materials, Philadelphia,
U.S.A., pp. 109-127, 1980

[10] KALTHOFF, J.F., "Zur Ausbreitung und Arretierung schnell laufender
Risse", Fortschritt-Berichte der VDI-Zeitschriften, Reihe 18, Nr.4,
VDI-Verlag, Düsseldorf, 1978.

[11] KALTHOFF, J.F., BEINERT, J., and WINKLER, S., "Influence of Dynamic
Effects on Crack Arrest", Final Report prepared for Electric Power
Research Institute, Palo Alto, Calif., under Contract No. RP 1022-1,
IWM-Report W 4/80, Fraunhofer-Institut für Werkstoffmechanik,
7800 Freiburg, West-Germany, 1980.

[12] CROSLEY, P.B., and RIPLING, E.J., "Crack Arrest Toughness of Pressure Vessel Steels", Nuclear Engineering and Design, Vol. 17,
No. 1, pp. 32-45, August 1971.

[13] CROSLEY, P.B., and RIPLING, E.J., "Towards Development of a Standard Test for Measuring K_{Ia}", ASTM STP 627 - Fast Fracture and
Crack Arrest, American Society for Testing and Materials, Philadelphia, U.S.A., pp. 372-391, 1977.

[14] HAHN, G.T., HOAGLAND, R.G., KANNINEN, M.F., and ROSENFIELD, A.R.,
"A Preliminary Study of Fast Fracture and Arrest in the DCB Test
Specimen", Proc. Int. Conf. Dynamic Crack Propagation, Ed. G.C.Sih,
Lehigh University, Bethlehem, Pa., U.S.A., July 10-12, 1972.

[15] HAHN, G.T., GEHLEN, P.C., HOAGLAND, R.G., KANNINEN, M.F., POPE-
LAR, C., ROSENFIELD, A.R., et al., "Critical Experiments, Measure-
ments and Analyses to Establish a Crack Arrest Methodology for
Nuclear Pressure Vessel Steels", BMI-1937, 1959, 1995, Battelle
Columbus Laboratories, Columbus, Ohio, Aug. 1975, Oct. 1976,
May 1978.

[16] BEINERT, J., KLEMM, W., and KALTHOFF, J.F., "Mechanik und Stoff-
verhalten bei der Arretierung von Rissen", IWM-Report V 16/83,
prepared for European Community on Steel and Coal, under Contract
No. 7210.Ke 107 Fraunhofer-Institut für Werkstoffmechanik, 7800
Freiburg, West-Germany, April 1983.

[17] KALTHOFF, J.F., BEINERT, J., and WINKLER, S., "Einfluß dynamischer
Effekte auf die Bestimmung von Rißarrestzähigkeiten und auf die
Anwendung von Rißarrestsicherheitsanalysen", 8. Sitzung, Arbeits-
kreis Bruchvorgänge im Deutschen Verband für Materialprüfung,
Köln, Germany, Oct. 6-7, 1976.

[18] BEINERT, J., and KALTHOFF, J.F., "The Development of a Crack Arrest
Test Specimen with Reduced Dynamic Effects", Int. Conf. Application
of Fracture Mechanics to Materials and Structures, Freiburg, Germany, June 20-24, 1983.

[19] KALTHOFF, J.F., WINKLER, S., and BEINERT, J., "The Influence of
Dynamic-Effects in Impact Testing", Int. Journ. of Fracture,
Vol. 13, pp. 528-531, 1979.

[20] KALTHOFF, J.F., BÖHME, W., WINKLER, S., and KLEMM, W., "Measurements of Dynamic Stress Intensity Factors in Impacted Bend Specimens", CSNI Specialist Meeting on Instrumented Precracked Charpy
Testing, EPRI, Palo Alto, Calif., U.S.A., Dec. 1-3, 1980.

[21] KALTHOFF, J.F., BÖHME, W., and WINKLER, S., "Analysis of Impact
Fracture Phenomena by Means of the Shadow Optical Method of
Caustics", VIIth Int. Conf. Experimental Stress Analysis, Society
for Experimental Stress Analysis, Haifa, Israel, Aug. 23-27, 1982.

[22] BÖHME, W., and KALTHOFF, J.F., "The Behavior of Notched Bend Specimens in Impact Testing", Int. Journ. of Fracture, Vol. 20,
pp. R139-143, 1982.

[23] KALTHOFF, J.F., WINKLER, S., BÖHME, W., and SHOCKEY, D.A., "Mechanical Response of Cracks to Impact Loading", Int. Conf. on Dynamic Mechanical Properties and Fracture Dynamics of Engineering Materials, Czechoslovak Academy of Sciences, Institute of Physical Metallurgy, Brno-Valtice, CSSR, June 16-18, 1983.

[24] ASTM E 24.03.03, "Proposed Standard Method of Test for Instrumented Impact Testing of Precracked Charpy Specimens of Metallic Materials", Draft 2c, American Society for Testing and Materials, Philadelphia, U.S.A., 1980.

[25] KALTHOFF, J.F., WINKLER, S., BÖHME, W. and KLEMM, W., "Determination of the Dynamic Fracture Toughness K_{Id} in Impact Tests by Means of Response Curves", 5th Int. Conf. Fracture, Cannes, March 29 - April 3, 1981, Advances in Fracture Research, Ed. D. Francois et al., Pergamon Press, Oxford, New York, 1980.

[26] SHOCKEY, D.A., and CURRAN, D.R., "A Method for Measuring K_{Ic} at Very High Strain Rates", ASTM STP 536 - Progress in Flaw Growth and Fracture Toughness Testing, American Society for Testing and Materials, Philadelphia, U.S.A., pp. 297-311, 1973.

[27] SHOCKEY, D.A., KALTHOFF, J.F., HOMMA, H., and EHRLICH, D.C., "Response of Cracks to Short Pulse Loading", Workshop on Dynamic Fracture, Ed. W.G. Knauss, California Institute of Technology, Pasadena, Calif., U.S.A., February 17-18, 1983.

[28] KALTHOFF, J.F., and SHOCKEY, D.A., "Instability of Cracks under Impulse Loads", J. Appl. Phys., Vol. 48, No. 3, pp. 986-993, March 1977.

[29] SHOCKEY, D.A., KALTHOFF, J.F., and EHRLICH, D.C., "Evaluation of Dynamic Crack Instability Criteria", Int. Journ. of Fracture, Vol. 22, pp. 217-229, 1983.

[30] HOMMA, H., SHOCKEY, D.A., and MURAGAMA, Y., "Response of Cracks in Structural Materials to Short Pulse Loads", submitted to J. Mech. Phys. Solids.

[31] SIH, G.C., EMBLEY, G.T., and RAVERA, R.J., "Impact Response of a Plane Crack in Extension", Int. Journ. Solids Structures, Vol. 8, pp. 977-993, 1982.

CONCLUSIONS OF SEMINAR

A. Janne Carlsson
Royal Institute of Technology
100 44 Stockholm
SWEDEN

ABSTRACT

The important aspects of fracture mechanics and fracture dynamics are
brought forth and discussed. Some important aspects which have not been
dealt with in the seminar papers are discussed in more detail. Comparison is made between different methods for engineering assessment of
failure. Finally possible areas for future seminar activity are proposed.

1. INTRODUCTION

For a cracked body there are in general several ways to failure as indicated in the introductory paper of this volume by L. H. Larsson. Which
way will be followed in a possible failure depends on material properties, loading conditions, geometry of the structure especially its compliance and also other conditions.
    The mechanisms of failure are indicated in the flow chart of Larsson
and are tearing collapse (high energy), shear decohesion and cleavage
(low energy). The designations of the mechanisms indicate great complexity of the material state at the crack tip and therefore great difficulty to determine and describe that state i.e. stress, strain and exact
geometry. In spite of this there are today attempts made to apply local
critera to the crack growth process i.e. criteria based on local stress
and strain.
    The predominant results followed in failure analysis are based on
global parameters as the stress intensity factor K, the path independent
integral J and the tearing modulus T (dJ/da). Much work is devoted to
making failure analysis based on these parameters accessible to engineers by developing simplified, approximate methods of analysis. It is
also interesting to note that dynamic crack problems involving consideration of rapid crack growth and crack arrest have become of interest to
engineers in design of large structures.

L. H. Larsson (ed.), Elastic-Plastic Fracture Mechanics, 509–519.

2   FUNDAMENTALS OF FRACTURE MECHANICS

2.1   Local versus global criteria

The basis of crack growth theories are the continuum mechanical analysis
of the state at the crack tip and physical studies of the mechanisms of
material damage and crack propagation.

It is of course essential to include into the modelling of the
material as many features as possible of the physical and mechanical
crack propagation process. In this respect the Kachanov damage model and
its improvement as described in the paper by François and the damage
modelling described by Mudry are interesting. Work of this type will
lead to an increased understanding of the fracture process. Whether it
can be brought to a state where it will be a useful tool for the designer and failure analyst seems so far uncertain. It is proposed to
lead to a local criterion for crack growth and thus requires detailed
calculations of the stress and strain state at the tip of the crack for
each geometric configuration. Even with very powerful finite element
programs such detailed analysis of crack problems seems to be out of
question other than in research situations.

The global crack parameters K and J, here offer the advantage of
being much easier to determine for the designer and engineer since they
are possible to determine from a normal structural analysis without
looking in great detail at the crack tip region.

Concerning the use of J there are however some fundamental problems
still to be solved which have not been dealt with very extensively in
this seminar:

2.2   Geometry dependence of J(Δa)

There is observed to be a geometry dependence in the traditional
J=J(Δa) curve showing that J for tension loading is higher than for
bending load cases. This dependence is predicted also from analytical-
numerical evaluations of the extension of the HRR-singularity at the
cracktip which is larger for bending than for tension loading, Shih and
Germain [2]. There is evidence indicating that this deviation appears
if the modified J-integral by Ernst [3] is used.

2.3   Non-proportional loading

The J-integral is based on a non-linear elastic material model which is
often used to describe elastic-plastic materials. This is also the case
in fracture mechanics. This would to be stringent require monotonically
increasing proportional loads. It is known that in this respect the J-
theory is very tolerant and allows e.g. crack growth. An important case
in engineering structures is the case where the external loads are controlled by several loading parameters varying independently. A simple
case with two loading parameters is a beam or a plate with an edge or
a surface crack exposed to tension (P) and bending (M) where P and M are
separately controlled. A common case is e.g. a pressure vessel wall with
a temperature gradient and a bending type stress which is displacement
controlled and with tensile, load controlled axial and circumferential

stresses due to internal pressure and thus load controlled.

Such problems give rise to several questions both of fundamental and of practacal importance:
- Is J at all possible to use under those circumstances?
- How is J to be determined?
- In cases of mixed boundary conditions-load and displacement control - what is the condition for instability?

These questions have been elaborated by Kaiser in several papers [4], [5]. The purpose of this paper is not to go into detail in any of these questions. It will only be concluded here from [4] and [5] that J also as concerns application to strongly non-proportional external loads seems to be much more tolerant than one would expect it to be considering its basis.

It also turns out to be possible to determine J and T for multi-parameter loading with methods similar to those for the Merkle - Corten - approach. The instability question for this case is also possible to handle. This latter point will be dealt with more extensively below.

2.4   Influence of residual and thermal stresses

Closely related to the previous paragraph is the problem of how to handle residual and thermal stresses in a crack growth analysis. A J-integral expression for such stresses is given by Ainsworth et al [6]. Applications of this integral are however scarce. It is verified that residual stresses have a decreasing influence on the fracture load of structures with increasing temperature or ductility. Although thermal and residual stresses are essentially displacement controlled they are not equivalent from a crack growth instability point of view to external, displacement controlled loads. Therefore the methods mentioned in paragraph 2.3 for stability analysis are not directly applicable to the present cases. It seems that no methods are available to treat crack growth instability for these cases at present.

2.5   Crack growth criteria for 3-dimensional cases

The most extensively treated 3-dimensional crack problem is the surface crack. Fundamental aspects of this problem are considered by Bakker and more practical view points by Munz.

There are evidently great difficulties in determining $J = J(s)$ along the crack front. As concerns conditions for growth our present knowledge is not sufficient to make good predictions.

Therefore good approximate methods are very valuable. The problem is of great practical importance.

2.6   Numerical methods in non-linear fracture mechanics

The inaccuracy of J-calculations is according to round robin calculations in Europe, L.H. Larsson[1] great as indicated in Fig 1. Partly, deviations are due to modelling of the problem - plain strain, plain stress or three-dimensional – but within each of these groups there are considerable deviations. These are suprising since the problem considered is simple and the reasons for the deviation are not clear.

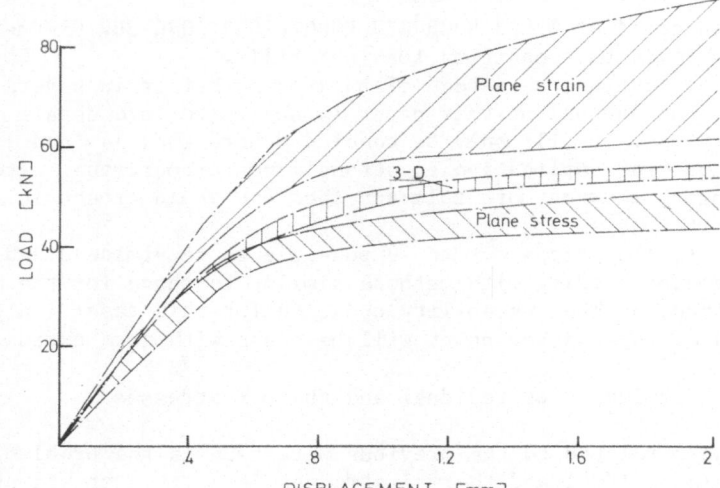

Figure 1.
Result of
Round Robin
FEM calculations.

Against this background one is tempted to draw the conclusion that FEM calculations for more complex 3-dimensional geometries must be very unreliable. One primary goal of fracture mechanics research must be to understand the difficulties involved in numerical treatment of crack problems and to improve the methods used.

3   ENGINEERING METHODS – APPLICATIONS

3.1   General viewpoints

One does not accept conditions in structures under which cracks grow stably. In spite of this it is of great interest to determine the true critical load for a structure and thus necessary to allow for stable crack growth in such determinations. This way true estimates can be made of the safety margin of a cracked structure. It is evident from what was said in chapter 2 that stringent, exact calculations can only be made for a minority of the applications one wants to consider. The reasons are as pointed out both fundamental and practical difficulties.
     For engineering applications several simple approximate and hopefully conservative methods of handling crack problems have therefore been developed.
     Fortunately there is for a structural crack problem two limiting cases which are often easy to handle in an accurate manner i.e. the linear elastic case and the limit load case. Especially the linear elastic case is reliable. The limit load is more difficult to determine in a conservative manner. For some structures e.g. pipes the failure

mode is not very well known and conseqently the limit load not so easy
to determine. It should also be said that the limit load itself is only
an engineering estimation of the load bearing capacity of a structure.
These limiting cases are easily recognized in the failure assessment
diagram (FAD) and as a matter of fact found the backbone of this
diagram.

Also in other engineering methods for failure assessment the
linear elastic case forms the basis and the extension of the
methods outside the LEFM regime can often be seen as an extrapolation. This guarantees in a sence that the methods give reasonable
results at least in the transition region from linear to non-linear
behaviour.

Determination of the instability point for a cracked structure
is a more difficult task and here more stringent methods are required. As will be seen in the following text general stability
concepts of solid mechanics can often be useful for instability
analysis in fracture mechanics.

An important part of applied fracture mechanics is determination or estimation of material toughness properties. For this
purpose other methods may be used than those suitable for failure
assessment of structures – especially for comparative toughness
measurements. The COD-method is extensively used in this connection
and serves this purpose excellently. As is shown in the paper by
Garwood procedures for critical CTOD measurements are today reliable and well established.

3.2   General stability criterion and its relation to the tearing modu-
       lus concept

One way of treating crack instability which has not been presented in
the seminar is the general instability condition for a structure containing a nonlinear component. Consider a system consisting of a non-
linear part and an elastic part in series Fig 2. The elastic part is
modelled with a spring C_{ext} and the non-linear part with a CT-specimen.
However the whole system in Fig 2 could also represent a piping system
with one crack. The non-linear part is then a short part of the pipe
containing the crack.

The non-linear part is assumed to have the force-displacement characteristic $P=P(\Delta)$, Fig 3, and the elastic part the compliance C_{ext}.
P and Δ may be generalized force and displacement.

The stability condition then reads (Kaiser & Carlsson [7])$d\Delta/dP>$-
C_{ext} and instability occurs for equality between the quantities. This
definition of instability can be shown [7] to be equivalent to the Paris
tearing modulus condition for instability $T_{appl} = T_{mat}$.

It must be observed that $P=P(\Delta)$ contains deformation due to both
crack growth and plasticity and its determination requires testing to
failure of the actual non-linear component under stiff loading conditions so that the $P(\Delta)$-curve, Fig 3, is determined down to low enough
values of P on the unstable branch of $P=P(\Delta)$.

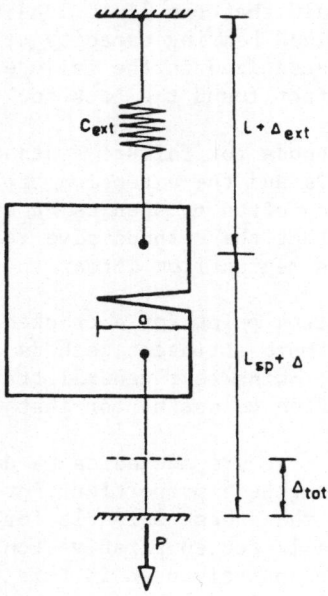

Figure 2   Non-linear component in elastic system (modelled by spring C_{ext})

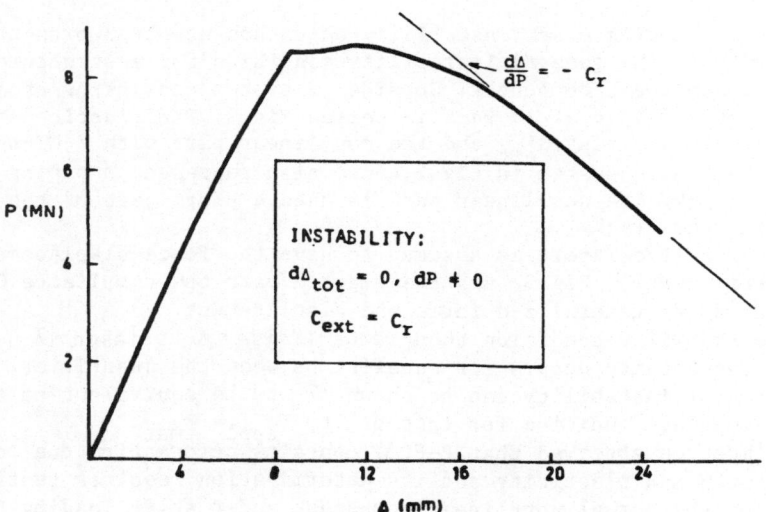

Figure 3   Load displacement curve for non-linear component of Fig. 2. Instability occurs for $C_{ext} = - C$.

3.3   Assessment methods in fracture mechanics

The procedures for engineering defect assessment discussed in this seminar are those based on the COD, R6, EnJ and EPRI methods. Other methods which have not been discussed but which merit being mentioned are the more general (than R6) failure assessment diagram, FAD, and the tearing modulus.

Failure assessment diagrams may be based on different theories as e.g. the simple Feddersen [8] approach or on more stringent theories as e.g. the Hutchinson – Shih [9] analysis of stable growth and instability. The original Feddersen approach is to define fracture strength from the Griffith-Irwin condition combined with a net section yield condition. It gives a FAD very close to the R6 curve, Fig.4.

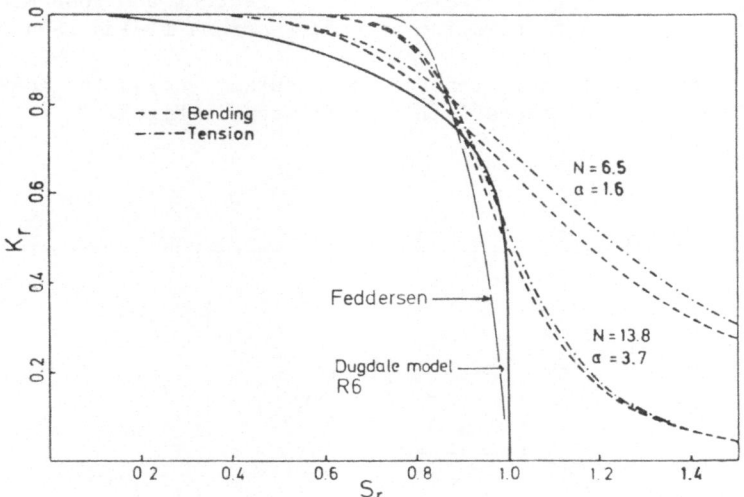

Figure 4.   Failure assessment diagram for powerlaw hardening
            material $\varepsilon/\varepsilon_\gamma = \alpha(\sigma/\sigma_\gamma)^N$.

In a more general theory workhardening can be accounted for and the FAD obtains a tail instead of a cut off at a limit load. An example given by Kaiser is shown in Fig 4. The FAD is derived for different constants in the Ramberg-Osgood stress strain relation.

The pros and cons of the different assessment criteria are given in Table 1. A full stable growth to instability analysis is possible to make with the T-modulus and EPRI-methods and in principle also with the general FAD method although in this case it requires more elaborate basic data than are generally available. These latter methods also appear to be the most exact ones and have a sound theoretical basis. They are also designer-friendly i.e practical to use in connection with traditional strength calculations.

The R6 and EnJ methods are more approximate. They should be conservative but it is not clear how much. The R6 method is very easy to use and its theoretical background is quite clear. The approximation of the method lies in the theoretical model itself. The EnJ method is in the linear elastic range based on linear elastic fracture mechanics with the Irwin plastic zone size correction. The non-linear range expression is of limit load type but the approximation it involves is difficult to estimate.

For all methods of assessment there is no sound basis to account for secondary (thermal and residual) stresses in a stringent way especially not as concerns instability. However there are rules how to handle these stresses in an approximate manner. It should be pointed out that not until lately have stress intensity factors and good methods for determination of such factors for secondary stress fields been available [10], [11], [12].

As concerns the sophistication and accuracy versus the cost and time requirement the methods fall as indicated in Fig 5.

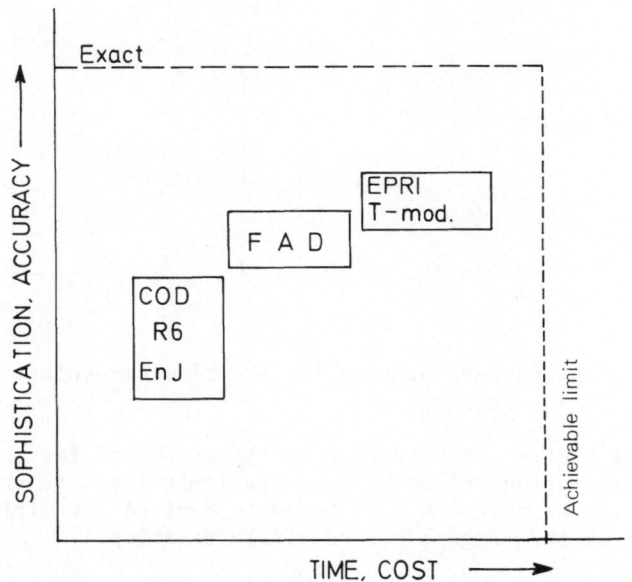

Figure 5. Sophistication and accuracy of different assessment methods versus the time they require and their cost.

A final viewpoint on the methods is obtained by dividing them into one category T, EPRI and the general FAD which is attractive to mechanical and structural engineers and one category COD and to some extent R6 which is attractive to materials specialists.

Dynamic fracture mechanics is a very important field since dynamic loads
are very common. In many cases of LEFM where we today use a static approach we should actually apply dynamic theory and dynamic material
data. It is only the complexity of the problem and the lack of data that
prevents us from using such a realistic approach. As seen from the
papers of this seminar dynamic fracture mechanics has today reached a
state both experimentally and theoretically where it can be used by
engineers.

One aspect which has already gained great attention in engineering
is crack arrest. Design for crack arrest has been used to guarantee
extra safety for structures like ships, pipelines and piping-vessel
systems.

The state of the art in linear fracture dynamics may be summarized
in the following points:
- The effect of impact loading on stationary cracks is well understood
  and can be studied with available methods. Material data are obtain-
  able.
- Crack propagation, arrest and reinitiation are all well understood
  and theoretical and experimental methods available.
- Wave loading effects on cracks are likewise understood.
- Experimental methods are well developed to obtain crack velocity å,
  thoughness data K_{pc} (å) and arrest data k_a.

 - Analytical solutions and FEM procedures exist.
In elastic plastic fracture dynamics the state of the art is in
summary:
- Analytical solutions exist but are few.
- FEM-procedures are not well developed.
- Fracture criteria are not well understood.
- Material relations are not well known as concerns strain rate effects
  and viscoplastic properties which are very important.
Since fracture dynamics is a rather new field one would like to urge for
future studies related to applications e.g. development of FEM-procedures, experimental studies of crack propagation in components and structures, production of material data for engineering materials. Development of engineering methods of analysis and assessment is also urgent.

5   CLOSURE

It is inappropriate to write conclusions to a contribution which is
in itself conclusions of the whole seminar. In closure of this paper it
seems however adequate to define areas of fracture mechanics suitable
for future ISPRA seminars.

Certainly the development in elastic-plastic fracture mechanics
will continue and it would be desirable to have in the future further
seminars in this important field. Perhaps one should then broaden the
outlook and have a more extensive presentation in the lectures of the
development outside Europe especially in US and Japan. One area of special importance is the effect of secondary stresses both in LEFM and
EPFM. Much work has been devoted to this problem during the last years.

In fracture dynamics the development is very strong. The area is
very important and worth strong attention perhaps even a special seminar
in the near future.

Probabilistic fracture mechanics and fatigue are other areas of
great interest to engineers and scientists and also suitable for future
seminars.

Table 1

Method	Pros	Cons
T-modulus	Sound theoretical basis Designer friendly Handles stable growth instability (s.g. is.)	Difficult for 3D-cases Secondary stresses? $T=T(\Delta a)$ not constant
EPRI	As for T. Handbook available	3D-cases? Secondary stresses
FAD (general)	Sound basis. Designer friendly.	3D-cases? Secondary stresses?
R6	Easy to use.	Approximate. Conservative? How much?
En J.	In principle easy to use.	As R6.
COD	Good for toughness assessment of materials.	Not designer friendly. Stable growth - instability?
Compliance	"Exact." Designer friendly.	Requires large scale testing.

REFERENCES

1   Larsson, L.H. EGF numerical round robins on EPFM in 3rd Int.Conf. on
    Numerical Methods in Fracture Mech. Ed. Luxmore, A.R. and Owen,
    D.R.J.   Swansea 1984.

2   Shih, C.A and Germain, M.D. (1978). Req. for One Parameter Charac-
    terization of Crack Tip Fields by the HRR Singularity. G.E.
    Techn.Rep.

3   Ernst, H.A. Material Resistance and Instability Beyond J-Controlled
    Crack Growth, ASTM ST P 803, Ed Shih, C.F. and Gudas, J.P.

4   Sönnerlind, H. and Kaiser, S. The J-integral for a SEN Specimen Under
    Non-Proportionally Applied Bend and Tension. To be published in J.
    Engn. Fr. Mech (1985)

5   Kaiser, S. An Extension of the Tearing Instability Theory to Multiple
    Loading Parameters, Report Nr. 55 Division Strength of Materials and
    Solid Mechanics, Stockholm. To be published in Int. J. of Fracture.

6   Ainsworth, R.A., Neale, B.K. and Price, R.H. (1978) Proc. Intern.
    Mech. Eng. Conf. on Tolerence of Flaws in Press. Comp., London, 197–
    204.

7   Kaiser, S. and Carlsson, A.J. (1983), ASTM-Symposium "Elastic-Plastic
    Fracture II – Fracture Resistance Curves and Engineering Applica-
    tions" AST; STP. 803 Eds. Shih, C.F. and Gudas, J.P.

8   Feddersen, C.E. Evaluation and Prediction of the Residual Strength of
    CC Tension Panels. ASTM STP 486, 1971.

9   Hutchinson, J. and Shih, C.F.   J. Engn. Mat. and Techn.,
    98, 1976 pp. 289.

10 Wu, X. (1984) The Effect of Residual Stress on Brittle Fracture of
    Plates with Surface Cracks, J. Engn. Fracture Mech. Vol 19, 427–
    439.

11 Wu, X. (1984) Stress Intensity Factors for Half-Elliptical Surface
    Cracks Subjected to Complex Crack Face Loadings, J. Engn. Fracture
    Mech, Vol 19, 387–405.

12 Wu, X. and Carlsson, A.J. (1984) Welding Residual Stress Intensity
    Factors for Half-Elliptical Surface Cracks in Thin and Thick Plates,
    J. Engn. Fracture Mech. Vol. 19. 407–425.

Crack initiation *35,77,85,89,94,101,121,128,374,417,483,490*
Crack length, crack size (corrected, effective, equivalent) *5, 180,341,380,400*
Crack mouth opening displacement *61,68*
Crack (Tip) Opening Displacement (COD, CTOD) *9,36,61,78,89,119, 173,182,251,398,399,470*
    COD for surface cracks *173*
    COD method *10,85,397,411,412,419,421,422,515,516,518*
    Critical values of CTOD *85,89,165,404,410*
    CTOD measurement *62,85,89,106,410*
    Determination of CTOD by FEM *78,251,252*
    Relationship between COD and J *36*
Crack propagation *436,443,445,473,480,483,485,487,517*
    Criterion for crack propagation and arrest *442,449,501-503*
Crack tip blunting *7,103,118,252*
    Crack tip deformation field *9,21*
    Crack tip elements *79,230*
    Crack tip opening angle *91,92*
    Crack tip singularity *22,230,449*
    Crack tip stress/strain field *4-6,18-22,230*
    State at a moving crack tip *428*
Crack velocity *442,443,485,487,517*
Cracked body: regimes of a cracked body *4*
Critical elongation *293*
Critical loads *193*
Critical non-dimensional cavity size *266*
Critical stress intensity factor *13,105,203*
Critical stress (for crack instability under short-pulse loading) *500-503*
Damage *9,145,268,285,510*
    Damage energy release rate *287*
    Damage function *267*
    Damage parameter *287*
    Damage theory *286*
    Damage threshold *290*
    Damaged zone *269,293*
Defect detection probability *2,3*
Deformation twins *209*
Design methods *9,422,512,515*. See also: COD method, EnJ procedure, EPRI method, R6 procedure
Detection and characterization of real defects in structures *2*
Detection of initiation *101,102,156*
Dislocation *205*
    Dislocation pile-up *205,206*
    Sessile dislocation *207*
Displacement control *43,56,73,98,101,510*
Double compliance method *387*
Driving force curve *107,108,374,377,382,390*
Drop weight loading *480,490-498*
Ductile-brittle transition *86,166,221,222,303,305*

Tearing: see also Ductile tearing, Maximum tearing load.
    Tearing instability *5,8,41,42,45,98,306,374,415,417,513,514*
    Tearing instability criterion *43,45,376,513,514*
    Tearing instability parameter J_{50} *138*
    Tearing modulus *43,44,138,373,376,509,513,515,516,518*
Temper embrittlement *303*
Temperature dependence of fracture toughness *310*
Testing standards *88*
Thermal shock *174,314*
Thermal stress *272,399,413,414,511,516*
Thermodynamic potential *287*
TITUS program *314*
Transferability (of experimental results from small specimens to
    real structures) *7,36,73-80*
Transient problems *433,439,470*
Transition from cleavage to ductile fracture *86,221.* See also
    Ductile-brittle transition
    Transition temperature *212,303,304*
Triaxiality factor *288,290*
Tunnelling: see Crack front t.
Ultrasonics *101*
Unloading compliance: see Single specimen...
Virtual crack extension technique *33,76,244,247,257*
Void nucleation, growth and coalescence *7,117,217,268,271,446*
Warm prestress effect *8,303,312*
Wave propagation *10,468,487,498,517.* Longitudinal waves, shear
    waves, Rayleigh waves *469*
Weibull's parameter *291,307*
    Weibull statistics *304,307*
    Weibull stress *308*
Weight function method *171*
Weldment *85,88,96,97,133,148,159,384,403,404,420*
    Testing weldments using CTOD procedures *96*
    Weldment toughness *96,133,159*
Work: elastic w., plastic w. *58,61*, total w. *57,63*
Work hardening *62,413,424.* See also Hardening
Yield: general y., gross (full ligament) y. *4,5,179*
    Yield stress *20,149,236,328,333,338,345,350,398,403,412*
Yielding: small scale y. *20,310,312,314,442,446*, net-section (li-
    gament) y. *179,515*